H. Ryssel, P. Pichler (eds.)

Simulation of Semiconductor Devices and Processes

Vol. 6

Springer-Verlag Wien New York

Univ.-Prof. Dr. Heiner Ryssel
Fraunhofer-Institut für Integrierte Schaltungen
Erlangen, Federal Republic of Germany

Institut für Elektronische Bauelemente
Universität Erlangen-Nürnberg, Federal Republic of Germany

Dr. Peter Pichler
Fraunhofer-Institut für Integrierte Schaltungen,
Erlangen, Federal Republic of Germany

This work is subject to copyright.
All rights are reserved, whether the whole or part of the material is concerned, specifically those of translation, reprinting, re-use of illustrations, broadcasting, reproduction by photocopying machines or similar means, and storage in data banks.

© 1995 Springer-Verlag/Wien
Printed in Germany

Typesetting: Camera ready by editors and authors
Printing and Binding: Druckerei zu Altenburg GmbH, D-04600 Altenburg

Printed on acid-free and chlorine-free bleached paper

With 567 Figures

ISBN 3-211-82736-6 Springer-Verlag Wien New York

EDITORIAL

This volume contains the Proceedings of the 6th International Conference on "Simulation of Semiconductor Devices and Processes" (SISDEP'95) which was held at the Campus of the University of Erlangen-Nuremberg in Erlangen on September 6-8, 1995. Over 200 participants from more than 20 different countries attended. SISDEP'95 continued a series of conferences which started in 1984 at the University College of Wales, Swansea, where it took place a second time in 1986. Subsequent conferences were organized at the University of Bologna in 1988, the Federal Institute of Technology in Zurich in 1991 and the Technical University of Vienna in 1993. SISDEP'95 is the last conference in this series because it was agreed with the International Committees of the International Workshop on "Numerical Modeling of Processes and Devices for Integrated Circuits" (NUPAD) and the International Workshop on "VLSI Process and Device Modeling" (VPAD), the other two major series of conferences on numerical simulation of semiconductor devices held in the USA and Japan, to combine the three conferences to the International Conference on "Simulation of Semiconductor Processes and Devices" (SISPAD). So, SISDEP'95 is the 6th and last SISDEP conference and at the same time the first SISPAD conference. The new series will be organized annually and will take place in turn in Europe, Japan, and the USA.

SISDEP'95 provided an international forum for the presentation of state-of-the-art research and development results in the area of numerical process and device simulation. Continuously shrinking device dimensions, the use of new materials, and advanced processing steps in the manufacturing of semiconductor devices requires new and improved software. The trend towards increasing complexity in structures and process technology demands advanced models describing all basic effects and sophisticated two and three dimensional tools for almost arbitrarily designed geometries. SISDEP/SISPAD is the major international forum for presenting the latest results and bringing together the simulation and modeling community, and the process as well as device engineers who need numerical simulation tools with high reliability for characterization, prediction, and development.

The conference committee of SISDEP'95 has prepard an excellent program with 6 invited papers, 75 papers for oral presentation and 39 posters, selected from a total of 191 abstracts. Their distribution reflects the international nature of the conference: 29 from the USA, 26 from Germany, 14 from Japan, 8 from each Austria and Switzerland, 5 from Italy, 4 from each Canada and France, 3 from the United Kingdom, 2 from Lithuania, Russia, Spain, Sweden, and The Netherlands, and 1 from Australia, Belgium, China, Czech Republic, Greece, Hungary, Korea, Poland, and Yugoslavia.

The proceedings were printed from the authors' camera-ready manuscripts. We would like to express our sincere appreciation to the authors for their high quality contributions, their cooperation and efforts. In addition, we would like to thank the members of the conference committee for carrying out the paper selection work with care and competence.

SISPAD'96 will be held on September 2-4, 1996 in Tokyo. Judging from the contents of this volume, the editors can foresee the presentation of a great deal of new and exciting research next year in Japan and we invision continuing strong interest in simulation of semiconductor processes and devices.

Heiner Ryssel
Peter Pichler
Editors

SIMULATION OF SEMICONDUCTOR DEVICES AND PROCESSES Vol. 6
Edited by H. Ryssel, P. Pichler - September 1995

SUPPORTING ORGANIZATIONS

Bayerische Verwaltung der Staatlichen Schlösser, Gärten und Seen
Bayerisches Staatsministerium für Wirtschaft, Verkehr und Technologie
VDE/VDI-Gesellschaft Mikroelektronik (GME)
IEEE Electron Devices Society
IEEE German Section
Informationstechnische Gesellschaft (ITG)
Siemens AG
Universität Erlangen-Nürnberg

CONFERENCE COMMITTEE

G. Baccarani	Università di Bologna	ITALY
K. de Meyer	IMEC	BELGIUM
W. Fichtner	ETH Zürich	SWITZERLAND
M. Fukuma	NEC	JAPAN
H. Jacobs	Siemens	GERMANY
S. Jones	GMMT	UNITED KINGDOM
S. Laux	IBM	USA
C. Lombardi	SGS-Thompson	ITALY
M. Orlowski	Motorla	USA
A. Poncet	CNET/CNS	FRANCE
H. Ryssel	Universität Erlangen-Nürnberg	GERMANY
W. Schilders	Philips	THE NETHERLANDS
S. Selberherr	Technische Universität Wien	AUSTRIA
T. Toyabe	Toyo University	JAPAN
H. Van der Vorst	Rijksuniversiteit Utrecht	THE NETHERLANDS

LOCAL ORGANIZING COMMITTEE

M. Ebner	S. List	P. Pichler	M. Schäfer
T. Klauser	F. Meyer	H. Ryssel	C. Scordo

The cover picture was reproduced with the kind permission of V. Axelrad, Technology Modeling Associates, Palo Alto, USA from Fig. 3 of his contribution on page 13 in these proceedings.

SIMULATION OF SEMICONDUCTOR DEVICES AND PROCESSES Vol. 6
Edited by H. Ryssel, P. Pichler - September 1995

Table of Contents

Numerical Modelling and Materials Characterisation for Integrated Micro Electro Mechanical Systems 1
    H. Baltes, J. G. Korvink, and O. Paul

Fast and Accurate Aerial Imaging Simulation for Layout Printability Optimization 10
    V. Axelrad

Efficient and Rigorous 3D Model for Optical Lithography Simulation 14
    K. D. Lucas, H. Tanabe, C.-M. Yuan, and A. J. Strojwas

Application of the Two-dimensional Numerical Simulation for the Description of Semiconductor Gas Sensors 18
    D. Schipanski, Z. Gergintschew, and J. Kositza

Analysis of Piezoresistive Effects in Silicon Structures Using Multidimensional Process and Device Simulation 22
    M. Lades, J. Frank, J. Funk, and G. Wachutka

Modeling of Magnetic-Field-Sensitive GaAs Devices Using 3D Monte Carlo Simulation 26
    C. Brisset, F.-X. Musalem, P. Dollfus, and P. Hesto

Quasi Three-Dimensional Simulation of Heat Transport in Thermal-Based Microsensors 30
    A. Nathan and N. R. Swart

Simulating Deep Sub-Micron Technologies: An Industrial Perspective 34
    P. Packan

An Improved Calibration Methodology for Modeling Advanced Isolation Technologies 42
    P. Smeys, P. B. Griffin, and K. C. Saraswat

Algorithms for the Reduction of Surface Evolution Discretization Error 46
    H. A. Rueda and M. E. Law

Polygonal Geometry Reconstruction after Cellular Etching or Deposition Simulation 50
    R. Mlekus, Ch. Ledl, E. Strasser, and S. Selberherr

A Data-Model for a Technology and Simulation Archive 54
    K. Wimmer, M. Noell, W. J. Taylor, and M. Orlowski

A Programmable Tool for Interactive Wafer-State Level Data Processing 58
    G. Rieger, S. Halama, and S. Selberherr

Layout Design Rule Generation with TCAD Tools for Manufacturing 62
    J. López-Serrano and A. J. Strojwas

ALAMODE: A Layered Model Development Environment 66
    D. W. Yergeau, E. C. Kan, M. J. Gander, and R. W. Dutton

TCAD Optimization Based on Task-Level Framework Services 70
    Ch. Pichler, N. Khalil, G. Schrom, and S. Selberherr

Cellular Automata Simulation of GaAs-IMPATT-Diodes 74
   D. Liebig

Two-Dimensional Simulation of Deep-Trap Effects in GaAs MESFETs with Different Types of Surface States 78
   K. Horio, K. Satoh, and T. Yamada

An Efficient Numerical Method to Solve the Time-Dependent Semiconductor Equations Including Trapped Charge 82
   L. Colalongo, M. Valdinoci, and M. Rudan

Advances in Numerical Methods for Convective Hydrodynamic Model of Semiconductor Devices 86
   N. R. Aluru, K. H. Law, and R. W. Dutton

An Advanced Cellular Automaton Method with Interpolated Flux Scheme and its Application to Modeling of Gate Currents in Si MOSFETs 90
   K. Fukuda and K. Nishi

Piezoresistance and the Drift-Diffusion Model in Strained Silicon 94
   A. Nathan and T. Manku

A Novel Approach to HF-Noise Characterization of Heterojunction Bipolar Transistors 98
   F. Herzel and B. Heinemann

Ge Profile for Minimum Neutral Base Transit Time in $Si/Si_{1-y}Ge_y$ Heterojunction Bipolar Transistors 102
   W. Molzer

Performance Optimization in Si/SiGe Heterostructure FETs 106
   A. Abramo, J. Bude, F. Venturi, M. R. Pinto, and E. Sangiorgi

On the Integral Representations of Electrical Characteristics in Si Devices ... 110
   S. Biesemans and K. De Meyer

Large Signal Frequency Domain Device Analysis Via the Harmonic Balance Technique 114
   B. Troyanovsky, Z. Yu, and R. W. Dutton

A Method for Extracting the Threshold Voltage of MOSFETs Based on Current Components 118
   K. Aoyama

2-D MOSFET Simulation by Self-Consistent Solution of the Boltzmann and Poisson Equations Using a Generalized Spherical Harmonic Expansion 122
   W-C. Liang, Y-J. Wu, H. Hennacy, S. Singh, N. Goldsman, and I. Mayergoyz

Ultra High Performance, Low Power 0.2 μm CMOS Microprocessor Technology and TCAD Requirements 126
   A. Nasr, J. Faricelli, N. Khalil, and C.-L. Huang

Viscoelastic Modeling of Titanium Silicidation 135
   S. Cea and M. Law

Multidimensional Nonlinear Viscoelastic Oxidation Modeling 139
   S. Cea and M. Law

Three-Dimensional Integrated Process Simulator: 3D-MIPS 143
    M. Fujinaga, T. Kunikiyo, T. Uchida, K. Kamon, N. Kotani, and T. Hirao

Effect of Process-Induced Mechanical Stress on Circuit Layout 147
    H. Miura and Y. Tanizaki

The Simulation System for Three-Dimensional Capacitance and Current Density Calculation with a User Friendly GUI 151
    M. Mukai, T. Tatsumi, N. Nakauchi, T. Kobayashi, K. Koyama, Y. Komatsu, R. Bauer, G. Rieger, and S. Selberherr

Numerical and Analytical Modelling of Head Resistances of Diffused Resistors ... 155
    U. Witkowski and D. Schroeder

New Spreading Resistance Effect for Sub-0.50μm MOSFETs: Model and Simulation 159
    M. Orlowski and W. J. Taylor

The Role of SEMATECH in Enabling Global TCAD Collaboration 163
    E. M. Buturla, J. Byers, A. Husain, M. Kump, P. Lloyd, R. Manukonda, S. Runnels, and D. Scharfetter

Three Dimensional Simulation for Sputter Deposition Equipment and Processes ... 166
    D. S. Bang, Z. Krivokapic, M. Hohmeyer, J. P. McVittie, and K. C. Saraswat

Comprehensive Reactor, Plasma, and Profile Simulator for Plasma Etch Processes 170
    J. Zheng, J. P. McVittie, M. J. Kushner, and Z. Krivokapic

Modeling the Wafer Temperature in a LPCVD Furnace 174
    A. Kersch and M. Schäfer

Determination of Electronic States in Low Dimensional Heterostructure and Quantum Wire Devices 178
    A. Abou-Elnour and K. Schünemann

An Exponentially Fitted Finite Element Scheme for Diffusion Process Simulation on Coarse Grids 182
    S. Mijalković

Achievement of Quantitatively Accurate Simulation of Ion-Irradiated Bipolar Power Devices 186
    P. Hazdra and J. Vobecký

Modeling of Substrate Bias Effect in Bulk and SOI SiGe-channel p-MOSFETs ... 190
    G. F. Niu, G. Ruan, and T. A. Tang

A Very Fast Three-Dimensional Impurity Profile Simulation Incorporating An Accumulated Diffusion Lenght and its Application to the Design of Power MOSFETs 194
    S. Kamohara, M. Sugaya, and H. Matsuo

Recovery of Vectorial Fields and Currents in Multidimensional Simulation ... 198
    D. C. Kerr and I. D. Mayergoyz

An Efficient Approach to Solving The Boltzmann Transport Equation in Ultra-fast Transient Situations 202
    M.-C. Cheng

Modeling of a Hot Electron Injection Laser 206
    V. I. Tolstikhin and M. Willander

Scaling Considerations of Bipolar Transistors Using 3D Device Simulation ... 210
    M. Schröter and D. J. Walkey

Three-Dimensional Monte Carlo Simulation of Boron Implantation into <100>
Single-Crystal Silicon Considering Mask Structure 214
    M.-s. Son, H.-s. Park, and H.-j. Hwang

A fully 2D, Analytical Model for the Geometry and Voltage Dependence of
Threshold Voltage in Submicron MOSFETs 218
    A. Klös and A. Kostka

On the Influence of Band Structure and Scattering Rates on Hot Electron
Modeling 222
    Chr. Jungemann, S. Keith, B. Meinerzhagen, and W. L. Engl

Finite Element Monte Carlo Simulation of Recess Gate FETs 226
    S. Babiker, A. Asenov, J. R. Barker, and S. P. Beaumont

Coupled 2D-Microscopic/Macroscopic Simulation of Nanoelectronic
Heterojunction Devices 230
    C. Pigorsch, R. Stenzel, and W. Klix

On the Discretization of van Roosbroeck's Equations with Magnetic Field 234
    H. Gajewski and K. Gärtner

Modeling of Impact Ionization in a Quasi Deterministic 3D Particle Dynamics
Semiconductor Device Simulation Program 238
    K. Tarnay, F. Masszi, T. Kocsis, and A. Poppe

Accurate Modeling of Ti/TiN Thin Film Sputter Deposition Processes 242
    H. Stippel and K. Reddy

Monte Carlo Simulation of InP/InGaAs HBT with a Buried Subcollector 246
    G. Khrenov and E. Kulkova

Design and Optimization of Millimeter-Wave IMPATT Oscillators Using a
Consistent Model for Active and Passive Circuit Parts 250
    M. Curow

Generalised Drift-Diffusion Model of Bipolar Transport in Semiconductors ... 254
    D. Reznik

Efficient 3D Unstructured Grid Algorithms for Modelling of Chemical Vapour
Deposition in Horizontal Reactors 258
    F. Durst, A. O. Galjukov, Yu. N. Makarov, M. Schäfer, P. A. Voinovich, and
    A. I. Zhmakin

Preventing Critical Conditions in IGBT Chopper Circuits by a Multi-Step Gate
Drive Mode 262
    W. Gerlach and U. Wiese

Control of Plasma Dynamics within Double-Gate-Turn-Off Thyristors
(D–GTO) 266
    U. Wiesner and R. Sittig

A Vector Level Control Function for Generalized Octree Mesh Generation .... *T. Chen, J. Johnson, and R. W. Dutton*	270
Comparison of Hydrodynamic Formulations for Non-Parabolic Semiconductor Device Simulations .... *A. W. Smith and K. F. Brennan*	274
Influence of Analytical MOSFET Model Quality on Analog Circuit Simulation *M. Miura-Mattausch, A. Rahm, and O. Prigge*	278
2-D Adaptive Simulation of Dopant Implantation and Diffusion .... *C.-C. Lin and M. E. Law*	282
Optimization of a Recessed LOCOS Using a Tuned 2-D Process Simulator ... *G. P. Carnevale, P. Colpani, A. Marmiroli, A. Rebora, and A. Tixier*	286
Simulation of Complex Planar Edge Termination Structures for Vertical IGBTs by Solving the Complete Semiconductor Device Equations .... *M. Netzel and R. Herzer*	290
Numerical Analysis of Hot-Electron Effects in GaAs MESFETs .... *Y. A. Tkachenko, C. J. Wei, J. C. M. Hwang, and D. M. Hwang*	294
Capacitance Model of Microwave InP-Based Double Heterojunction Bipolar Transistors .... *C. J. Wei, H.-C. Chung, Y. A. Tkachenko, and J. C. M. Hwang*	298
Estimation of the Charge Collection for the Soft-Error Immunity by the 3D-Device Simulation and the Quantitative Investigation .... *Y. Ohno, T. Kishimoto, K. Sonoda, H. Sayama, S. Komori, A. Kinomura, Y. Horino, K. Fujii, T. Nishimura, N. Kotani, M. Takai, and H. Miyoshi*	302
D. C. Electrothermal Hybrid BJT Model for SPICE .... *J. Zarebski and K. Górecki*	306
Alpha-Particle Induced Soft Error Rate Evaluation Tool and User Interface .. *P. Oldiges*	310
Hydrodynamic Modeling of Electronic Noise by the Transfer Impedance Method *P. Shiktorov, V. Gružinskis, E. Starikov, L. Reggiani, and L. Varani*	314
Monte Carlo Simulation of S-Type Negative Differential Conductance in Semiconductor Heterostructures .... *E. Starikov, P. Shiktorov, V. Gružinskis, L. Reggiani, and L. Varani*	318
Two-Barrier Model for Description of Charge Carriers Transport Processes in Structures with Porous Silicon .... *S. P. Zimin, V. S. Kuznetsov, and A. V. Prokaznikov*	322
Monte-Carlo Simulation of Inverted Hot Carrier Distribution Under Strong Carrier-Optical Phonon Scattering .... *I. Nefedov and A. Andronov*	325

Algorithms and Models for Simulation of MOCVD of III-V Layers in the
Planetary Reactor 328
    T. Bergunde, M. Dauelsberg, Yu. Egorov, L. Kadinski, Yu. N. Makarov,
    M. Schäfer, G. Strauch, and M. Weyers

An Approach for Explaining Drift Phenomena in GTO Devices Using Numerical
Device Simulation 332
    S. Eicher, F. Bauer, and W. Fichtner

Parallel 3D Finite Element Power Semiconductor Device Simulator Based on
Topologically Rectangular Grid 336
    A. R. Brown, A. Asenov, S. Roy, and J. R. Barker

Investigation of Silicon Carbide Diode Structures via Numerical Simulations
Including Anisotropic Effects 340
    E. Velmre, A. Udal, F. Masszi, and E. Nordlander

A New Physical Compact Model of CLBTs for Circuit Simulation Including
Two-Dimensional Calculations 344
    D. Freund and A. Kostka

Combining 2D and 3D Device Simulation with Circuit Simulation for Optimising
High Efficiency Silicon Solar Cells 348
    G. Heiser, P. P. Altermatt, and J. Litsios

A New Quasi-two Dimensional HEMT Model 352
    C. G. Morton, C. M. Snowden, and M. J. Howes

Simulations of the Forward Behaviour of Hybrid Schottky-/pn-Diodes 356
    U. Witkowski and D. Schroeder

HFET Breakdown Study by 2D and Quasi 2D Simulations: Topology Influence .. 360
    Y. Butel, J. Hédoire, J. C. De Jaeger, M. Lefebvre, and G. Salmer

Investigation of GTO Turn-on in an Inverter Circuit at Low Temperatures Using
2-D Electrothermal Simulation 364
    Y. C. Gerstenmaier, and E. Baudelot

Large Scale Thermal Mixed Mode Device and Circuit Simulation 368
    J. Litsios, B. Schmithüsen, and W. Fichtner

Scaling of Conventional MOSFET's to the 0.1μm Regime 372
    M. J. van Dort, J. W. Slotboom, and P. H. Woerlee

Monte Carlo Simulation of Carrier Capture at Deep Centers for Silicon and
Gallium Arsenide Devices 380
    A. Palma, J. A. Jiménez-Tejada, A. Godoy, and J. E. Carceller

A New Statistical Enhancement Technique in Parallelized Monte Carlo Device
Simulation 384
    K. Shigeta, K. Tanaka, T. Iizuka, H. Kato, and H. Matsumoto

Stability Issues in Self-Consistent Monte Carlo-Poisson Simulations 388
    A. Ghetti, X. Wang, F. Venturi, and F. A. Leon

The Path Integral Monte Carlo Method for Quantum Transport on a Parallel Computer 392
    C. Schulz-Mirbach

A Monte Carlo Transport Model Based on Sperical Harmonics Expansion of the Valence Bands 396
    H. Kosina, M. Harrer, P. Vogl, and S. Selberherr

Full-Band Monte Carlo Transport Calculation in an Integrated Simulation Platform 400
    U. Krumbein, P. D. Yoder, A. Benvenuti, A. Schenk, and W. Fichtner

On Particle-Mesh Coupling in Monte Carlo Semiconductor Device Simulation ... 404
    S. E. Laux

T^{2}CAD: Total Design for Sub-μm Process and Device Optimization with Technology-CAD 408
    H. Masuda

Modelling Impact-Ionization in the Framework of the Spherical-Harmonics Expansion of the Boltzmann Transport Equation with Full-Band Structure Effects 416
    M. C. Vecchi and M. Rudan

Impact Ionization Model Using Second- and Fourth-Order Moments of Distribution Functions 420
    K. Sonoda, M. Yamaji, K. Taniguchi, and C. Hamaguchi

An Accurate NMOS Mobility Model for 0.25μm MOSFETs 424
    S. A. Mujtaba, M. R. Pinto, D. M. Boulin, C. S. Rafferty, and R. W. Dutton

A 2-D Modeling of Metal-Oxide-Polycrystalline Silicon-Silicon (MOPS) Structures for the Determination of Interface State and Grain Boundary State Distributions 428
    A.-C. Salaün, H. Lhermite, B. Fortin, and O. Bonnaud

Sensitivity Analysis of an Industrial CMOS Process Using RSM Techniques .. 432
    M. J. van Dort and D. B. M. Klaassen

Process- and Devicesimulation of Very High Speed Vertical MOS Transistors . 436
    F. Lau, W. H. Krautschneider, F. Hofmann, H. Gossner, and H. Schäfer

Two-Dimensional Transient Simulation of Charge-Coupled Devices Using MINIMOS NT 440
    M. Rottinger, T. Simlinger, and S. Selberherr

Determination of Vacancy Diffusivity in Silicon for Process Simulation 444
    T. Shimizu, Y. Zaitsu, S. Matsumoto, E. Arai, M. Yoshida, and T. Abe

Precipitation Phenomena and Transient Diffusion/Activation During High Concentration Boron Annealing 448
    A. Höfler, T. Feudel, A. Liegmann, N. Strecker, W. Fichtner, Y. Kataoka, K. Suzuki, and N. Sasaki

Modelling of Silicon Interstitial Surface Recombination Velocity at Non-Oxidizing Interfaces 452
    C. Tsamis and D. Tsoukalas

Efficient Hybrid Solution of Sparse Linear Systems 456
    A. Liegmann, K. Gärtner, and W. Fichtner

Mesh Generation for 3D Process Simulation and the Moving Boundary Problem 460
    S. Bozek, B. Baccus, V. Senez, and Z. Z. Wang

Three-Dimensional Grid Adaption Using a Mixed-Element Decomposition Method 464
    E. Leitner and S. Selberherr

Unified Grid Generation and Adaptation for Device Simulation 468
    G. Garretón, L. Villablanca, N. Strecker, and W. Fichtner

Platinum Diffusion at Low Temperatures 472
    M. Jacob, P. Pichler, H. Ryssel, and R. Falster

Lattice Monte–Carlo Simulations of Vacancy-Mediated Diffusion and Implications for Continuum Models of Coupled Diffusion 476
    S. T. Dunham and C. D. Wu

A New Hydrodynamic Equation for Ion-Implantation Simulation 480
    S. Kamohara and M. Kawakami

Monte Carlo Simulation of Multiple-Species Ion Implantation and its Application to the Modeling of 0.1μ PMOS Devices 484
    A. Simionescu, G. Hobler, and F. Lau

Analytical Model for Phosphorus Large Angle Tilted Implantation 488
    A. Burenkov, W. Bohmayr, J. Lorenz, H. Ryssel, and S. Selberherr

Statistical Accuracy and CPU Time Characteristic of Three Trajectory Split Methods for Monte Carlo Simulation of Ion Implantation 492
    W. Bohmayr, A. Burenkov, J. Lorenz, H. Ryssel, and S. Selberherr

Author Index 496

Numerical Modelling and Materials Characterisation for Integrated Micro Electro Mechanical Systems

Henry Baltes, Jan G. Korvink and Oliver Paul

Physical Electronics Laboratory,
ETH Zurich, CH-8093 Zurich, Switzerland
Email: baltes@iqe.phys.ethz.ch

Abstract

Integrated micro electro mechanical systems *(iMEMS)* include sensors, actuators and circuits made by silicon IC technology combined with micromachining, deposition or electroplating. We present two essential *iMEMS* development tools:
(i) the data base **ICMAT** of material parameters obtained from measuring process-dependent IC thin film electrical, magnetic, thermal and mechanical properties by using dedicated materials characterisation microstructures and
(ii) the toolbox **SOLIDIS**, providing coupled numerical modelling of the electrical, magnetic, thermal and mechanical phenomena and their boundary and interface conditions occurring in *iMEMS* devices in a uniform and consistent environment.

1. Introduction

Micro Electro Mechanical Systems *(MEMS)* include micromachined structures with mechanical, thermal, magnetic and fluid effects for sensors and actuators. Integrated Micro Electro Mechanical Systems *(iMEMS)* combine *MEMS* with integrated circuits. Our approach to *iMEMS* is to combine industrial CMOS or bipolar IC technology with bulk or surface micromachining, thin film deposition or electroplating [1]. Integrated thermoelectric, magnetic, mechanical and chemical microtransducers have been demonstrated in this way [2]-[7].

Design and simulation of *iMEMS* devices requires precise knowledge of process-dependent material parameters, such as resistivity and its temperature coefficient, thermal conductivity, heat capacity, Seebeck coefficient, stress, Young's modulus, and Poisson's ratio of all IC layers (which include the metallisation and polysilicon layers, the field, contact and via oxides, the standard passivation) and any additional layers such as silicon nitride passivation for sensors. In Section 2, we report how these material properties are measured for a variety of industrial CMOS IC processes by designing material characterisation microstructures fabricated by the corresponding IC process.

In efficient *iMEMS* devices, magnetic, thermal or mechanical vector fields couple strongly (and often nonlinearly) to "ordinary" IC electrical properties. Thus *iMEMS* simulation goes beyond semiconductor device modelling based on some approximation of Boltzmann's equation [8]. In Section 3 we report the *iMEMS* simulation toolbox SOLIDIS, which includes equations describing the various reversible and irreversible transducer effects and their coupling [9]. Specific simulation results are reported for a thermoelectric characterisation device, a CMOS micromirror and a micromagnetic flux concentrator. There is a "bootstrap" connecting Section 2 and 3: Modelling allows the validation of the materials characterisation microstructures. In turn, the simulation can be validated by comparison with the measured material properties. The latter serve as input parameters for realistic *iMEMS* device simulation and optimisation.

2. IC Material Properties

Thermal properties of IC thin films depend on their fabrication process. The layers being characterised must be produced with the same process as the corresponding microdevice layer. We therefore use characterisation microstructures fabricated with the IC process used for the fabrication of microtransducers [10]. Such IC compatible microstructures are smaller than 500 μm and have been successfully used for the determination of thermal conductivities [11, 12] and the heat capacities [13] of CMOS layers, and of transport coefficients of CMOS polysilicon [14], such as its Seebeck coefficient.

IC compatible characterisation structures exploit ideas implemented in microsensors, such as micromachined dielectric beams and integrated CMOS polysilicon resistors and thermistors. In contrast to sensors optimised to transduce an external signal into the electrical domain, the structures have to sense a property of one of their constitutive materials. Cross-sensitivities to other properties are suppressed by suitable design optimisation. Validation and optimisation of a thermal structure with numerical simulation is described in [15]. A further case is detailed in Section 3.2.

2.1. Thermal and Electronic Properties of CMOS Layers

As an example we describe the measurement of the Seebeck coefficient of CMOS polysilicon, a parameter of thermoelectric CMOS sensors [1]. The test structure is shown schematically in Figure 1. It consists of a micromachined cantilever and lateral arms composed of the sandwich of dielectric CMOS layers. Micromachining is made possible by suitable layout design of the field, contact, via, and pad masks. Cuts through the corresponding layers allow the silicon to be anisotropically etched away and the structure to be undercut. A micrograph is shown in Figure 2.

At its tip the cantilever contains two CMOS polysilicon resistors. The one closer to the end serves as a heater. The other resistor is used as a thermistor to determine the temperature at the free end. A rectangular polysilicon film to be characterised is integrated into the dielectric sandwich between the hot cantilever tip and the bulk silicon die. The polysilicon is contacted at its hot and cold ends with the lower CMOS metal. When electrical power is dissipated in the heater, a thermoelectric voltage $U_{th} = \alpha \Delta T$ is measured between these two contacts, where α is the Seebeck coefficient sought and ΔT is the temperature difference between the hot and the cold contacts of the polysilicon. The electrical measurement of U_{th} is straightforward. For the accurate determination of ΔT we improved the geometry of earlier versions of the test structure [14]. One improvement is the integration of a temperature smoothing CMOS metal layer, that extends the hot polysilicon contact beyond the temperature monitoring resistor. With this test structure we have shown that the

Seebeck coefficient of the standard gate poly of three commercial CMOS processes is in the range between 108 and 120 $\mu V/K$ [10].

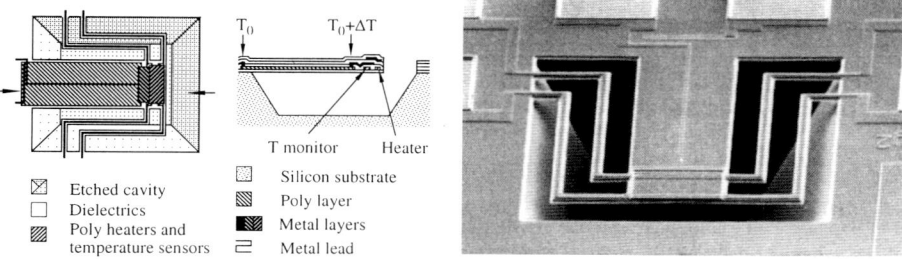

Figure 1. Schematic top view (left) and cross-section (right) of CMOS characterisation structure to determine the Seebeck coefficient of the gate or capacitor polysilicon for CMOS microsensors. The Seebeck coefficient is measured between T_o and $T_o + \Delta T$.

Figure 2. SEM micrograph of the micromachined Seebeck test structure. Heater and temperature monitoring resistors are integrated between the lateral connection arms.

Property Layer	ρ_{sq} ($10^{-3}\Omega cm$)	TC of ρ ($ppm K^{-1}$)	$\alpha - \alpha_{Metl}$ ($\mu V/K$)	κ_{th} (W/mK)
n^+-Poly	25 ± 1	860 ± 30	-120 ± 5	24 ± 1.5
p^+-Poly	215 ± 5	-140 ± 5	190 ± 10	17 ± 1.5
Metal 1	65 ± 0.7	2960 ± 100	0	181 ± 8
Metal 2	36 ± 0.4	3010 ± 100		166 ± 8

Table I. Sheet resistance ρ, temperature coefficient of ρ, Seebeck coefficient α and thermal conductivity κ_{th} of conducting layers of the 1.2μm CMOS process of Austria Mikrosysteme (AMS) at 300K.

Property Layer	κ_{th} (W/mK)	c ($Jcm^{-3}K^{-1}$)
Field oxide	1.28 ± 0.18	
Contact ox.	1.32 ± 0.18	1.05 ± 0.1
Via oxide	1.16 ± 0.24	
Passivation	1.5 ± 0.25	2.7 ± 0.25

Table II. Thermal conductivity κ_{th} and heat capacity c of dielectric thin films of the AMS 1.2μm CMOS process at 300K.

Similar thermal test structures provide the thermal conductivities and heat capacities of CMOS layers. Sheet resistances and temperature coefficients (TC) of resistance are measured with van-der-Pauw structures. In our laboratory, measurements of material properties are systematically performed between 100 and 420 K to obtain temperature coefficients and to allow better understanding of the process dependence of the properties. Experimental thermal and electronic properties of films of the CMOS process of EM Microelectronic Marin, Austria Mikrosysteme (AMS) and European Silicon Structures (ES2) were measured. As an example, results of AMS' 1.2μm CMOS process are shown in Tables I and II. The thermal conductivities of polysilicon and metal layers deviate strongly from bulk values. Such deviations show the importance of characterising thermal material properties relevant for *iMEMS*.

Thin film	σ (MPa)	E (GPa)	ν
LPCVD silicon nitride	1040 ± 30	195 ± 6	0.13 ± 0.02
Contact oxide	-40 ± 10	20 ± 10	$0.5^{+0.0}_{-0.1}$
Intermetal dielectric standard passivation	-37 ± 6	65 ± 5	0.2 ± 0.2
PECVD silicon nitride passivation for CMOS sensors	82 ± 5	97 ± 6	0.13 ± 0.07

Table III. Stress σ, Young's modulus E and Poisson's ration ν of the dielectric layers of the $2\mu m$ CMOS process of EM Microelectronic Marin, used for the fabrication of microsensors and actuators.

2.2. Mechanical Properties of CMOS Layers

Mechanical thin film properties of the $2\mu m$ CMOS process of EM Microelectronic Marin (EM), Switzerland, were determined using the bulge test. This method has previously been applied to measure the residual stress σ and the elastic modulus E of polymer films [16, 17], and of CVD silicon nitride and polysilicon [18]. It exploits the load-deflection behavior of thin film membranes, as shown in Figure 3. The mid-point deflection W of thin membranes under unilateral pressure P is described by

$$P = C_1 \sum_i \frac{h_i}{a^2}\sigma_i W + C_2 \sum_i f(\nu_i)\frac{h_i}{a^4}\frac{E_i}{1-\nu_i}W^3,$$

where h_i, σ_i, ν_i, and E_i denote the thicknesses, the stresses, the Poisson's ratios and the elastic moduli, respectively, of the membrane component layers, and a denotes a linear dimension of the membrane. The constants C_1 and C_2 and the function $f(\nu)$ for square, rectangular, and circular membranes have been calculated with finite element and analytical methods [19]. Implicit in the equation is the assumption that the contribution of membrane bending to the total energy is negligible. This condition is fulfilled by sufficiently thin and tensile membranes.

We produced membranes containing the layers to be characterised using the standard CVD deposition conditions of the CMOS process. The membranes are composed of a 1680 Å thick, tensile base layer of LPCVD silicon nitride, on top of which standard thickness laysers of either the contact oxide, the intermetal oxide, or a $2.08\mu m$ thick nitride passivation layer optimised for CMOS microsensors were deposited [2]. After the CVD steps, rectangular membranes with sizes up to $2mm$ and side ratios of 1:1, 1:2, and 1:4 were produced by silicon micromachining in KOH. During the etch the wafer-fronts were mechanically protected. Pressure-dependent deflections of the membranes were measured with a contactless surface profiler.

Figure 4 shows the mid-point responses of various LPCVD nitride membranes with a side ratio of 1:4. First the silicon nitride properties were obtained with single layer membranes. Based on these values, the mechanical properties of the second thin film in the bilayer membranes were then determined. Results are listed in Table III. The contact and intermetal dielectrics are under compressive stress, whereas the nitride layers are under tension. Poisson's ratios of both nitrides and the contact oxide show a significant deviation from the values of 0.25 to 0.3 usually assumed in the literature. Clearly, the determination of Poisson's ratio of thin films deserves more attention than it has hitherto received.

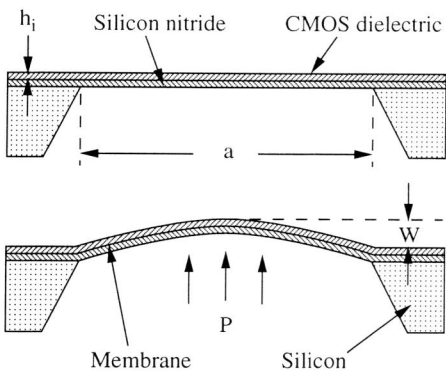
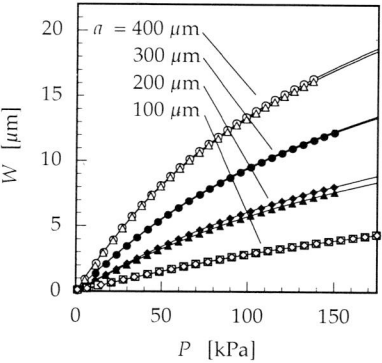

Figure 3. Micromechanical membrane to determine mechanical thin film parameters (a) After fabrication and micromachining, (b) Deflection under pressure load.

Figure 4. Deflection of rectangular membranes with side length ratio 1:4 as a function of applied pressure. The membranes consist of 1680 Å silicon nitride.

3. *iMEMS* Modelling and Simulation

Sensor and actuator development using industrial IC processes places tight limits on device performance. Modelling and simulation is important for the purposes of device optimisation, and understanding the detailed device operation principles. Simulations must be representative of the physical processes and provide sufficiently accurate predictions of sensitivity. High performance simulation requires three key elements: (i) precise knowledge of material properties and device geometries, (ii) accurate (self-correcting) numerical solution methods and (iii) correct model systems (physical model equations, geometry, boundary and interface conditions). The operation of *iMEMS* is governed by one or more of a large variety of coupled physical effects. A functional tool establishes consistent solutions, and supports extensions when required [9]. With these prerequisites, simulation significantly reduces the number of prototypes required for the development process.

3.1. The *iMEMS* Simulation Toolbox SOLIDIS

SOLIDIS addresses the simulation of coupled effects that arise in sensor chips and in packaged solid-state sensors and actuators. It comprises a set of computational *kernels* and interactive visualisation *front ends*. Each *kernel* implements a hybrid finite element discretisation [20] of a tailored set of partial differential equations. A scalar field $\psi(\vec{x})$ and its associated vector field $\vec{\Phi}(\vec{x})$ are separately approximated over the domain covered by a mesh of finite elements and nodes:

$$\psi(\vec{x}) = \sum_{i=1}^{n} p_i S_i(\vec{x}); \quad \vec{\Phi}(\vec{x}) = \sum_{j=1}^{m} f_j \vec{V}_j(\vec{x}).$$

The subscripts are associated with mesh nodes. The $S_i(\vec{x})$ are the scalar basis functions and $\vec{V}_j(\vec{x})$ are vector basis functions. The p_i and f_i are scalar weights. For an equation of the form $\vec{\nabla} \cdot \vec{\Phi} = \rho(\vec{x})$, with $\vec{\Phi} = \beta(\vec{x})\vec{\nabla}\psi$, where $\rho(\vec{x})$ and $\beta(\vec{x})$ are parameters[1], we write two weighted residual equations, each integrated over the simulation domain Ω,

$$\int_\Omega \vec{V}_i \cdot \left[\vec{\Phi} - \beta\vec{\nabla}\psi\right] d\Omega = 0; \quad \int_\Omega S_i \left[\vec{\nabla} \cdot \vec{\Phi} - \rho\right] d\Omega = 0.$$

After substitution of the expansions for ψ and $\vec{\Phi}$ and some standard manipulations we obtain two discrete matrix residual equations $A\,f + B\,p = 0$ and $B^T f - r = 0$. After eliminating f we obtain $B^T A^{-1} B\,p + r = 0$. This is the equation to be assembled and solved for the nodal degrees of freedom p, from which f is computed. This hybrid method ensures better compatibility between ψ and $\vec{\Phi}$ than with a conventional finite element method, and bears fruit when used in conjunction with mesh adaptivity converging to spatially accurate solutions, and when discretising coupled effects that require accurate distributions of $\vec{\Phi}$.

3.2. Characterisation Device Modelling

We next consider a device for measuring the Seebeck coefficient of a BiCMOS polysilicon layer, as described in Section 2.1 and Figure 2. One modelling goal is to minimise external influences and stray effects. Accurately resolved field distributions and fluxes in 3D are therefore essential.

For the simulation we model half of the geometry of the beam structure, taking advantage of the beam's symmetrical layout (Figure 5). The device is operated in vacuum. Its small size ensures negligible heat loss through radiation. The thermal conductivities are taken from Tables I and II. The power dissipated is accounted for in all conductors with resistivities taken from Table I.

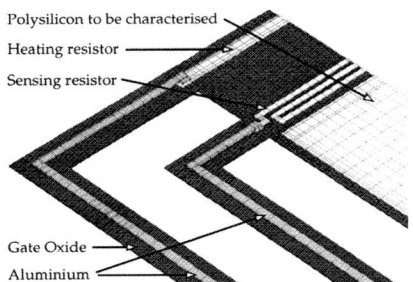

Figure 5. Part of the 3D mesh used for the modelling. The passivation, intermetal oxide, contact oxide, and the aluminium covering the heater and temperature sensor, is removed for clarity. The mesh has 15,869 finite elements and 20,112 nodes.

Figure 6. The temperature contours on the upper surface of the mesh depicted in Figures 5. δT between contours is 0.3K. On the polysilicon $T_h = 310.2K$ and $T_o = 300K$. Note the uniform temperature distribution over the temperature sensor.

---

[1]Note that we are not restricted to this particular partial differential equation.

Modelling was used to verify the device function, and quantifyd stray effects arising from the current-carrying arms connected to the free end of the beam. Figure 6 indicates the effectiveness of the aluminium sheets in providing a low-resistance path for the generated heat into the polysilicon beam, and for averaging the temperature profile over the meandering thermometer. The computed temperature drop over the beam is $10.2\ K$. The resistor is only $0.12\ K$ hotter than the hot end of the polysilicon, and uniform to within $0.05\ K$. Thus all assumptions regarding the temperature measurement (see Section 2.1) hold.

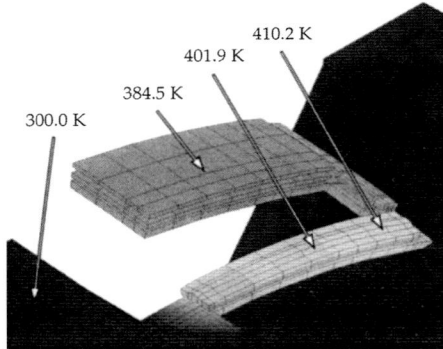

Figure 7. The mesh of the integrated micromirror, with the temperature distribution over the surface (most oxides and nitrides are removed) and in the mechanically deformed ($\times 30$) state.

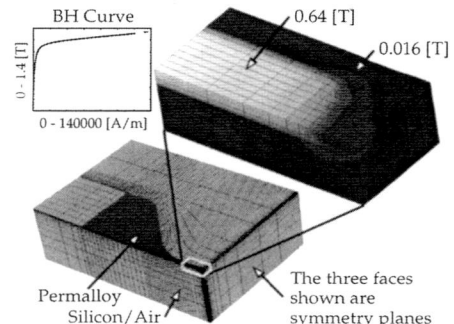

Figure 8. The mesh of 1/8 of the integrated microconcentrator (dark shading). Above right is the magnetic induction flux distribution over the tip of the magnetoconcentrator. Above left is the measured BH curve of the permalloy.

3.3. Sensor and Actuator Modelling

In the thermomechanical micromirror actuator [6], Joule heat is dissipated in a polysilicon resistor within the mirror's suspension beams to generate a local temperature rise. This results in thermal expansion of the beam's materials. The difference in thermal expansion coefficients causes bending of the beam (also called the multimorph effect) and hence deflection of the mirror. A principal target is to obtain maximum mirror deflection for given input power. It is also essential that the mirror does not warp during actuation (see Figure 7). Finally, effects due to mask misalignment - causing asymmetry in the deflection - must be minimised. The simulations rely heavily on knowing the thermoelectric and mechanical properties listed in Tables I, II and III accurately. A full study is presented in [21].

For the integrated magnetic flux concentrator [3] in Figure 8, a thin layer ($5\mu m$) of permalloy is electroplated onto the surface of a chip and serves to focus the magnetic flux onto a magnetotransistor. The saturative nature of the permalloy is a source of nonlinearity. Saturation is a local effect — the concentrator causes large variations in magnetic field intensity. A coarse mesh of $20,000$ finite elements underestimates the concentrated flux density and smoothens the saturation curve of the device. Numerically, large finite elements are driven uniformly into saturation, whereas this should only happen for a small sub-region. Adaptive mesh refinement leads to meshes in the region of $80,000$ finite elements. A full numerical study is presented in [22].

4. Conclusions

In this paper we demonstrated a variety of characterisation microstructures for measuring process-dependent electrical, thermal and mechanical properties of IC layers used in *iMEMS* devices. The resulting data base ICMAT in combination with the simulation toolbox SOLIDIS, also presented here, allows precise and efficient numerical modelling of a wide variety of *iMEMS* devices and thus provides a crucial tool for resource-efficient prototyping.

Future measurements will aim at additional coupling parameters, e. g. thermomechanical and piezoresitive coefficients. The extension to further CMOS processes involved in *iMEMS* manufacturing is another obvious task. This will lead to a set of standard characterisation microstructures to be added to the conventional IC process test structures. Likewise, modelling will be extended to further coupling effects such as convection, the piezo-electric coupling and the Peltier effect. It is imperative that the mutual validation of characterisation structures and modelling tools is maintained for any reliable extension of ICMAT and SOLIDIS.

5. Acknowledgements

It is our pleasure to thank Dr. Arokia Nathan, visiting Professor at ETH Zurich, on leave from the University of Waterloo, Ontario, Canada, for enlightening discussions.

References

[1] H. Baltes, O. Brand, J. G. Korvink, R. Lenggenhager, O. Paul, "IMEMS - integrated micro electro mechanical systems by VLSI and micromachining," *ESSDERC'94 Proc. 24th European Solid State Device Research Conference, (Editions Frontires, Gif-sur-Yvette, France)*, pp. 273-280 (1994).

[2] R. Lenggenhager, D. Jaeggi, P. Malcovati, H. Duran, H. Baltes, E. Doering, "CMOS membrane infrared sensor and improved TMAHW etchant," *IEDM Technical Digest, (IEEE)*, pp. 531-534 (1994).

[3] M. Schneider, R. Castagnetti, M. G. Allen, H. Baltes, "Integrated flux concentrator improves CMOS magnetotransistor," *Proc. IEEE MEMS*, pp. 151-156 (1995).

[4] M. Hornung, R. Frey, O. Brand, H. Baltes, C. Hafner, "Ultrasound barrier based on packaged micromachined membrane resonators," *Proc. IEEE MEMS*, pp. 151-156 (1995).

[5] O. Paul, A. Häberli, P. Malcovati, H. Baltes, "Novel integrated thermal pressure gauge and read-out circuit by CMOS IC design," *IEDM Technical Digest, (IEEE)*, pp. 131-134 (1994).

[6] J. Bühler, H. Baltes, "Thermally actuated CMOS micro mirrors," *Sensors and Actuators 47*, pp. 525-575 (1995).

[7] F.-P. Steiner, A. Hierlemann, C. Cornila, G. Noetzel, M. Bächtold, J. G. Korvink, W. Göpel, H. Baltes, "Polymer coated capacitive microintegrated gas sensor," *Transducers'95 Digest of Technical Papers*, in press (1995).

[8] S. Selberherr, "Analysis and Simulation of Semiconductor Devices," *(Springer-Verlag, Wien)* (1984).

[9] J. G. Korvink, J. Funk, H. Baltes, "IMEMS modelling," *Sensors and Materials 6*, pp. 235-243 (1994).

[10] O. Paul, M. von Arx and H. Baltes, "Process-dependent thermophysical properties of CMOS IC thin films," *Transducers'95 Digest of Technical Papers*, in press (1995).

[11] F. Völklein and H. Baltes, "A microstructure for measurement of thermal conductivity of polysilicon thin films," *J. of Microelectromech. Systems*, *1*, pp. 193-196 (1993).

[12] O. Paul, M. von Arx, and H. Baltes, "CMOS IC layers: complete set of thermal conductivities," *Proc. Intl. Workshop on Semiconductor Characterization, (NIST, Gaithersburg)*, in press (1995).

[13] M. von Arx, O. Paul, H. Baltes, "Determination of the heat capacity of CMOS layers for optimal CMOS sensor design," *Sensors and Actuators A*, *47*, pp. 428-431 (1995).

[14] O. Paul and H. Baltes, "Measuring thermogalvanomagnetic properties of polysilicon for the optimization of CMOS sensors," *Transducers'93 Digest of Technical Papers (IEE, Japan, Tokyo)*, pp. 606-609 (1993).

[15] O. Paul, J. Korvink and H. Baltes, "Determination of the thermal conductivity of CMOS IC polysilicon," *Sensors and Actuators A*, *41-42*, pp. 161-164 (1994).

[16] P. Lin, "The in-situ measurement of biaxial modulus and residual stress of multilayer polymeric thin films," *Mat. Res. Soc. Symp. Proc.*, *188*, pp. 41-46 (1990).

[17] M. G. Allen, M. Mehregany, R. T. Howe and S. D. Senturia, "Microfabricated structures for the in situ measurement of residual stress, Youngs modulus, and ultimate strain of thin films," *Appl. Phys. Lett.*, *51*, pp. 241-243 (1987).

[18] D. Maier-Schneider, A. Ersoy, J. Maibach, D. Schneider and E. Obermeier, "Influence of annealing on elastic properties of LPCVD silicon nitride and LPCVD polysilicon," *Sensors and Materials*, *7*, pp. 121-129 (1995).

[19] J. Pan, P. Lin, F. Maseeh and S. D. Senturia, "Verification of FEM analysis of load-deflection methods for measuring mechanical properties of thin films," *Tech. Digest of IEEE Solid-State Sensor and Actuator Workshop, (Hilton Head SC)*, pp. 70-73 (1990).

[20] J. G. Korvink, "An Implementation of the Adaptive Finite Element Method," *(Verlag der Fachvereine, Zurich)* (1993).

[21] J. Funk, J. Bühler, J. G. Korvink and H. Baltes, "Thermomechanical Modelling of an Actuated Micromirror," *Eurosensors VIII Conf. Book of Abstracts, (LAAS, Toulouse)*, p.192 (1994).

[22] M. Schneider, J.G. Korvink, H. Baltes, "Magnetostatic Modelling of an Integrated Microconcentrator," *Transducers'95 Digest of Technical Papers*, in press (1995).

Fast and Accurate Aerial Imaging Simulation for Layout Printability Optimization

V. Axelrad

Technology Modeling Associates,
Palo Alto, CA, USA

1 Introduction

Optical lithography has been a major force in the continuing reduction of feature size in VLSI. Pushing the limits of lithography by using advanced lenses (high NA, in-lens filtering, etc.), light sources (annular, etc.) and mask designs (phase shift masks, optical proximity corrections, etc.) allowed to extend its life span far beyond what was predicted only 10 years ago.

Since lithographic image quality is a major limiting factor in VLSI processing, exact understanding of the printability of a certain layout is crucial to detect possible product quality problems. In certain cases it is possible to improve the image quality by optimizing the mask to compensate for non-local optical interaction effects (so-called optical proximity correction [1]). The ability to rapidly and accurately evaluate the expected image quality has therefore attained the status of considerable practical significance for industrial applications.

This work discusses the application of highly efficient algorithms based on the Fast Fourier Transform to achieve aerial image calculations many orders of magnitude faster than conventional lithography simulators such as DEPICT [2] and SPLAT [3]. An algorithm is presented which compares the original mask image with the calculated aerial image to estimate printability. The algorithm has essentially linear dependence of CPU time on image size and linear dependence of memory on image size, therefore full-chip applications are feasible. The current implementation requires approximately $1KB/\mu m^2$ of memory and 40 ms/μm^2 CPU time on a SPARC Station 10. For a 500μm by 500μm image this translates into 250MB and 3 hours. Utilization of redundancy in regular layouts and windowing techniques can substantially reduce the requirements.

The techniques described here have been previously developed in connection with closely related wave propagation problems in acoustics [4], [5]. Applications of related techniques in lithography have also been reported [6].

2 The Method

Monochromatic wave propagation is governed by the well-known Helmholtz equation as well as appropriate boundary conditions:

$$\Delta\varphi + k^2\varphi = 0, \quad \text{with} \quad k = \omega/c = 2\pi/\lambda \quad (1)$$

A general solution to this equation can constructed as an integral over all plane waves with the same wavelength $1/\lambda$ and amplitudes $\Phi(k_x, k_y)$ traveling in all directions given by the wave vector $\vec{k} = (k_x, k_y, k_z)$:

$$\varphi(x, y, z) = \int\int \Phi(k_x, k_y) \cdot e^{i(k_x x + k_y y + k_z z)} \cdot dk_x \cdot dk_y \quad (2)$$

Since the length of the wave vector k is given by equation (1), its z-direction component is the projection of the k-vector upon the z-axis:

$$k_z = \pm\sqrt{k^2 - (k_x^2 + k_y^2)} \qquad (3)$$

It is interesting to note that real values of k_z are only possible for $k_x^2 + k_y^2 \leq k^2$. This means that only spectral components up to a certain cut-off frequency given by the wavelength are propagated, higher-frequency components are represented by so-called evanescent modes with imaginary k_z and are not present in the far field [4]. These modes are usually quite important in acoustics but of no significance in optics, where typical propagation distances through lenses are very much larger than the wavelength of light.

A physical interpretation of the Ansatz in eq. (2) is shown in Figure 1.: as the length of the projection of k onto the x-y plane approaches the total length of the k-vector given by the wavelength λ, its projection upon the z-axis decreases until it reaches zero. At this point the plane wave is traveling in the x-direction, orthogonally to the optical axis z. Beyond this point, higher values of $\sqrt{k_x^2 + k_y^2}$ do not correspond to real values of k_z:

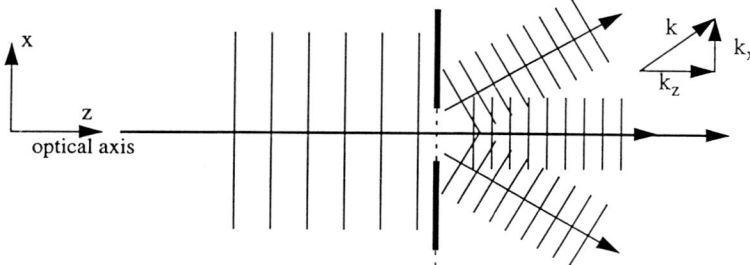

Figure 1. Physical interpretation of the Fourier Integral approach.

A two-dimensional inverse Fourier transform is recognized at the core of equation (2). We can therefore write the Fourier transform of the field in a plane z as a combination of forward and backward propagating waves for the two signs in equation (2):

$$\Phi(k_x, k_y; z) = \Phi^{forward}(k_x, k_y) \cdot e^{ik_z z} + \Phi^{back}(k_x, k_y) \cdot e^{-ik_z z} \qquad (4)$$

The aerial imaging problem as well as certain problems in acoustics are described by the special case of forward propagation only, i.e. reflections are not taken into account. For this case we skip the superscript *forward* and obtain the solution as:

$$\Phi(k_x, k_y; z) = \Phi(k_x, k_y; z_0) \cdot e^{ik_z(z - z_0)} \qquad (5)$$

Equation (5) is of central importance to the algorithm. It means that the Fourier transform of the field in the plane z can be calculated by multiplying the Fourier transform of the field in another plane z_0 by a linear space-invariant transfer function. In lithography, this transfer function can be used to model defocus, since a defocused image can be constructed as a propagation problem from the focus plane.

$$G(k_x, k_y; z - z_0) = e^{i\sqrt{k^2 - (k_x^2 + k_y^2)} \cdot (z - z_0)} \qquad (6)$$

Lens effects are taken into account by another transfer function. A perfectly focused ideal lens creates a controlled phase delay for each plane wave direction causing the wave front to converge in the focus. The image thus created is the superposition of the original plane waves minus certain high spatial frequency components. The ideal lens can thus be described by an ideal low-pass filter. Its cutoff frequency is determined by the numerical aperture NA of the lens and the wavelength λ as shown in Figure 2. due to the fact that the lens can only capture plane waves with $k_x/k_z < X/f$.

Lens aberration effects lead to an additional phase error, which renders the lens transfer function to:

$$G_{Lens} = e^{iL(k_x, k_y)}, \quad \forall \ \sqrt{k_x^2 + k_y^2} < \frac{2\pi NA}{\lambda}, \quad 0 \text{ otherwise} \qquad (7)$$

As a result, the calculation of the field in the image plane is reduced to the calculation of the field in

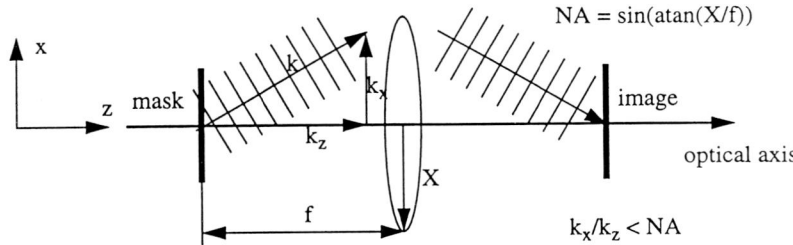

Figure 2. Derivation of the lens transfer function.

the mask plane and applying the two linear filters (6) and (7). The field in the mask plane given a single monochromatic light source is a product of a linear phase function times the mask transmittance and phase delay, which is spatially dependent for a phase-shift mask. The linear phase function is described by the complex exponential in the equation below:

$$\varphi(x, y; 0) = Mask(x, y) \cdot e^{i(\alpha_x x + \alpha_y y)} \qquad (8)$$

The aerial image for a single monochromatic point source is thus given by:

$$\varphi(x, y; z) = F^{-1} \{ G(\Delta z) \cdot G_{Lens} \cdot F \{ \varphi(x, y; 0) \} \} \qquad (9)$$

Fourier transforms are evaluated numerically using the Fast Fourier Transform algorithm with the number of operations given by $O(N \log N)$, with N being the total number of sampling points. An interesting and practically important aspect of evaluating eq. (9) is that while the low-pass nature of the lens transfer function ensures that the Fourier transform of the aerial image is band-limited with no spectral components for frequencies higher than $2\pi NA/\lambda$, the transform of the mask image is not band-limited. As a consequence, special care must be paid to calculating the forward transform in eq. (9) to keep aliasing errors sufficiently small.

Typical light sources in optical lithography are not coherent. To calculate the light intensity resulting from a larger light source we thus have to perform a numerical integration over the source. In other words the light source is represented by a number of point sources and the total light intensity is calculated as a weighted sum of the light intensities produced by each point source.

The total numerical effort involved is one FFT to calculate the transform of the mask image and a number of inverse transforms according to the discretization of the source. For different locations of the point source the Fourier Transform of the field in the mask plane is simply shifted by a distance in frequency domain according to the location of the point source.

3 Extraction of Printability

Calculating the aerial image numerically according to eq. (9) leaves us with a sampled image. Its sampling density is determined by the cut-off frequency $2\pi NA/\lambda$. The image can be output in a standard format (GIF) for display. Since the aerial image is band-limited, trigonometric interpolation can be used to generate additional sampling points if desired. The sampled aerial image can be directly compared to the sampled mask information to determine the quality of the aerial image. Degradation of quality is

defined as image details which are likely not to print as desired. Two cases of quality loss can be distinguished:
  i) image intensity is below a threshold value in an open mask region.
  ii) image intensity is above a threshold in a dark mask region.

This comparison is carried out locally for each sampling point in the image. Following [8] and others we use a value of 0.3 for both thresholds, with 1 being the aerial image intensity at the center of a large open mask region

4 Application

A test mask layout of the size 15x15 µm was chosen as an example of application. The layout is a combination of the test pattern reported in [7] and some additional mask features. Figure 3. (left) shows the aerial image calculated for $\lambda=0.365\mu m$, NA=0.55, $\Delta z=0.4\mu m$ and a single centered point light source. The line thickness is 0.35 µm, which is very close to the theoretical resolution limit of $0.5 \cdot \lambda/NA \approx 0.33\mu m$. As a consequence, loss of resolution is clearly visible. Figure 3. (right) displays the results of a printability analysis, where mask areas which are likely not to print as desired are marked in gray.

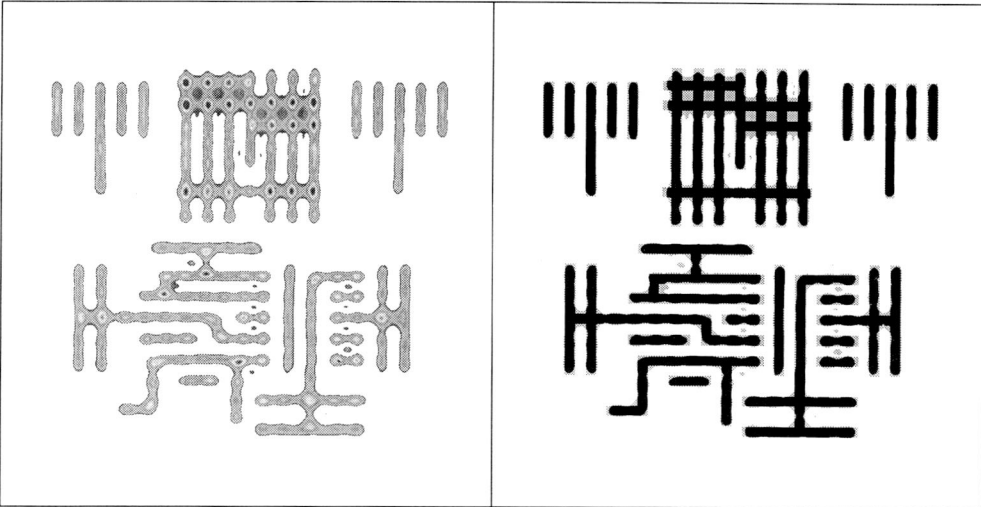

Figure 3. Application example: Aerial image and Printability analysis of a test mask pattern.

The author acknowledges helpful discussions on the subject of optical lithography with Dr. D. Bernard of TMA.

References

[1] O.W. Otto, J.G. Garofalo, K.K. Low, C.-M. Yuan, R.C. Henderson, C. Pierrat, R.L. Kostellak, S. Vaidya, P.K. Vasudev, "Automated Optical Proximity Correction - A Rules-Based Approach", *SPIE* Vol. 2197, 1994
[2] Technology Modeling Associates, DEPICT User Manual, 1994
[3] UC Berkeley, SPLAT
[4] H. Fleischer and V. Axelrad, "Fourier-Acoustics: An Approach to Acoustic Field Analysis", *Acustica*, vol. 57, 1985
[5] H. Fleischer and V. Axelrad, "Restoring an Acoustic Source from Pressure Data Using Wiener Filtering", *Acustica*, vol. 60, 1986
[6] C. Spence, J. Nistler, E. Barouch, U. Hollerbach and S. Orszag, "Automated Determination of CAD Layout Failures Through Focus: Experiment and Simulation", *SPIE* Vol. 2197, 1994
[7] N. Shiraishi, S. Hirukawa, Y. Takeuchi and N. Magome, Proc. *SPIE* 1674, p. 741, 1992
[8] D.C. Cole, E. Barouch, U. Hollerbach, S.A. Orszag, "Derivation and Simulation of Higher Numerical Aperture Scalar Aerial Images", *Jpn. J. Appl. Phys.*, Vol. 31, Pt. 1, No. 12B, 1992

Efficient and Rigorous 3D Model for Optical Lithography Simulation

Kevin D. Lucas[1,3], Hiroyoshi Tanabe[2], Chi-Min Yuan[1] and Andrzej J. Strojwas[3]

[1]Advanced Products Research & Development Lab, Motorola Inc., Austin TX 78721, USA
[2]Opto-Electronics Research Lab, NEC Corporation, Kawasaki, Kanagawa 216, Japan
[3]Electrical & Computer Engin. Dept., Carnegie Mellon Univ., Pittsburgh, PA 15213, USA

Abstract

A new workstation-based rigorous model for 3D vector lithography simulation is introduced. The model extends a successful 2D lithography model, and has been applied to the simulation of 3D photomasks. The theory behind the new model is presented, and examples are given of the model's results and computational efficiency. The procedures for extending the model to the simulation of 3D optical alignment, metrology and photoresist bleaching problems are also given.

1. Introduction

Decreasing dimensions and increasing non-planarity of devices are creating complicated problems for the optical lithography process in semiconductor manufacturing. The large cost and time necessary for experiments make rigorous photolithography simulation increasingly cost effective in the solution of these problems. These problems include the modeling of realistic corners, contacts, vias, alignment marks and defects. We are presenting a new vector 3D model which runs quickly on common engineering workstations. This model has been implemented into a simulator, METROPOLE-3D, for the study of 3D photomasks.

The basis for this work was the extension of the fast and rigorous 2D *waveguide method*, a spatial frequency solution to Maxwell's equations[1]. The 2D method was previously implemented into a simulator, METROPOLE, and applied to the study of optical alignment and metrology[1], substrate bleaching[2] and phase shifting masks[3]. A major benefit of the *waveguide method* is its ability to simultaneously solve for all incoherent incoming light orders. Traditional models must repeat their simulations for each of these orders, which in 3D modeling may number in the hundreds.

2. 3D Theory

The potential for a 3D waveguide theory using vector potentials was shown by Tanabe[5]. The E and H fields are described using vector potential A and scalar potential ϕ as: $H = \nabla \times A$ and $E = ikA - \nabla\phi$. Inserting into Maxwell's equations provides the characteristic equation:

$$\nabla^2 A + k^2 \varepsilon A - \nabla(\nabla \bullet A) + ik\varepsilon\nabla\phi = 0 \cdot$$

Because of gauge freedom, the vector and scalar potentials can be changed simultaneously by $\quad A \to A - \nabla \Lambda \quad\quad \phi \to \phi - ik\Lambda \quad$ and observables E and H are unaffected.

Implementing the Lorentz gauge, which provides $\nabla \bullet A = ik\varepsilon\phi$, allows writing the characteristic equation as $\nabla^2 A + k^2 \varepsilon A - \nabla (\log \varepsilon)(\nabla \bullet A) = 0$.

Tanabe used the *waveguide method* approximation where non-planar structures are described by (often thin) layers in which the dielectric function is constant vertically (z-direction) but varies in the horizontal x and y directions (See Figure 1). This allows simplifying the z component of the characteristic equation in a layer to be $\nabla^2 A_z + k^2 \varepsilon A_z = 0$.

However, gauge freedom still exists. This allows using the transformation $\frac{\partial \Lambda}{\partial z} = A_z$ to set $A_z = 0$. Significantly, the characteristic equation, which is Maxwell's equations in vector potential form in each layer, is reduced to two coupled partial differential equations:

$$\nabla^2 A_x + k^2 \varepsilon A_x - \frac{\partial}{\partial x}\log\varepsilon\left(\frac{\partial A_x}{\partial x} + \frac{\partial A_y}{\partial y}\right) = 0 \qquad \nabla^2 A_y + k^2 \varepsilon A_y - \frac{\partial}{\partial y}\log\varepsilon\left(\frac{\partial A_x}{\partial x} + \frac{\partial A_y}{\partial y}\right) = 0$$

This formulation has only two variables to be solved for as opposed to three for traditional methods. In this work, we solve these equations, beginning by using the relations, $A_x = f(x,y)Z(z)$, $A_y = g(x,y)Z(z)$ and $Z(z) = C \cdot \exp(\alpha z) + C' \cdot \exp(-\alpha z)$.

Substituting truncated Fourier series for ε, $\log \varepsilon$, $f(x,y)$ and $g(x,y)$, such as

$$f(x,y) = \sum_{l=-N}^{N}\sum_{m=-N}^{N} B_{l,m} \cdot \exp\left(i2\pi\{b_1 lx + b_2 my\}\right) \qquad g(x,y) = \sum_{l=-N}^{N}\sum_{m=-N}^{N} D_{l,m} \cdot \exp\left(i2\pi\{b_1 lx + b_2 my\}\right)$$

where N is the # of approximating orders, we create a coupled eigenvalue problem for each layer which may be solved for the α, B and D coefficients

$$[J_1]B + [J_2]D = -\alpha^2 B \quad \text{and} \quad [J_3]D + [J_4]B = -\alpha^2 D \quad \text{or} \quad \begin{bmatrix} J_1 & J_2 \\ J_4 & J_3 \end{bmatrix}\begin{bmatrix} B \\ D \end{bmatrix} = -\alpha^2 \begin{bmatrix} B \\ D \end{bmatrix}$$

where each J_i is a full complex matrix of size $4N^2 \times 4N^2$.

3. Boundary Conditions

The C and C' coefficients still need to be calculated. These can be obtained from the continuity of parallel components of the E and H fields between layers. Inside any layer j, we are able to write A_x and A_y, using $\Psi_{l,m} = \exp\left(i2\pi\{b_1 lx + b_2 my\}\right)$, at a relative z position (See Figure 1), as

$$A_x^j = \sum_{h=1}^{2N}\left[C_h^j \cdot \exp\left(\alpha_h^j(z-z_j)\right) + C_h^{j'} \cdot \exp\left(-\alpha_h^j(z-z_j)\right)\right]\sum_l\sum_m B_{h,l,m}^j \Psi_{l,m}$$

$$A_y^j = \sum_{h=1}^{2N}\left[C_h^j \cdot \exp\left(\alpha_h^j(z-z_j)\right) + C_h^{j'} \cdot \exp\left(-\alpha_h^j(z-z_j)\right)\right]\sum_l\sum_m D_{h,l,m}^j \Psi_{l,m}$$

In layer 0, where $\Omega_{l,m}^0 = ik\sqrt{\varepsilon - (lb_1\lambda)^2 - (mb_2\lambda)^2}$, A_x and A_y are given by:

$$A_x^0 = \sum_l\sum_m \{X_{l,m}^0 e^{-z\Omega_{l,m}^0} + a_{x(l,m)} e^{z\Omega_{l,m}^0}\}\Psi_{l,m} \qquad A_y^0 = \sum_l\sum_m \{Y_{l,m}^0 e^{-z\Omega_{l,m}^0} + a_{y(l,m)} e^{z\Omega_{l,m}^0}\}\Psi_{l,m}$$

In the substrate A_x and A_y are defined relative to the substrate depth, z_s, as:

$$A_x^S = \sum_l\sum_m \{X_{l,m}^S e^{ik(z-z_s)\sqrt{\varepsilon_s - (lb_1\lambda)^2 - (mb_2\lambda)^2}}\}\Psi_{l,m} \quad A_y^S = \sum_l\sum_m \{Y_{l,m}^S e^{ik(z-z_s)\sqrt{\varepsilon_s - (lb_1\lambda)^2 - (mb_2\lambda)^2}}\}\Psi_{l,m}$$

At the interface of layers 0 and 1, we impose the boundary conditions and eliminate X^0 and Y^0 to obtain: $G^1\begin{bmatrix} C^1 \\ C'^1 \end{bmatrix} = R\begin{bmatrix} a_x \\ a_y \end{bmatrix}$ which can be written as $T^0\begin{bmatrix} C^1 \\ C'^1 \end{bmatrix} = \begin{bmatrix} a_x \\ a_y \end{bmatrix}$.

We use boundary conditions between layers j and $j+1$ to relate C coefficients

$$G^j \begin{bmatrix} C^j \\ C'^j \end{bmatrix} = G^{j+1} \begin{bmatrix} C^{j+1} \\ C'^{j+1} \end{bmatrix} \quad \text{or} \quad \begin{bmatrix} C^j \\ C'^j \end{bmatrix} = T^j \begin{bmatrix} C^{j+1} \\ C'^{j+1} \end{bmatrix}$$

and we use boundary conditions between the last layer and the substrate to write the expression:

$$G^q \begin{bmatrix} C^q \\ C'^q \end{bmatrix} = H \begin{bmatrix} X^S \\ Y^S \end{bmatrix} \quad \text{or} \quad \begin{bmatrix} C^q \\ C'^q \end{bmatrix} = T^q \begin{bmatrix} X^S \\ Y^S \end{bmatrix}.$$

H and T have size $16N^2 \times 8N^2$. For a structure with q layers, we combine T matrices to obtain $U^j = T^j T^{j+1} T^{j+2} \ldots T^q$ allowing us to write $\begin{bmatrix} X^S \\ Y^S \end{bmatrix} = (U^0)^{-1} \begin{bmatrix} R_1 \\ R_2 \end{bmatrix}$ and $\begin{bmatrix} C^j \\ C'^j \end{bmatrix} = U^j \begin{bmatrix} X^S \\ Y^S \end{bmatrix}$ where R contains the illumination information, X^S & Y^S are the amplitudes of transmitted light and X^0 & Y^0 are the amplitudes of light reflected from the structure. Thus, for modeling 3D photomasks we can compute X^S and Y^S, for 3D alignment and optical metrology we can compute X^0 & Y^0 from C^1 and C'^1. and for modeling 3D photoresist bleaching, the internal light amplitudes in layers can be known by using the C and C' of a layer to solve for A_x and A_y.

4. Results

The model was implemented into a photomask simulator, METROPOLE-3D, which runs on engineering workstations. We have simulated examples of 3D binary and phase shifting masks for 1X and 5X reduction imaging systems. Rigorous simulation is necessary for masks with small structure sizes (approaching the wavelength of illuminating light) and for masks with vertical topography. These masks have light scattering effects which are not taken into account by approximate, scalar simulators. To verify the results of our model, we compared aerial images generated by METROPOLE-3D versus experimental images and those generated by a scalar model, SPLAT[5](See Figure 2). The imaging system had a 5X reduction factor, I-line illumination (λ=0.365 micron), a numerical aperture of 0.6, and a partial coherence of 0.6. The images are cross sections for isolated contact openings of 0.49 and 0.39 microns, after reduction, in a 10% transmission embedded attenuating phase shifting mask. The experimental aerial images were provided by SEMATECH. As can be seen, with the larger 0.49 micron opening, the results of METROPOLE-3D and the scalar model both match the experimental data well. However, for the 0.39 micron opening, where light scattering effects are more pronounced, only our rigorous vector model is able to accurately predict the experimental light intensity results. Modeling of these small mask features is important for the quick and accurate design of masks in the critical layers of next generation devices. We also implemented theoretical checking routines for comparing the model's results against the conservation of energy and reciprocity theorems. Our model's error from the theoretical predictions is very slight, typically much less than 1%.

5. Run Time and Memory Usage

Traditional workstation-based rigorous lithography models may take days to complete complex simulations. The run time of our model on an IBM RS6000 model 550 workstation is typically under a few hours for non-symmetric simulations. Simulations of fully symmetric structures typically run in under an hour[6]. The memory usage is also moderate for today's desktop workstations, typically using less than 120MB.

6. Summary

Using a vector potential extension to an existing 2D electromagnetic solution method, we have created a new rigorous 3D lithography model. This model has been implemented in a photomask simulator for engineering workstations and has been shown to run quickly and accurately. In addition, the model is easily extendable to the efficient solution of other non-planar 3D lithography problems in optical alignment, metrology and photoresist bleaching.

References

[1] C. M. Yuan, A. J. Strojwas. *Proc. of SPIE Microlithography* Vol. 1264, 1990.
[2] K. D. Lucas, C. M. Yuan, A. J. Strojwas. *SISDEP IV,* 1991.
[3] K. D. Lucas, A. J. Strojwas. *Proc. of SPIE Microlithography* Vol. 1809, 1992.
[4] H. Tanabe. *Proc. of SPIE Microlithography* Vol. 1674, 1992.
[5] K. K. H. Toh, A. R. Neureuther. *Proc. of SPIE Microlithography* Vol. 1463, 1991.
[6] K. D. Lucas, H. Tanabe, A. J. Strojwas. *Proc. of SPIE Microlithography* Vol. 2440, 1995.

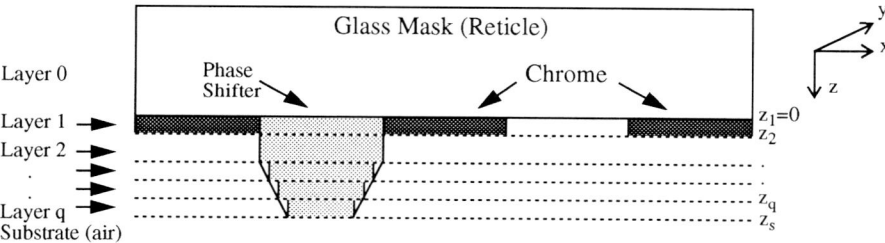

Figure 1. 2D cross section of multiple layer approximation for slanted shifter phase shifting mask. The approximation is valid if the distance errors introduced are much less than the wavelength of light.

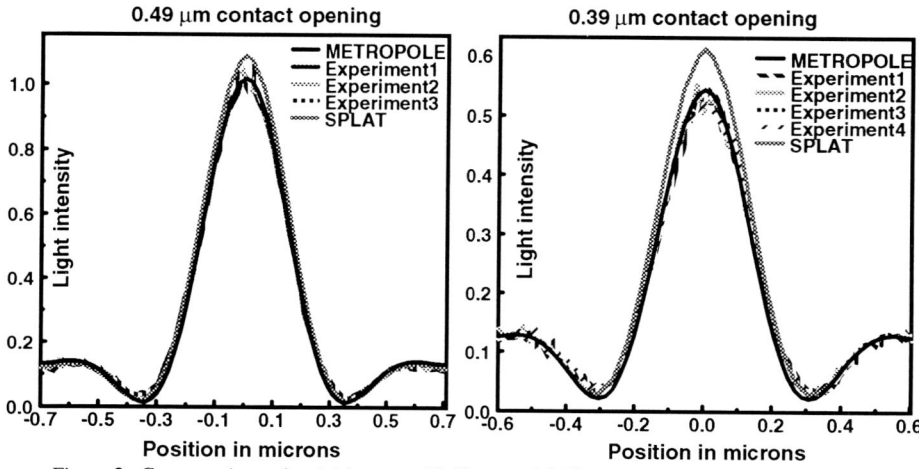

Figure 2. Cross sections of aerial images of 0.49 μm and 0.39 μm contacts in 10% attenuated PSM.

Application of the Two-dimensional Numerical Simulation for the Description of Semiconductor Gas Sensors

D.Schipanski, Z.Gergintschew, J.Kositza

Technical University of Ilmenau, Institute for Solid State Electronics
PF 0565 Kirchhoffstr.1, 98684 Ilmenau, Germany

Abstract

The implementation of models for the gas adsorption on metal-oxide-semiconductors in a two-dimensional device simulator is the subject of this paper. Further we give an application example for the simulation of the sensitive reactions and the electronic device in their exchange. For that we use a conductivity type gas sensor with a polycrystalline ZnO active layer.

1. Introduction

The usage of electronic devices like transistors and resistors is nowadays a common approach to realize the transducer function of modern chemical gas sensors. Moreover the development of gas sensors with a metal-oxide (like ZnO, SnO_2 or Ga_2O_3...) as sensitive layer is mainly empirical. That's why, it is useful to employ numerical simulation methods for the development, description and optimization of semiconductor gas sensors. For that it is necessary to use in addition to the equations for electrical behaviour description, models for the gas adsorption and for the corresponding electrical parameter modification of the sensitive layer and the device behaviour.

2. Model for the gas chemosorption and use in the simulator PROSA

The description of the adsorption effects is based on the 'electron theory of catalysis on semiconductors' by Wolkenstein [1], which has been used already by Geistlinger [2] for the description of gas sensitive effects. The main properties of this model are discussed in [2][3], so we will describe it here briefly by means of the oxygen adsorption on n-type metal-oxide-semiconductor. We consider the steady-state case with the following assumptions for temperatures under 700 K: adsorption without dissociation of the oxygen molecule and no reaction of the O-vacancies with the O_2 from the gas.

$$O_2 + e^- \rightleftharpoons O_2^- \qquad (1)$$

In [1] Wolkenstein considered the neutral 'weak' chemosorption as precursor of the charged 'strong' chemosorption. The first one acts as an acceptor state on the metal-oxide surface. The particle is strong chemosorbed after an electron exchange with the

metal-oxide and a surface charge arise due to the strong chemosorption.
As a result of the steady-state chemosorption consideration by Wolkenstein we have for the fractional surface coverage Θ with O_2^- the following isotherm, which describes the chemosorption in dependence on the semiconductor properties (E_f) of the sensitive surface

$$\Theta = \frac{N}{N_0} = \frac{\beta P_{O_2}}{\beta P_{O_2} + 1} \qquad (2)$$

with

$$\beta = \frac{s_0}{\nu\sqrt{2\pi MkT}exp\left(\frac{-Q^0}{kT}\right)} \frac{1}{f^0\left[1 + exp\left(\frac{E_f - E_c}{kT}\right)\right]} \qquad (3)$$

where N is the density of occupied surface states; N_0 the density of the maximal available surface states; P_{O_2} the oxygen partial pressure; s_0 the adhesion coefficient; M the mass of the adsorbed particle; k the Boltzmann constant; T the lattice temperature; ν the phonon frequency of the adsorbed particle; Q^0 the weak chemosorption energy; E_c the energy of the conduction band edge; E_f the Fermi energy; f^0, f^- the occupation probabilities for the weak or strong chemosorption.
The charging of the metal-oxide surface due to the strong chemosorption is given by the following expression:

$$Q'_{ad} = -eN_0\Theta^- = -eN_0 f^- \Theta \qquad (4)$$

with e as elementary electronic charge.
For the simulation of the complete sensor devices the Poisson equation (5), the equation for the adsorption charge (like Wolkenstein) (6) and the continuity equations for electron and holes (7)(8) have to be solved

$$div[Q'_{ad} - \epsilon grad\phi] = e(V_0^+ + 2V_0^{++} - n + p - N_A^- + N_D^+) \qquad (5)$$
$$Q'_{ad} = -eN_0 f^- \Theta(P_{O_2}, \phi...) \qquad (6)$$
$$div J_n = eR \qquad (7)$$
$$div J_p = -eR \qquad (8)$$

where ϕ is the electrical potential; V_0^+, V_0^{++} the density of the single and double ionized oxygen vacancies; N_A, N_D the acceptor and donator doping density; n, p the electron and hole density; J_n, J_p the current density of electrons and holes.
The equation system (5)..(8) is solved self-consistently with the 2d device simulator PROSA [4], which was developed at the TU Ilmenau. PROSA uses the finite difference method.

3. Application on a semiconductor gas sensor

Up to now there are two types of semiconductor gas sensors. The first one uses the effect of the work function change on the metal-oxide sensitive layer surface due to the band bending caused by the adsorption charge. An example for such sensor device is the Suspended Gate Field Effect Transistor (SGFET). A detailed description of the SGFET and a modelling of the influence of such parameters like sensitive layer doping and thickness are presented in [3]. A second type of a semiconductor gas sensor is the conductivity one [5]. As shown in Fig.1 it consists of a insulating substrate, a polycrystalline metal-oxide semiconductor layer and two electrodes. The charging of

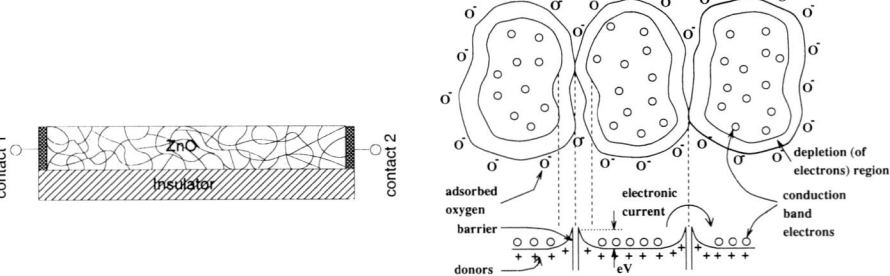

Figure 1: Schematic plot of a conductivity type gas sensor

Figure 2: Mechanism of the conductivity change due to O_2-chemosorption

the grain surface due to the chemosorption has two effects on the layer conductivity. The first one is the influence of the potential barriers on the grain boundaries and the second one is the conductivity changing in the grain volumina due to the space charge regions (Fig.2).

For the simulation we divide the metal-oxide layer in several regions with regular shape, which represent the polycrystalline grains. We distinguish between compact sputtered metal-oxide sensitive layers, where the gas adsorption can take place only on the interface to the gas phase and the sintered metal-oxide layers, where the gas can diffuse around the grains. The schematic structures for the first case is shown in Fig.3. The current transport across the grain boundaries is considered to be controlled by the potential barriere, that means by the trap distribution or by the adsorption charge density at the boundaries. Effects like the thermionic emission (9) and recombination at the grain boundaries (10) are also taken into account.

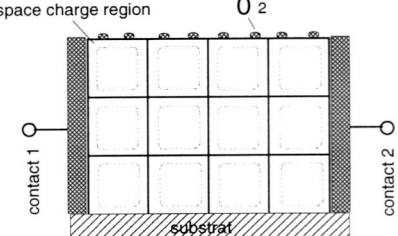

Figure 3: Schematic structure of the simulate conductivity type gas sensors with a compact metal-oxide layer

$$J = -\frac{A^*T^2}{N_C}(n_m - n_0) \quad \text{with} \quad n_0 = N_C exp\left(-\frac{e\phi_B}{kT}\right) \quad (9)$$

$$R_s = \frac{np - n_i^2}{\frac{1}{s_p}(n + n_i) + \frac{1}{s_n}(p + n_i)} \quad (10)$$

Here are A^* the effective Richardson constant, ϕ_B the barrier height, n_m the electron density at nonequilibrium, n_i the intrinsic density and s the grain boundary recombination velocity.

In Fig.4 the current density distribution in a ZnO layer with 5x3 grains is shown for the case of gas adsorption only on the top-interface. The average grain size is 200nm, the O-vacancy density amount 10^{14} cm^{-3} and the applied voltage amount 2 Volt. The current flows always within the grains. It is smaller near at the surface and larger in the bulk of the ZnO layer. A cause for this current density distribution is on the one hand the depletion region at the ZnO-surface as a result of the oxygen chemosorption and on the other hand effect of the steady surface state density at the grain surface.

For the same case the relative resistance change by O_2-adsorption is shown in Fig.5. One can see a nearly linear dependence between the relative resistance and the logarithm of oxygen partial pressure over a large region.

4. Summary

The chemosorption considered in dependence on the semiconductor properties of the sensitive surface and the compatibility with the equations of the drift-diffusion model are the main advantages of the adsorption description by Wolkenstein for our application. The first applications of the 2d-simulator PROSA show that it is a suitable tool for modelling of semiconductor gas sensors. PROSA offers the possibility to model the sensor devices completely that means the sensitive chemo-physical reactions and the transducer function of the electron devices in their exchange. For the application for sensors with other metal-oxide sensitive layers and different gases or gas mixtures one have to determine some of the needed input parameters.

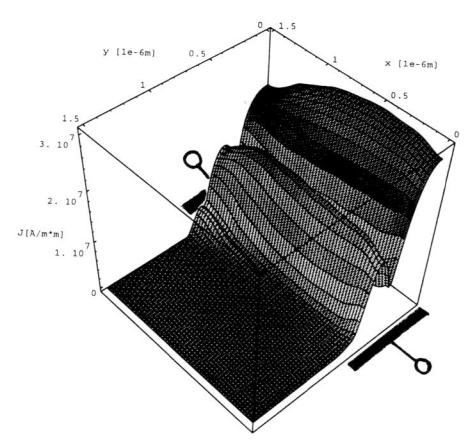

Figure 4: Current density distribution in a polycrystalline ZnO layer

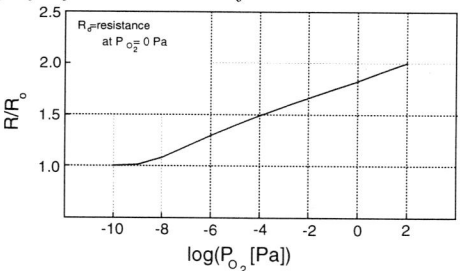

Figure 5: Relative resistance change of the ZnO layer by O_2-adsorption

Acknowledgement

This work has been supported by the German Federal Ministry for Research and Technology under Project Number 13 MV 0331 0. Further we acknowledge the support of Mr.H. Foerster and Mr.D. Nuernbergk with the basic version of the simulator PROSA.

References

[1] T. Wolkenstein, "Electronic Processes on Semiconductor Surfaces during Chemisorption," *Consultants Bureau, New York*, 1991.

[2] H. Geistlinger, "Electron theory of thin-film gas sensors," *Sensors and Actuators B*, vol. 17, no. 1, pp. 47-60, 1993.

[3] Z. Gergintschew, H. Foerster, J. Kositza and D. Schipanski, "Two-dimensional numerical simulation of semiconductor gas sensors," *Sensors and Actuators B*, vol. 26, no. 1-3, pp. 170-173, 1995.

[4] H. Foerster and D. Nuernbergk, "PROSA-3.0," Technische Universitaet Ilmenau, Institut fuer Festkoerperelektronik, 1995.

[5] N. Tagutchi, "UK Patent specification 1280809," 1970.

Analysis of Piezoresistive Effects in Silicon Structures Using Multidimensional Process and Device Simulation

M. Lades, J. Frank[a], J. Funk[b], G. Wachutka

Chair for Physics of Electrotechnology
Technical University Munich
Arcisstrasse 21, D-80290 Munich, GERMANY
[a] Corporate Research and Development, Siemens AG
Otto-Hahn-Ring 6, D-81730 Munich, GERMANY
[b] Physical Electronics Laboratory
Swiss Federal Institute of Technology
ETH-Hönggerberg, HPT, CH-8093 Zurich, SWITZERLAND

Abstract

With the view to analyzing piezoresistive effects in silicon microstructures we implemented a rigorous physically-based model in the multidimensional general purpose device simulator DESSIS^{ISE}. In this model, the dependence of the piezoresistive coefficients on temperature and doping concentration is included in a numerically tractable way. Using a commercial TCAD system (ISE), the practicability of the approach is demonstrated by performing a complete simulation sequence for realistic microdevices ranging from the layout design up to the analysis of the device operation.

1. Introduction

Mechanical distortion of silicon microstructures results in a change in the electric conductivity. In modern semiconductor technology this effect is employed in realizing smart integrated micromechanical sensors. On the other hand, piezoresistivity arises as undesired parasitic effect in silicon devices due to mechanical stress induced by thermal treatment or packaging. Up to now, a predictive numerical analysis of the performance of piezoresistive elements integrated in semiconductor microdevices was restricted to idealized structures with simplified geometry, assuming spatially uniform piezoresistive coefficients along high-symmetric crystal orientations without local dependence on the doping concentration or temperature distribution. Of course, with these quite coarse approximations a quantitative analysis of realistic devices is hardly possible. For accurate results a full multidimensional numerical simulation is required, which is based on a coupled field description of the piezoresistive effects. A reliable numerical approach is demonstrated in this work.

2. Modelling

A rigorous physically-based model [1] describing the strain-induced changes in the electric conductivity of single-crystalline silicon has been implemented in the multidimensional general purpose device simulator DESSIS^{ISE} [2]. The basic part of the model is a linear extension of the constitutive current relations for electrons and holes,

$$J_\alpha = -\sigma_\alpha \cdot (\bar{I} + \bar{\Pi}_\alpha \cdot \bar{X}) \cdot \nabla \varphi_\alpha, \quad (\alpha = n, p)$$

where σ_α denotes the isotropic electric conductivity in the absence of stress, $\bar{I}$ the identity tensor, $\bar{X}$ the mechanical stress tensor, φ_α the quasi-Fermi potential and $\bar{\Pi}_\alpha$ the tensor of piezoresistive coefficients which depend on the doping concentration and the temperature distribution.

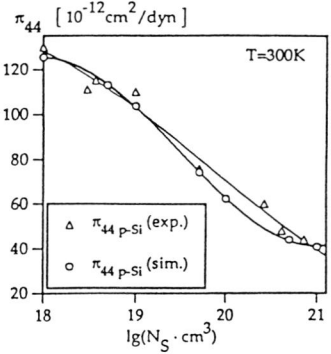

Fig.1:
Comparison of measured [3] and calculated piezoresistive shear coefficient Π_{44} in p-doped silicon versus concentration N_s of the diffused piezoresistive layer.

The implemented model was validated with reference to experimental piezoresistive coefficients measured in diffused n- and p-type silicon layers [3]. As shown in Fig. 1 for the shear coefficient Π_{44} in p-silicon, for instance, we achieved good agreement between simulation and measurement of the test structures.

3. Simulation of integrated silicon piezoresistive sensors

The capabilities of the interfaced simulation system were demonstrated by a complete simulation sequence ranging from layout and process simulation up to piezoresistive device simulation. Two realistic microtransducer structures fabricated by means of industrial silicon IC technology were investigated. Each of them is basically a square silicon diaphragm (10x1000x1000 µm³) with integrated piezoresistors connected in a Wheatstone bridge. The first structure represents the conventional layout of a pressure sensor (Fig. 2) as proposed in [4]. The second device (Fig. 3) is an electrothermally excited microresonator with piezoresistive readout, which is used as test structure for determining thermoelastic material properties. The four piezoresistors of the Wheatstone bridge probe the thermoelastic deformations caused by a heating resistor placed at the diaphragm centre. Obviously the two structures are similar with respect to the mechanical behavior, but differ in the arrangement of the piezoresistors.

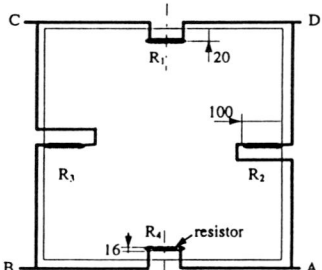

Fig. 2: Top view of a sensor structure for pressure measurement [4] (lengths in µm).

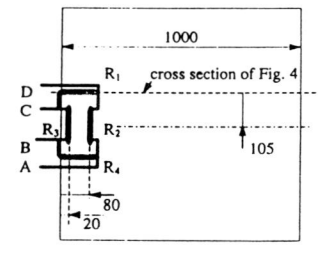

Fig. 3: Top view of a microresonator structure with piezoresistive readout (lengths in µm).

For easy comparison of the respective sensitivities, we assumed equal technological and geometrical parameters for both structures (namely those measured on the microresonator). The piezoresistors were fabricated using a boron implantation with subsequent drive-in diffusion. The maximum doping concentration amounts to 2×10^{18} cm^{-3} and is located at a depth of about 0.6 µm. In our simulations we assumed that a pressure difference $\Delta p = p_1 - p_0$ between top and bottom side of the diaphragm caused the mechanical deformation (cf. Fig. 4).

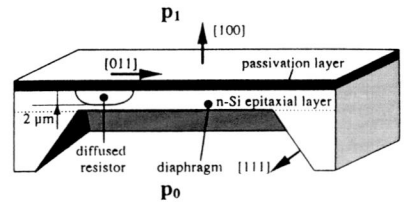

Fig. 4:
Cross section of the simulated pressure sensor, showing crystal orientations and the location of the piezoresistors. p_0 is the reference pressure, p_1 the measurand.

The sequence of simulation steps is shown in Fig. 5. Using the two-dimensional technology simulator DIOS^{ISE} [5] all fabrication steps of the diaphragm and the piezoresistors can be simulated, yielding their doping profile.

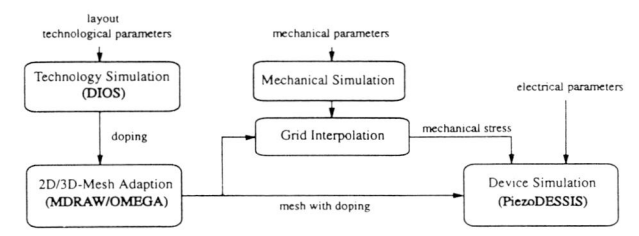

Fig. 5: Schematic simulation sequence and data flow.

The DIOS output mesh is adapted for the use in the piezoresistive simulation step by means of the automated grid manipulating programs MDRAW^{ISE} (2D) and OMEGA^{ISE} (3D). Multi-dimensional structural analysis is interfaced by projecting the resulting stress field onto the adapted grids. In order to achieve high accuracy in the calculation of the diaphragm deformation, the entire transducer structure (i.e., diaphragm and its suspension on bulk silicon) was taken into account. Also the passivation layer on top, consisting of 0.6 µm SiO$_2$ and 0.3 µm Si$_3$N$_4$, was included in the simulation domain, since it is known that the mechanical behavior can significantly be influenced by that. Fig. 6 illustrates the different areas underlying the mechanical and the electrical simulations.

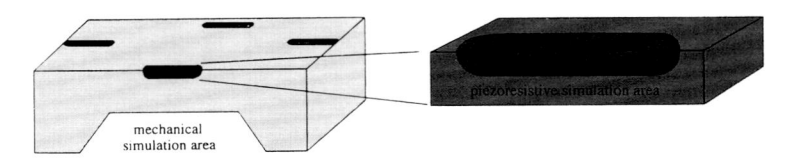

Fig. 6: Embedding of the electrical simulation domains in the full diaphragm structure used in the mechanical analysis.

Lastly, with the device structure and the field of mechanical stress as input, the extended DESSIS version computed the relative change in electrical resistivity $\Delta R_i / R_0$ ($i=1,...,4$) for each of the piezoresistors. From this data, the output voltage of the Wheatstone bridge is determined according to [6]

$$V_{BD}/V_{AC} = \Delta V_{BD}/V_{AC} = \left(-\Delta R_1 + \Delta R_2 + \Delta R_3 - \Delta R_4\right)/R_0$$

where R_0 is the basic reference resistance and V_{AC} denotes the supply voltage. A reasonable measure of the pressure sensitivity S can be defined as

$$S = \Delta V_{BD}/(V_{AC} \cdot \Delta p)$$

Fig. 7 displays the calculated relative change of the bridge voltage versus the applied pressure difference. For the first example (pressure sensor), a sensitivity of S = 6.1 mV / (V bar) was obtained. This value falls in a range typical of such an arrangement of piezoresistors, as it can easily be estimated from the functional dependence of the sensitivity on the diaphragm side length [6]. The sensitivity of the second structure (microresonator) was calculated to be S = 1.1 mV / (V bar), which is significantly smaller than the measured value. Presumably the difference results from a calibration problem. In experiment, the deformation of the diaphragm was caused by the thermoelastic effect as mentioned above, and not by a uniform pressure difference as simulated. Reference for the calibration was the measured elongation at the diaphragm center which, of course, is not a linear measure of the overall deformation. A comprehensive simulation of the coupled thermo-mechanical effects inside the structure, planned as future work, should provide clarity.

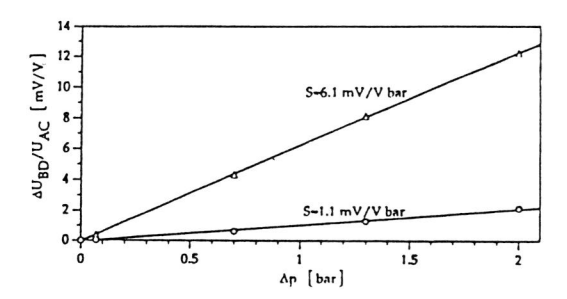

Fig. 7: Sensor response of the analyzed diaphragm structures (Δ structure from Fig. 2, o structure from Fig. 3).

4. Conclusion

The presented approach constitutes a practicable method for the numerical analysis of realistic piezoresistive structures. The practicability of a full simulation sequence ranging from layout and process simulation through structural analysis up to piezoresistive device simulation has been demonstrated with reference to realistic microtransducer structures.

5. Acknowledgement

The authors are grateful to U. Krumbein, Dr. N. Strecker and Dr. A. Schenk from the Institute of Integrated Systems, ETH Zurich, for providing helpful assistance and technical support.

References

[1] Z. Z. Wang, "Modélisation de la Piézoresistivité du Silicium", *PhD thesis* No. 9929, Univ. of Science and Technology, Lille, France (1994).
[2] DESSIS[ISE] Reference Manual Version 1.3 (1994), Institute of Integrated Systems, ETH Zurich (1993).
[3] O. N. Tufte, E. L. Stelzer, "Piezoresistive Properties of Silicon Diffused Layers", *J. Appl. Phys.* **34** (1963) 313-318.
[4] K.W. Lee, K. D. Wise, "SENSIM: A Simulation Program for Solid-State Pressure Sensors", *IEEE Trans. on ED*, vol. **ED-29** (1982) 34.
[5] N. Strecker, DIOS[ISE] User's Manual, Institute of Integrated Systems, ETH Zurich (1993).
[6] P. Ciampolini et al, "Electro-Elastic Simulation of Piezoresistive Pressure Sensors", in *Simulation of Semiconductor Devices and Processes*, vol. 5, ed. by S. Selberherr et al, Springer, Vienna (1993) 381-384.

Modeling of magnetic-field-sensitive GaAs devices using 3D Monte Carlo simulation

C. Brisset [a,b], F.-X. Musalem [a], P. Dollfus [a], and P. Hesto [a]

[a] Institut d'Électronique Fondamentale, Bâtiment 220
Université Paris-Sud, 91405 Orsay cedex, France
[b] present adress: CEA, Service Électronique, BP 12
91680 Bruyères-le-Châtel, France

Abstract

The action of the Lorentz force on carrier motion is now included in MONACO, our 3D Monte Carlo device simulator. As examples we model two typical magnetic-field-sensitive devices, in which the detected signal is either the Hall voltage, or the induced current at two contacts. The results are in good agreement with expected characteristics.

1. Introduction

The trend towards the size reduction of integrated magnetic sensors makes necessary to account for geometrical effects and inhomogeneity of fields in theoretical prediction of galvanomagnetic phenomena governing device operation. Simulation tools are usually based on 2D numerical solution of macroscopic drift-diffusion equations augmented by magnetic-field-dependent terms [1]. Using the powerful Monte Carlo technique, studies of electron transport under homogeneous electric and magnetic fields [2] or in a gyrotron oscillator [3] have been also reported. This modeling technique can be applied to galvanomagnetic device simulation in order to gain insight into the description of physical phenomena involved in device operation. In this work, we present analyses of magnetic-field-sensitive devices using 3D Monte Carlo simulation.

2. The model

The main characteristics of MONACO, a 3D Monte Carlo device simulator, have been previously described [4,5]. The 3D Poisson equation is solved using a finite-elements scheme. The action of the Lorentz force on the carrier motion is now implemented in this software. This implementation is restricted to the case of spherical band structure leading to quite tractable equation of carrier motion. The magnetic field is uniformly applied in the device.

An important aspect of the algorithm used for the calculation of carrier trajectories is the detection of cell changes that allows to update in "real time" the electric field experienced by particles. This detection requires, before each free-flight, the calculation of the flight-time needed to reach a cell boundary, *i.e.*, the inversion of $\vec{r}(t)$ function. In this order, the carrier real-space motion is simplified in the following manner. First the conductivity mass is assumed to be constant during the free flight and calculated as a function of the non parabolicity coefficient and the initial energy. This is all the more justified that duration of free-flight is smaller, i.e., the energy change is lower. Second, by denoting ω_c the cyclotron frequency, the functions $\sin(\omega_c t)$ and $\cos(\omega_c t)$ are expanded in series terminated with the $(\omega_c t)^3$-term, which is perfectly justified for usual values of magnetic field and mean free time between collisions. This yields a third degree power function $\vec{r}(t)$ from which the flight-time needed to reach a cell boundary is easily extractable. However, by using this simplified approach, the energy gained by the particle in real space may be slightly inconsistent with its motion in $\vec{k}$-space. To ensure consistency, the final components of wave vector are adjusted *a posteriori* without significant error.

3. Simulation of magnetic sensors

We have simulated the two structures schematized in Fig.1, consisting of a uniformly doped GaAs bar ($N_D=10^{16}$cm^{-3}) with two electrodes (DEV-a) and four electrodes (DEV-b). In both devices the length L separating the bias electrodes numbered 1 and 2 is 3μm, the width W is 6μm, and the contact width W_c is 2μm. DEV-b differs from DEV-a by the presence of two lateral electrodes numbered 3 and 4.

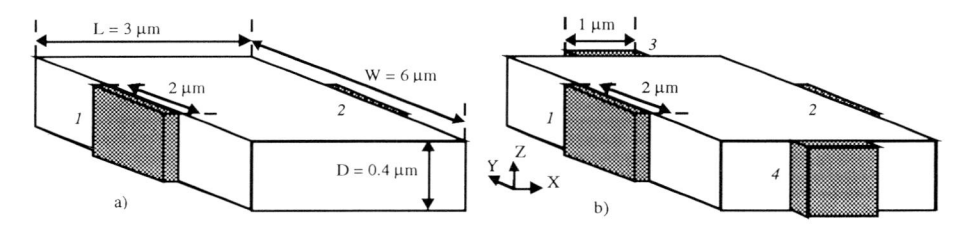

Figure 1: Simulated stuctures (GaAs, $N_D=10^{16}$cm^{-3}) ; a) DEV-a ; b) DEV-b

In both devices a voltage $V_X = 0.3$V is applied between the contacts numbered 2 and 1 to set the electric field E_X about 1kV/cm. A vertical magnetic field B_Z is uniformly applied in the whole structure. In DEV-a, it results in the building of an electric field E_Y and then of a Hall voltage V_H across the device width. In DEV-b the contacts 3 and 4 are short-circuited and the combination of electric field E_X and magnetic field B_Z induces a current I_H between these contacts.

Fig.2 illustrates the equipotential lines in a XY-plane resulting from an applied magnetic field $B_Z = 0.8$ T in DEV-a and Fig.3 shows the potential profiles along the width of the structure for several X positions. Under these field conditions the Hall voltage developed across the device is $V_H = 95$ mV. Fig.4 is a plot of V_H as a function of applied B_Z. For $B_Z \leq 0.6$ T the linearity of $V_H(B_Z)$ is excellent (S = 127 mV/T), in agreement with the usual expression of Hall field E_Y:

$$E_Y = -\omega_c \frac{\langle \tau_m^2 \rangle}{\langle \tau_m \rangle} E_X = -\frac{q}{m^*} \frac{\langle \tau_m^2 \rangle}{\langle \tau_m \rangle} E_X B_Z \quad (1)$$

By defining the Hall mobility as $\mu_H = R_H \sigma = V_H / E_X B W_c$, the obtained value of 6350 cm^2/Vs is in good agreement with experimental ones.

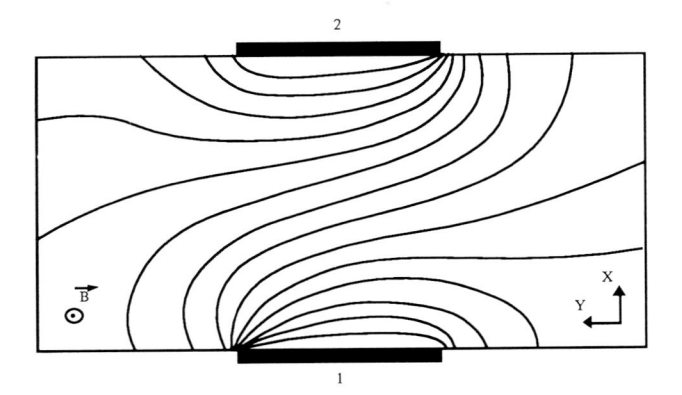

Figure 2: Equipotential lines in DEV-a.($V_x=0.3V$, $B_z=0.8T$)

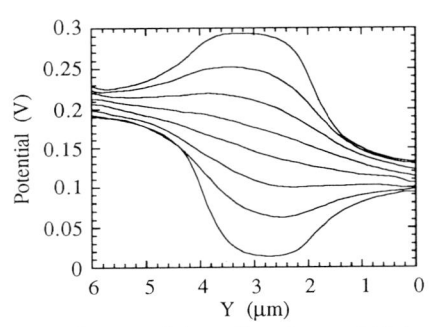

Figure 3: potential profiles along width in DEV-a for different X values. ($V_x=0.3V$, $B_z=0.8T$).

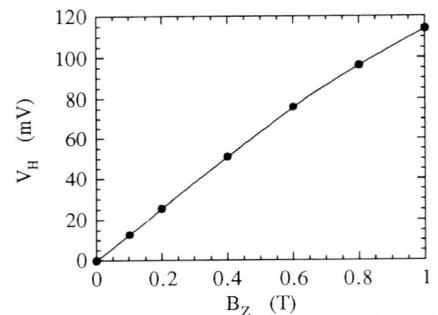

Figure 4: Hall voltage as a function of magnetic field in DEV-a. ($V_x=0.3V$).

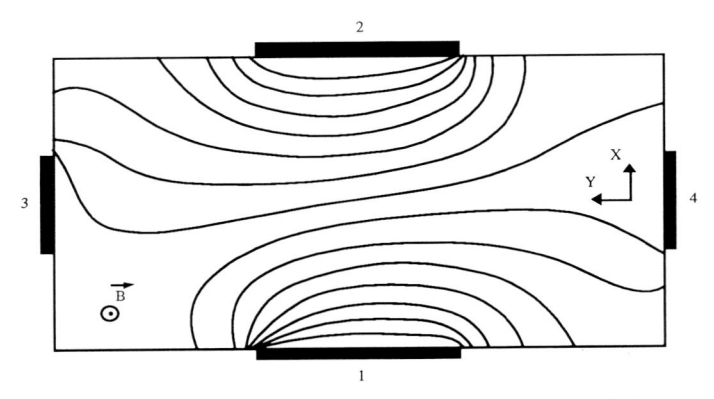

Figure 5: Equipotential lines in DEV-b.($V_x=0.3V$, $B_z=0.8T$)

Fig.5 shows the equipotential lines in a XY-plane of DEV-b for $B_Z = 0.8$ T, with electrode 3 and 4 short-circuited. The magnetic field-induced-current $I_Y = I_H$ is plotted as a function of B_Z in Fig.6. This plot is very well fitted (solid line in Fig.6) by a function similar to the theoretical expression of the current density j_Y given by:

$$j_Y = \frac{n\,q^3}{m^{*2}} \langle \tau_m^2 \rangle E_X \frac{B_Z}{1 + \frac{q^2 \langle \tau_m^2 \rangle}{m^{*2}} B_Z^2} \qquad (2)$$

The best fit is obtained for $\langle \tau_m^2 \rangle = 3.9 \times 10^{-26}$ s. Introducing this value in eq. (1) yields a scattering factor $r_H = \langle \tau_m^2 \rangle / \langle \tau_m \rangle^2$ about 1.3.

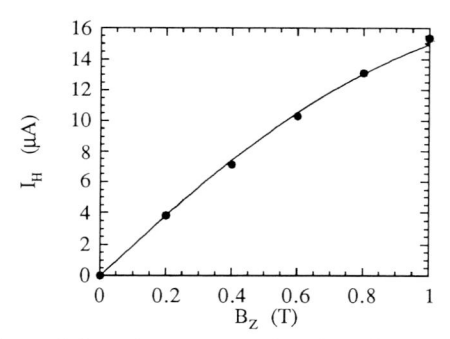

Figure 6: lateral current as a function of magnetic field in DEV-b. (V_x=0.3V).

4. Conclusion

MONACO is now a powerful 3D simulation tool for investigation of magnetic sensors and devices (magnetotransistors, Hall structures,...). Qualitative and quantitative results on typical magnetic-field-sensitive devices shown in this work validates our approach. Complex geometries including heterostructures may be now taken into account.

References

[1] C. Riccobene, G. Wachutka, J. Bürgler, H. Baltes, *IEEE Trans. Electron Devices*, vol. ED 41, pp. 32-43, 1994

[2] D. Chattopadhyay, *J. Appl. Phys.*, vol. 45, pp. 4931-4933, 1974

[3] A.K. Ganguly, B.H. Hui, K.R. Chu, *IEEE Trans. Electron Devices*, vol. ED 29, pp. 1197-1209, 1982

[4] C. Brisset, P.Dollfus, N.Chemarin, R.Castagné, P.Hesto, Proc. SISDEP'93, pp. 189-192, 1993

[5] C.Brisset, P.Dollfus, O.Musseau, J.-L.Leray, P.Hesto, IEEE Trans. Nucl. Sci. vol. 41, pp. 2297-2303, 1994

Quasi Three-Dimensional Simulation of Heat Transport in Thermal-Based Microsensors

A. Nathan[a] and N.R. Swart[b]

[a]Electrical and Computer Engineering, University of Waterloo
Waterloo, Ontario N2L 3G1, Canada
on leave at: Physical Electronics Laboratory, ETH Hoenggerberg
CH-8093 Zurich, Switzerland

[b]Institut National d'Optique, Sainte-Foy, Quebec G1P 4N8, Canada

Abstract

Results based on quasi three-dimensional numerical solutions of electrothermal behaviour in thermally isolated microstructures are presented. Here, we solved the two-dimensional system of electrothermal equations with heat loss to the surrounding (due to natural convection) incorporated as a mixed boundary condition. The convective heat loss was calculated based on a three-dimensional solution to the heat conduction equation using a boundary element method. The technique, employed in the analysis of heat transfer in a µ-Pirani gauge, yields numerical soultions which provide good agreement with measurement data.

1. Introduction

Thermal-based microsensors are required in a variety of sensing applications including detection of flow rate [1], pressure (vacuum) [2], and gas species [3]. In flow sensing, the heat loss from a resistively heated microstructure (due to forced convection) is modulated by the flow rate. In gas sensing, an isothermal heat surface (microhotplate) can be used to raise the temperature of the sensing film to increase absorption or reaction of gas on the film surface. With pressure sensing (based on the Pirani principle), heat transfer from a heating element is modulated by the mean free scattering length of molecules (or gas thermal conductance) which is pressure-dependent.

For insight into device operation including underlying heat transport mechanisms and for optimization of device design to meet key requirements of fast response time, low operating power and temperature, fast thermal response, it is necessary to solve the coupled system of equations governing electrothermal behaviour. Of particular importance in all of the above applications, is the modeling of heat loss from the microstructure to the ambient. For example, with the microstructure we have considered here (see Figs. 1 and 2), the heat loss due to natural convection can be as much as 99% of the input power at standard temperature and pressure (STP) [4]. Thus accurate modeling of the boundary condition accounting for the convective heat loss is crucial in a two-dimensional simulation.

2. Assumptions, Model Equations, and Numerical Procedure

In view of the planar nature of thermally isolated microstructures and because of small film thicknesses (relative to other linear dimensions), the electrothermal behaviour within the structure can be adequately described in two-dimensions. The approximation, although reasonable in terms of electrical behaviour, may not be intuitively obvious from a thermal standpoint and requires justification. In most structures, under conditions of natural convection in the diffusive limit (i.e. stagnant fluids), the Biot number is relatively small; this number describes the ratio of the surface heat conductance to the internal heat conductance across the microstructure thickness. Measurements performed on heat transfer test structures show that the gradient across the thickness at STP is negligible compared to the lateral gradients [5]. Given these conditions, electrical and heat transport within the microstructure can be described by the following 2-D system:

$$\nabla_{x,y} \cdot [\sigma(T) \nabla_{x,y} \psi] = 0 \qquad (1)$$

$$\nabla_{x,y} \cdot [\kappa(T) \text{ grad } T] = \sigma(T) (\nabla_{x,y} \psi)^2. \qquad (2)$$

Here, $\sigma(T)$ denotes the temperature-dependent electrical conductivity of the polysilicon regions (see Fig. 2), ψ is the electric potential, $\kappa(T)$ is the temperature-dependent thermal conductivity, and T is the temperature. The term on the right-hand-side of eqn. (2) denotes Joule heat.

In the presence of a flow stream, the steady-state heat transport from the surface of the microstructure, assuming negligible viscous dissipation, is governed by the energy equation:

$$\rho C_p [u \, \partial T/\partial x + v \, \partial T/\partial y + w \, \partial T/\partial z] = \nabla \cdot (\kappa(T) \nabla T) \qquad (3)$$

where u, v, and w are the components of the velocity field, ρ is the density, and C_p is the specific heat. With zero flow, we can assume that the natural convective currents are negligible. Despite the excessively high (≈ 400 °C) in-situ heated temperatures, the resulting Grashof number, which describes the ratio of buoyancy to viscous fluid forces, is small. Measurements show that the heat loss from such microstructures is independent of its orientation with respect to the gravity vector [5]. Thus we can assume that the fluid is stagnant (diffusive) and that the temperature distribution obeys $\nabla \cdot (\kappa(T) \nabla T) = 0$ where $\kappa(T)$ denotes the thermal conductivity of the fluid expanse (air) and is pressure-dependent. For given surface temperature of the microstructure, $T_s(x,y)$, the temperature of the micromachined cavity walls, and the temperature of the fluid far from the microstructure surface (free stream temperature, T_∞), the equation can be solved to determine the convective heat loss from the microstructure surface to the surrounding [5]. This heat loss constitutes the boundary condition required for eqn. (2). The technique to solve this equation is based on a boundary element method along the lines reported in [6]. However, it can be reduced to Laplace's equation by employing Kirchoff transformation and this intrinsically accounts for the temperature dependence of the thermal conductivity [7]. Based on the solution, we calculate a heat transfer function $G(x,y)/(\kappa°C)$ on the membrane surface. The surface heat density at position (x,y) on the membrane surface

$$q_s = \kappa\, G(x,y)\, [T_s(x,y) - T_\infty] \qquad (4)$$

is then employed as a mixed boundary condition for the two-dimensional solution of eqn. (2). Since the heat transfer function $G(x,y)$ is a function of $T_s(x,y)$, it is recalculated within the iterative loop used in the solution of eqns. (1) and (2). Radiative heat losses, if necessary, can also be incorporated as a mixed boundary condition. In our case, in view of the given device active area and input powers involved, such losses are negligible.

Equations (1) and (2) are discretized using a control area approximation with node count of 1600 and 4700, respectively, for the structure shown in Fig. 1. The procedure starts with the calculation of $G(x,y)$, following which a system of equations for the electrical conduction equation is generated. Based on the solution, the Joule heat terms are calculated and eqn. (2) is solved. Since (2) can be potentially nonlinear, an inner iterative loop is employed. The solution of the systems of equations is based on a conjugate gradient scheme. Solution verification is based on electrical and heat flux conservation checks.

3. Results and Discussion

The simulated temperature distribution in the device (see Figs. 1 and 2) in air at STP is illustrated in Fig. 3. The peaks and valleys correspond to the temperature distribution of the active (current carrying) and passive (temperature sensing) coils, respectively. The thermal behaviour of both coils was measured for both air and helium at STP. The temperature values shown were based on measurements of coil resistance from which an average coil temperature is retrieved following prior temperature coefficient characterization of the polysilicon layer. The temperature difference between active (T_{ac}) and passive coils is illustrated in Fig. 4 as a function of the input power. With maintaining the active coil average temperature at 70°C, we clearly see that the input power (heat loss) is consistent with the thermal conductivities of the two gases implying an almost 100% transduction efficiency. The larger temperature difference in the case of helium is due to the convective heat loss to the surrounding which predominates over the inter-coil lateral heat transfer.

4. Conclusions

In this paper, we have presented the necessary assumptions, resulting model equations, and boundary conditions pertinent to two-dimensional numerical simulation of heat transport in thermal-based microsensors taking into account convective heat losses. The simulations provide good agreement with measurement data.

References

[1] R.G. Johnson and R.E. Higashi, Sensors and Actuators, vol. 11 (1987) 63.
[2] A.W. van Herwaarden and P.M. Sarro, J. Vac. Sci. Tech., vol. A5 (1987) 2454.
[3] J.S. Suehle, R.E. Cavicchi, M. Gaitan, and S. Semancik, IEEE Electron Dev. Letts., vol. 14 (1993) 118.
[4] N.R. Swart and A. Nathan, Tech. Digest, IEEE IEDM, 1994, p. 135.
[5] N.R. Swart, Heat Transport in Thermal-Based Microsensors, Ph.D. dissertation, University of Waterloo, 1994.
[6] K. Nabors and J. White, IEEE Trans. CAD, vol. 10 (1991) 1447.
[7] W. Allegretto, B. Shen, Z. Lai, and A.M. Robinson, Sensors and Materials, vol. 6 (1994) 71.

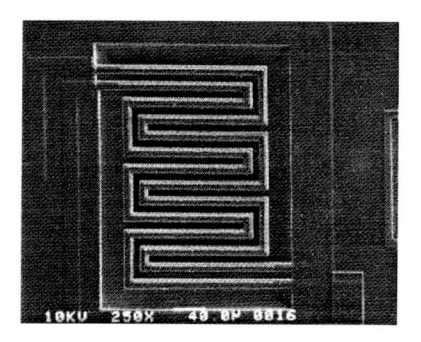

Fig. 1 Photomicrograph of device used in simulations.

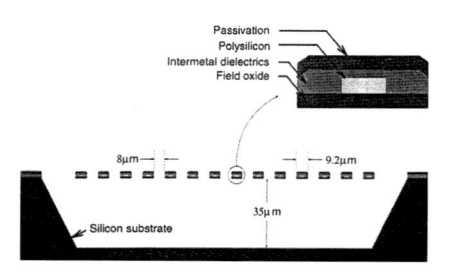

Fig. 2 Device cross-section.

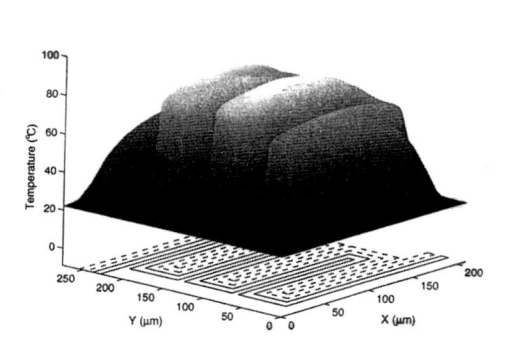

Fig. 3 Temperature distribution in device at STP (dashed line: active coil, solid line: passive coil).

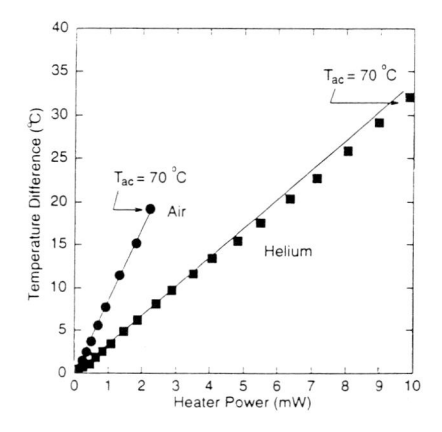

Fig. 4 Temperature difference between coils as a function of input power (solid lines: simulations, points: measured values)

Simulating Deep Sub-Micron Technologies: An Industrial Perspective

P. Packan

Intel Corporation,
5200 NE Elam Young Pkwy,
Hillsboro, Oregon, 97124-6497

Abstract

Meeting the performance goals necessary to be competitive in the semiconductor industry will force novel process and device designs to be evaluated and optimized. Process and device simulators can be a valuable tool in the evaluation and optimization process. As device dimensions approach the 0.10 μm regime, device and process simulators will be pushed to new levels. In device simulations, non-local hot electron effects and mobility modeling are crucial for predictive simulations. The shallow, highly doped junctions required to improve short channel effects forces accurate diffusion models for extremely low energy implants as well as predictive modeling of extended defect interactions. Current areas of application for device and process simulation tools including development, optimization and manufacturing are discussed.

1. Introduction

It is becoming increasingly difficult to maintain the current pace of transistor performance improvement. Not only are these improvements expected, but they are expected with a concomitant decrease in power consumption. As transistor gate lengths of MOS devices approach the 0.10 μm regime, short channel effects are extremely difficult to control. In order to meet these challenges, aggressive scaling options will be required. It is not obvious which, if any, of the current scaling options will provide the best means of meeting the performance goals of the next generation devices. Because of these uncertainties, many possible solutions need to be evaluated. However, due to time and material constraints, it is not always possible to evaluate all options. A wide range of inputs are used to focus resources in the most promising directions. Increasingly, process and device simulation tools are being used to assist in this process. These simulators are being used in a variety of ways from technology evaluations to split lot design and optimization. However, all of these applications share one thing in common: they are critically dependent on the accuracy of the simulation tools.

Development of a new technology goes through a number of distinctive phases. In the initial phase, there are a number of process and device design options from which to choose. Once this is done, the second phase, optimization, is entered. In this phase, the process flow and device characteristics are refined and optimized to attain the best device performance. Finally, the technology is ready for manufacturing. In this phase, impacts of manufacturing changes on performance and yield are evaluated. In addition, simulation tools can be used to quickly determine causes of process excursions. In each of these phases, process and device simulation tools can play a key role in decreasing the time to develop and maintain a new device process.

2. Architecture Development

During the initial stages of technology development, architectural options for both the process and device must be evaluated. Methods to suppress short channel effects (retrograde wells, halo implants, etc.), isolation schemes, metalization, etc., must all be addressed. As expected, the starting point for this evaluation is the previous generation technology. It may be that a clear direction for scaling the previous technology exists. Decreasing implant energies, reducing thermal cycles and printing smaller gate dimensions may be the simplest and most direct path to the next generation technology. However, as device dimensions continue to scale, this approach may not be feasible. Effects that were second and third order in older generations may now dominate transistor performance. Transient enhanced diffusion, implant channeling, salicide resistance, velocity saturation, etc. may play such a large part in determining transistor performance that conventional scaling may not be possible. In this case, new architectural options must be evaluated.

During this early phase of technology development, simulation tools can have their largest impact. At this time, absolute accuracy is not as critical as the ability to accurately assess the relative trade-offs between different device and process options. This requires a more global calibration, even though local accuracy may be sacrificed. Since there may be a wide range of device and process options, models must be physically based to account as much as possible for unforeseen interactions. These simulations can then be used to identify the most promising options for experiments in silicon.

The push in the semiconductor industry for MOS devices is toward shallow, highly doped S/D extensions (tips) with deeper S/D's for contacting the device [1],[2]. The reason for this is two-fold. First, shallow junctions decrease short channel effects by decreasing charge sharing effects. This allows the gate lengths to be decreased without undo source-drain leakage. Second, highly doped, abrupt profiles are needed to decrease the external resistance or at least to offset any resistance increase due to the shallower junctions.

Perhaps the most straightforward way to decrease the junction depths is to scale the implant energies. This is a methodology adopted by many companies and explains the current push to implant energies below 5 keV for both n and p-type dopants [3]. It is well known that TED is strongly dependent on the dopant species and implant dose [4]-[6]. In addition, at these extremely low energies, there is also a strong energy dependence that can differ significantly for different dopant species. The diffusion in most processes currently used in industry are dominated by TED. Therefore, simulation of trade-offs in junction depth, lateral underdiffusion, resistance and ultimately drive currents depend on the accuracy of these TED models.

Figure 1 shows simulations of diffused boron profiles for 5e14 boron tip implants of 5, 10, 20 and 40 keV. These tip profiles were implanted through a screen oxide and underwent an anneal of 900 C for 10 min. The depths of the profiles at a concentration of 5e17 are shown in table 1. Also shown is the percent change in junction depth as the energy is scaled. It is seen that scaling the implant energy alone will not continue to lead to a constant scaling of the junction depth. It is becoming increasingly difficult

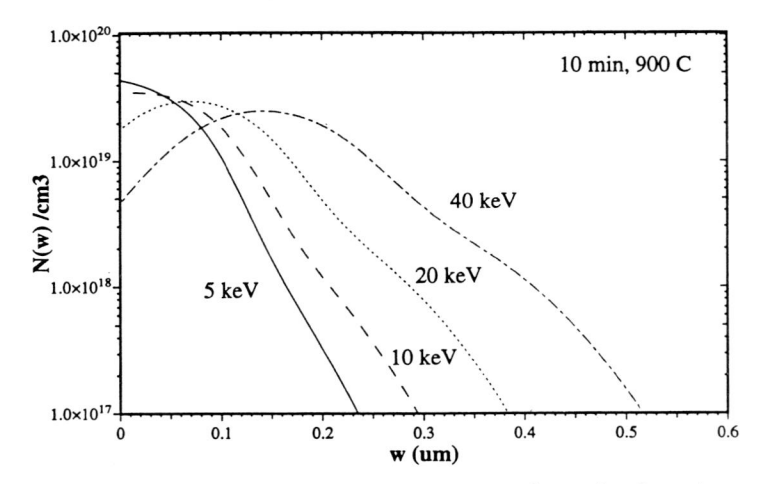

Figure 1: Simulated boron concentration profiles for 5e14 boron implanted at energies from 5 to 40 keV through a 10 nm oxide and annealed at 900°C for 10 min.

arsenic energy	Junction Depth (μm)		Scaling Percentage	
	900°C, 10 min	1000°C, 30 sec	900°C, 10 min	1000°C. 30 sec
5	0.17	0.16	19%	20%
10	0.24	0.22	28%	27%
20	0.32	0.30	28%	30%
40	0.45	0.43	—	—

Table 1: Effects of energy scaling on junction depth. Scaling the junction depth by decreasing the implant energy is becoming more difficult as the implant energies are reduced. TED and extended defects play a large role for this trend.

to scale tip junction depths merely by decreasing the energy. At these low energies, the improvement in the "as implanted" profile are being offset by TED effects.

A complicated interaction of implant and TED effects determines final junction depths. By changing the anneal conditions as well as the implant conditions, these scaling relationships can be altered. Table 1 also shows junction scaling factors if the anneal cycle is changed to 1000 C, 30 sec. For this case, the scaling factors vary even more. It will become increasingly difficult to scale junction depths and alternate doping techniques may need to be addressed.

These results become even more complicated when the damage from the source-drain implants is included. The interactions between point defects, extended defects and dopant atoms must all be taken into account for accurate, predictive process simulation. Even seemingly insignificant changes in thermal cycles can have large consequences on final device characteristics. Through the routine use of process and device simulations, these effects can be monitored to assure that any changes will not adversely affect device performance.

In conjunction with lowering implant energies, thermal cycles are being decreased to limit TED effects. It is commonly accepted that short time, high temperature anneals help to minimize TED [4]. Almost every major semiconductor company uses RTA processing in some capacity. Many companies are using very low temperature processing with strategically placed RTA anneals to help minimize TED. In these flows, the majority of dopant diffusion is controlled by these few RTA steps. Since enhanced diffusion effects due to TED are not localized, many process steps will

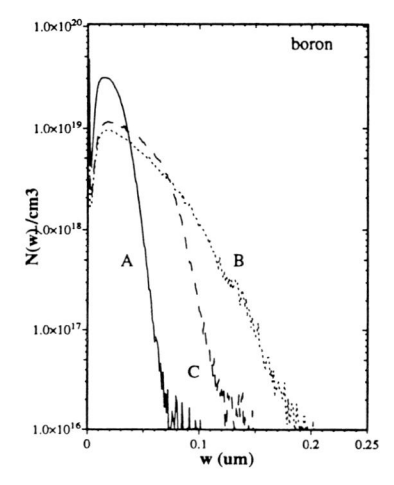

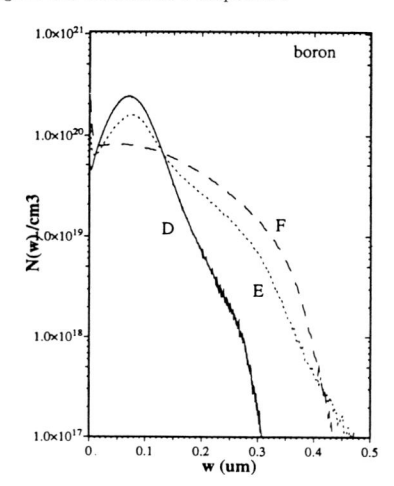

Figure 2: Boron profiles for a 1e14, 20 keV BF2 implant: A) as implanted, B) 800°C, 120 min anneal and C) 1000°C, 10 sec + 800°C, 120 min anneal. A silicon pre-amorphizing implant was used. Also shown are boron profiles for a 2e15, 20 keV boron implant: D) as implanted, E) 900°C, 10 min anneal and F) 1000°C, 30 sec + 900°C, 10 min. Although the addition of an RTA cycle will often help to minimize TED effects, this is not always the case. Extended defect-dopant interactions can strongly influence diffusion.

interact to define the final dopant profiles. Tip implants will affect channel profiles. S/D implants will affect tip profiles. It is critical to understand these interactions when developing an integrated process.

Figure 2 shows the effect of adding an RTA step immediately after ion-implantation. For low concentration implants (curves A,B,C), the addition of an RTA step can decrease dopant diffusion. For this experiment, a silicon pre-amorphization 1e15, 60 keV implant was used, followed by a 1e14, 20 keV BF2 implant. In one case, an 800°C, 120 min furnace anneal was performed. In the second case, a 1000°C, 10 sec anneal was performed prior to the 800°C furnace anneal. In this case, the RTA anneal reduced the total amount of diffusion. However, figure 2 also shows (curves D,E,F) that the addition of an RTA step immediately following an implant will not always lead to a reduction in junction depths. In this figure, the effect of inserting a 1000°C, 30 sec RTA step prior to a 900 C, 10 min anneal is shown. The boron was implanted at 20 keV to a dose of 2e15. For this case, the addition of an RTA step does not lead to a shallower junction. Extended defects and clustering effects interact to change both the TED dependency and mobile dopant concentrations. It is extremely difficult to simulate these effects using current models and can be done locally at best. For these high concentration profiles, modeling active concentrations is necessary, but difficult. As thermal cycles decrease, modeling the kinetics of activation and deactivation will be become even more important since the active concentrations determine device performance. This will mean that extended defect-dopant interactions will become even more critical as S/D extensions continue to increase in concentration and decrease in depth.

Velocity saturation and velocity overshoot are extremely important in very small MOS devices. NMOS devices, in particular, show a very strong velocity saturation effect in the sub-micron regime. Accurate evaluation of the improvement in device performance for different device options requires predictive simulation of these non-local hot carrier effects. If these effects are not taken into account, incorrect evaluations of de-

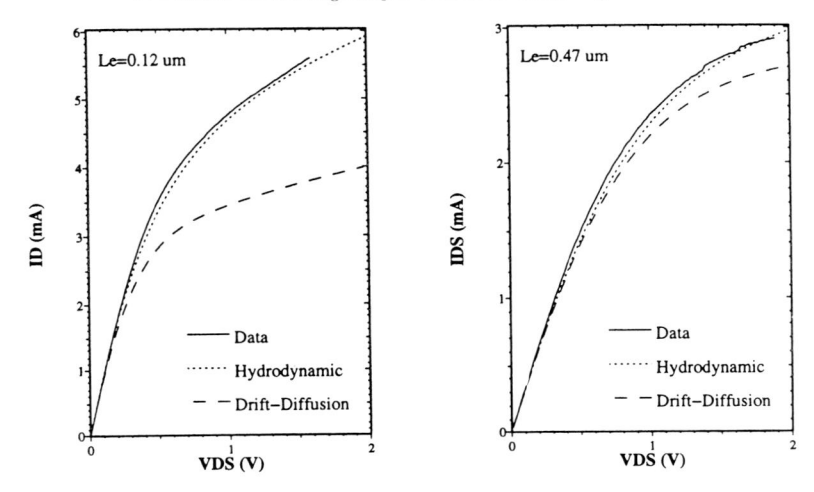

Figure 3: IV characteristics for a thick gate SOI structure. To accurately simulate these devices, non-local hot electron effects must be modeled. Using a hydrodynamic model with an energy relaxation time of 0.5 ps results in excellent matches.

vice improvement will be made. Hydrodynamic, Monte Carlo and other deterministic solution methods for solving the Boltzmann transport equation such as the scattering matrix approach and spherical harmonic expansion can capture non-local effects such as velocity overshoot [7]-[9]. The accuracy and ability to globally calibrate each of these methods is still debated.

Figure 3 shows data and simulated IV device characteristics for thick gate SOI structure developed by Assaderaghi, et. al. to evaluate velocity saturation and overshoot effects [10]. Two different size devices are shown. Drift diffusion simulations alone cannot be used to model both devices. Hydrodynamic simulations with an energy relaxation time of 0.5 ps match the data quite well. It should be noted that the value of the relaxation time is sensitive to the mobility model used. These results show that drift diffusion models can grossly underpredict saturation currents. When device architectural options such as halo implants and retrograde wells which strongly effect electric fields are evaluated, accurate simulation of these non-local effects are critical.

3. Architecture Optimization

Once an architecture has been determined, it must be optimized. During this portion of technology development, the simulation tools can be used to save time and silicon by effectively targeting split conditions for experiments. The simulations can also be used to better understand underlying physical mechanisms responsible for different phenomena. The requirements on accuracy of the simulations increase during the optimization phase. This may force locally calibrated models and may even necessitate the addition of new models to comprehend specific effects that may have been considered second order during the initial definition stage.

The optimization of one aspect of a trench isolation process can be used to illustrate these points. It has been reported that narrow channel effects for trench isolated devices can be very different than LOCOS isolated devices. For the LOCOS structure, gate edge lifting increases the gate oxide thickness at the trench edge. This leads to an increase in threshold voltages for narrow channel devices. However, the opposite trend has been observed for trench isolated devices, namely, a parasitic transistor

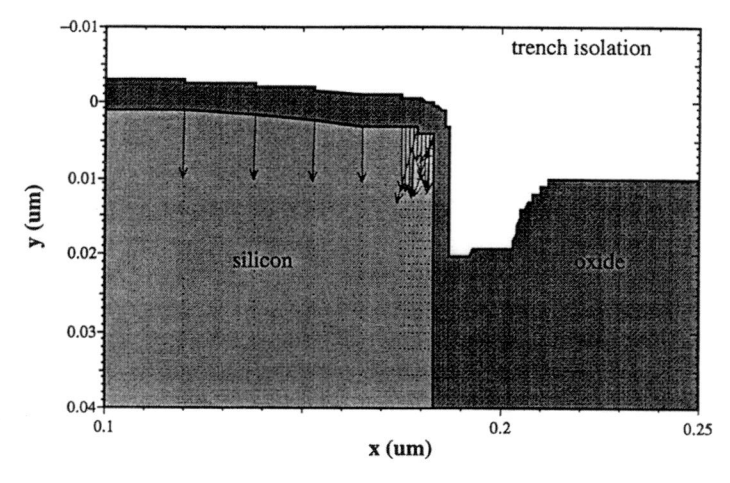

Figure 4: Electric field lines near the trench edge corner showing the importance of the corner geometry. High electric fields at the corner create a parasitic edge device.

with a decreased threshold voltage which can lead to a kink in the subthreshold IV characteristics [11],[12]. The origin of these effects were evaluated using process and device simulations. Figure 4 shows the electric field in the corner region of the device. As has been reported, the amount of curvature in this corner region strongly affects the electric field. This higher electric field will permit this edge region to invert at a lower voltage. Another important effect is seen in figure 5. This figure shows that dopant dose loss from the silicon substrate can occur during annealing of implanted dopants even though the surface is capped with an oxide. Figure 5 shows this effect for phosphorus. Dose loss is also seen for other dopants. The exact mechanism for this phenomenon is still under debate, but it is generally believed that a dopant rich interfacial layer forms between the silicon and the oxide. The dopant atoms in this layer are electrically inactive and can be removed by etching the oxide. At the trench corner, two dimensional effects increase the dose loss. This effect can be as important as the electric field effect. In addition to these effects, oxide edge thinning can play a role. Possible solutions to the narrow device Vt decrease include sidewall implants to increase the edge dopant concentrations and careful modification of sidewall oxidation conditions to change the edge geometry. The relative importance of the electric field crowding and dopant loss can be isolated and the appropriate solution determined through these types of simulations.

4. Manufacturing

After the device and process have been optimized in a development environment, it must be transferred to manufacturing. Simulations can aid in determining if the process conditions chosen are precariously close to a process window cliff. Response surface modeling can be used to evaluate the impact of process skews on device performance. These simulations can help target any problem areas where particularly tight control specifications may be needed. This requires extremely accurate simulations since the effect of very small changes are being evaluated.

The simulations can also be used to evaluate the impact of process changes that could make the process more manufacturable. Changes which affect throughput such as stabilization times and temperatures, ramp rates, etc. can be evaluated. However, these changes can adversely affect device performance and must be monitored.

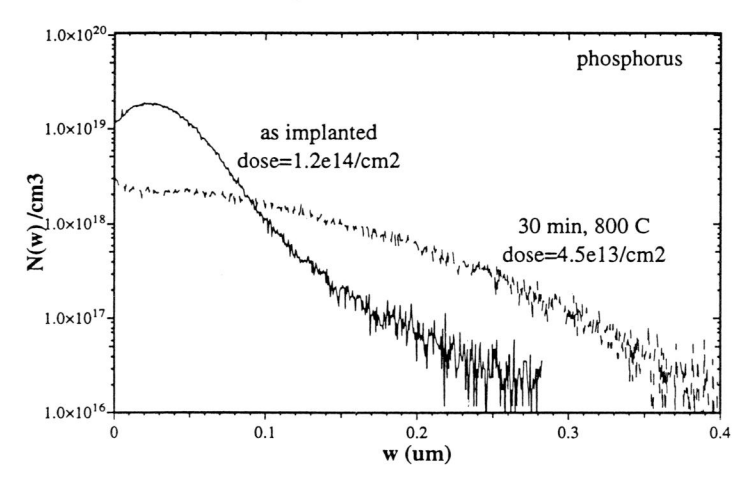

Figure 5: Phosphorus doping profiles showing extensive loss of dose during anneals in which TED effects dominate. This effect leads to a depletion of dopant at the trench edge.

Matching results from fabrication facility to fabrication facility can also benefit from simulations. Equipment and environmental differences must be compensated for and simulations can be used to identify possible solutions.

Knowledge base systems can be constructed to help to determine the cause for excursions in electrical characteristics. Electrical characteristics can be compared through the knowledge base systems and related back to particular process steps. Quick determination of the causes of process excursions is extremely valuable and because of this is drawing increasing attention as a priority area for simulation tools.

Dislocation loops caused by the interaction of point and extended defects are affected by implant and anneal conditions and can be strong yield limiters in a manufacturing environment. Relatively small changes in process conditions can strongly affect these extended defects and cause heavy leakage and even catastrophic electrical shorts. It is not possible to model these types of interactions in any global way at this time. However, the ability to model such effects would be extremely useful to manufacturing groups so that potential yield limiting process windows can be avoided.

5. Summary

Process and device simulations are used in a wide variety of ways within industry. Each application has specific requirements on accuracy, reproducibility and predictivity. As the cost of development continues to rise at an accelerated rate, the benefit of predictive process and device simulations increases. The trend in device design is toward shallow highly doped S/D extensions to decrease both resistance and short channel effects. To attain these extremely shallow highly doped junctions, implant energies are decreasing to the sub-5 keV range as doses increase to the 1e15 range. This puts added importance on low energy TED and extended defect interactions. Solubility and clustering kinetics determine the electrical profiles which are necessary for accurate device simulations. As the device dimensions continue to decrease, non-local hot carrier effects become very important for novel device evaluation.

There are still many interactions in the process and device simulations which are not well modeled. Extended defect formation and dopant interactions are locally modeled with empirical models at best. Hydrodynamic and Monte Carlo device simulations focusing on non-local effects are not well calibrated and thus are not readily applied to problems of interest. However, as device dimensions continue to decrease, non-local effects must be accurately addressed to predictively assess and evaluate new technology options. Simulations currently provide an advantage in developing a technology generation. However, the promise of future impact is much larger if more predictive simulations become a reality.

References

[1] Y. Taur et al.,"High Performance 0.1μm CMOS Devices with 1.5 V Power Supply,"*IEDM Tech. Dig.*, pp. 127-130, 1993.

[2] T. Hori,"A 0.1μm CMOS Technology with Tilt-Implanted Punchthrough Stopper (TIPS),"*IEDM Tech. Dig.*, pp. 75-78, 1994.

[3] A. Hori et al.,"A 0.05μm CMOS with Ultra Shallow Source/Drain Junctions Fabricated by 5keV Ion Implantation and Rapid Thermal Annealing," *IEDM Tech. Dig.*, pp.485-488, 1994.

[4] P. Packan and J. Plummer,"Transient enhanced diffusion of low-concentration B in Si du to ^{29}Si implantation damage,"*Appl. Phys. Lett.*, vol. 56, pp. 1787-1789, 1990.

[5] S. Solmi et al.,"High-concentration diffusion in silicon: Simulation of the precipitation phenomena,"*J. Appl. Phys.*, vol. 68, no. 7, pp.3250-3258, 1990.

[6] P. Griffin et al.,"Species, Dose and Energy Dependence of Implant Induced Transient Enhanced Diffusion,"*IEDM Tech. Dig.*, pp. 295-298, 1993.

[7] C. Jacoboni and L. Reggiani,"The Monte Carlo method for the solution of charge transport in semiconductors with application to covalent materials," *Rev. Mod. Phys.*, vol. 55, pp645-705, 1983.

[8] N. Goldsman et al.,"A physics-based analytical/numerical solution to the Boltzman transport equation for use in device simulation," *Solid-State Elec.*, vol. 34, no. 4, 1991.

[9] A. Das and M. Lundstrom,"A scattering matrix approach to device simulation,"*Solid-State Elec.*, vol. 33, no. 10, pp. 1299-1307, 1990.

[10] F. Assaderaghi et al.,"Saturation Velocity and Velocity Overshoot of Inversion Layer Electrons and Holes,"*IEDM Tech. Dig.*, pp. 479-482, 1994.

[11] K. Ohe et al.,"Narrow Width Effects of Shallow Trench-Isolated CMOS with n+-Polysilicon Gate,"*IEEE Trans. Electron Devices*, Vol. 36, pp. 1110-1116, 1989.

[12] K. Ishimaru et al.,"Trench Isolation Technology with 1μm Depth n- and p-wells for a Full-CMOS SRAM Cell with a 0.4μm n+/p+ Spacing," *Symp. on VLSI Tech.*, pp. 97-98, 1994.

An Improved Calibration Methodology for Modeling Advanced Isolation Technologies

Peter Smeys, Peter B. Griffin and Krishna C. Saraswat

Stanford University
Department of Electrical Engineering, Stanford, CA 94305-4055

Abstract

An improved calibration methodology for simulating advanced isolation technologies using SUPREM-IV is presented. Based on the experimental determination of the material properties of silicon nitride, an improved parameter set for the stress dependent oxidation models is derived. The calculated substrate stress using this new parameter set is compared with micro-raman spectroscopy stress measurements to validate the calibration methodology.

Introduction

With the scaling of IC technologies to deep-submicron dimensions, the inter-device isolation becomes increasingly critical. Local Oxidation of Silicon (LOCOS) is still the dominant isolation technology due to its process simplicity and superior isolation characteristics. However, pushing the scalability limits of LOCOS technology has become virtually impossible without relying on accurate process simulators. Accurate simulations and design for manufacturability require knowledge of the material properties of the layers involved. Typically, these parameters are determined by fitting complete isolation structures after field oxidation using stress dependent oxidation models where SiO_2 and Si_3N_4 are treated as non-linear viscous and linear viscous materials, respectively. While the experimental boundary shapes correlate well with SUPREM-IV simulations over a wide temperature range, the absolute values of the model parameters are still uncertain. In this paper, we determine experimentally the viscosity of thin LPCVD silicon nitride films as a function of temperature and derive the other model parameters based on these experimental data. A comparison of the stress levels in the substrate with those obtained by micro-raman spectroscopy [1] is presented as well and is used to validate the calibration procedure.

Experimental results and calibration methodology

LPCVD silicon nitride films were deposited on silicon substrates at a temperature of 780 °C. After deposition, the film thickness was recorded. Subsequently, the wafers were annealed in argon and steam at 900, 1000 or 1100 °C for 60 minutes. The length of the anneal was chosen so that it would be of the order of the film relaxation time. The initial film thickness was about 140 nm. After the 900 °C, 1000 °C and 1100 °C anneals, a decrease in film thickness was observed (figure 1). The film stress, measured at room temperature using a wafer curvature measurement set-up, was found to increase with anneal temperaure from 1070 MPa to 1180

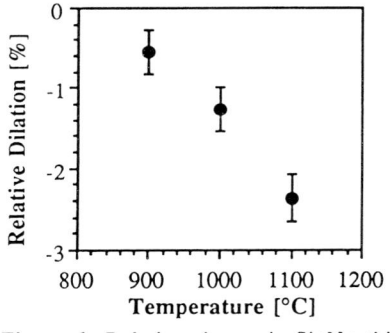

Figure 1: Relative change in Si_3N_4 thickness with anneal temperature

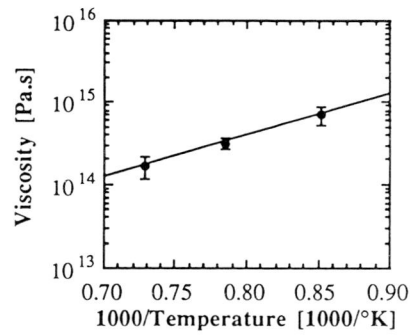

Figure 2: Arrhenius plot of the experi-mental Si_3N_4 viscosity

MPa. This increase is believed to be due to film densification [2]. The high stresses observed in as-deposited silicon nitride films are primarily due to non-equilibrium deposition conditions and to a lesser extent to the thermal mismatch between the film and the substrate [2]. A very similar behavior is observed for samples annealed in steam, indicating that the annealing ambient does not affect the densification or the viscous flow behavior. If we assume that film shrinkage occurs in an isotropic manner, the viscosity of the silicon nitride as a function of temperature can be approximated using a simple linear viscous model described by $-(\Delta d/\Delta t) = \sigma/\eta$, where $\Delta d/\Delta t$ is the magnitude of the film shrinkage rate, σ is the film stress and η is the nitride viscosity. The experimental results for the viscosity of silicon nitride estimated using this model are summarized in figure 2. The results fit an Arrhenius expression extremely well. It is worthwhile noting that the Si_3N_4 viscosity extracted by this method is independent of any fits to experimental LOCOS boundary shapes and forms the foundation for the calibration strategy to be discussed in the next section.

In SUPREM-IV, high temperature oxide deformation is described by a non-linear viscous flow model with stress dependent oxidation parameters. This appears to give excellent fits to oxide thinning data on convex and concave structures of different radii for temperatures between 900 °C and 1100 °C [3,4]. Silicon nitride is treated as a purely viscous material, an approximation we will show to be valid. The viscosity, oxidant diffusivity and reaction rate have the following stress dependency:

$$\eta = \eta_0 \frac{\tau V_c/kT}{\sinh \tau V_c/kT} \quad (1), \quad D = D_0 \, e^{-PV_d/kT} \quad (2), \quad k_s = k_{s0} \, e^{-\sigma_n V_r/kT} \quad (3),$$

where τ is the shear stress, P the hydrostatic pressure and σ_n the normal stress. V_c, V_d and V_r are the activation volumes for critical stress, stress-dependent diffusion, and stress-dependent reaction rate respectively. This leaves us with a set of five fitting parameters to model any isolation boundary shape: η_0 the low stress viscosity, which must be specified for both nitride and oxide and the three activation volumes V_c, V_d and V_r. These five parameters form a quintuplet and cannot be chosen independently. The quintuplet is not unique in that identical boundary shape solutions can be obtained when η_0 is multiplied and V_c, V_d and V_r are divided by a constant. In the previous section, we have experimentally determined the viscosity of silicon nitride and will scale the default SUPREM-IV quintuplet according to the ratio of the original fitted values [5] and the experimental values of the Si_3N_4 viscosity. In order to verify the quintuplet scalability, simple LOCOS structures with 10 nm pad-oxide and 100 nm silicon nitride were simulated using the two parameter sets which are presented in table 1. It is clear from figure 3 that there is no significant difference in the boundary shapes when the model parameters are scaled appropriately. When analyzing the simulated stress levels in the substrate however, there is a distinct difference between the two quintuplets (figure 4). The ratio between the stress components, calculated with the two different quintuplets is approximately equal to the nitride viscosity scale factor, irrespective of the stress component and the temperature.

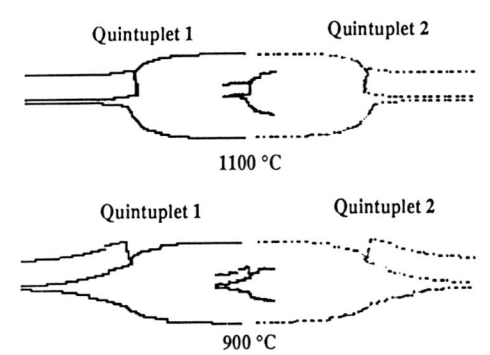

Figure 3: SUPREM-IV boundary shapes for the two quintuplets

Figure 4: Plot of maximum magnitude of the different stress components along Si-SiO$_2$ interface as a function of temperature (Open: Quintuplet 1, Solid: Quintuplet 2).

Quintuplet	Parameter	900 °C	1000 °C	1100 °C
1 (default)	$\eta_{nitride}$[Pa.s]	1.15E14	4.7E13	2.3E13
	η_{oxide}[Pa.s]	2.5E14	2.8E13	4.5E12
	V_c [Å^3]	300	522	1000
	V_d [Å^3]	65	65	65
	V_r [Å^3]	12.5	12.5	12.5
2 (scaled)	$\eta_{nitride}$[Pa.s]	7.1E14	3.12E14	1.67E14
	η_{oxide}[Pa.s]	1.67E15	1.87E14	3.01E13
	V_c [Å^3]	44.8	78	149
	V_d [Å^3]	9.7	9.7	9.7
	V_r [Å^3]	1.9	1.9	1.9

TABLE I: Overview of the two parameter sets used in the simulations.

Micro-Raman spectroscopy is one of the few techniques that allows the direct measurement of stress levels in the silicon substrate with small enough resolution so that an estimate of the stresses can be made [1,6]. In order to compare our results with those obtained by micro-raman, the same LOCOS structures De Wolf et al. [1] used in their experiments were simulated. Micro-raman spectroscopy measures the average stress over a depth of about 0.3 μm [6]. On wide Si$_3$N$_4$ lines (> 4 μm) away from the bird's beak, a tensile stress σ_{XX} of 0.52 GPa was estimated from the raman shift measurements for this particular LOCOS structure[1]. Calculating the average stress σ_{XX} over a depth of 0.3 μm away from the bird's beak region from the SUPREM-IV simulations yielded an average stress σ_{XX} of 0.02 GPa when using quintuplet 1 to simulate the structure and 0.14 GPa when using the experimentally derived quintuplet. Clearly, the parameter set derived from the experimental nitride viscosity gives the best agreement and it can probably be improved by a more careful interpretation of the micro-raman results.

Conclusion

We have shown that the fitting parameters for the oxidation models in SUPREM form a quintuplet that can be scaled and yield identical boundary shapes of LOCOS structures for many different parameter sets. The calculated stresses in the substrate however do not remain constant but scale as well. This has serious implications when designing new isolation structures. Based on the scalability information and the experimentally extracted nitride viscosity, we designed a strategy that allows for an improved calibration of SUPREM-IV.

References

[1]   I. De Wolf, et al., J. Appl. Phys., **71**, 898 (1992).
[2]   P. H. Townsend, PhD Dissertation, Stanford University (1987).
[3]   C. S. Rafferty, PhD Dissertation, Stanford University (1989).
[4]   D-B. Kao, et al., IEEE Trans. Elec. Dev., ED-35,25 (1988)
[5]   P. B. Griffin, et al.,Proceedings IEDM, 741 (1990).
[6]   K. Kobayashi, et al., J. Electrochem. Soc. **137**, 1987 (1990)

Algorithms for the Reduction of Surface Evolution Discretization Error

Hernan A. Rueda and Mark E. Law

Department of Electrical Engineering, University of Florida
Gainesville, FL 32611
Phone: 904-392-6276 Fax: 904-392-8381
har@tcad.ee.ufl.edu law@tcad.ee.ufl.edu

Abstract

This paper investigates numerical error due to time and spatial discretization. The relative error contributions of each discretization scheme is examined. Two algorithms are then introduced to control the error due to each modelling discretization scheme.

1. Introduction

Surface evolution algorithms[1-4] all suffer from some numerical error as a function of both the time and spatial discretization. We have studied both components of the error, and conclude that the spatial discretization is the largest component. We suggest techniques that can control the error from both time and spatial discretization.

2. Time Discretization

The velocity of etch or deposition of any node on a surface may be dependent on the node's current position on the surface and the current time in the process, V(x,t). In general terms, this may be summarized by the following equations:

$$\vec{V} \propto x, t$$

$$x_1 - x_0 = \int_{t_0}^{t_1} \vec{V}(x, t) \, dt$$

When solving this integration, one may use different approaches computationally. Traditionally, in most surface evolution simulators[1-4], a "forward euler approach (FE)" has been implemented:

$$x_1 = x_0 + \Delta t \cdot \vec{V}(x_0, t_0)$$

$$\Delta t = t_1 - t_0$$

The new position of the discretized surface node is computed by adding the old position to the product of the time step and the velocity vector. This algorithm has the advantage of simplicity since all the parameters may be determined at any given initial position for a surface node. However, the error introduced by the equation depends on the rate of change of the velocity. For an isotropic etch or deposition it is accurate, but inaccurate in an anisotropic simulation, where the velocity vector is changing in magnitude and/or direction. For this algorithm, the error may be controlled by choosing a very small time step continuously throughout the simulation, in hopes that the change in velocity vector for the given node changes negligibly during that time interval. However, this wastes CPU time, since at times throughout the simulation, the velocity vectors of all the surface nodes may not change. Also, the correct size of the time step to take is difficult to compute.

Two algorithms have been investigated to reduce this quantization error. The first is the "trapezoidal rule (TR)" algorithm:

$$x_1 = x_0 + \Delta t \cdot \left[\frac{\vec{V}(x_0, t_0) + \vec{V}(x_1, t_1)}{2} \right]$$

This procedure calculates the new node position by computing the average of the initial and final velocities in the given time interval. It then also uses the initial and final velocity vectors to calculate an appropriate succeeding time step. This adaptive time step algorithm can then control the time discretization error. The disadvantage of using this algorithm is that it is more computationally intensive because the new velocity must be solved iteratively.

The differences between these methods greatly depends on the grid resolution of the test structures. For fine grids the differences between these methods is much smaller than coarse grid surfaces. Figure 1 summarizes these results for different surface evolutions. In general, the error from the time discretization is less than the error from the spatial discretization.

3. Spatial Discretization

This led to the development of a surface grid refining technique (SGR) for etching and deposition modeling. This technique uses the radius of curvature at each given node on the surface to determine whether new nodes should be placed next to that node to help define the surface local to the node. When the radius is less than a specified length, new neighboring nodes are then added. This technique allows more defined corners to be evolved. It proves to be very helpful in both isotropic and anisotropic process modelling.

When the grid refining technique is coupled with the adaptive time step procedure, an initial coarse grid will not affect the surface development of the simulation. An initially coarse grid simulation will then eventually result in the similar final surface structure and therefore less discretization error as compared to a very fine grid.

Figure 2 shows how the result of an anisotropic deposition process, using both error reduction algorithms, with an initial grid spacing of 0.10um can resemble that of a 0.01um grid.

Figure 3 shows the relative differences among the different surface evolution algorithms at 0.10um spacing and how they compare to the 0.01um grid simulation using a fixed very small time step.

4. Conclusion

From this investigation of discretization error, we conclude that both spatial and time discretization error can be minimized by the use of the trapezoidal rule and the surface grid refining algorithms.

References

1. E.W. Scheckler and A.R. Neureuther, NUPAD Workshop, Seattle, p. 9, 1992.
2. J.J. Helmsen, E.W. Scheckler, A.R. Neureuther and C.H. Sequin, NUPAD Workshop, Seattle, p. 3, 1992.
3. E. Strasser and S. Selberherr, A General Simulation Method for Etching and Deposition Processes, SISDEP, Vol. 5, p. 357, Sept 1993.
4. M. Seifert, F. Richter, and R.G. Spallek, Simulation of Sputter Deposition Process by DUPSIM, SISDEP, Vol. 5, p. 197, Sept 1993.

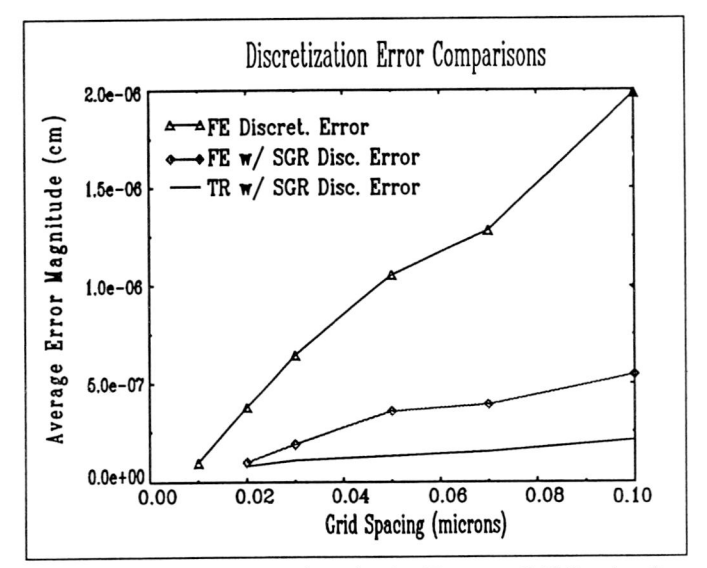

Figure 1. Comparisons of Average Discretization Error vs. Grid Spacing for each algorithm examined.

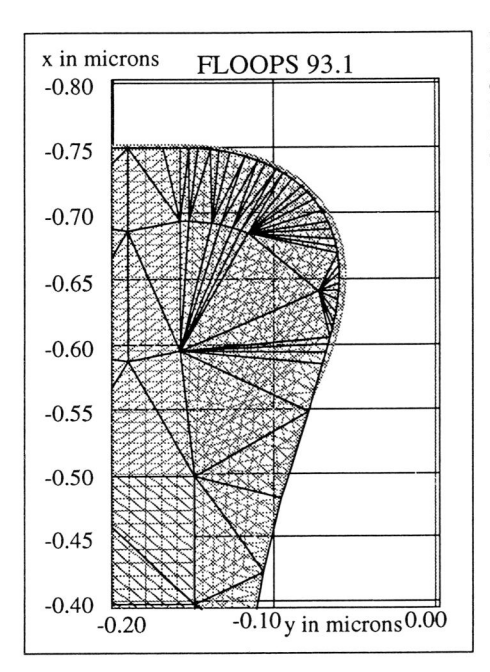

Figure 2.
An anisotropic LPCVD with initial 0.10um grid spacing simulation with both grid refining and TR time-step control resembles that of an initial 0.01um grid spacing simulation.

Figure 3.
A surface profile comparison showing how the trapezoidal rule and surface grid refining algorithms reduce the discretization error for coarse grids.

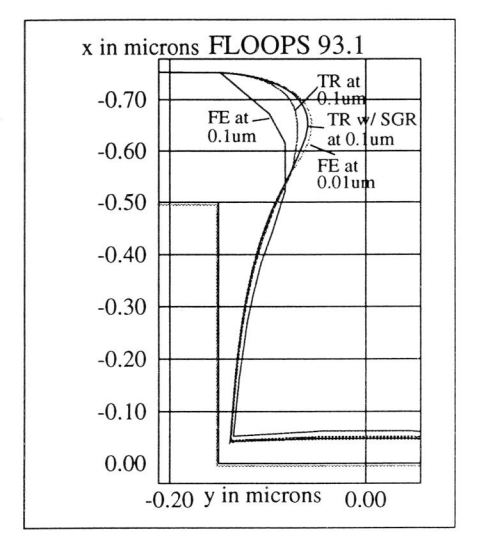

Polygonal Geometry Reconstruction after Cellular Etching or Deposition Simulation

R. Mlekus, Ch. Ledl, E. Strasser, and S. Selberherr

Institute for Microelectronics, TU Vienna
Gusshausstrasse 27–29, A-1040 Vienna, Austria

Abstract

This paper provides a new algorithm for the recalculation of a polygonal geometry representation after the computation of etching and deposition simulations based on a cellular geometry representation. The purpose of that algorithm is to avoid totally any discretization errors in those parts of the geometry which were not affected by the surface movements resulting from the simulation.

1. Introduction

In two-dimensional process simulation, etching and deposition simulations are central steps. The thereby required surface advancement algorithms are often performed on a cellular geometry representation, e.g. [1]. During the simulation each of the cells contains one material type. Etching and deposition is modeled by changing the material type of some cells, leaving their geometric extensions unchanged.

Therefore it is necessary for each etching or deposition simulation step during the process simulation, to discretize the original polygonal geometry (OPG), run the simulation and recalculate a final polygonal geometry (FPG) representation. Former algorithms, e.g. [2], use only the final discrete geometry description to compute the FPG. Discretization errors occur all over the geometry, which demand regriding of every geometry conform grid defined on the original geometry. In addition discretization errors of subsequent etching or deposition simulation steps might accumulate and under certain circumstances endanger the accuracy of the whole process simulation.

To minimize these problems a cellular algorithm was developed which generates the FPG by combining informations from the OPG, the original discrete geometry and the final discrete geometry. This algorithm totally avoids any discretization errors in those parts of the geometry which were not affected by the surface movements resulting from the simulation. Therefore the extensions of the cells giving the accuracy of the discretization must only be adjusted to the minimum extensions of the affected parts of the geometry. Structures much smaller than the resolution of the discretization will keep their original shape when they were not affected by the etching or deposition simulation.

2. The Algorithm

The computation of the FPG starts with a copy of the OPG: Firstly, in the main part of the algorithm, a provisional polygonal geometry is assembled by the following three steps which are performed on the cells of the discrete geometry:

1. **Classification:** Every cell of the discrete geometry is classified depending on the original and final materials of the cell itself and all it's neighboring cells. These have to be taken into account because the regions of the OPG usually do not correspond to the borders of the cells. Five different types of cells are distinguished:

   Etched Cells: a cell is classified as etched if the material changed to vacuum due to the simulation, or if it is originally vacuum and one of its neighboring cells changed to vacuum.

   Partially Etched Cells: the original and final material of the cell is not vacuum and at least one of the neighboring cells changed to vacuum in the final discrete geometry. This category is necessary to describe accurately etching at etch stops.

   Deposited Cells: the material changed from vacuum to the deposited material, or the cell was originally containing some material and one of its neighboring cells changed to the deposited material.

   Original Vacuum Cells: the original and final material is vacuum and the cell is not classified as etched, partially etched or deposited before.

   Original Material Cells: the original and final material is not vacuum and the cell is not classified as etched, partially etched or deposited before.

2. **Geometry-Extraction:** For etched, partially etched and deposited cells a polygonal description of the original geometry is computed. This description contains every part of the OPG which is located inside of the cell, informations about the material types inside and outside of the borders of the cell and the classifications of the cell and all its neighboring cells. (Fig. 2 shows an example for this geometry extraction using quadratic cells.)

3. **Geometry-Correction:** the FPG inside of the recent cell is computed by modifying the polygonal description of the cell dependent on the classification of the cell:

   Etched Cells: Any region of the geometry which does not contain vacuum is removed from the provisional polygonal geometry.

   Partially Etched Cells: Every region of such a cell which is not containing the material itself and is bordering to an etched cell is removed from the provisional polygonal geometry. Remaining parts of the borders to neighboring cells which are classified as *Original Material Cell* or *Etched Cell* are added to the provisional polygonal geometry to ensure a consistent description of the geometry.

   Deposited Cells: Regions of the geometry containing vacuum are replaced by regions containing the deposited material and added to the provisional polygonal geometry like newly created borders to cells which were classified as *Original Material Cell* or *Original Vacuum Cell*.

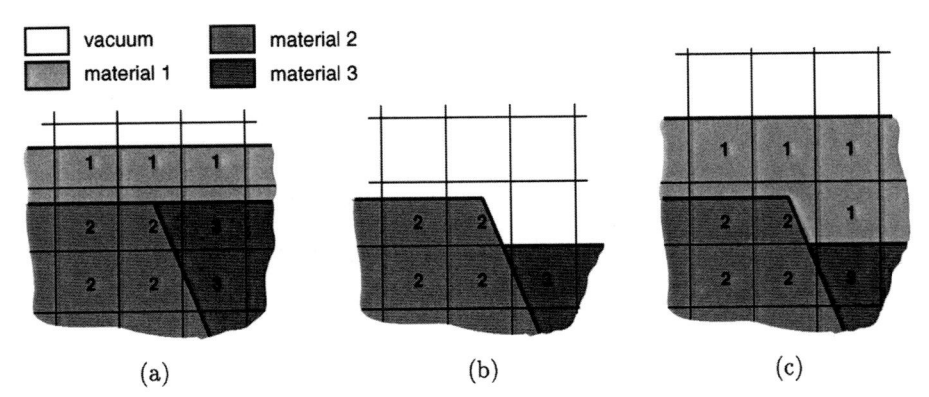

Figure 1: Discrete and polygonal geometries. (a) original, (b) etched, (c) redeposited

Fig. 1 shows a two-dimensional example for the construction of the FPG after an etching and deposition simulation. In Fig. 1a the discretization of th OPG is demonstrated. In this example the material of a cell is determined by the material type of the OPG at the center of the cell. The resulting discrete and polygonal geometries after removing some cells by an etching simulation are presented in Fig. 1b. Fig. 1c shows the resulting discrete and polygonal geometries after the redeposition of material 1 on top of the geometry of Fig. 1b.

In Fig. 2 the extraction and modification of the polygonal geometry is demonstrated for this simulation step for the cell in the center of Fig. 1b. The cell contains the material type 2 and is classified as a deposited cell. The redeposition of material 1 is simulated by changing the material type of the face in the northeast of the cell, and correcting the material references at the borders of the cell. After that the local geometry is inserted into the provisional polygonal geometry.

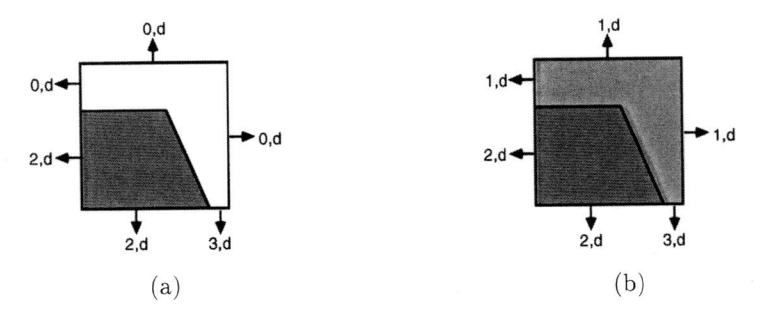

Figure 2: Extraction and modification of the polygonal geometry for the redeposition simulation of the cell in the center of figure 1b and 1c. The vectors at the borders give the material types outside of the cell and the classifications of the neighboring cells. (d...deposited)

The structure which is thereby created contains a high number of segments. Therefore in a second step the face structure is simplified and locally smoothed as far as it is not defined by parts of the OPG. The extent of reduction can be controlled, and the number of segments is often drastically reduced. Fig. 3 shows the polygonal geometry

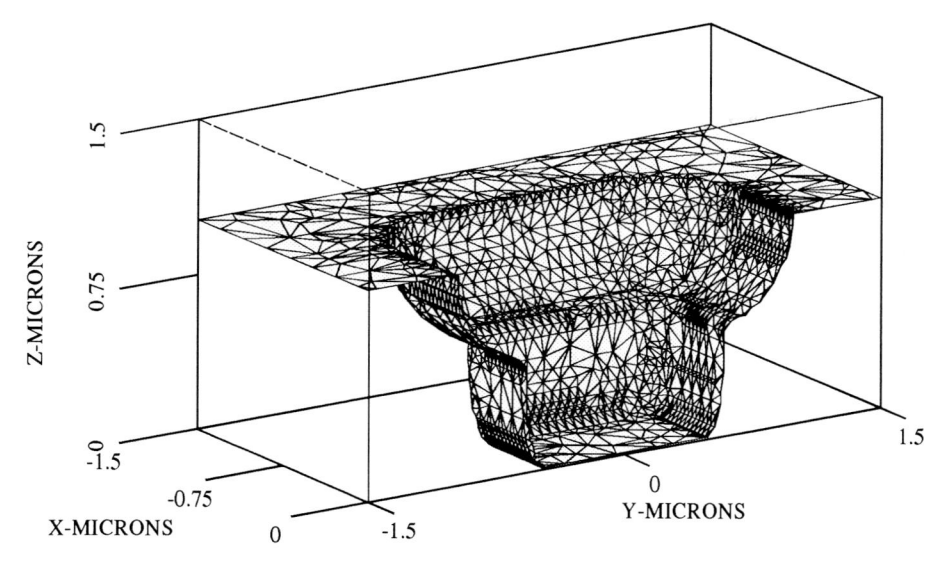

Figure 3: A square contact hole.

which was obtained after the simulation of a complete three-dimensional contact hole etching process.

3. Conclusions

The algorithm is highly independent from the dimension and shape of the discretization cells. Possible restrictions arise only out of numerical and algorithmic problems during the computation of the inner geometry of the cells. Therefore it is applicable to a large group of problems which require temporary conversions from polygonal to discrete geometry representations.

The increased computational effort of this new algorithm can be justified by considerable savings of calculation time in following regriding algorithms, because these have only to be applied in those parts of the geometry which actually changed during the etching or deposition simulation.

Acknowledgements

Our work is significantly supported by Digital Equipment Corporation, Hudson, USA; and Siemens AG, Munich, Germany.

References

[1] E. Strasser, G. Schrom, K. Wimmer, and S. Selberherr, "Accurate Simulation of Pattern Transfer Processes Using Minkowski Operations", *IEICE Trans. Electronics*, vol. E77-C, pp. 92–97, 1994.

[2] W.E. Lorenson and H.E. Cline, "Marching Cubes: A High Resolution 3D Surface Construction Algorithm", *Computer Graphics*, vol. 21, pp. 163–169, 1987.

A Data-Model for a Technology and Simulation Archive

K. Wimmer, M. Noell, W. J. Taylor, and M. Orlowski

Advanced Product Research and Development Laboratory, Motorola Inc.
3501 Ed Bluestein Blvd., Austin, TX 78721, USA

Abstract

In this paper, we describe software aimed at achieving a significant reduction in the time required to develop new semiconductor technologies, by facilitating simulation sharing, and recognition and reuse of technology modules. We have created a technology and simulation archive based on the World-Wide Web (WWW) data-model. The WWW data-model is very flexible in capturing the wide variety of data formats of technology and simulation related information, and it proves to be sufficiently fast for frequent direct data retrieval by simulation tools.

1. Introduction

In a typical design environment for semiconductor technology, engineers of many different skill levels have to navigate through a wide variety of data (simulations, experiments, process flows, etc.) from many different sources.[1]

A rigid data-base approach for storage and retrieval of predefined objects is too inflexible to capture our needs:

- facilitate storage, retrieval and reuse of calibrated simulation know-how (input decks, experimental/simulation results, hints and documentation)
- make process flows and recipes directly accessible by simulators
- allow engineers of all levels to acquaint themselves with the technologies and simulation capabilities.

Here we present a complete departure from traditional data bases: An **archive** which is based on the **World-Wide Web** (WWW) **data-model** [1]. The WWW data-model is a distributed hypermedia system which allows many possible relations (*using hypertext links* and *index searches*) between any individual objects and others. Each object can be any type of data file that can be stored on a computer.

---

[1]Although there are excellent proposals for a unified data exchange format (e.g. [2]) such a format has not become generally established.

2. Implementation

To use the WWW data-model, one has to set up servers for the `http` [1] and `wais` [3] protocols and to install clients (commonly refered to as browsers) for requesting data from the servers. In this section we focus on selected features in the the WWW data-model which significantly improve its capabilities for a technology and simulation archive.

A browser (most popular browsers are Mosaic and Netscape) needs to know what to do with the data being requested. In order to handle this, each file in the archive is assigned an explicit **content-type** [4]. When a browser requests an object, the server returns the object's content together with its content-type.[2] Depending on the content-type the browser performs **actions** to view the object. This action, in many cases, is just the presentation in the document window of the browser, but we have found it to be very beneficial to include transformation of the data-formats and spawning of external viewers.

The archive is easily extended to include additional data types by creating a mail-cap [5] entry in the browser's `.mailcap` file to associate the content type with the program to view it. For instance, as shown in Figure 1, `tonyplot` is used to view results from Silvaco [6] simulators.

For each object in the archive a **description** is entered. This description is displayed during directory-based browsing to help the user to identify the object. Furthermore we plan to index the descriptions to allow searches.

During check-in of an object into the archive, both the content-type and the description are entered into a local `.htaccess` file. Additionally, new hyperlinks need to be created and indexes need to be updated by corresponding routines on the server.

Uploading of new information and other administrative functions are performed by `http-POST`s to WWW-agents (CGI-script [7]) from either the browser (using HTML-forms [8]) or separate programs (Figure 2).

```
# Use 'tonyplot' to view ssf.
application/x-ssf;      tonyplot %s
# Convert to ssf and use 'tonyplot'
application/x-pep-sav;  pep2ssf -s %s\; tonyplot %s
application/x-pbf;      tifwrap -s %s\; tonyplot %s
application/x-tif;      tif2ssf -s %s\; tonyplot %s
....
```

Figure 1: In the browsers' mailcap file content-types and corresponding viewing actions are specified.

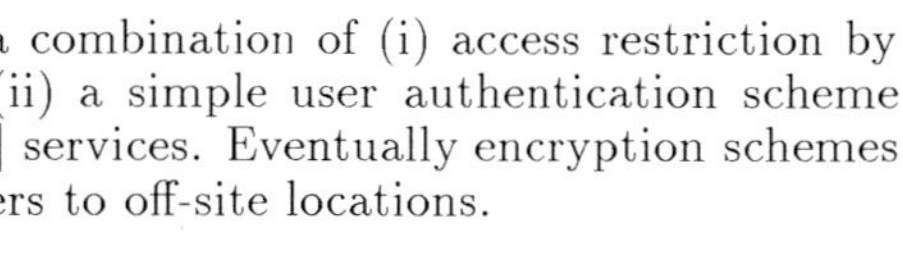

Figure 2: Uploading data to the server is performed by `http-POST` to WWW-agents (CGI-scripts).

Proprietary information is **protected** by a combination of (i) access restriction by domain (Internet address of client), and (ii) a simple user authentication scheme (userid, password) using X.500 directory [7] services. Eventually encryption schemes such as PGP [9] will be included for transfers to off-site locations.

3. Utilization

We have used the WWW data-model to build up a technology and simulation archive which we call the "Simulation and Technology Information Discovery and Access System" (Figure 3). Currently our system provides for browsing (hyperlinked, directory based, and search-able through free text and fielded indexes) of the following kind of semiconductor technology related data:

---

[2]Actually, the browser negotiates the acceptable content-types with the server

- Calibrated process and device simulations (input files, results, etc.) for, e.g., PowerPC, fast SRAM (4Mbit, 16Mbit, 64Mbit), non-volatile memory (Flash, EE, 1T, 1.5T cell). These simulations are intended for reuse and to demonstrate current simulation capabilities.
- For simulation automation tools (see below) as well as for training and discovery learning, process flows hyperlinked to miniflows and to technology cross-sections at key steps in the process; individual recipes (miniflows) and masksets.
- Solutions to recurrent simulation issues (e.g. shallow trench and PELOX isolation)
- Manuals, examples and helpful hints
- WAIS index of U.S. patent abstracts; to follow projects complete index of Sematech and SRC abstracts; technical reports, table of contents of selected journals, etc.

We store process flow representation (**PFR**) and recipe (**miniflow**) information in the original format as generated by the **CIM-system** [10], which is **familiar to the engineers**. We have developed a set of supporting flow-to-<simulator> tools which translate the PFRs, mask information, and the miniflows into valid simulation input files (Figure 4). The individual tools can be used from the command-line or from VISTA [11] to perform standard device characterizations. The fact that the application engineers now have only to deal with the familiar PFR greatly **enhanced** the **acceptance** of advanced simulators.

Each recipe referenced by the PFR (Figure 5) is stored in a **miniflow** file (Figure 6). The miniflows from the CIM-environment are represented together with process simulator commands to capture knowledge from previous experience. To account for the variety of process simulators (1D vs. 2D, different vendors) the miniflow's command lines are translated to the individual simulator syntax. To adopt to specific process simulator requirements we use cpp-like extensions in the miniflows. The masks referenced in the PFR are stored in a maskset file. For each mask and each device cross-section the information about covered and exposed areas is stored in parameterized form (Figure 7).

An issue is the average response time of the server during **frequent accesses** to the archive. The above mentioned translation process by the the flow-to-<simulator> tools requires to access several hundred objects from the archive server. We experienced, that the response time of the server is short enough to translate typical PFRs into simulator-input files within a few seconds. I.e., the WWW data-model is **sufficiently fast** for frequent direct data retrieval by simulation tools.

4. Conclusion

We have created a technology and simulation archive based on the WWW data-model. The WWW data-model proved its efficiency and flexibility in capturing the wide variety of data formats of technology and simulation related information. Technology knowledge, including simulation, can be **collectively** built up and made available to **individuals**. Due to its open configuration with hyperlinks and index searches, the archive is the perfect environment for **discovery learning** for users of all skill levels. Additionally, the WWW provides a convenient platform for technology transfer while complying with security directives.

References

[1] T. Berners-Lee, et al.: *Electronic Networking*, 2(1), pp. 52-58, Meckler Publ. 1992.

[2] F. Fasching, et al.: *IEEE Trans. CAD*, 13(1), pp. 72–81, 1994.
[3] B. Kahle, et al.: *Electronic Networking*, 2(1), pp. 59–68, Meckler Publ. 1992.
[4] N. Borenstein: Internet Network Working Group, *RFC 1521*, 1993.
[5] N. Borenstein, et al.: Internet Network Working Group, *RFC 1521*, 1993.
[6] Silvaco International: *Virtual Wafer Fab User's Manual*, 1994.
[7] C. Liu, et al.: *Managing Internet Information Services*, pp. 357–380, O'Reilly & Associates 1994.
[8] J.E. Tilton: *Composing Good HTML*. http://www.willamette.edu/html-composition/strict-html.html, 1994.
[9] S. Garfinkel: *PGP: Pretty Good Privacy*. O'Reilly & Associates 1994.
[10] PROMIS Systems Corp.: *PROMIS User's Manual*, 1992.
[11] S. Halama, et al.: *Technology CAD Systems*, pp. 197–236, Springer-Verlag 1993.
[12] R.W. Dutton, et al.: *Technology CAD: Computer Simulation of IC Processes and Devices*. Kluwer 1993.

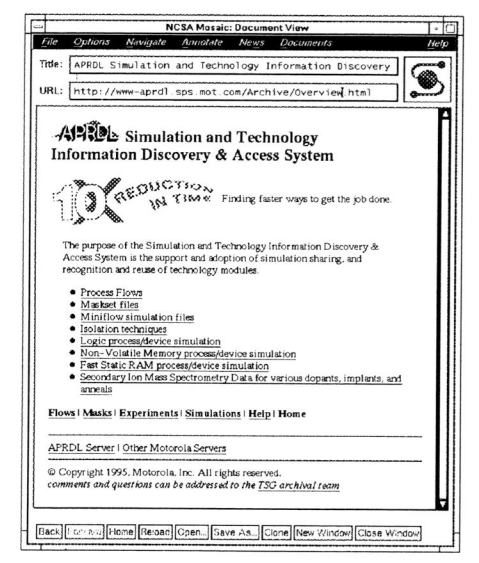

Figure 3: Example *home page*, the front door to APRDL Simulation and Technology Information Discovery & Access System.

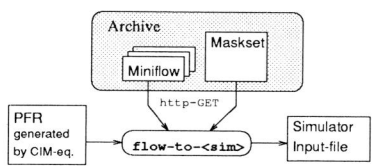

Figure 4: `Flow-to-<simulator>` tools create input files for various simulators from original CIM-syntax.

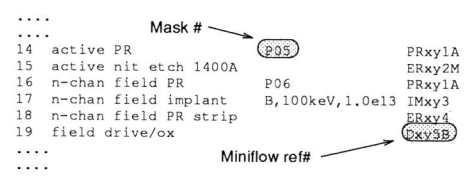

```
....
....                Mask #
14    active PR                  P05          PRxy1A
15    active nit etch 1400A                   ERxy2M
16    n-chan field PR            P06          PRxy1A
17    n-chan field implant   B,100keV,1.0e13  IMxy3
18    n-chan field PR strip                   ERxy4
19    field drive/ox                           Dxy5B
....
....                  Miniflow ref#
```

Figure 5: Example PFR generated by the CIM-equipment [10]. The process shown is from [12].

```
### Miniflow: Dxy5B - Field Drive & Oxidation
#       Push       0:10:00    800 degC  N2+2%O2
#       Ramp       0:12:00    1000 degC N2+2%O2
#       Drive      2:00:00    1000 degC N2+2%O2
#       Oxidize    3:10:00    1000 degC O2+H2
#       Ramp       0:40:00    800 degC  N2
#       Pull       0:15:00    800 degC  N2
....
diff    time=120   temp=1000     dry   pressure=0.0196
#ifdef Athena
  method viscous oxide.rel=5e-3 grid.oxide=0.05
#endif
diff    time=190   temp=1000     wet   pressure=0.85
....
```

Figure 6: Example of a miniflow file, showing recipe [12], generic simulator commands and simulator specific extensions.

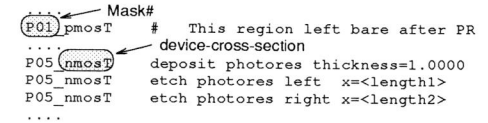

```
....        Mask#
P01  pmosT    #    This region left bare after PR
....              device-cross-section
P05  nmosT        deposit photores thickness=1.0000
P05_nmosT         etch photores left  x=<length1>
P05_nmosT         etch photores right x=<length2>
....
```

Figure 7: Maskset: specifies the covered and exposed areas for each mask and device.

A Programmable Tool for Interactive Wafer-State Level Data Processing

G. Rieger, S. Halama, and S. Selberherr

Institute for Microelectronics, TU Vienna
Gusshausstrasse 27-29, A-1040 Vienna, Austria

Abstract

This paper presents a tool for initial and intermediate data processing at wafer-state level. Geometric operations and analytical attribute calculations are combined with the graphical user interface and with some TCAD shell functionality to reduce the effort of both manual editing and automated operations within complex simulation flows.

1. Introduction

With the automated simulation of complex process flows some new requirements arose for TCAD frameworks. One of these is the creation of scalable input data for the first simulation step. Others are to perform simple tasks on the actual wafer-state description, like geometry modifications, or to emulate prototype processing steps like mask deposition. As these tasks must be performed in both interactive and batch mode, a highly flexible and versatile implementation is required.

2. The PIF editor

The PED (PIF Editor) has been developed within the VISTA (Viennese Integrated System for TCAD Applications, [4]) framework which is based on the PIF (Profile Interchange Format, [3]). The PED allows creation and modification of device data in batch mode as well as interactively via an X11 Window based graphical user interface. One-, two-, and three-dimensional device geometries can be handled. The basic operations include geometry modeling, attribute handling, and simple visualization. External executables, like, e.g., grid generators, solid modelers, or even simulators can be directly invoked from the PED. This provides an alternative environment for task-level tool integration to the VISTA shell, which is typically used when an immediate interactive access to the output or input of the integrated too is desired.

3. Extension language

A CAD tool like the PIF editor provides functionality for a wide spectrum of tasks. Nevertheless, instead of restricting the user to the functionality provided, it is desirable to adopt the editor to ones needs either by combining basic functions to

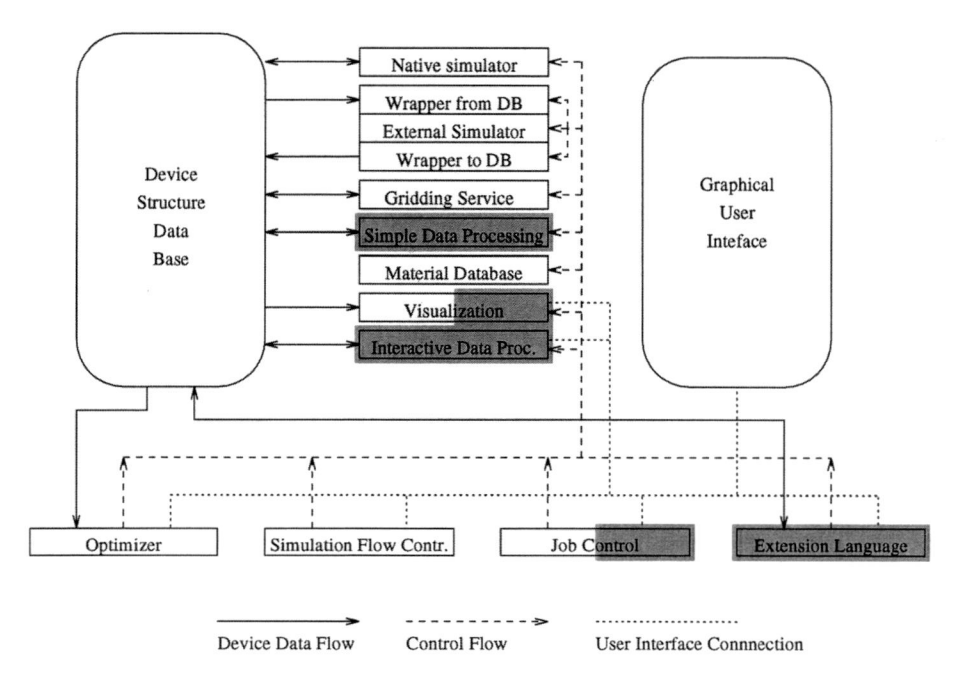

```
                    ┌──────────────────┐
                    │ Native simulator │
                    ├──────────────────┤
                    │ Wrapper from DB  │
                    │ External Simulator│
                    │ Wrapper to DB    │
     Device         ├──────────────────┤    Graphical
    Structure       │ Gridding Service │      User
      Data          │Simple Data Processing│  Interface
      Base          │ Material Database│
                    │  Visualization   │
                    │Interactive Data Proc.│
```

Figure 1: Functional diagram of a TCAD framework and topics met by the PIF editor (shaded)

comfortable editing commands or by generating application–specific facilities. An interpreting extension language is a suitable approach to meet these wishes.

The PED is linked to the VLISP (VISTA LISP) interpreter, a version of the public domain XLISP interpreter, extended with a variety of framework-oriented functions including generic operating system bindings, a comprehensive user interface, and functional access to the PIF database. Utilizing this interpreter makes available a large collection of task level LISP functions already developed for the VISTA shell.

3.1. Configuration and styles

The control and configuration parts of the PED are written in LISP, while low level and time-critical functions are implemented in C, but are bound to the LISP interpreter. Thus the PIF editor can read LISP style- and command-files. Loading the required files makes it possible to build custom versions of the PED that support all geometric elements and attributes required for a specific simulator, whilst allowing the user to focus on only the necessary tasks instead of becoming lost in menus and commands.

3.2. Integration of external executables

The interface for the invocation of the grid generator TRIGEN [2] is shown in Fig. 2. After selection of the program via a menu and filling in the parameters, the TRIGEN executable is started for the device geometry currently displayed. After successful termination of TRIGEN, the result is displayed immediately, allowing for verification and manual improvements of the new grid.

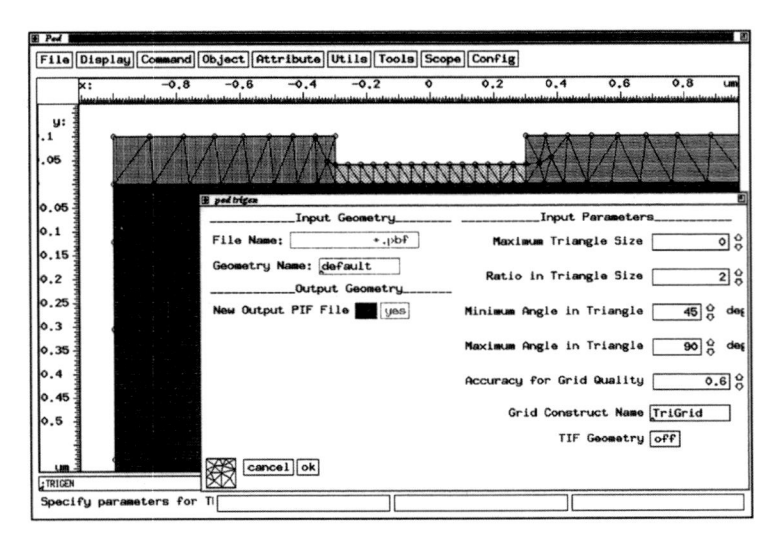

Figure 2: PIF editor with the panel for the TRIGEN grid generator, and the resulting grid

3.3. User interface

The usual elements of a graphical user interface - mouse, keyboard, menus and, pop-up windows - are controlled from LISP. All user actions trigger evaluation of LISP expressions which depend on the current style and mode. Operations that require more than one user event are controlled by an infinite state machine [1] that processes an appropriate rule to switch the mode of the PIF editor, to check the input, and to invoke the processing functions.

4. Example Application

A typical example for a PED application is the deposition of a contact and the specification of its electric potential for use by a device simulator.

First a LISP function "contact" is written with parameters for contact geometry, position, material, and contact voltage. It generates a rectangular segment and creates appropriate MaterialType and ContactVoltage attributes. This function can be used, as is, in batch mode. The parameters can be provided by invoking other functions, while "contact", which returns a reference to the new segment, might be used in LISP-programs.

"contact" can be invoked with the following line:

```
(contact 0.0  -1.0  0.1  0.05 "Al" 0.0)
```

which will generate a rectangular Aluminum segment.

```
(defun contact
  (ybase xmid width height mat voltage
  &aux point1 point2 point3 point4
       segment1 segdesc1 mattype1 contvolt1)

  (setq point1 (point (- xmid (/ width 2)) ybase))
  (setq point2 (point (+ xmid (/ width 2)) ybase))
  (setq point3 (point (+ xmid (/ width 2)) (+ ybase height)))
  (setq point4 (point (- xmid (/ width 2)) (+ ybase height)))

  (setq segment1
        (segment (rectangle)))

  (setq segdesc1  (attribute NIL "SegmentDescription" segment1))
  (attribute segdesc1 "MaterialType"    segment1 mat)
  (attribute segdesc1 "ContactVoltage"  segment1 voltage
             :unit "V")
  segment1)
```

This function can be integrated to the user interface by specification of a rule. This rule lets the state machine inquire the parameters from the user and then invoke the above defined function. The user can specify the values with mouse, keyboard, or by activation of other state machine rules that return appropriate values or objects.

5. Conclusion

The PIF editor is an important tool for the VISTA framework. It allows data processing either program-controlled (for use within extended simulation flows) or interactively via a graphical user interface. It can be tailored to specific requirements with an integrated LISP interpreter.

Acknowledgements

Our work is significantly supported by Sony Corporation, Atsugi, Japan; and Digital Equipment Corporation, Hudson, USA.

References

[1] A.V. Aho, R. Sethi, and J.D. Ullmann, *Compilers*, Addison-Wesley, 1986.
[2] R.E. Bank, "PLTMG: A Software Package for Solving Elliptic Partial Differential Equations", *Frontiers in Applied Mathematics*, vol. 15, SIAM, 1994.
[3] F. Fasching, W. Tuppa, and S. Selberherr, "VISTA-The Data Level", *IEEE Trans. Computer-Aided Design*, vol. 13(1), pp. 72–81, 1994.
[4] S. Halama, F. Fasching, C. Fischer, H. Kosina, E. Leitner, Ch. Pichler, H. Pimingstorfer, H. Puchner, G. Rieger, G. Schrom, T. Simlinger, M. Stiftinger, H. Stippel, E. Strasser, W. Tuppa, K. Wimmer, and S. Selberherr, "The Viennese Integrated System for Technology CAD Applications", In F. Fasching, S. Halama, and S. Selberherr, editors, *Technology CAD Systems*, pp. 197–236, Springer, 1993.

Layout Design Rule Generation with TCAD Tools for Manufacturing

José López-Serrano[*], Andrzej J. Strojwas

Electrical and Computer Engineering Department, Carnegie Mellon University,
Pittsburgh, PA 15213, USA

Abstract

This paper presents a methodology for estimating the effects of changes in the layout design rules on the manufacturability of a VLSI technology. 2-D process and device simulations were used to estimate parametric yield, while functional yield was predicted with state-of-the-art yield modeling tools. A spectrum of TCAD tools was therefore capable of estimating the resulting number of good chips per wafer for different sets of VLSI layout design rules.

1. Introduction

Minimizing the circuit layout feature sizes can lead to improved performance and packing density, but it may also reduce the manufacturing yield. The smaller dimensions increase the relative variability of the process and make the circuits sensitive to smaller particles, which can degrade the manufacturing yield. If that possible reduction in yield is not taken into account properly, it may negate the gains due to greater packing densities.

Layout design rules, at present, are developed by taking the minimum feature size that guarantees process repeatability and no electrical parasitics, while allowing for the tolerance of photolithography steps. Functional yield loss due to particles is usually not considered, and considerable R&D resources are often spent on reducing design rules that do not impact final circuit layout size the most. It is common to treat the probability of failure for each design rule independently, which is not always correct. For example, treating the design rules for minimum metal-1 width and minimum metal-1 spacing independently may underestimate the joint probability of failure because they are inversely correlated, i. e., metal-1 bridging shorts are more likely for wider metal-1 lines.

We present a methodology for the development of layout design rules in a more statistically rigorous way, considering their joint probability of failure, and also taking into account functional yield. Our approach uses TCAD simulation tools and parametric and functional yield loss estimation techniques, and is intended to fit within an integrated process synthesis system [1].

2. Methodology

Figure 1 shows the proposed methodology. Given a process flow and an initial set of design rules,

---

[*] At present with the Product Engineering Department, AT&T Microelectronics- Spain, Polígono Industrial Tres Cantos s/n (Zona Oeste), 28760 Tres Cantos (Madrid), SPAIN

the effects of different sets of design rules on manufacturability and electrical were analyzed.

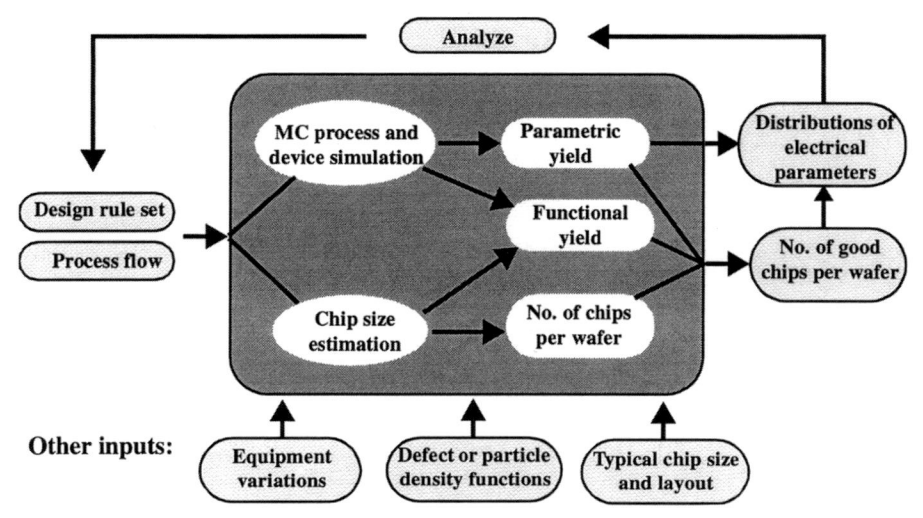

Figure 1: Proposed methodology for design rule development

The process flow for the technology was assumed to be fixed. It was also assumed that it was possible to obtain precise information about the variability of the processing equipment, and measured distribution functions for the density and size of particles or defects. The optimum design rules might be different for different types of circuits (ASICs, DRAMs, etc.,), so the layout of a typical circuit, of the type that the technology was intended for, was used in the analysis.

The measure for performance was the simulated distribution of device characteristics. The metric for manufacturability was the estimation of the number of good chips per wafer for a typical circuit, even though more elaborate cost measurements might be used [2].

2.1. Chip Size Estimation and Number of Printed Chips per Wafer

We have implemented a C program to estimate the impact of the different layout design rules on the size of a typical chip. This program allowed us to approximate the change in size for a small chip or for a typical cell, without having to determine the new layout that the change in design rules required. It would not have been computationally viable to attempt that with a layout compactor.

The approach consisted in decoupling the two-dimensional problem into two one-dimensional ones. The program considered many horizontal and vertical slices, and for every design rule instance in those slices, it decided whether a local shrink or expansion would have been needed to accommodate the new design rules, without calculating the actual changes in the layout features. The total accumulated shrinks or expansions for each slice were then studied and a final change was determined for both horizontal and vertical dimensions. The estimation of the number of printable chips per wafer was then straightforward.

2.2. Monte Carlo Process and Device Simulations and Parametric Yield.

Parametric yield estimation was obtained by introducing equipment variations into the process flow for the 2-D process simulations. The resulting Monte Carlo simulation produced a distribution of electrical device characteristics, instead of single values. Those distributions, with a set of specification limits, gave the estimated parametric yield [3]. In every case, the electrical parameters were simulated under worst-case operating conditions. Figure 2 shows how the yield loss was calculated, by identifying those points that did not meet the specifications for any of the electrical parameter simulations. Scatter-plots like these only show the interactions between two electrical parameters, while in reality many more parameters were considered.

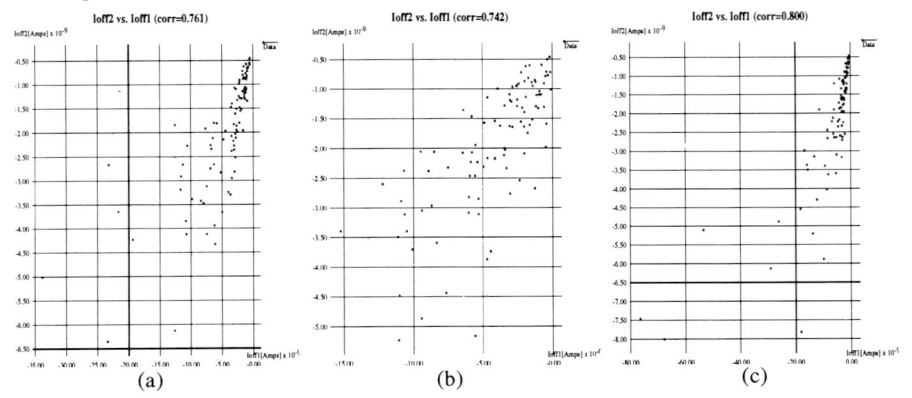

Figure 2: Estimation of parametric yield from simulations, for three changes in design rules: (a) nominal case, (b) reduction in channel length, and (c) reduction in drain spacing. The Ioff for PMOS transistors and the leakage current between two P+ drains are plotted.

2.3. Functional Yield

Functional yield was estimated by integrating the product of the layout critical area function and the defect size distribution function [4]. The critical area for the circuit under study was calculated with the CREST program [5]. If particle size distributions functions were available (form laser-scan and digital image processing equipments), instead of defect size distribution functions, there would be a need to translate the particle size to final defect size -if defects are created at all. That transformation is possible with precise lithography/topography simulators, such as METROPOLE [6]. Finally, the changes in the critical area function due to changes in design rules were calculated by simple shifts in the defect-radius-axis.

3. Application to a CMOS Process

This methodology was applied to a modified 0.8 micron CMOS technology taken from industry. Since we decided to generate design rules for logic circuits, we used a typical cell for those circuits: a 2500 transistor 8-bit multiplier, to study the changes in chip size and the critical area functions. We extrapolated the results to a 500K transistor circuit by multiplying the area by 200.

The effects on electrical performance were estimated using 2-D process and device simulations. We used TMA SUPREM-IV [7] for the process simulations, and SIMOS, a 2-D device simulator within PDFAB [8]. PDFAB, an environment for statistical process, device and circuit simulations, controlled the Monte Carlo runs, linked the different simulators, and extracted the final electrical characteristics.

Table 1 shows the results for the manufacturing aspects of the problem. The metric chosen was the number of good chips per wafer, which, for a fixed process flow, can directly be translated into circuit cost. The effects of three changes in three different design rules were analyzed. In reality, of course, many more rules would have to be considered, and the process flow could also be modified to accommodate the new rules. We were just trying to show instances of situations where not considering functional yield or interactions could produce very negative results. In this case, only the reduction in drain spacing could be justified from the manufacturing point of view. The increase in packing density for metal spacing reduction did not compensate the associated yield degradation due to larger sensitivity to smaller particles. The defect distribution data came from actual measurements [4]. Finally, the shrinking of the transistor channel length did result in faster devices, but the reduction in parametric yield (due to PMOS transistor leakage) was unacceptable.

Table 1: Results Summary	Nominal case	1 λ reduction in drain spacing	1 λ reduction in metal spacing	.5 λ reduction in transistor length
Parametric yield	94%	92%	94%	9%
Functional yield	86.6%	86.6%	78%	86.6%
Total yield:	81.4%	79.7%	73.3%	7.8%
Chip size (as% of initial size):	100%	97.4%	95.4%	96.5%
Number of chips per wafer	200	205	210	207
Good chips per wafer:	**162.8**	**163.4**	**153.9**	**16.1**

4. Conclusions

Functional yield loss due to particles must be considered for layout design rule development: while it might be technologically possible to shrink a technology, it might not be advisable in certain instances, because it would end up producing fewer good chips per wafer. The interactions between different design rules must also be taken into account by studying the joint probability of failure. 2-D TCAD simulation tools and the most advanced yield estimation techniques can be used to study the manufacturability of different sets of layout design rules, by estimating the number of resulting good chips per wafer. The use of TCAD tools could offer significant development cost savings, and help focus R&D investments into the most profitable areas.

5. Acknowledgments

We gratefully acknowledge the help and data from K. Lucas, P. K. Nag, J. Khare, H. Heineken, D. Gaitonde, and H. Read (from Carnegie Mellon University), and J. Kibarian, K. Michaels, and T. Cobourn (from PDF Solutions Inc., Pittsburgh). The financial support from the Fundación Barrié de la Maza (La Coruña, Spain) is also greatly appreciated.

References

[1] Z. Lemnios, "Beyond MMST: the Virtual Factory (IC Manufacture)", Solid State Technology, vol. 37-2, p. 25, February, 1994.
[2] W. Maly, "Cost of Silicon Viewed from VLSI Design Perspective", ACM/IEEE DAC-1994.
[3] J. López-Serrano, S. Koh, T. Crandell, J. Delgado, H. Nicolay, T. Haycock, A. Strojwas, "Yield Enhancement Prediction with Statistical Process Simulations in an Advanced Poly-Emitter Complementary Bipolar Technology", Proc. of IEEE CICC, May, 1994.
[4] J. Khare, W. Maly, S. Griep, D. Schmitt-Landsiedel, "SRAM-Based Extraction of Defect Characteristics", Proc. IEEE ICMTS, Vol. 7, March, 1994.
[5] P.K. Nag, W. Maly, "Fast Critical Area Extraction for Shorts in Very Large IC's", SRC report, CMUCAD-93-26, April, 1993.
[6] K.Lucas, A. Strojwas, "A New Vector 2D Lithography Simulation Tool", Proc. IEDM, Dec-92.
[7] TMA Inc., "TSUPREM-4 V.5.2 User's Manual", 1992.
[8] PDF Solutions Inc., "pdFab V.2.1 User's Manual", 1994.

ALAMODE: A Layered Model Development Environment

D.W. Yergeau, E.C. Kan, M.J. Gander, R.W. Dutton

Integrated Circuits Laboratory, Stanford University,
Stanford, CA 94305, USA

Abstract

To accurately simulate modern semiconductor process steps, TCAD tools must include a variety of physical models and numerical methods. Increasingly complex physical formulations are required to account for effects that were not important in previous generation of technology. As a specific example, the impurity diffusion mechanisms owing to point defects and damage kinetics are not well understood, and thus flexibility in definition of models is highly desirable. An object-oriented approach has been applied to implementing a 1-2-3D finite-element *dial-an-operator* PDE solver. The control interface is based on Tcl and allows layered access to model definitions and solution techniques.

1. Configurable TCAD Framework

To achieve a configurable and resuable TCAD framework, we have analyzed process simulation information flow carefully and separated the common functionalities such as the geometry/field server, the parameter/model library, and the visualization tools from the physical definition and numerical solvers (Figure 1) [3]. In this open framework, one can choose to use individual tools and in any combination. For example, one has the choice of using SUPREM-IV diffusion code or the new ALAMODE dial-an-operator solver for impurity diffusion simulation. All tools communicate wafer representation through the geometry/field servers, and additional information for individual tools such as model and boundary condition setup can come from either the model library or the front-end controller.

This environment also provides several layers of access to the variety of users. A process engineer may need only to select a calibrated model appropriate for the process step (e.g. diffusion using the SUPREM "fermi" model) and specify the process conditions (e.g. time, temperature, and ambient). For calibration, one would need to modify the parameters used in a model to find the optimal set. At a lower level, the model developer may need to modify the equations used to implement the model, either by adding terms or even equations. Previous generation software provided user access to the first two levels, but any changes beyond altering parameters required access to and modification of the source code. By utilizing a *dial-an-operator* [1] paradigm, we have provided access to the model definition layer.

2. Design of the ALAMODE PDE Solver

A portion of the information model for the ALAMODE PDE solver is shown in Figure 2 (see [4] for discussion of information model). An *operator* is a dimensionally independent discretized representation of a term in a PDE such as $\nabla \cdot D \nabla C$ or $K_r(VI - V^{eq}I^{eq})$. Operators generate residual vectors and tangent matrices for use in a nonlinear solver. Grouping operators via *equation* and *system in region* provides the coupled system of equations describing the homogeneous physical model within a *region*. Boundary and interface constraints are handled similarly by treating boundaries and interfaces as distinct regions and attaching operators to meet the constraint equations. Operators may be arbitrarily attached to any *field* and can utilize derived field quantities via *functions* such as is needed for the diffusion coefficient for the more compact diffusion models. Element interpolation is organized in the inheritance hierarchy shown in Figure 3, which provides the dimensional independence of the operators as well as the ability to solve equations mesh and field from a variety of sources.

The object-oriented data structure provides an efficient and versatile means to represent coupled systems of PDEs. To provide the user with access to the data structure, we have extended a Tcl interpreter [2] with commands which mirror the information model objects. This allows the user to completely describe the physical model in a compact script, with nearly direct correspondence between the terms in the PDEs and the commands in the script. We will illustrate this mapping using the first example below.

3. Examples

For the first example, we use a previous generation model for boron diffusion with segregation at a stationary silicon/oxide interface to illustrate the Tcl interface in ALAMODE, and to show applicability to models in current simulation technology. This model is roughly equivalent to SUPREM's "fermi" diffusion model (without the electric field) and is described by $C_{B,t} = \nabla \cdot D_B \nabla C_B$, where D_B is a constant in the oxide and $D_B = D_B(C_B) = D_B^x + D_B^+(p/n_i)$ in the silicon. Segregation of dopant between the regions is modeled by a segregation flux (flux = $\text{Trn}(C_B^{\text{outer}} - C_B^{\text{inner}}/m)$). The Tcl script in Figure 4 directly maps the model's PDEs, interface conditions, and functional relationships into *equations* composed of *operators* and *functions* applied in the appropriate *region*. Figure 5 shows the initial and final boron profiles for a 15 minute, 1100°C anneal using SUPREM's defaults for diffusivities and the transport coefficient and value of 2.2 for the equilibrium segregation ratio (m).

Note that the model for the silicon diffusion coefficent is drawn from a function library while the oxide diffusion coefficient is a constant. Additional enhancements, such as vacancy and interstitial supersaturation effects (if the user adds equations for interstitials and/or vacancies), can be chained into the diffusion coefficient.

In the second example, we solve a system of reactive-diffusive equations for phosphorus diffusion [5]. The five-species model and the parameters used are shown in Figure 6. The profiles developed during a 10 minute, 900°C anneal are shown in Figure 7. Multi-species reactive-diffusive systems, similar to the one shown, are becoming increasing important to model low temperature and rapid thermal annealing as well as implant damage enhanced diffusion.

Acknowledgement: This work was supported by ARPA under contract DABT63-93-C-0053. Also, the mentorship of Drs. M. Orlowski (Motorola) and R. Goossens (National) is gratefully acknowledged.

References

[1] B.J. Mulvaney, et. al., *PEPPER 1.2 User Manual*, MCC Tech. Rep. CAD-239-90, 1989.

[2] J.K. Ousterhout, *Tcl and the Tk Toolkit*, Addison-Wesley, 1994.

[3] S. Bebee, et. al., "Next-generation Stanford TCAD – PISCES 2ET and SUPREM OO7, " *IEDM Technical Digest*, 1994, p. 213.

[4] D.W. Yergeau, R.W. Dutton and R.J.G. Goossens, "A General OO-PDE Solver for TCAD Applications," *OONSKI 94*, Mar. 1994, Oregon.

[5] W.B. Richardson and B.J. Mulvaney, "Plateau and kink in P profiles diffused into Si: A result of strong bimolecular recombination?," *Appl. Phys. Lett.*, vol. 53, no. 20, pp. 1917-1919, 1988.

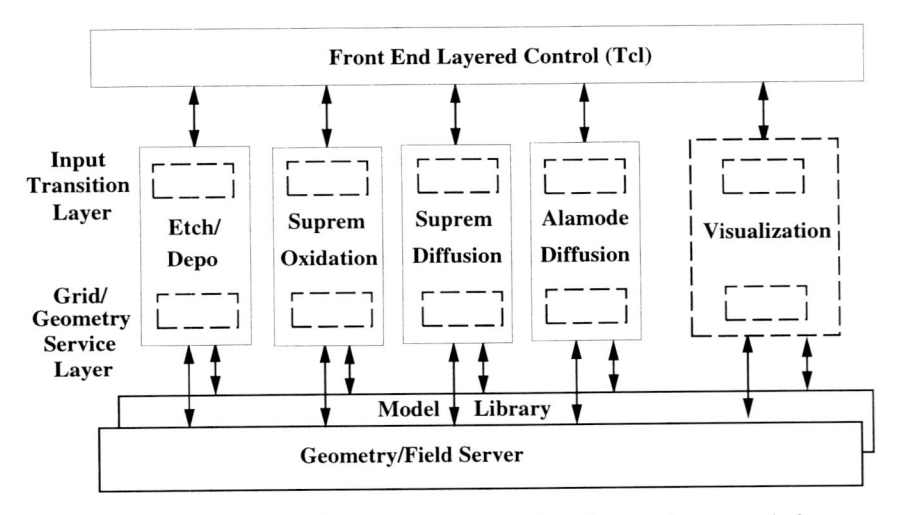

Figure 1: SUPREM OO7 architecture: an agent-based open framework for process simulation.

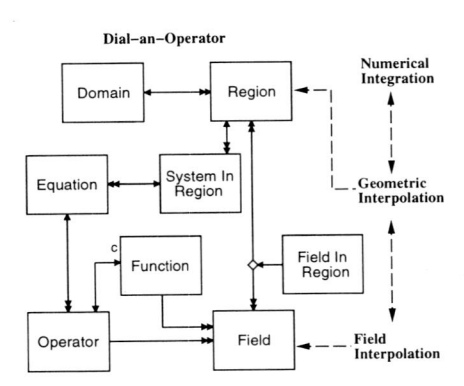

Figure 2: Information model for ALAMODE dial-an-operator solver.

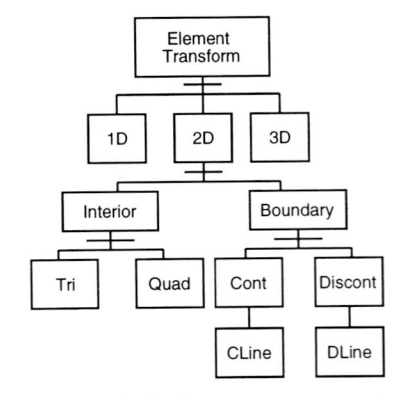

Figure 3: Inheritance structure of element transforms.

```
set DBSi [function fermi
  -parameters "B ni dbsi0 dbsip"]
set m 2.2
set modelDiff [model
 -systems [list
  [systemInRegion -region Si
   -equations [equation -parabolic
    -operators [operator diffusion
     -parameters "$DBSi B"]]]
  [systemInRegion -region Ox
   -equations [equation -parabolic
    -operators [operator diffusion
     -parameters "dbox B"]]]
  [systemInRegion -region {Ox/Si}
   -equations [equation
    -operators [operator segregation
     -parameters "B Trn m"]]]]
```

Figure 4: TCL script for boron segregation example.

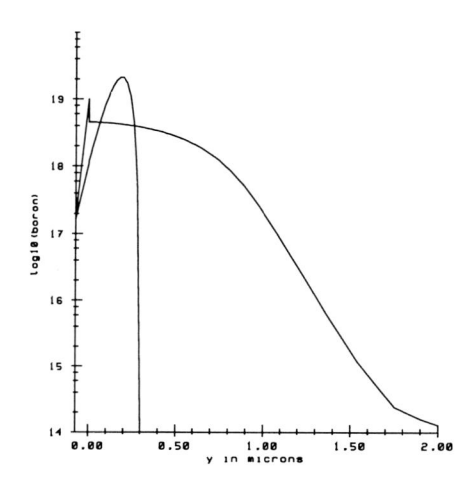

Figure 5: ALAMODE simulation of equations described by Figure 4.

$V + I \leftrightarrow <0>, \ P + V \leftrightarrow E, \ P + I \leftrightarrow F$

$V_{,t} = \nabla \cdot D_V \nabla V - k_{for}^E PV + k_{rev}^E E$
$\quad - k_{bi}(VI - V^{eq}I^{eq})$

$I_{,t} = \nabla \cdot D_I \nabla I - k_{for}^F PI + k_{rev}^F F$
$\quad - k_{bi}(VI - V^{eq}I^{eq})$

$E_{,t} = \nabla \cdot D_E \nabla E + k_{for}^E PV - k_{rev}^E E$

$F_{,t} = \nabla \cdot D_F \nabla F + k_{for}^F PI - k_{rev}^F F$

$P_{,t} = -k_{for}^E PV + k_{rev}^E E - k_{for}^F PI + k_{rev}^F F$

	C_{eq} (cm^{-3})	D_{def} $(\frac{cm^2}{s})$	D_{pair} $(\frac{cm^2}{s})$	k_{for}^{pair} $(\frac{cm^3}{s})$	k_{rev}^{pair} (s^{-1})
V	10^{14}	10^{-10}	10^{-13}	10^{-14}	10
I	10^{14}	10^{-9}	$2 \cdot 10^{-13}$	10^{-14}	12

Figure 6: Five species reactive-diffusive model for phosphorus.

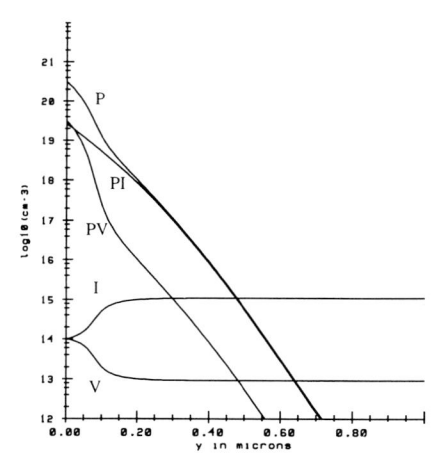

Figure 7: ALAMODE simulation of phosphorus plateau and kink as described by Figure 6. The results reproduce those in [5].

TCAD Optimization Based on Task-Level Framework Services

Ch. Pichler, N. Khalil*, G. Schrom, and S. Selberherr

Institute for Microelectronics, TU Vienna
Gusshausstrasse 27-29, A-1040 Vienna, AUSTRIA
Phone: +43-1-58801/5239, FAX: +43-1-5059224
e-mail: pichler@iue.tuwien.ac.at
* Digital Equipment Corporation, Hudson, USA

Abstract

This paper presents the integration of an external optimizer into the VISTA TCAD framework. A collection of task-level framework services is used by the optimizer to request the execution of process flow and device simulation tasks. All aspects of tool control and simulation data management are taken care of by these services, allowing for an easy implementation of a variety of task-level applications such as sensitivity analysis tasks, optimization, and RSM extraction. An example shows the calibration of MINIMOS' mobility parameters.

1. Introduction

The automatic optimization of semiconductor devices by means of computer simulation requires the repeated execution of multistep process flows to compute a set of response variables as functions of a set of control variables. For example, the implant dose and energy are chosen as control variables to optimize the LDD implant of an n-channel MOSFET, with the response variables being the drive current $I_{D,max}$ and the substrate current $I_{B,max}$, and the goal is to minimize $I_{B,max}$ and to maximize $I_{D,max}$. Finding the optimum design involves two independent tasks: the *selection* of sampling points in the design space, and the *evaluation* of the design at these points. The former is provided by an optimization algorithm, i.e., the optimizer, which, in general, determines the location of sampling points from the results of previously evaluated points in an iterative fashion until an optimum is reached. The latter returns response values for given input settings, obtained, e.g., by running a process flow simulation for each evaluation, or by using a previously established response surface model (RSM) of the process. The operations required to determine the responses as functions of the settings are not relevant to the optimizer. They can be treated as being hidden inside of a black box that takes care of executing the appropriate tasks to produce the desired outputs.

In the present work, the Vienna Integrated System for TCAD Applications (VISTA) [1] provides a collection of task-level services for the definition and simulation of process flows, catering to an external optimization tool that acts as a client and uses these services to evaluate and optimize the design. An evaluation may lead to the start of simulation tools, the computation of a response surface model, or simply the retrieval of a previously computed result; from the client's point of view, all these cases are identical.

2. Framework Service Layer

The service layer provides access to a set of high-level framework services based on VISTA's simulation flow representation [3]. It allows for the creation, modification, and execution of process flow instances, the submission of tasks for execution, the retrieval of responses, and the persistent storage of results. Automatic split generation and scheduling minimize the number of simulator runs required for iterative as well as parallel optimization techniques. Independent split branches are executed simultaneously over the network to quickly obtain results.

The service layer is implemented in VISTA's extension language VLISP, a superset of XLISP. An instance of a process flow together with its run-time data is called an *experiment* and is represented by a VLISP object. Table 1 gives examples of available services to create and manipulate experiments.

Service	Description
Define Experiment	Defines experiment attributes, e.g., process flow, initial wafer, etc.
New Experiment	Creates new instance of existing experiment.
Edit Step Parameter	Modifies parameter values at step in process flow.
Submit Experiment	Requests execution of process flow or retrieves previously computed results.
Inquire Step Data	Returns responses, current wafer data, etc.

Table 1: Examples of framework services to create and access experiments.

By means of these services, high-level TCAD applications like design-of-experiments (DoE), sensitivity analysis, and optimization can be conveniently implemented with all tool invocation details, etc., hidden.

3. Framework – Optimizer Interface

When an optimization task is initiated by the framework, an *agent* is assigned to the optimizer tool, which establishes a connection between the task-level services and the optimizer. The agent is realized as a VLISP object. It takes care of passing messages between the optimizer and the framework by means of a callback-based, asynchronous connection, allowing for the execution of multiple optimization tasks at the same time. Fig. 1 shows the interaction between the optimizer agent and the service layer on the one hand, and between the agent and the external optimizer on the other hand. The framework passes a description of the model to the optimizer, defining the model's type and its control and response variables. During the course of the optimization, the optimizer requests the evaluation of the model for a certain set of control values by sending a message to the framework. Messages between the optimizer and the framework rely on VISTA's operating-system independent standard-input/standard-output redirection capabilities. Depending on the internal operation of the optimizer, evaluation requests may be sent synchronously, or a number of requests may be sent at a time. Upon termination of the optimization, the result found and diagnostic information are passed back to the framework.

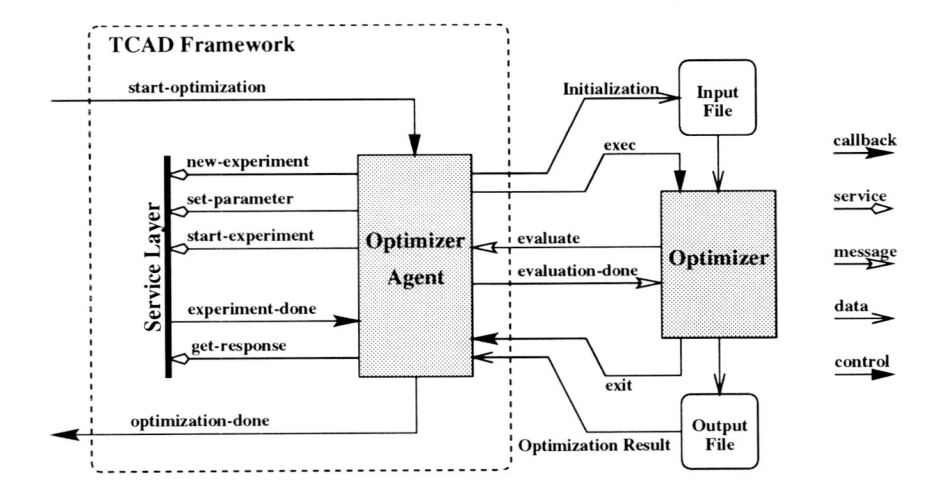

Figure 1: Communication between the service layer, the optimizer agent, and the external optimizer.

4. Application Example

To verify the feasibility of the approach presented above, the MINIMOS [4] mobility model equation parameters were calibrated using p-channel data from devices fabricated with a retrograde n-well, salicided dual-gate CMOS process. The calibration was done using nonlinear least-squares optimization to adjust physical model parameters to minimize the errors between calculated and experimental values. The two-dimensional doping profile was determined experimentally [2]. Process and device simulation were performed by the TCAD framework upon request of the optimizer. With the calibrated mobility parameters, MINIMOS simulation accurately reproduces experimental I–V data over a wide range of biases and lengths. The width of all the simulated and measured devices is 64 μm, the oxide thickness (t_{ox}) is 72.7 Å. The polysilicon gate doping concentration (N_p) is equal to 2.7×10^{19} cm^{-3} as determined from gate capacitance measurement with the device biased in the inversion region. Fig. 2 shows measured and simulated results for the linear region currents for three gate lengths. Good agreement is found in all regions of bias for all three lengths.

5. Conclusion

Using an external optimizer as a client in the simulation environment separates the evaluation of the *model* to be optimized from the optimization task, thereby effectively liberating the optimizer from dealing with tool invocation intricacies, error handling provisions, and user interface requirements. Different optimizer tools can be used,

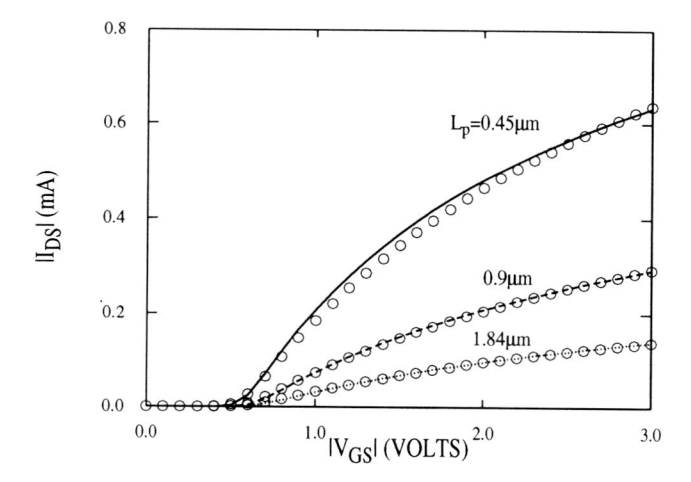

Figure 2: Comparison of measure and simulated I–V characteristics in the linear region ($V_{DS} = -50$ mV) for three gate lengths ($L_p = 0.45$, 0.9 and 1.84 μm).

as they access the evaluation services in a standardized way. Other task level applications such as sensitivity analysis or RSM generation can be easily implemented using the services presented. The framework's parallel execution and split generation capabilities provide fast responses to multiple model evaluation requests.

6. Acknowledgment

Part of this work was carried out in cooperation with ADEQUAT (JESSI project BT11) and has been funded by the EU as ESPRIT project No. 8002.

References

[1] S. Halama et al., The Viennese Integrated System for Technology CAD Applications, *Technology CAD Systems*, 197-236, Springer 1993.
[2] N. Khalil, J. Faricelli, D. Bell, and S. Selberherr, The Extraction of Two-Dimensional MOS Transistor Doping via Inverse Modeling, *IEEE Electron.Dev.Lett.*, 16:17-19, 1995.
[3] Ch. Pichler and S. Selberherr, Process Flow Representation within the VISTA Framework, *SISDEP V*, 25-28, 1993.
[4] S. Selberherr, W. Hänsch, M. Seavey, and J. Slotboom, The Evolution of the MINIMOS Mobility Model", *Solid-State Electron.*, 33(11):1425-1436, 1990.

Cellular Automata Simulation of GaAs-IMPATT-Diodes

D. Liebig

Technische Universität Hamburg–Harburg
Arbeitsbereich Hochfrequenztechnik,
D-21071 Hamburg, Germany.

Abstract

A new 3-dimensional cellular automata method is presented which improves the numerical efficiency of standard Monte-Carlo codes by keeping the physical accuracy. With both methods calculated results of stationary and time-dependent bulk transport quantities of GaAs and stationary transport characteristics of a GaAs-IMPATT-diode for D-Band applications will be presented and compared in detail.

1. Introduction

The study of the microscopical transport properties of IMPATT-devices is of great importance for a successful design of mm-wave GaAs-based transit-time devices for frequencies beyond 150 GHz. The Monte-Carlo (MC) simulation of these devices offers physical insight with high accuracy, but the calculation of the avalanche processes are numerically extremely demanding, what makes a full numerical optimization impossible. To overcome this problem, a new Cellular Automata (CA) technique has been developed and applied to investigate the operation of mm-wave IMPATT-diodes.

2. Method

The CA method stochastically calculates the semi-classical movement of an ensemble of pseudo-particles like in the known MC method, but it treats the dynamics of scattering and motion on an underlying grid in wave-vector-space iteratively with constant time steps Δt (typically 1-5 fs). The structure of each conduction band valley (or valence band) is described in polar coordinates and in energy space as shown in Fig. 1. In each time step the energy, the polar and azimuthal angles of the pseudo-particles are changed by stochastically scattering into another angular cell and into another energy-level of the same or another valley (or band). These super-scattering events include the effective change of states whithin the time step Δt and include also many-scattering events with a certain probability. These probabilities are stored in "connection tables". In contrast to an earlier published CA method [1], the electric field is treated deterministically here, and the motion in real space is given by means of finite distances which are assigned to each cell when particles underlie free flights, or by mean values assigned to a certain cell combination ($\vec{r}_{scat}$ in Fig. 1) when a particle super-scatters into another cell.

The super-scattering probabilities are calculated by a deterministic integration scheme, which allows the inclusion of generalized bandstructure, adjustable grid-sizes, and arbitrary number of valleys (or bands). Nevertheless, a non-parabolic and isotropic standard 3-valley-(band) representation is used here to describe the bandstructure of GaAs. This allows a direct comparison of the CA calculations with the results of a standard MC code. Both methods incorporate the same ionized impurity, intra- and intervalley phonon scattering, and impact ionization processes and are self-consistently coupled to a one-dimensional Poisson solver for the simulation of IMPATT-diodes. The CA method is also coupled to a one-dimensional drift-diffusion model which describes the stationary transport in the contact regions even more efficient whereas all processes in the active region are modelled by the CA method. The impact ionization rate is calculated by invoking the random-k approximation of Kane [2].

3. Results

In Fig. 2 stationary and time-dependent bulk quantities of GaAs calculated with both the CA and the MC method are presented. Fig. 2a shows the convergence of the CA method for the calculation of the autocorrelation function versus the angular mesh size. The MC result is well reproduced when 32 cells in angular space are used (leading to a necessary memory of 3.5 MByte for one connection table).

The stationary drift velocity and the longitudinal diffusivity calculated with the MC method are nicely reproduced by the CA method in all important electric field ranges (Fig. 2b, 2c). At electric fields beyond 250 kV/cm, the drift velocity and the longitudinal diffusivity increase again because impact ionization occurs which scatters carriers back into low-energy regions where scattering rates are lower.

The CA and the MC calculations also show very good agreement when non-stationary conditions are considered. As an example, the time-dependent velocity overshoot arising when thermal carriers are exposed to a sudden switch-on of an electric field is presented in Fig. 2d.

The CA method shows an at least factor of 10 higher computational speed on a HP 735 RISC workstation compared to efficient standard MC methods. At high electric fields or high temperatures (these are typical conditions in transit-time devices like IMPATT-diodes), the CA technique is about 40 times faster than MC codes because the increase of scattering rates at high energies increases the numerical effort in MC codes more than in the CA technique.

With the MC and the CA method, the Read-type GaAs-IMPATT-diode of Fig. 3 has been simulated. The investigated structure has theoretically been proposed for application at oscillation frequencies higher than 150 GHz and was experimentally realized and characterized in [3]. The diode shows an extremely short injection region of only a few 10 nm and a short drift region of approx. 130 nm. As can be seen in Fig. 3a, the electric field strengh reaches 1 MV/cm in the injection region. Consequently, the electron and hole average energies reach maximum values of more than one eV near the end of the $p^+ - i$ resp. $i - n^+$ injection regions. Related to the peak structure of the carrier energy, the generation rate also shows maximum values at the end of the injection region. This behaviour leads to a drastic *dead space* effect, that is a low ionization rate of electrons in the $p^+ - i$ and of holes in the $i - n^+$ regions, respectively. As discussed above, one finds velocity overshoot of electrons and holes in the injection region (Fig. 3d) where the transport is dominated by impact ionization scattering processes. As shown in Fig. 3a-d, the results of the CA method are in very good agreement with those of the MC method for all physical quantities.

The typical cpu time to simulate 100 picoseconds of IMPATT-operation on a HP 735 RISC workstation is one hour when 2 fs time steps and 20.000 pseudoparticles are used in the active region.

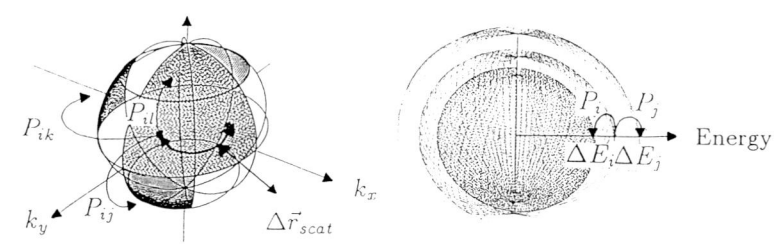

Figure 1: Discretization and super-scattering probabilities (P) in the CA method

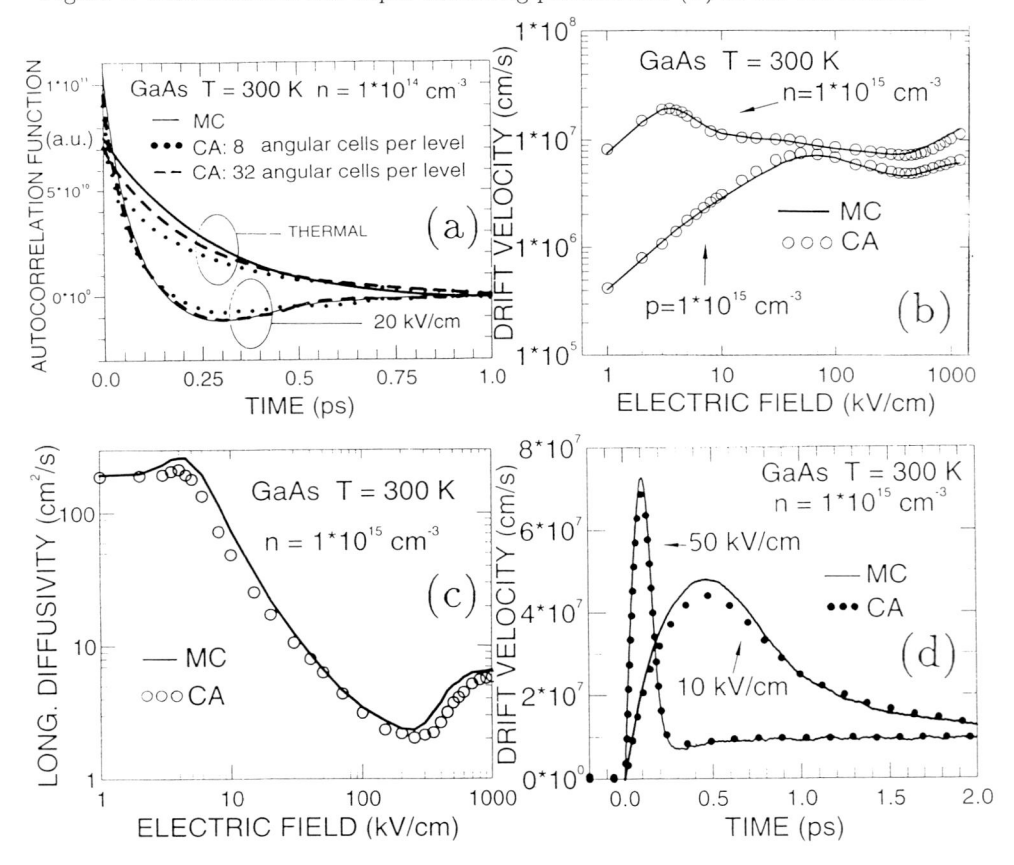

Figure 2: Comparison of stationary autocorrelation function (a), drift velocity (b), and longitudinal diffusivity (c), and of time-dependent velocity overshoot (d) calculated with the CA and the MC method

4. Conclusion

A new 3-dimensional Cellular Automaton method was presented. This method gives an accurate description of all relevant physical quantities and shows a factor of 10 - 40 higher computational speed compared to efficient standard Monte-Carlo codes. This makes the present method suitable to characterize the operation of mm-wave devices and to optimize their structure.

Acknowledgement

The author gratefully acknowledges the Deutsche Forschungsgemeinschaft for financial support.

References

[1] K. Kometer, G. Zandler and P. Vogl, *Phys. Rev. B*, **46**, 1382 (1992).
[2] E. O. Kane, *J. Phys. Chem. Solids* **12**, 181 (1959); *Phys. Rev.* **159**, 624 (1967).
[3] M. Tschernitz et al, "GaAs Read-type IMPATT-diode for D-Band", *Electronics Letters*, **30**, 1070 (1994).

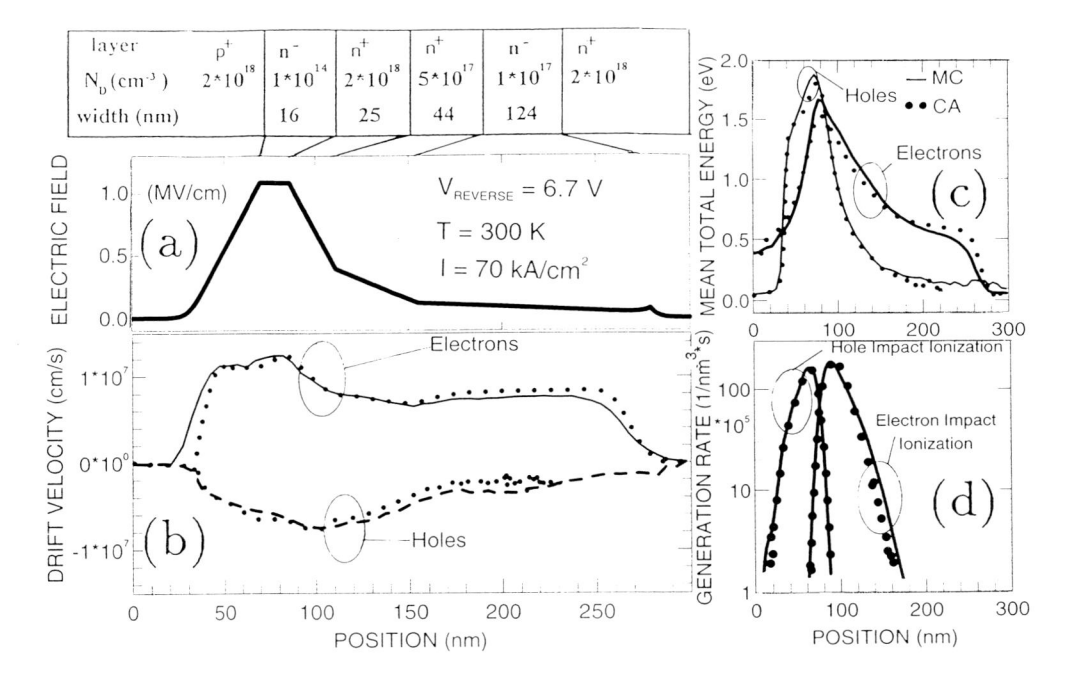

Fig.: 3 Comparison of the calculated distributions of electric field (a), drift velocity (b), mean energy (c), and impact ionization rate (d) for stationary operating conditions of the GaAs-IMPATT-diode

Two-Dimensional Simulation of Deep-Trap Effects in GaAs MESFETs with Different Types of Surface States

K. Horio, K. Satoh and T. Yamada

Faculty of Systems Engineering, Shibaura Institute of Technology
307 Fukasaku, Omiya 330, JAPAN

Abstract

Effects of surface states on I-V curves and turn-on characteristics in GaAs MESFETs are studied by 2-D simulation. These characteristics are essentially determined by deep-acceptor-like state. Depending on whether it acts as an electron trap or a hole trap, the turn-on characteristics change drastically. Physical mechanism of slow transients due to surface states is discussed.

1. Introduction

Many performance instabilities such as drain-current drifts, hysteresis in I-V curves, sidegating effects, and frequency-dependence of small-signal parameters have been reported experimentally in GaAs MESFETs. These were supposed to occur due to deep levels in the semi-insulating substrate or surface states on the active layer. However, the detailed mechanisms were not well clarified. As to the effects of semi-insulating substrate, many theoretical works have been made since a numerical model including deep levels was proposed [1], and clarified to some extent how the deep levels affect device characteristics. As to the effects of device surface conditions, only a few theoretical works were recently reported [2],[3]. So, in this work, we have systematically simulated GaAs MESFETs considering surface states, and found that drastic change of device characteristics arises depending on the nature of surface states.

2. Physical Model

We consider a GaAs MESFET where the active-layer thickness is 0.2 μm and its doping density is 10^{17} cm^{-3}. The gate lenght is typically 0.3 μm. For a substrate, we consider undoped semi-insulating LEC GaAs where deep donors "EL2" (N_{EL2}) compensate shallow acceptors (N_{Ai}) [1]. For the surface-state model, we assume that the surface states consist of a pair of deep donor and deep acceptor [4], and the following two cases are considered for GaAs surface [4],[5].
    a) Sample 1: $E_{SD} = 0.925$ eV, $E_{SA} = 0.8$ eV [3],[4]
    b) Sample 2: $E_{SD} = 0.87$ eV, $E_{SA} = 0.7$ eV [2],[5]

where E_{SD} is energy difference between the bottom of conduction band and deep donor's energy level, and E_{SA} is energy difference between deep acceptor's energy level and the top of valence band. The surface states are assumed to distribute uniformly within 5 Å from the surface. Their density and capture cross-section for carriers are typically set to 10^{13} cm^{-2} and 10^{-15} cm^2, respectively. Basic equations are the Poisson's equation, continuity equations for electrons and holes, and three rate equations for the deep levels. These are solved numerically in two dimension.

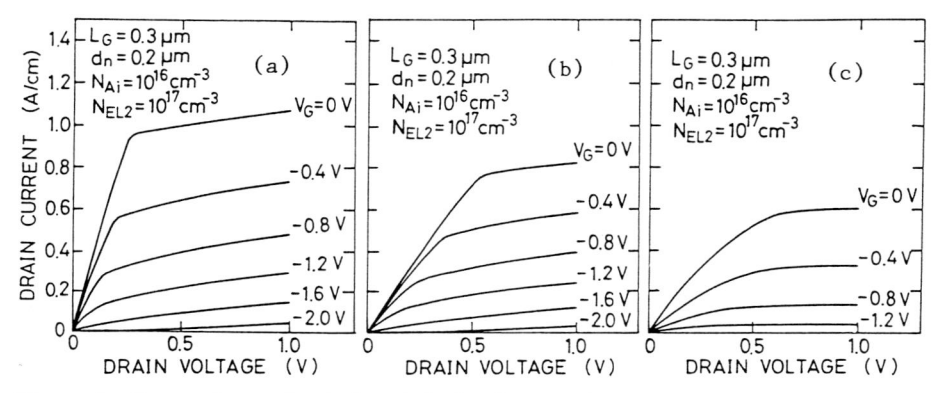

Figure 1: Comparison of calculated drain characteristics of GaAs MESFETs with and without surface states. (a) Without surface states, (b) Sample 1, (c) Sample 2.

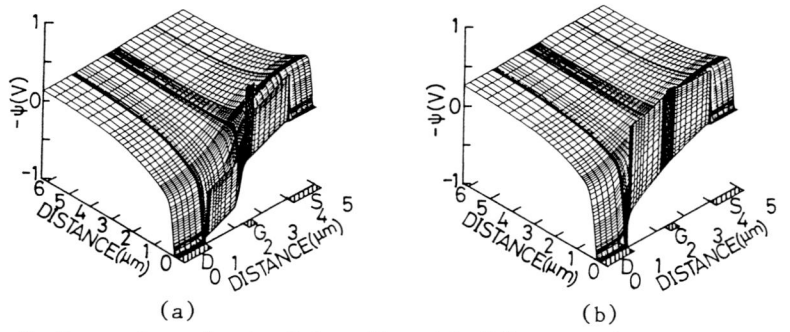

Figure 2: Comparison of potential profiles with different surface states. $V_G = 0$ V and $V_D = 1$ V. (a) Sample 1, (b) Sample 2.

3. I-V Characteristics

Figure 1 shows an example of calculated drain characteristics. When considering surface states, drain currents are estimated lower because the Fermi level is pinned around the mid-gap and the depletion layer is formed (Figure 2). It is also seen that the drain currents are lower for the case of Sample 2. This is because the dominant surface state is the deep acceptor in these cases and its energy level for Sample 2 is nearer to the valence band. In Figure 2, it should be noted that in Sample 1, the drain voltage is applied along surface-state layer, but in Sample 2, the drain voltage is entirely applied along the interface between drain electrode and surface-state layer. This is because in Sample 1 the deep acceptor acts as an "electron trap", while in Sample 2 the deep acceptor acts as a "hole trap". This difference of trap nature does not strongly affect I-V characteristics, but leads to dramatic difference in transient characteristics as described in the next section.

4. Turn-on Characteristics

Figure 3 shows comparison of turn-on characteristics with and without surface states when the gate voltage changes abruptly. Without surface states, the drain current becomes a steady-state value around $t = 10^{-11}$ s. For Sample 1 (with electron trap), the drain current shows fast response, too, and becomes constant temporarily around $t = 10^{-11}$ s, but decreases slightly during $t = 10^{-5}$ to 10^{-1} s. This current decrease starts when the deep acceptor (electron trap) begins to capture electrons. However, the difference of ionized-trap density between OFF and ON states is small in this electron-trap case, so the surface state does not strongly affect the characteristics. A dramatic feature arises in the case of Sample 2 (with hole trap). The drain current remains a low value for some period and begins to increase slowly around $t = 10^{-2}$ s. (This sort of slow transient is sometimes observed experimentally and called "gate-lag".) This slow transient is due to slow response of the deep acceptor (hole trap). In this case, as seen in Figure 4(a), the depletion layer exists along the entire region from source to drain. So, even if the gate voltage is changed, the drain current remains low (Figures 4(b),(c)) until the deep acceptors capture or emit carriers to change their ionized density much. The ionized deep-acceptor density should decrease when the deep acceptors (hole traps) begin to capture holes, which can be supplied from the Schottky contact, as is understood from the potential profiles in Figures 4(b)-(e). Thus the width of depletion region under the surface-state layer begins to decrease, leading to the slow increase in drain current.

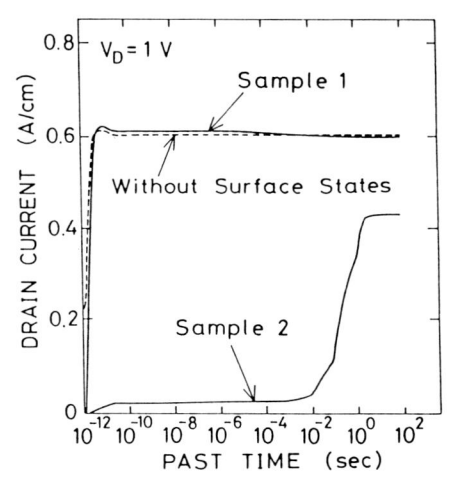

Figure 3: Comparison of turn-on characteristics of GaAs MESFETs with and without surface states. $V_D = 1$ V.

(a) OFF (b) t = 10^{-10} s (c) t = 10^{-4} s

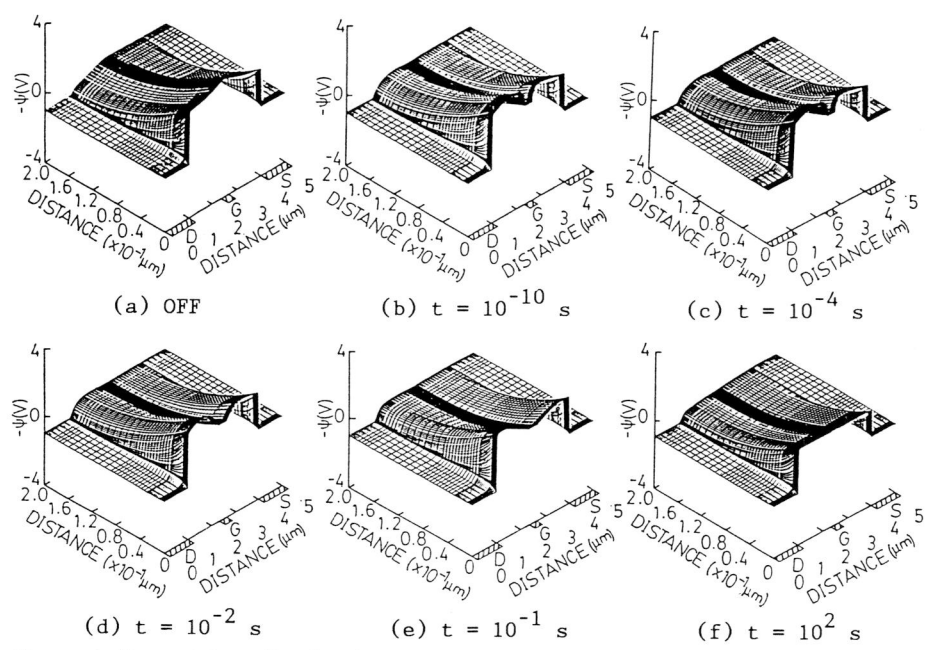

(d) t = 10^{-2} s (e) t = 10^{-1} s (f) t = 10^{2} s

Figure 4: Potential profiles for the case of Sample 2. (b) to (f) are profiles during the turn-on process. t is the past time after the gate voltage is changed abruptly.

5. Conclusion

2-D simulation of surface-state effects in GaAs MESFETs has been made. The characteristics are essentially determined by the deep-acceptor-like state. Depending on whether it acts as an electron trap or a hole trap, the turn-on characteristics can change drastically. Physical mechanism of the slow transients due to surface states has been discussed and clarified.

References

[1] K. Horio, H. Yanai and T. Ikoma, "Numerical simulation of GaAs MESFET's on the semi-insulating substrate compensated by deep traps", *IEEE Trans. Electron Devices*, vol.35, pp.1778-1785, 1988.

[2] C. L. Li, T. M. Barton and R. E. Miles, "Avalanche breakdown and surface deep-level trap effects in GaAs MESFET's", *IEEE Trans. Electron Devices*, vol.40, pp.811-816, 1993.

[3] S. H. Lo and C. P. Lee, "Analysis of surface state effect on gate lag phenomena in GaAs MESFET's", *IEEE Trans. Electron Devices*, vol.41, pp.1504-1512, 1994.

[4] W. E. Spicer *et al.*, "New and unified model for Schottky barrier and III-V insulator interface states formation", *J. Vac. Sci. Technol.*, vol.16, pp.1422-1433, 1979.

[5] H. H. Wieder, "Surface Fermi level of III-V compound semiconductor-dielectric interfaces", *Surface Sci.*, vol.132, pp.390-405, 1983.

An Efficient Numerical Method to Solve the Time-Dependent Semiconductor Equations Including Trapped Charge

L. Colalongo, M. Valdinoci, M. Rudan

Dipartimento di Elettronica, Università di Bologna,
viale Risorgimento 2, 40136 Bologna, ITALY

Abstract

In recent years, the increasing interest for Thin-Film Transistor (TFTs) has made the modeling of semiconductor devices with localized states increasingly important. In transient conditions, the dynamic change of trapped charge must be properly accounted for and two continuity equations ought to be considered in addition to the standard semiconductor equations. We propose here a novel methodology to solve this problem without increasing the number of resulting equations, which takes advantage of the locality of the trapped-charge conservation equations. In this way, the solution is achieved without resorting to approximation in the description of the trap-states dynamics.

1. Introduction

In recent years increasing attention has been paid to the design, fabrication and electrical characterization of amorphous and polycrystalline-silicon thin film transistors, which are being used in active-matrix flat-panel displays as addressing devices. Hence a growing interest is now devoted to the modeling and simulation of such devices which, due to the presence of large, energy-distributed, bulk (or grain-boundary) states, pose a few challenging simulation problems. In steady-state conditions the charge trapped in the gap states may correctly be accounted for by redefining the generation or recombination rate (e.g., [1]). In transient conditions, in order to take the dynamic variation of trapped charge into account, two more continuity equations must be added to the system describing the transport in the semiconductor. Such additional equations turn out to be differential in time but purely algebraic in space; thanks to this, a suitable manipulation can be found such that the model is solved without increasing the number of equations with respect to the drift-diffusion one. This makes the implementation easy while maintaining the efficiency of the drift-diffusion scheme. It is worth adding that no approximations in the description of the trap-state dynamics are involved here; in fact, opposite to other approaches, no particular assumption on the features of the intra-gap transitions is made.

2. Theory and Implementation

Taking the time dependence of both free and trapped charge into account, the complete system of device equations in transient conditions turns out to be

$$- \text{div}\,(\varepsilon_s \,\text{grad}\,\varphi) = q\,(p - n + N_D^+ - N_A^- + p_t - n_t), \tag{1}$$

$$\frac{\partial n}{\partial t} - (1/q)\,\text{div}\,\boldsymbol{J}_n = -U_n, \qquad \frac{\partial p}{\partial t} + (1/q)\,\text{div}\,\boldsymbol{J}_p = -U_p, \tag{2}$$

$$\frac{\partial n_t}{\partial t} = -U_{nt}, \qquad \frac{\partial p_t}{\partial t} = -U_{pt}, \tag{3}$$

where n_t, p_t are the concentrations of trapped charge and the remaining symbols have the usual meaning. The current densities $\boldsymbol{J}_{nt}$, $\boldsymbol{J}_{pt}$ associated to the traps are set to zero owing to the negligible mobility of trapped carriers. The system (1,2,3) is made of 5 equations in the unknowns φ, n, p, n_t, p_t. As anticipated in the Introduction, the number of equations can be reduced to 3; it is shown below that this result is achieved without approximations by incorporating the two continuity equations (3) in a modified expression of the recombination formula. It is worth adding that, for the sake of generality, the donor and acceptor states are treated separately; the corresponding concentrations of states are indicated by N_{tD}, N_{tA}. Combining (2) and (3) and observing that $q\,\partial(n + n_t - p - p_t)/\partial t = \text{div}(\boldsymbol{J}_n + \boldsymbol{J}_p)$ one obtains $U_n - \partial n_t/\partial t = U_p - \partial p_t/\partial t$ which, after straightforward manipulation, leads to two linear, first-order equations in n_t and p_t:

$$\frac{\partial n_t}{\partial t} + D_A n_t = N_{tA}\,(\alpha_{nA} n + e_{pA}), \qquad \frac{\partial p_t}{\partial t} + D_D p_t = N_{tD}\,(\alpha_{pD} p + e_{nD}). \tag{4}$$

In (4) it is $D_A = \alpha_{nA} n + \alpha_{pA} p + e_{nA} + e_{pA}$, $D_D = \alpha_{nD} n + \alpha_{pD} p + e_{nD} + e_{pD}$, while e_{nA}, e_{pA}, e_{nD}, e_{pD} are the emission probabilities and $\alpha_{nA} = \sigma_{nA} u_{th}$, $\alpha_{pA} = \sigma_{pA} u_{th}$, $\alpha_{nD} = \sigma_{nD} u_{th}$, $\alpha_{pD} = \sigma_{pD} u_{th}$. The time discretization of (4), implemented for instance using the Backward-Euler method and referring to the ith node, yields

$$n_{ti} = \left[\frac{n_t^{old} + N_{tA}\,(\alpha_{nA} n + e_{pA})\Delta t}{1 + D_A \Delta t}\right]_i, \qquad p_{ti} = \left[\frac{p_t^{old} + N_{tD}\,(\alpha_{pD} p + e_{nD})\Delta t}{1 + D_D \Delta t}\right]_i. \tag{5}$$

Eqs. (5) could also be obtained by first integrating (4) analytically and then taking Δt small. One sees that n_{ti}, p_{ti} are decoupled from each other; in particular, n_{ti}^{old}, p_{ti}^{old} in (5) are the values of the trapped charge calculated and stored at the previous time step. Since the probability of direct band-to-band transition is negligibly small, the net recombination rate U_n is given only by the transition of electrons between the conduction band and the gap states; similarly, the net recombination rate U_p is given by the transitions of holes between the valence band and the gap states. It follows that U_n, U_p can be written as functions of n, p, n_t, p_t:

$$U_n = \alpha_{nA}\,(nN_{tA} - nn_t) + \alpha_{nD} n p_t + e_{nD}(p_t - N_{tD}) - e_{nA} n_t, \tag{6}$$

$$U_p = \alpha_{pD}\,(pN_{tD} - pp_t) + \alpha_{pA} p n_t + e_{pA}(n_t - N_{tA}) - e_{pD} p_t. \tag{7}$$

Eq. (5) is replaced into the discrete form of (6,7); the result is then used to calculate the RHS of (2) at the current iterate. As a consequence, Eqs. (1) and (2) thus modified are fully equivalent to the original system (1,2,3) but, on the other hand, retain the same discretization scheme as in the trap-free case. Since no simplifying assumption is introduced, all the possible transitions are considered in (1) and (2); among these, in particular, are the intra-gap transitions between acceptor and donor states, which are seldom considered in the literature.

3. Results

Numerical simulations have been carried out using a two-dimensional version of the device-analysis program HFIELDS, supplemented with the method described above. The turn-off transient of an n-p polycrystalline-silicon diode is shown by way of example. The diode is biased with a 1 ns linear voltage ramp starting at 10^{-7} s, which brings the anodic voltage from 1 to -3 V. The program accounts for the large number of defects in polycrystalline silicon (grain boundaries, intra-grain defects etc.) assuming the density of states in the semiconductor proposed in [2]: the states located in the lower half of the gap are donor-like while those located in the upper half are acceptor-like. The energy distribution of each set of donor-like (acceptor-like) states is approximated by the sum of two exponential functions. Following [3], the distribution of states is assumed uniform in space; this is of course irrelevant as far as the scheme proposed here is concerned. The current across the two contacts of the diode is shown in Fig. 1. The calculation has been carried out accounting for the displacement component of the current: the perfect balance of the currents indicates that the charge is correctly conserved by the transient analysis. The corner at $t = 1$ ns corresponds to the end of the ramp. As a comparison, the same simulation has been repeated using the steady-state expression of p_t and n_t in Poisson's equation (1) and the corresponding values of U_p, U_n in the continuity equations (2), that is, one of the commonly-accepted approximations at low trap concentrations; the result is shown in Fig. 2 (on a different time scale from Fig. 1) and demonstrates the importance of taking the dynamics of trapped charge into account. One sees in fact that the currents flowing through the two contacts begin to differ as soon as the voltage drop across the diode changes. This means that the total charge within the device is not conserved. The importance of trap dynamics is also evident from Fig. 3, where the current is calculated at the same contact using the full model (continuous line) and the approximation of Fig. 2 (dashed line). The full model exhibits an intermediate transient corresponding to the release of the trapped charge, which is instead missing in the approximate model. The latter, in fact, exhibits a much sharper peak due to the displacement current only.

4. Acknowledgements

The activity of L. Colalongo and M. Valdinoci has been supported by fellowships provided by ST-CO.RI.M.ME, which are gratefully acknowledged.

References

[1] M. Valdinoci, A. Gnudi, M. Rudan and G. Fortunato, "Analysis of Amorphous Silicon Devices," *Proc. NUPAD V*, pp. 19-22, 1994.

[2] M. Shur and C. Hyun, "New high field-effect mobility regimes of amorphous silicon alloy thin-film transistor operation," *J. Appl. Physics*, vol. 59, pp. 2488-2497, 1986.

[3] G. Fortunato and P. Migliorato, "Model for the above-threshold characteristics and threshold voltage in polycrystalline silicon transistors" *J. Appl. Physics*, vol. 68, pp. 2463-2467, 1990.

[4] S. M. Sze, "Physics of Semiconductor Devices", *John Wiley & Sons*, 1981.

Fig. 1

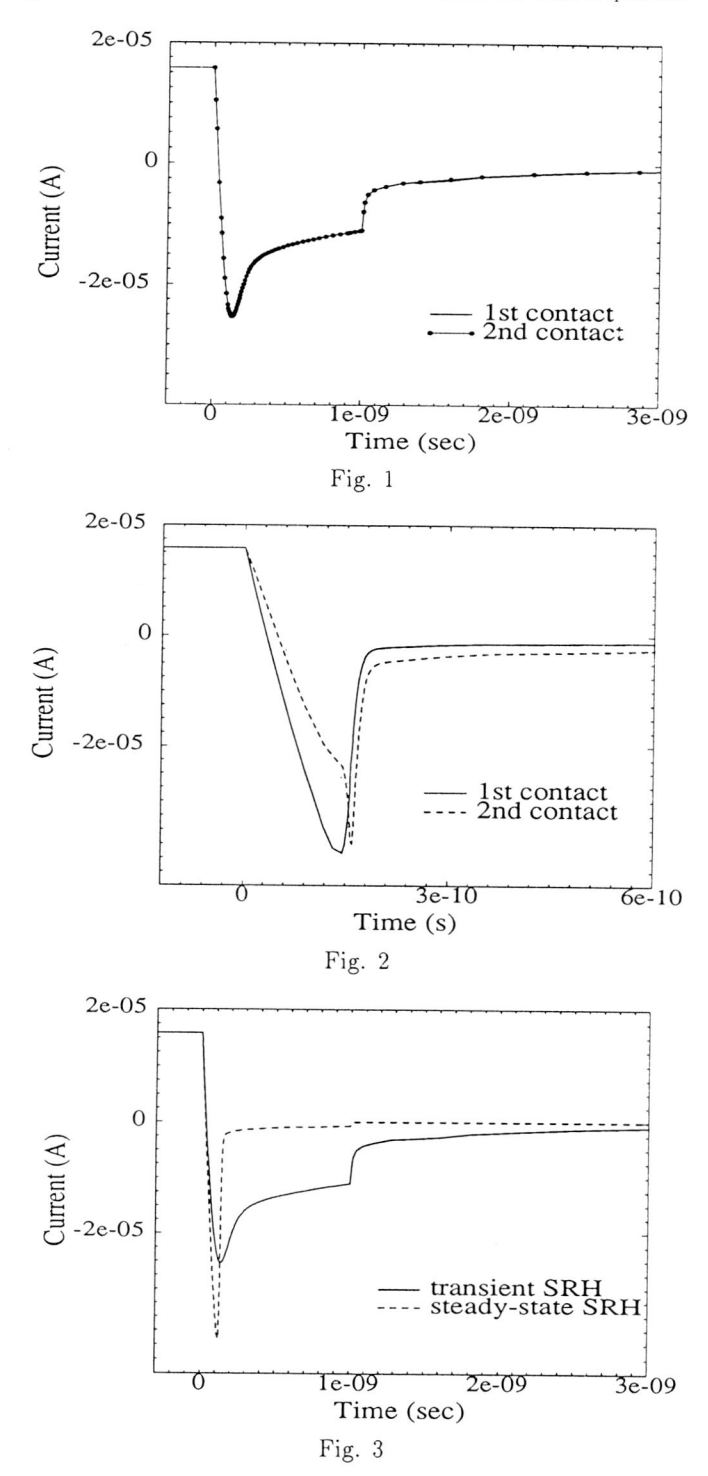

Fig. 2

Fig. 3

Advances in Numerical Methods for Convective Hydrodynamic Model of Semiconductor Devices

N. R. Aluru, K. H. Law and R. W. Dutton

Integrated Circuits Laboratory, Stanford University,
Stanford, California 94305 USA

Abstract

The convective hydrodynamic model of semiconductor devices is analyzed employing parallel and stabilized finite element methods. The stabilized finite element method for the two-carrier hydrodynamic equations and the parallel computational model are briefly described. Numerical results are shown for a bipolar transistor. A comparison of drift-diffusion, energy-transport and the hydrodynamic models is presented for a $0.1\mu m$ channel n^+-n-n^+ silicon diode.

1. Introduction

Comprehensive semiconductor device simulations employing the hydrodynamic model involve the solution of the Poisson equation, electron and hole hydrodynamic conservation laws and the lattice thermal diffusion equation [1]. The electron and hole hydrodynamic equations are derived by considering the zeroth, first and second moments of the Boltzmann transport equations and are summarized as follows:

$$\frac{\partial c_\alpha}{\partial t} + \nabla \cdot (c_\alpha \mathbf{u}_\alpha) = [\frac{\partial c_\alpha}{\partial t}]_{col} \tag{1}$$

$$\frac{\partial \mathbf{p}_\alpha}{\partial t} + \mathbf{u}_\alpha (\nabla \cdot \mathbf{p}_\alpha) + (\mathbf{p}_\alpha \cdot \nabla)\mathbf{u}_\alpha = (-1)^\alpha \epsilon c_\alpha \mathbf{E} - \nabla(c_\alpha k_b T_\alpha) + [\frac{\partial \mathbf{p}_\alpha}{\partial t}]_{col} \tag{2}$$

$$\frac{\partial w_\alpha}{\partial t} + \nabla \cdot (\mathbf{u}_\alpha w_\alpha) = (-1)^\alpha \epsilon c_\alpha (\mathbf{u}_\alpha \cdot \mathbf{E}) - \nabla \cdot (\mathbf{u}_\alpha c_\alpha k_b T_\alpha) - \nabla \cdot \mathbf{q}_\alpha + [\frac{\partial w_\alpha}{\partial t}]_{col} \tag{3}$$

where $\mathbf{u}_\alpha$, $\mathbf{p}_\alpha$, T_α, w_α and $\mathbf{q}_\alpha$ are the velocity vector, momentum density vector, temperature, energy density and heat flux vector of the carrier α. (For electrons, α = n or $\alpha = 1$; for holes, α = p or $\alpha = 2$). The terms $[]_{col}$ represent the rate of change in the particle concentration, momentum and energy due to collision of the carriers; the collision terms can be approximated by their respective relaxation times and the expressions can be found in [1].

Numerical studies employing the convective hydrodynamic model cannot be performed trivially since conventional numerical methods often fail when convective terms are included. The classical Scharfetter-Gummel (SG) method for discretization, that can be extended for the simplified hydrodynamic (neglecting convective terms) and the energy-transport models, does not work well for the hydrodynamic model. We have developed new stabilized finite element methods and they are summarized in this paper. This paper also presents a parallel computational model for the finite element method on distributed memory parallel computers and provides a comparison of drift-diffusion, energy-transport and hydrodynamic models for a $0.1\mu m$ channel n^+-n-n^+ silicon diode.

2. Stabilized Finite Element Methods

The Poisson and the lattice thermal diffusion equations are elliptic in nature and standard Galerkin finite element discretization can be shown to be stable. Galerkin finite element formulation is, however, unstable for electron and hole hydrodynamic equations as the solutions to these equations contain steep layers. Hence, advanced Galerkin/least-squares finite element formulations are developed. The temporal behavior of the equations is accounted by employing a discontinuous Galerkin method in time. With a discontinuous Galerkin in time and a Galerkin/least-squares in space, the discretization technique is referred to as a space-time Galerkin/least-squares finite element method. The space-time Galerkin/least-squares finite element method can be shown to be stable for the electron and hole hydrodynamic equations. Hence this method is also referred to as a stabilized finite element method. The important steps in the space-time Galerkin/least-squares formulation are summarized as follows:

1. A least-squares term of a residual type is introduced to the weak form of the hydrodynamic equations so that the numerical stability of the system is enhanced. Furthermore, a discontinuity-capturing term is added to overcome the undershoot and overshoot phenomena near steep gradients. The least-squares and discontinuity capturing terms vanish when the exact solution is substituted, thus making the method consistent.

2. Within each space-time slab, the trial and test functions are approximated by linear in space and constant in time basis functions.

3. The nonlinear system is solved using a Newton iterative scheme by linearizing the nonlinear equations with respect to the unknown trial solution.

Comprehensive semiconductor device simulations are performed employing the stabilized methods discussed above. The boundary conditions required for the electron and hole hydrodynamic equations are more complicated compared to the drift-diffusion or the energy-transport model. They are discussed in greater detail in [1]. Figure 1 and Figure 2 show, respectively, the electron concentration and the lattice temperature for a bipolar transistor. The numerical results indicate that stabilized finite element methods produce extremely robust and accurate solutions.

3. Parallel Computational Model

The convective hydrodynamic model demands enormous computations due to both the advanced nature of the model and the numerical method. A single-program-multiple-data (SPMD) programming model is designed and implemented on distributed memory parallel computers such as iPSC/860, Touchstone Delta and IBM SP-1. The power of the programming model lies in the quick adaptation of the serial code to the parallel code. The parallel algorithms are shown to be scalable and excellent efficiencies are reported. Figure 3 is a description of the parallel programming model and the various features available in the element library. A pre-processor reads/generates a mesh, reads boundary conditions and several other input parameters for the program. In a parallel program the pre-processor includes a domain decomposition algorithm, which partitions the domain into several subdomains. The pre-processor also prepares the input data for each subdomain and sends it to the corresponding processor. The same serial program is then executed on each processor with changes made only to accomodate inter-processor communication. Inter-processor communication is needed when iterative solvers perform vector-vector and matrix-vector products. The calculation of currents is also done in parallel and the final step is visualization which is done on a host. Figure 4 presents a comparison of the CPU times on serial and parallel computers. The CPU time on serial machines increases exponentially as the mesh size increases. The CPU time is shown to be significantly lower employing 32 processors of iPSC/860. The hydrodynamic simulations are shown to be very efficient employing just 8 processors of SP-1. These results indicate that serial computers are bottlenecks for grand challenge device applications and parallel computers can enable solution of the hydrodynamic model in a reasonable time.

4. Matrix-Free Techniques

The non-symmetric system of equations obtained from discretizing the convective hydrodynamic equations are solved using a matrix-free generalized minimal residual (GMRES) algorithm. The GMRES algorithm primarily involves vector-vector and matrix-vector products. The matrix-vector products can be evaluated by matrix-free algorithms to reduce storage costs and improve convergence of the fractional-step solution algorithm. In a matrix-free algorithm, the Jacobian is never formed and a matrix-vector product involving the Jacobian, $\mathbf{J}(\mathbf{v})$, and a vector, $\mathbf{u}$, can be obtained directly using the residual $\mathbf{R}$ as shown below

$$\mathbf{J}(\mathbf{v})\mathbf{u} = \lim_{\varepsilon \to 0} \frac{\mathbf{R}(\mathbf{v} + \varepsilon \mathbf{u}) - \mathbf{R}(\mathbf{v})}{\varepsilon} \tag{4}$$

where ε is taken as a small but finite value. The choice of ε is important for accurate determination of the matrix-vector product $\mathbf{J}(\mathbf{v})\mathbf{u}$ and is discussed in greater detail in [1].

5. Comparison of Transport Models

The significance of the convective terms in the transport equations can be revealed by a comparison of currents (see Figure 5) obtained from the drift-diffusion, energy-transport and the hydrodynamic models for a $0.1\mu m$ channel n^+-n-n^+ silicon diode. The results indicate that the convective hydrodynamic model produces non-negligible effects on submicron terminal characteristics. For large applied voltages the currents obtained from the hydrodynamic model are almost twice greater than the currents obtained from the energy-transport model. While the accuracy of the transport models could not be established because of the lack of experimental results for the test structure, an analysis of the energy resulted in the hydrodynamic model shows that the kinetic energy is approximately 50% of the total energy in the channel region (see Figure 6). The non-negligible amount of kinetic energy in the channel region could indicate that the results obtained from the hydrodynamic model are more accurate since the simplified hydrodynamic and the energy-transport models neglect convective effects and assume that the kinetic energy is negligible compared to the thermal energy.

References

[1] N. R. Aluru, *Parallel and Stabilized Finite Element Methods for the Hydrodynamic Transport Model of Semiconductor Devices*, Ph.D Thesis, Stanford University, June 1995.

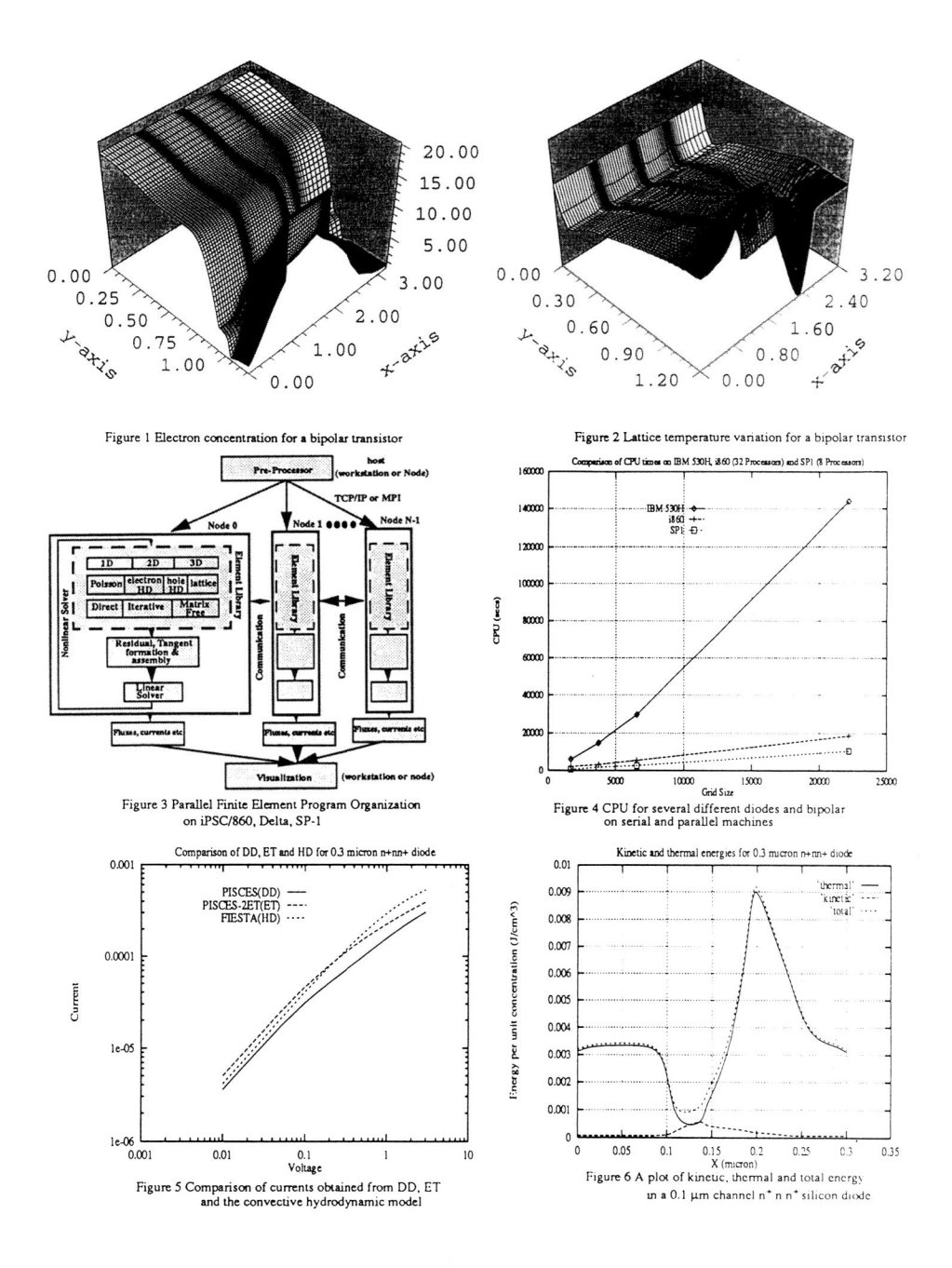

Figure 1 Electron concentration for a bipolar transistor

Figure 2 Lattice temperature variation for a bipolar transistor

Figure 3 Parallel Finite Element Program Organization on iPSC/860, Delta, SP-1

Figure 4 CPU for several different diodes and bipolar on serial and parallel machines

Figure 5 Comparison of currents obtained from DD, ET and the convective hydrodynamic model

Figure 6 A plot of kinetic, thermal and total energy in a 0.1 μm channel n$^+$ n n$^+$ silicon diode

An Advanced Cellular Automaton Method with Interpolated Flux Scheme and its Application to Modeling of Gate Currents in Si MOSFETs

K. Fukuda, K. Nishi

VLSI R&D Center, Oki Electric Industry Co., Ltd.,
550-1, Higashiasakawa-cho, Hachioji-shi, Tokyo 193, Japan

Abstract

An improved cellular automaton(CA) method is proposed in which an interpolated flux concept is introduced to suppress a well-known artifitial diffusion problem. The new method allows larger mesh sizes both in real and momentum space without losing numerical accuracy. Consequently, it becomes a practical modeling tool of non-linear carrier transport in semiconductor devices. In its application to Si MOSFETs, obtained results are extremely stable beyond the advanced weighted Monte Carlo method. Furthermore, making the best use of the stability, gate currents are studied in detail. 2 peaks of gate currents on gate bias, so-called drain avalanche hot carrier(DAHC) and channel hot electron(CHE), are well explained by thermionic emission and Fowler Nordheim(FN) tunneling of hot carriers respectively.

1. Introduction

Solving Boltzmann transport equation(BTE) by Monte Carlo method(MC), carrier distribution functions(DFs) which are essential information for hot carrier analysis of Si MOSFETs can be obtained. Because MC requires a huge amount of computer resources to get stable solutions, some new approaches to solve BTE based on a cellular automaton(CA) concept have been developed by several groups[1-4]. Although CA produces fairly stable solutions, CA still needs a significant calculation time because of a lot of data arizing from full momentum mesh points on every real space mesh points. In this paper, a new CA method is proposed in which interpolated flux scheme has been introduced to reduce the number of meshes drastically by allowing much larger mesh sizes both in momentum space and real space. The efficiency of the present method is demonstrated through applications to gate currents analysis of Si MOSFETs.

2. Numerial and Physical Models

As other CA approaches[1-4], carrier momentum distribution functions are stored in numerical tables defined on each geometrical mesh points (fig.1). To keep numerical accuracy, not logical type but real type data are used in these tables and updated by each time steps. Potential is also updated to be consistent to carrier distribution functions(fig.2). In these steps, flux is calculated assuming interpolated distribution fucntions between adjacent mesh points which suppress the artificial diffusion problem mentioned in ref.[4]. The suppression of the artificial diffusion is schematically explained in fig.3. Interpolation is adopted also to momentum space. By the interpolation of physical quantities between adjacent meshes, much larger mesh sizes can be used without losing accuracy, compared with the conventional method without interpolation.

As for physical models,followings are used.
(1) Energy band considering the shape of high energy electron's density of states[5].
(2) Phonons and impact ionization scatterings consistent with the band model[5, 6].
(3) Brooks-Herring model of impurity scattering modified in high doping conditions[7].
(4) Surface scattering model considering the universal mobility in inversion layers[8].
(5) Gate currents of thermionic emission and of tunneling with WKB approximation[9].

2-dimensional impurity profiles of the process of coded gate length of $0.5\mu m$ and gate oxide thickness of 9nm are obtained from our process simulator OPUS[10].

3. Application Results

The present method is applied to Si n-MOSFETs where steep slopes of carrier concentration, highly doped source/drain and rare events such as impact ionization and gate injection will cause difficulties even to the advanced weighted Monte Carlo method. Drain, substrate and gate currents versus gate voltage(V_{GS}) characteristics for drain biases(V_{DS}) of 3,4 and 5 volts are shown in fig.4. It is remarkable that all these currents from subthreshold region to saturation region can be obtained in the same CPU time(3 hours for each on EWS of 200-Spec92fp) while MC needs more CPU time in low current conditions. This is because these current values are not averaged ones as in MC, but snapshots where obtained distribution functions contain no fluctuation. Distributions of carrier concentration, carrier temperature and generation rates are shown in fig.5-7 which are again stable snapshots of the $V_{DS} = 5V, V_{GS} = 2V$ case. The peak position of the generation rate is more drain side than the peak of the eletron temperature which is a reasonable non-local effect. Back to fig.4, 2 peaks of gate currents can be observed (at $V_{GS} = 3$ and $V_{GS} = 5$ on $V_{DS} = 5$ for examples) which are conventionally explained as DAHC($V_{GS} = 3$) and CHE($V_{GS} = 5$). Distributions of gate injection along the channel both for 2 peaks are shown in fig.8 as $V_{GS} = 3$ in a) and $V_{GS} = 5$ in b), where solid lines represent thermionic emission current while dashed lines represent FN tunneling. Such FN tunneling is of hot electrons just below the barrier height of thermionic emission because they feel narrower tunneling distances of triangular potential than low energy electrons. In case b) with $V_{GS} = 5$, electrons beyond the barrier are less than in case a), but below the barrier , more tunneling is induced by the higher gate bias. As a conclusion, it is well explained that DAHC is attributed to thermionic emission and CHE to FN tunneling of hot electrons respectively.

References

[1] T. Iizuka and M. Fukuma, *Solid-St. Electron.*, vol. 33, no. 1, pp. 27-34, 1990.
[2] A. Das and M.S. Lundstrom, *Solid-St. Electron.*, vol. 33, no. 10, pp. 1299-1307, 1990.

[3] M.G. Ancona, *Solid-St. Electron.*, vol. 33, no. 12, pp. 1633-1642, 1990.
[4] K. Kometer, G. Zandler and P. Volg, *Phys. Rev. B*, vol. 46, no. 3, pp. 1382-1394, 1992.
[5] R. Brunetti, C. Jacoboni, F. Venturi, E. Sangiorgi and B. Ricco, *Solid-St. Electron.*, vol. 32, no. 12, pp. 1663-1667, 1989.
[6] R. Thoma, H.J. Peifer and W.L. Engl, W. Quade, R. Brunetti and C. Jacoboni, *J.Appl.Phys.*, vol. 69, no. 4, pp. 2300-2311, 1991.
[7] K. Fukuda and K. Nishi, *IEICE Trans. Electron.*, vol. E78-C, no. 3, pp. 281-287, 1995.
[8] E. Sangiorgi and M.R. Pinto, *IEEE Trans. Electron Devices*, vol. ED-39, no. 2, pp. 356-361, 1992.
[9] C. Huang, T. Wang, C.N. Chen, M.C. Chang and J. Fu, *IEEE Trans. Electron Devices*, vol. ED-39, no. 11, pp. 2562-2568, 1992.
[10] K. Nishi, K. Sakamoto, S. Kuroda, J. Ueda, T. Miyoshi, and S. Ushio, *IEEE Trans. Computer-Aided Design.*, vol. CAD-8, no. 1, pp. 23-32, 1989.

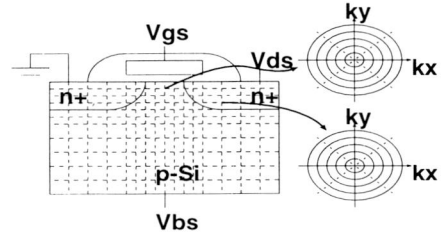

Figure 1: A schematic representation of momentum tables on each geometrical mesh points are shown. Each tables contain carrier distribution functions as real type data.

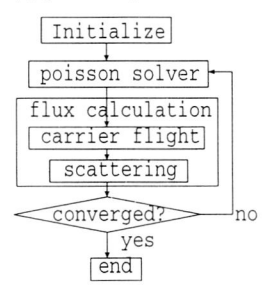

Figure 2: Flow chart of the presented simulation method. In the flux calculation, interpolated flux between adjacent mesh points is assumed.

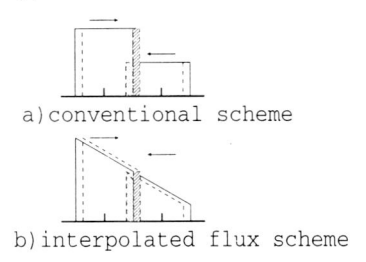

Figure 3: A schematic explanation of the suppression of artifitial diffusion problem. In the conventional scheme a), flux is constant in assumed cells and the hatched boxes clearly show that flux is much more from the left cell to right. Such artifitial diffusion is suppressed in the present method b).

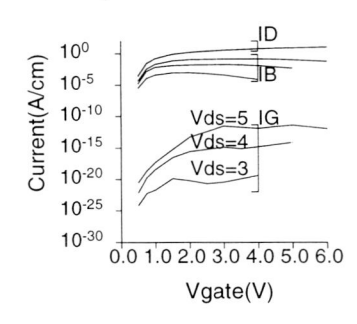

Figure 4: I_D, I_B and I_G versus V_{GS} on $V_{DS} = 3, 4$ and $5V$ which are stable over wide range.

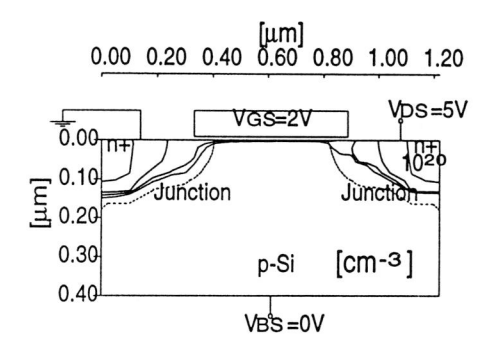

Figure 5: Carrier concentration of the $V_{DS} = 5$ and $V_{GS} = 2V$ case. Neighbouring lines differ by a factor of 10, referring to $10^{16}, \ldots, 10^{20} (1/\text{cm}^3)$.

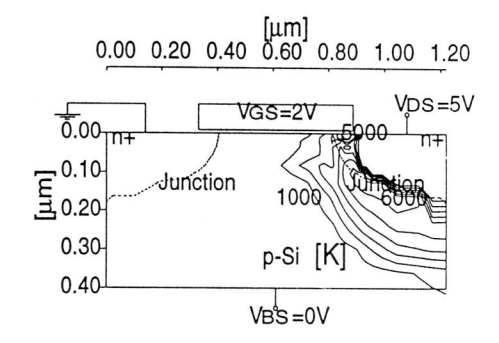

Figure 6: Carrier temperature of the same condition with fig.5. Neighbouring lines differ by 1000K, referring to $1000, \ldots, 6000K$.

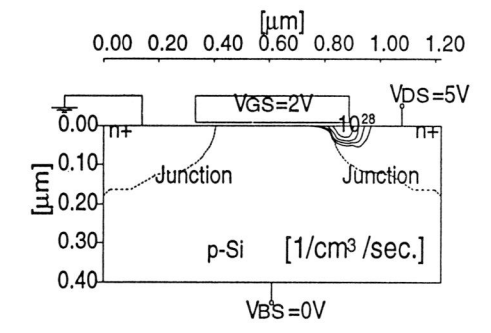

Figure 7: Generation rates of the same condition with fig.5. Neighbouring lines differ by a factor of 10, referring to $10^{25}, \ldots, 10^{28} (1/\text{cm}^3/\text{sec.})$.

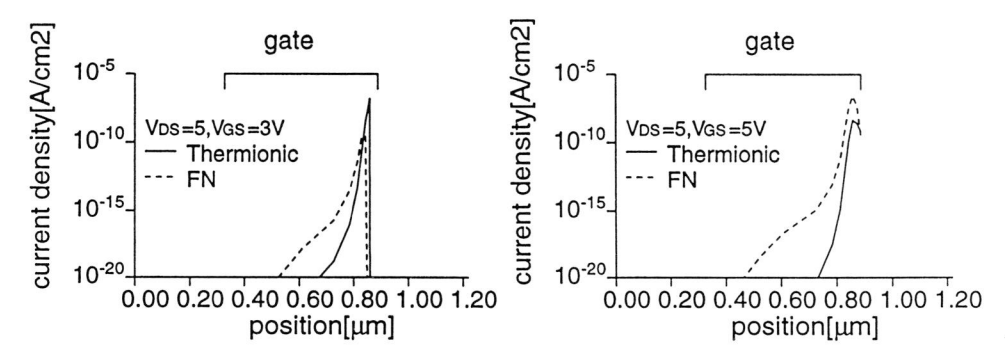

Figure 8: Carrier injection current density into SiO_2 along the channel for a) $V_{DS} = 5, V_{GS} = 3$ and b) $V_{DS} = 5, V_{GS} = 5V$ cases respectively where solid lines represent themionic emission currents while dashed lines represent FN tunneling currents. In case b), FN tunneling of hot carriers dominates the gate current.

Piezoresistance and the Drift-Diffusion Model in Strained Silicon

A. Nathan[a] and T. Manku[b]

[a]Electrical and Computer Engineering, University of Waterloo
Waterloo, Ontario N2L 3G1, Canada
on leave at: Physical Electronics Laboratory, ETH Hoenggerberg
CH-8093 Zurich, Switzerland

[b]Electrical Engineering, Technical University of Nova Scotia
P.O. Box 1000, Halifax, Nova Scotia B3J 2X4, Canada

Abstract

We have computed the strain-dependent tensorial mobility values for p-type Si based on a rigorous analysis of the valence band structure taking into account spin-orbit coupling effects. The mobilities, computed for not too large strain levels, are in agreement with the well known measured piezoresistance coefficients. Thus we now have a very convenient form of description for the drift-diffusion current density which can be incorporated in any device simulator.

1. Introduction

Mechanical strain can noticeably alter the carrier transport properties of silicon. Whilst on the one hand this can be usefully exploited for the realization of MEMS-based devices, it can have adverse effects on material characterization and on the operation of other microsensors as well as VLSI bipolar and MOS devices. For example, in Hall-based material characterization experiments, because of the strain-induced material anisotropy (resistivity, Hall coefficient) in the device, the values retrieved for the Hall mobility may not be meaningful. In magnetic sensors, the encapsulation-induced mechanical stresses and intrinsic stresses in the overlying conducting and dielectric thin films can affect the accuracy and long-term stability of the output response [1,2]. In trench-isolated VLSI technologies, there are regions of high stresses (see [3]) that can have undesirable effects on device characteristics and on device matching. Thus it has become imperative to account for the dependence of the various transport coefficients on strain for reliable prediction of device behavior.

2. Effects of strain on band structure in silicon

In general, the presence of mechanical strain alters the conduction and valence band structures by either shifting it in energy, distorting it, removing degeneracy effects, or any combination of the three. With n-type silicon, the interpretation of the effects of strain are less involved since the conduction band to first order, only shifts in energy. Since the shape of the ellipsoids remain unaltered, the effective mass of each conduction band remains unaltered but the concentration of electrons in each band can be altered due to energy shifts. The resulting change in resistivity can be reduced to an effective mobility change [4].

The effects of strain on the valence band of silicon, however, are quite different due to its strongly degenerate nature. The shape of the valence band is altered resulting in change of the effective masses and mobility. In addition, the symmetry is reduced and the bands are shifted in energy. In view of the multitude of different effects, the piezoresistance in p-silicon is more dominant than in n-type silicon. Using **k.p** perturbation theory coupled with deformation potential theory, we have extended the analysis of [5] to include spin-orbit coupling effects in computation of the valence band structure [6]. The computed band structure is shown in Figs. 1 and 2. Figure 1 illustrates the energy spectra (E-k) in the [100] and [111] directions for the heavy hole (HH) band at zero stress, uniaxial stress (10^9 dynes/cm^2), and shear stress (10^9 dynes/cm^2). As expected, the stress introduces significant asymmetry and the bands become nondegenerate at **k** = 0; the HH band moves upward and the light hole band (LH), although not shown, moves downwards. The energy shift associated with the light hole band is less pronounced in comparison to the HH band implying a larger proportion of the hole concentration in the HH band. The constant energy surfaces (taken at −15 meV from the top of the corresponding band) at zero stress and shear stress of 10^9 dynes/cm^2, are shown in Fig. 2. The band distortion with shear stress is the largest leading to the largest change in mobility, yielding also the largest piezoresistance coefficient (π_{44}).

3. Relation of strained mobility with piezoresistance

The dependence of mobility on stress can be computed using the expression

$$\mu_{ij} = [(q/h^2) \int d^3k \, \tau(E) \, (\partial E/\partial k_i)(\partial E/\partial k_j)(\partial f/\partial E)] / [\int d^3k \, f(E, E_f, T)] \quad (1)$$

derived from the Boltzmann transport equation at low fields. Here, f is the distribution function, E is the distorted energy spectrum of the valence band, and τ is the relaxation time due to scattering events which are a function of the wavevector, k. Since the valence band is made up of predominantly the HH and LH bands, the computed mobility is a weighted sum of the individual mobilities in the corresponding bands. The mobility becomes highly directional-dependent and its variation with applied stress can be easily related to the well known piezoresistance coefficients that have been measured by Smith [7]. Here, the normalized change in resistivity (Δρ) per unit stress was measured, and assuming a first order change in the resistivity with stress, the piezoresistance coefficients were retrieved. Since, the sample was uniformly stressed, the global change in hole concentration can be assumed unaltered ($\Delta p = 0$) and the change in mobility can thus be expressed in terms of the piezoresistance coefficients, viz.,

$$\Delta \mu_{pij} / \mu_{po} = - \Delta \rho_{ij} / \rho_o = - \Sigma_{k,l} \, \pi_{ijkl} \, T_{kl} \qquad (2)$$

where $\Delta \mu_{pij} / \mu_{po}$ is the normalized change in the mobility, π_{ijkl} is the tensor of piezoresistance coefficients, and T_{kl} are the components of the stress tensor. In view of cubic symmetry of silicon, there are only three distinct piezoresistance coefficients. Table 1 illustrates a comparison of the calculated (using eqns. (1) and (2)) and measured piezoresistance coefficients [7].

coefficient	calculated	measured
π_{11}	8	6.6
π_{12}	-2	-1.1
π_{44}	140	138.1

Table 1 Values shown are in units of 10^{-12} cm^2/dyne.

The calculated values shown in Table 1 are consistently larger in magnitude and this can be attributed to numerical errors arising from evaluation of the integrals over the entire (whole) energy surface that are required to compute the difference in mobility, $\Delta \mu_p$; at zero stress, the computations are simplified resulting in evaluation of integrals over just 1/8 of the total energy surface.

4. The strained drift-diffusion model

The diffusion coefficient, in the presence of stress, becomes a tensor and using Einstein's relation, $D_{pij} = (kT/q) \mu_{pij}$, we can express the standard isothermal drift-diffusion relation for the hole current density as

$$J_p = - q \, D_p \, [\text{grad } p + p \text{ grad } (q\psi/kT)] \qquad (3)$$

which becomes modified, in the presence of a magnetic field, to read

$$J_p - q \, p \, \mu_p \, [(R_H \, B) \times J_p] = - q \, D_p \, [\text{grad } p + p \text{ grad } (q\psi/kT)]. \qquad (4)$$

In eqns. (3) and (4), the transport coefficients D_p, μ_p, and R_H are tensors, and **B** denotes the magnetic field vector. Equations (3) and (4) are valid only within the cubic cell coordinate system. They can be transformed to the Cartesian system of arbitrary orientation using a transformation matrix expressed in terms of the Euler's angles. In the absence of stress, the mobility, diffusion coefficient, and the Hall coefficient in eqn. (4) become scalars, and we recover the usual form of the galvanomagnetic transport equation in terms of the Hall mobility, μ_{Hp} ($= |q p \mu_p R_H|$). A relation for the electron current density can be obtained along similar lines.

References

[1] A. Nathan and T. Manku, Appl. Phys. Letts., vol. 62 (1993) 2947.
[2] H.P. Baltes and R.S. Popovic, Proc. IEEE, vol. 74 (1986) 1107.
[3] J.L. Egley and D. Chidambarrao, Sol.-St. Electron., vol. 36 (1993) 1653.
[4] C. Herring and Vogt, Phys. Rev., vol. 78 (1950) 173.
[5] G.E. Pikus and G.L. Bir, Soviet Physics-Technical Physics, (1958) 2194.
[6] T. Manku and A. Nathan, J. Appl. Phys., vol. 73 (1993) 1205.
[7] C.S. Smith, Phys. Rev., vol. 94 (1954) 42.

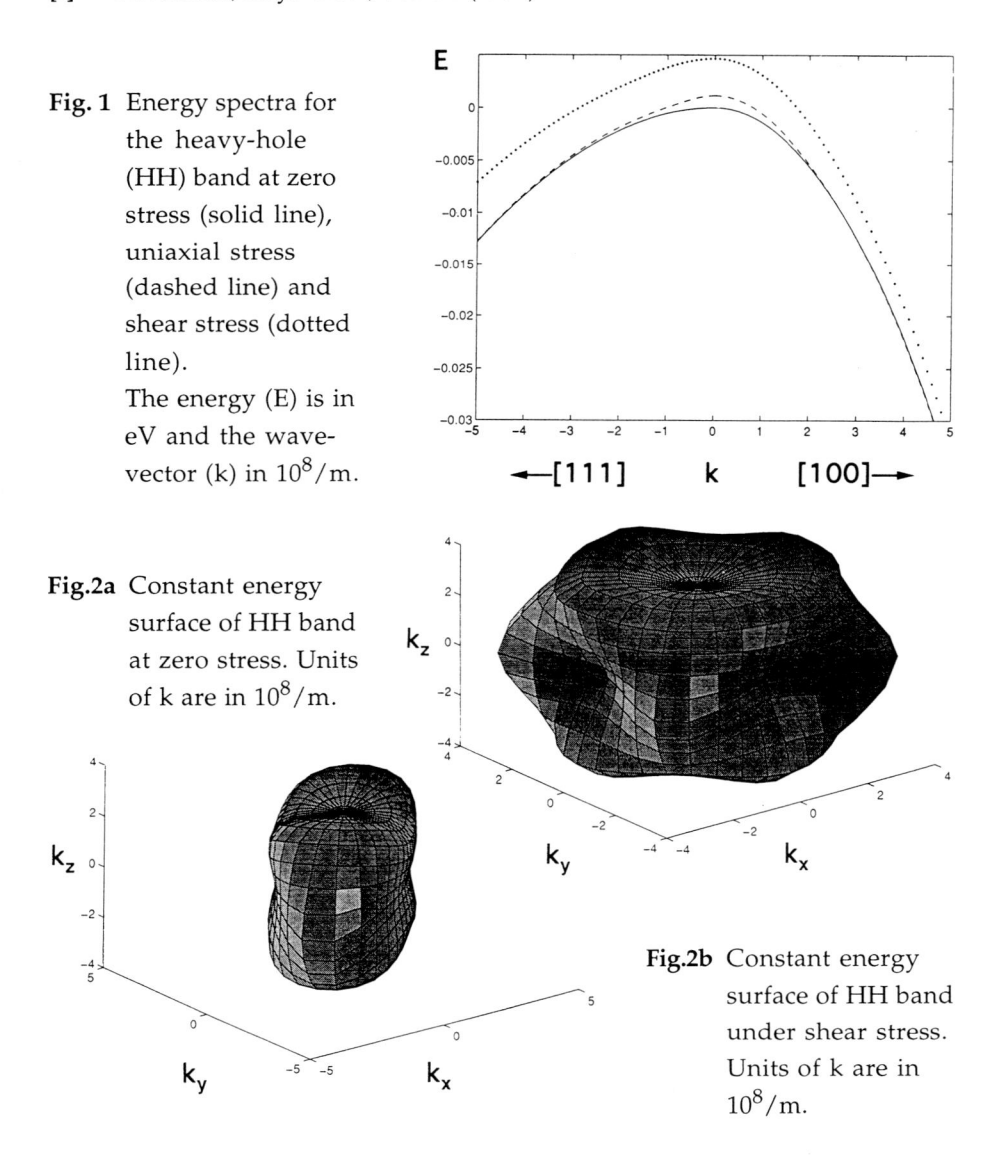

Fig. 1 Energy spectra for the heavy-hole (HH) band at zero stress (solid line), uniaxial stress (dashed line) and shear stress (dotted line). The energy (E) is in eV and the wave-vector (k) in 10^8/m.

Fig. 2a Constant energy surface of HH band at zero stress. Units of k are in 10^8/m.

Fig. 2b Constant energy surface of HH band under shear stress. Units of k are in 10^8/m.

A Novel Approach to HF-Noise Characterization of Heterojunction Bipolar Transistors

Frank Herzel, Bernd Heinemann

Institut für Halbleiterphysik
Walter-Korsing-Straße 2, D-15230 Frankfurt (Oder), GERMANY

Abstract

We present a numerical HF noise analysis of $Si/Si_{1-x}Ge_x/Si$ heterojunction bipolar transistors based on the time-dependent solution of the drift-diffusion equations. For the MHz range an analytical expression for the noise figures is derived.

1. Introduction

$Si/Si_{1-x}Ge_x/Si$ heterojunction bipolar transistors (HBTs) are expected to become important in certain HF applications, where the use of a single technology for the complete microwave part of the system may reduce cost [1]. Therefore, the investigation of HF noise figures (NF) of these devices is an important task. The aim of our paper is to calculate the dependence of the minimum NF on frequency and current using a two-dimensional solver for the drift-diffusion equations.

2. Theory

In [2] the authors developed a quantum statistical approach to thermal noise in semiconductor devices for arbitrary space-dependent carrier temperatures including low ones. Especially, a general expression for the NF of bipolar transistors was derived which relates the NF to the y-parameters and effective carrier temperatures. We will simplify the result of [2] by some reasonable assumptions: First, quantum statistical corrections are neglected since at room temperature the classical limit is a very good approximation. Secondly, electron heating in the collector is neglected since this effect turns out to be small for $f < f_T$. Finally, the noise generator is assumed to be at ambient temperature T. Then we obtain the NF defined as the ratio of input to output signal-to-noise ratios

$$F = 1 + \frac{e|I_B| + 2k_B T \mathrm{Re}(y_{11})}{2k_B T \mathrm{Re}(y_G)} + \frac{e|I_C| + 2k_B T \mathrm{Re}(y_{22})}{2k_B T \mathrm{Re}(y_G)} \frac{1}{|G|^2} \tag{1}$$

with e being the elementary charge, G the current gain of the terminated fourpole, and $y_G = g_G + ib_G$ the complex generator admittance. The crosscorrelation between the noise sources i_B and i_C has been neglected since this contribution turns out to be small for $f < f_T$. If the condition $b_G = -\text{Im}(y_{11})$ is fulfilled we arrive at the tuned-out NF for optimum load

$$\tilde{F} = 1 + \frac{e|I_B| + 2k_BT\text{Re}(y_{11})}{2k_BTg_G} + \frac{e|I_C| + 2k_BT\text{Re}(y_{22})}{2k_BTg_G}\left|\frac{\text{Re}(y_{11}) + g_G}{y_{21}}\right|^2 \quad (2)$$

In this formula the tuned-out NF is expressed by the y-parameters and the direct currents. The minimum NF F_{min} is defined as the minimum of $\tilde{F}$ with respect to g_G. It is important to note that, according to the three-dimensional derivation independent of transistor geometry [2], the y-parameters refer to the whole transistor, that is, series resistances are included. Furthermore, the distributed nature of the base resistance is properly taken into account.

3. Noise Figure at High Frequencies

The calculation of the y-parameters is based on the idea to solve the drift-diffusion equations in the time domain for sufficiently small voltage perturbations applied to the base or collector contact, respectively. We use the two-dimensional device simulation code TOSCA [3]. The Fourier decomposition of small-signal voltages and currents is performed as postprocessing. This approach may give an advantage over solving the equations in the frequency domain, especially, if a great number of points in the frequency domain is required as in our case. In contrast to many other HF noise models, the input for the simulation are device geometry and doping profiles instead of circuit element values. Thus, our approach is closely related to the technological process. Geometry and doping profile of our $Si_{1-x}Ge_x$-HBT model device are represented in Fig. 1 and Fig. 2.

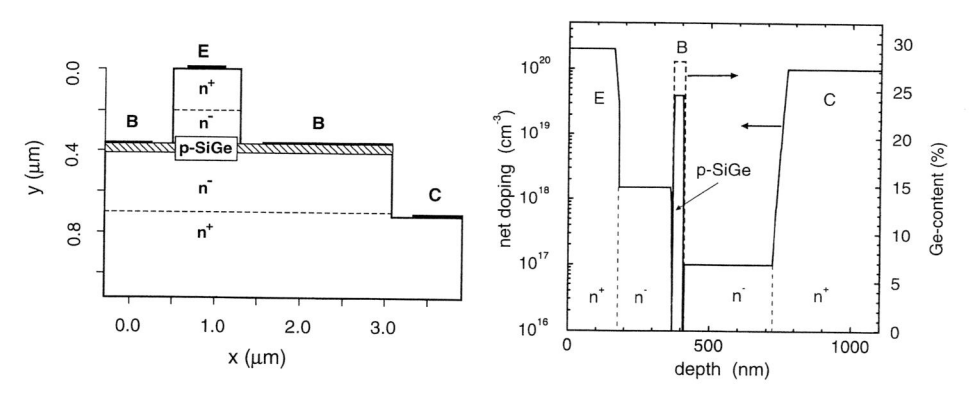

Figure 1: Geometry of the model device.   Figure 2: Vertical profile.

The sheet resistance of the base amounts to 700 $\Omega/\square$. Maximum transit frequency of f_T=46 GHz is reached at collector current of about 10 mA and f_{max} amounts to 120 GHz. In Fig. 3 F_{min} and associated gain are plotted versus frequency for different

Ge contents.

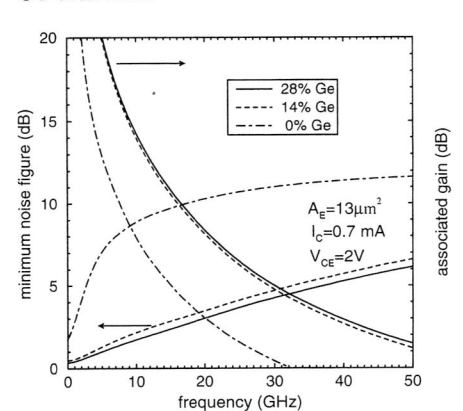

Figure 3: Minimum NF and associated gain versus frequency

Figure 4: Minimum NF versus collector current.

This figure illustrates the noise reducing influence of Ge both at medium and high frequencies. Furthermore, we realize that for our reference device (28% Ge) the static limit applies for frequencies f <1GHz. Looking at Fig. 4 we see that the noise optimum collector current for $f \approx 10$ GHz is about 1 mA, i.e., one order of magnitude lower than the f_T peak position. This agrees well with experimental experiences.

4. Noise Figure at Medium Frequencies

Now we are going to discuss medium frequencies (MF) where the NF is independent of frequency. In this case the NF can be calculated from the static characteristics. In order to obtain a basic understanding we use the ideal characteristics $I_{C,B} \propto exp(\Delta V_{BE}/V_T)$ with $V_T = (kT)/e$ and neglect the thermal noise contribution from the collector since it is small. With these assumptions the minimum NF in the MF range $\tilde{F}^0_{min}$ can be calculated analytically as a function of the static differential current gain β_0:

$$F^0_{min} = F^0_{opt} = 1 + \frac{1}{\beta_0} + \sqrt{\frac{3}{\beta_0} + \frac{1}{\beta_0^2}} \approx 1 + \sqrt{\frac{3}{\beta_0}}. \qquad (3)$$

Note, that F^0_{min} does not depend on I_C. Hence, it equals the MF optimum NF which represents the minimum of the NF with respect to b_G, g_G and I_C as well. From Eq. (3) we conclude that at medium frequencies the static current gain is the only criterion for the optimum NF. Since the current gain of HBTs can be increased by orders of magnitude in comparison to BJTs, they are suited for low-noise amplification in the MHz range. The optimum NF as a function of current gain is shown in Fig. 5.

Additionally, we plotted the values corresponding to Hawkins' model [4].

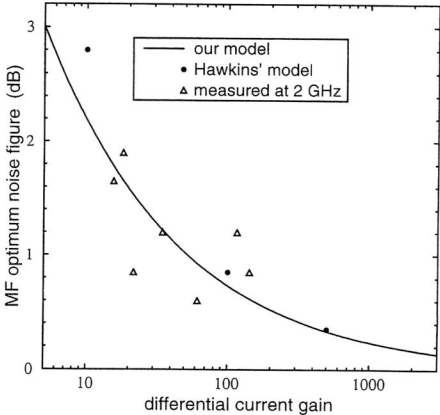

Figure 5: Optimum NF at medium frequencies as a function of static differential current gain. The circuit model values (circles) and the experimental values (triangles) are taken from [1].

The two models show a relatively good agreement. The small difference is due to the assumption of an ideal Gummel plot. Our result is identical to that resulting from Hawkins' theory for a base resistance small compared to the generator resistance.

Finally, we want to remark that our approach is not confined to simulation but also allows to extract HF noise figures at relatively little effort from small-signal parameters only.

References

[1] H. Schumacher and U. Erben, Heterojunction Bipolar Transistors for Noise-Critical Applications, *Proc. of 18. SOTAPOCS*, vol. 93, no. 27, pp. 345-352, 1993.

[2] F. Herzel and B. Heinemann, High-Frequency Noise of Bipolar Devices in Consideration of Carrier Heating and Low Temperature Effects, accepted for publication in *Solid-St. Electron.*

[3] H. Gajewski, K. Zacharias, H. Langmach, G. Telschow, and M. Uhle, *TOSCA user's guide*, Weierstraß-Institute for Mathematics, Berlin, 1986.

[4] R. J. Hawkins, Limitations of Nielsen's and Related Noise Equations Applied to Microwave Bipolar Transistors, and a New Expression for the Frequency and Current Dependent Noise Figure, *Solid-St. Electron.*, vol. 20, pp. 191-196, 1977.

Ge Profile for Minimum Neutral Base Transit Time in Si/Si$_{1-y}$Ge$_y$ Heterojunction Bipolar Transistors

Wolfgang Molzer

Corporate Research and Development, SIEMENS AG, ZFE T ME 25
Otto-Hahn-Ring 6, D-81739 Munich, GERMANY
Institut für Theoretische Physik, Leopold-Franzens-Universität Innsbruck
Technikerstraße 25, A-6020 Innsbruck, AUSTRIA

Abstract

A simple but effective numerical method for the determination of Ge profiles leading to minimum neutral base transit time τ_B is presented. The profiles under consideration have been taken from a large general class of functions. The resulting profiles show that significant reductions in τ_B can be achieved when they are compared to other investigated profile types. Moreover the position dependence of these optimum profiles has a simple structure. Thus there are no additional technological difficulties for their realization.

1. Introduction

The increasing interest in Si/Si$_{1-y}$Ge$_y$ heterojunction bipolar transistors (HBTs) raises the question of how to design the vertical Ge profile y in the base layer. For high speed applications, a minimum base transit time τ_B is extremely important. These considerations lead to the specific question of which Ge profile minimizes τ_B. As we will see later, for a negligible dependence of electron diffusivity D on y, base transit time can be minimized with respect to intrinsic carrier concentration profile n_i.

A mathematically equivalent question arises in the context of bandgap narrowing caused by high doping. It has been treated so far mostly by restricting the class of admissible profiles to a small set of trial functions with one parameter each, like exponential or gaussian doping profiles [1, 2, 3]. In those cases, τ_B has been minimized with respect to the corresponding parameter for each of the functions separately and the result has been compared to that of a uniform doping. To the author's knowledge there has been only one attempt which tries to solve the problem in a much more general sense by using the analytical methods of Variational Calculus [4]. But the differential of the considered functional is not zero anywhere on a reasonable domain. Thus the Euler-Lagrange equation leads to contradictory results [5] and the original question in its general sense has not been answered yet. For the first time in this investigation the problem is successfully solved by utilizing a numerical approach and by posing reasonable additional constraints on the considered functions.

2. Procedure

Similar to the work of McGregor et al. [4] we use the formula by Kroemer [6] to define base transit time τ_B in terms of n_i^2, hole concentration p and electron diffusivity D.

$$\tau_B\left(n_i^2, p, D\right) = \int_{[0,W]} \frac{n_i^2}{p}(x) \int_{[x,W]} \frac{p}{Dn_i^2}(z)\, dz\, dx$$

All three of them are functions of position through their dependence on base doping profile N and Ge profile y. The space coordinate is chosen to have a value of zero at the emitter side of the base and that of W at the collector side. In the neutral base, p can be approximated by N in the low injection regime. For any given N and for negligible dependence of D on y, τ_B can be minimized with respect to n_i^2, thus solving the problem for any one-to-one correspondence between n_i^2 and y.

This procedure has been carried out for the special case of constant N in the neutral base for which D and p can be assumed to be also constant functions of position. Additionally we fix the value of W as well as the values of n_i^2 at the boundaries of the base. Furthermore we limit the minimum and maximum value of n_i^2 to the same values. These additional constraints are not only technologically and physically reasonable but moreover avoid the problems which can be expected by the mentioned features of the differential of the functional for τ_B. Since it is nowhere equal to zero the solutions might come to lie at a boundary. For the discrete approximation this means some interpolation points should take the limiting values.

For the numerical treatment, a library routine for constraint optimization from the numerical package MATLAB® is employed. Using the above a discrete approximation of τ_B is minimized with respect to the function values of about 50 evenly spaced interpolation points of n_i^2 in the base interval.

The solution is expressed in terms of y using the common model for the dependence of n_i^2 on y as given in [7, 8].

$$n_i^2(y) = n_{i,\mathrm{Si}}^2 \frac{2 + \exp\left(-\frac{E_s(y)}{E_{th}}\right)}{3} \left(\frac{m_v(y)}{m_{v,\mathrm{Si}}}\right)^{\frac{3}{2}} \exp\left(\frac{E_{g,\mathrm{Si}} - E_g(y)}{E_{th}}\right)$$

The six conduction band minima are split into two groups of four and two. The difference in energy E_s is a function of the Ge content y. The curvatures of the two valence bands are also functions of y. Their combined effect on n_i^2 is represented by the effective hole mass m_v. Furthermore there is the dependence of the energy gap E_g on y, E_{th} being the value of thermal energy. The minimum value of Ge content is chosen to be 0%, maximum values have been selected from an interval between 0% and 25%.

3. Results

Four examples of the solutions in terms of Ge profiles can be seen in Fig. 1 with maximum values of 5, 10, 15 and 20%. They consist of three clearly discernible parts. In the border regions of the base, the Ge content is limited by the inequality constraints. These plateaus are connected by a region where the intrinsic carrier concentration n_i^2 is exponentially varying, giving an almost exactly linear Ge profile. For small differences in the limiting values, the positions where the plateaus meet the linear part are near the middle of the base and for large differences they come to lie near the boundaries of the base resulting in a linear profile over most of the neutral base. The dependence of the location and meeting points of the three parts on the maximum Ge content are shown by the shades and curves in Fig. 2.

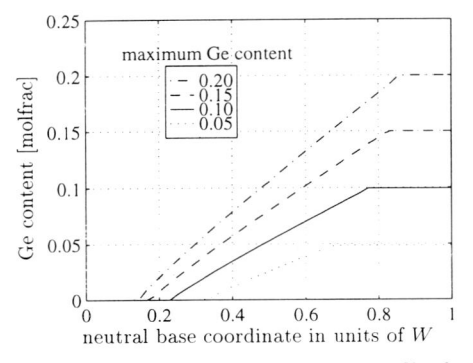

Fig. 1: Four examples of optimum Ge profiles for different values of maximum Ge content which are given in the legend.

Fig. 2: Location of the three different parts of optimum Ge profiles and the points where they meet with their dependence on maximum Ge content.

In Fig. 3 the dependence of minimum neutral base transit time on maximum Ge content for the calculated optimum profile is compared to two other profile types. The first one has no Ge in the emitter sided half of the base and a constant value towards the collector. The location of the step is placed in the centre of the base, which is the optimum position for this type of profile. The second one has an exponentially varying n_i^2 over the whole neutral base, i.e. a linear Ge grading from zero to the maximum value. The values of τ_B are normalized to a value obtained with constant n_i^2, which corresponds to a constant Ge content. The detailed comparison of the linear case and the calculated optimum in Fig. 4 shows an improvement of more than 10% in τ_B over a range from 5% maximum Ge content to well above 20%. Compared to the step profile the reduction in τ_B is more than 10% for any value of maximum Ge content above 10%. This reduction is increasing with increasing maximum Ge content and τ_B can even be lower by a factor of two at 25% maximum Ge content.

Fig. 3: Dependence of neutral base transit time τ_B on maximum Ge content for different Ge profile types.

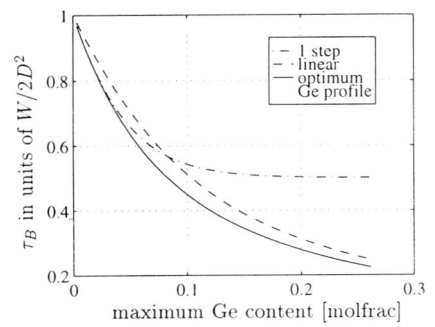

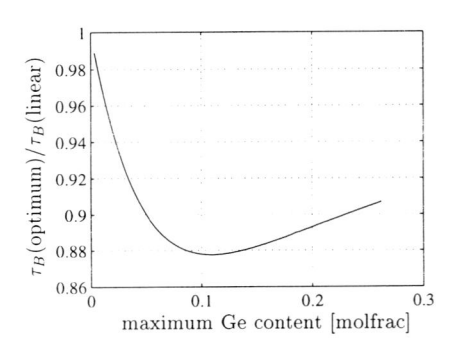

Fig. 4: Ratio of τ_B for optimum and linear profile for maximum Ge contents from 0 to 25%.

4. Conclusions

We have calculated the Ge profile in the base of a HBT that minimizes base transit time. By this we have demonstrated that the optimum Ge profile can lead to a significant reduction in neutral base transit time compared to profile types employed in existing HBTs and those usually investigated in modelling and device simulation. In addition the result clearly shows that the optimum profile is a simple function of position which is not without practical relevance. On the contrary it is as technologically feasible as the above 'ad hoc' profiles using advanced processing techniques like selective epitaxial growth or ultrahigh vacuum chemical vapor deposition [9].

References

[1] S. Szeto and R. Reif, "Reduction of f_t by nonuniform base bandgap narrowing", *IEEE Electron Device Letters*, vol. EDL-10, no. 8, pp. 341–343, 1989.

[2] P. J. van Wijnen and R. D. Gardner, "A new approach to optimizing the base profile for high-speed bipolar transistors", *IEEE Electron Device Letters*, vol. EDL-11, no. 4, pp. 149–152, 1990.

[3] K. Suzuki, "Optimum base doping profile for minimum base transit time", *IEEE Transactions on Electron Devices*, vol. ED-38, no. 9, pp. 2128–2133, 1991.

[4] J. M. McGregor, T. Manku, and D. J. Roulston, "Bipolar transistor base bandgap grading for minimum delay", *Solid-State Electronics*, vol. 34, no. 4, pp. 421–422, 1991.

[5] J. M. McGregor, T. Manku, and D. J. Roulston, "Retraction: Bipolar transistor base bandgap grading for minimum delay", *Solid-State Electronics*, vol. 35, no. 9, p. 1383, 1992.

[6] H. Kroemer, "Two integral relations pertaining to the electron transport through a bipolar transistor with a nonuniform energy gap in the base region", *Solid-State Electronics*, vol. 28, no. 11, pp. 1101–1103, 1985.

[7] R. People, "Physics and applications of Ge_xSi_{1-x}/Si strained-layer heterostructures", *IEEE Journal of Quantum Electronics*, vol. QE-22, no. 9, pp. 1696–1710, 1986.

[8] S. Marksteiner, "Bauelement-Simulation von $Si_{1-y}Ge_y$-Hetero-Bipolar-Transistoren", Master's thesis, Institut für Theoretische Physik, Universität Innsbruck, Innsbruck, 1992.

[9] E. F. Crabbé, J. H. Comfort, W. Lee, J. D. Cressler, B. S. Meyerson, J. Y.-C. Sun, and J. M. Stork, "73-GHz self-aligned SiGe-base bipolar transistor with phosphorus-doped polysilicon emitters", *IEEE Electron Device Letters*, vol. EDL-13, no. 5, pp. 259–261, 1992.

Performance optimization in Si/SiGe heterostructure FETs

A. Abramo[a], J. Bude[b], F. Venturi[c], M.R. Pinto[b], and E. Sangiorgi[d]

[a] DEIS, University of Bologna, Viale Risorgimento 2, 40136 Bologna, Italy
[b] AT&T Bell Labs., Murray Hill, NJ USA
[c] DII, University of Parma, Parma, Italy
[d] DIEGM, University of Udine, Udine, Italy

Abstract

In this paper we investigate the role of structure design in determining high effective mobility (μ_{eff}) values in Si/SiGe FETs. To this purpose we have developed a one-dimensional self-consistent Schrödinger-Poisson simulator and applied it to the study of the mobility behavior of different Si/SiGe FET structures. As a result we propose a structure which, despite to its simple design, shows improved theoretical performance.

1. Introduction

High μ_{eff} is crucial to obtain high-speed, high-performance transistors [1]. Since record high μ_{eff} values in Si/SiGe heterostructures were demonstrated both at low [2, 3] and room temperature [4, 5, 6], the Si/SiGe system is a promising candidate to carry on the improvement of performance standards in modern technologies. However, the strengths of this novel technology for future ULSI still have to be demonstrated.

In Si/SiGe structures the channel quantum well forms inside a Si layer lying on a fully-relaxed SiGe buffer, thus it is subject to tensile strain. The quantization and strain are responsible for the improved mobility with respect to bulk silicon, since they induce an energy splitting between the fourfold degenerate conduction band belonging to the growth plane and the twofold one along the growth direction. Consequently, since electrons inside the twofold band show the lower effective mass to the transport direction and since their quantized levels are lower in energy, the 2D electron gas (2DEG) μ_{eff} is improved. In addition, the spatial separation of the transport layer from the insulator tends to avoid the presence of surface scattering mechanisms limiting electron μ_{eff}.

2. The simulator

To address the problem of the simulation of μ_{eff} in Si/SiGe structures, we have developed a one-dimensional self-consistent Schrödinger-Poisson simulator which fully accounts for the 2D nature of the carrier gas.

First, the simulator computes the strain-induced energy splitting in the frame of the model-solid theory, following [7]. Then, eigenvalues and corresponding envelope functions are determined as result of a self-consistent Newton iteration loop between Schrödinger and non-linear Poisson equations.

From energy eigenvalues and eigenfunctions the scattering rate for the 2DEG are computed. We have included optical, acoustic elastic phonons, and surface roughness to evaluate the impact of surface vicinity. Optical and elastic acoustic phonon scatterings among subbands were included in a conventional way [8]. Surface roughness scattering was implemented as a many-subband generalization of the model described in [9]. The transport parameters we used are those of [10] for phonon scattering and of [11] for surface roughness.

3. Simulation results

As already mentioned, the theoretical peak performance of this technology is very promising, but practical requirements, such as bias compatibility, impose large gate voltage swings that eventually degrade μ_{eff}. In fact, as gate bias is increased, an inversion layer is progressively formed at the surface, thus competing with the channel 2DEG in determining μ_{eff}. In these conditions, a strong limiting effect on μ_{eff} due to surface roughness is expected. In order to fully exploit the possibilities of the material, this effect must be avoided by means of a proper device design, or at least pushed towards higher gate bias.

In the preliminary stage of our work, we applied the simulator to reproduce available experimental data for electron μ_{eff} in real devices [12]. We report here the behavior of μ_{eff} both as a function of temperature (Fig. 1) and of gate bias (Fig. 2) for a structure like the one described in [11], with a 5 nm SiGe cap layer.

In a second step we exploited simple modifications to the structure design in order to improve the confinement of the 2DEG inside the quantum well. The improved structure we propose in this paper is shown in Fig. 3.

Compared to the one in [11], the modulation doping layer has been removed, the n^+-poly has been replaced with the p^+ one, and n doping has been added to the SiGe cap layer and to a small fraction of the SiGe relaxed layer. The presence of the p-poly determines a band bending at the surface of the device that tends to keep the carriers away from the Si-SiO$_2$ interface, while the n-doped cap layer, as in buried channel MOSFETs, spreads the electrons of the surface channel deeper in the device when the gate voltage is increased, thus displacing the centroid of the electron charge away from surface. Instead, the strained-Si quantum well region was kept intrinsic in order to avoid ionized impurity scattering.

The behavior of the bottom of the conduction band of the new structure as a function of the gate bias is reported in Fig. 4. As in buried channel devices, the n-doped surface layer shows a smooth band bending, thus spreading the electrons of the channel deeper in the device.

For this structure a linear extrapolated threshold voltage of approximately 1 V is obtained (inset of Fig. 4). Although slightly high, this value is positive (as opposed to the negative V_T of the structure of [11]), and can be adjusted by changing the doping dose of the n layers.

As expected, the μ_{eff} behavior as a function of V_{GS} is also improved, as shown in Fig. 5. The peak value (above 2800 cm^2/Vs) is similar to the one reported for the structure of [4, 5] (2600 cm^2/Vs) that we also simulated to check the prediction ability of the simulator.

Finally, as a test on device performance, we report in Fig. 6 the linear transconductance behavior of the new structure. Since we are solving the one-dimensional equilibrium problem, no information on the saturation properties of the electron gas can be directly extracted, as well as 2D effects such as the short channel ones. Nevertheless, the g_m in the linear regime can be estimated. The g_m curve of Fig. 6 has been obtained multiplying the sheet concentration of the inset of Fig. 4 by the mobility curve of Fig. 5, assuming L_{eff} =0.5 μm and V_{DS}=0.2V (corresponding to an average longitudinal electric field of 4 kV/cm).

4. Conclusions

We have investigated the role of structure design in determining high μ_{eff} values in Si/SiGe FETs. A one-dimensional self-consistent Schrödinger-Poisson simulator has been developed and applied to the study of the mobility behavior of different Si/SiGe FET structures. As a result, we have proposed a relatively simple structure design showing improved theoretical performance.

5. Acknowledgments

A.A. gratefully acknowledge Dr. L. Selmi for many helpful discussions.

References

[1] M. Pinto et al., *IEEE Electron Device Lett.*, vol. 14, p. 375, 1993.
[2] F. Schäffler et al., *Semicond. Sci. Technol.*, vol. 7, p. 260, 1992.
[3] Y.-H. Xie et al., *J. Appl. Phys.*, vol. 73, p. 8364, 1993.
[4] S. Nelson et al., *Appl. Phys. Lett.*, vol. 63, p. 367, 1993.
[5] K. Ismail et al., *IEEE Electron Device Lett.*, vol. 14, p. 348, 1993.
[6] S. Verdonckt-Vandebroek et al., *IEEE Electron Device Lett.*, vol. 12, p. 447, 1991.
[7] C. V. de Walle, *Phys. Rev. B*, vol. 39, p. 1871, 1989.
[8] M. Fischetti et al., *Phys. Rev. B*, vol. 48, p. 2244, 1993.
[9] M. Ishizaka et al., in *IEDM Tech. Dig.*, p. 763, 1990.
[10] C. Jacoboni et al., *Rev. Mod. Phys.*, vol. 55, p. 645, 1983.
[11] A. Abramo et al., in *IEDM Tech. Dig.*, p. 731, 1994.
[12] E. Fitzgerald et al., *J. Vac. Sci. Technol. B*, vol. 10, p. 1807, 1992.

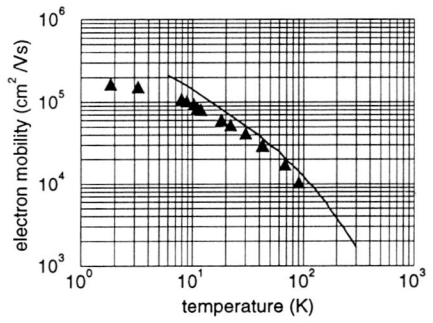

Fig. 1 Zero field electron mobility versus temperature for the 5 nm cap layer device of Ref. [11]. Triangles: experimental Hall mobility data from [12].

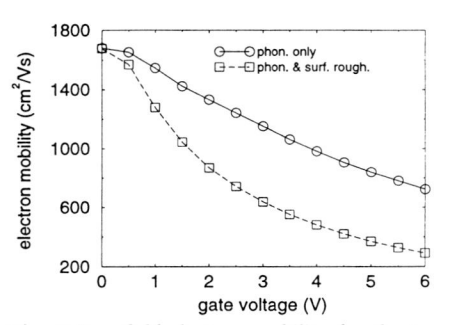

Fig. 2 Zero field electron mobility for the 5 nm cap layer device of Ref. [11].

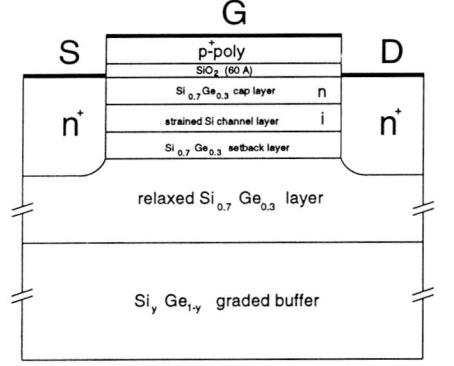

Fig. 3 The proposed structure. The channel resides inside the strained intrinsic silicon layer.

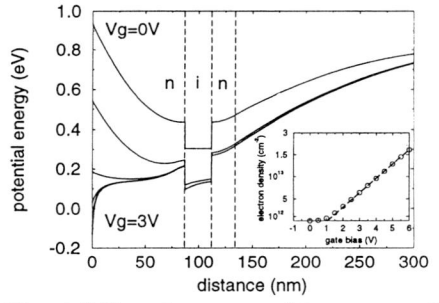

Fig. 4 Self-consistent potential energy profile at different gate voltage of the simulated structure. The voltage step is 0.5V. Inset: integrated sheet charge versus gate voltage and extrapolated threshold voltage.

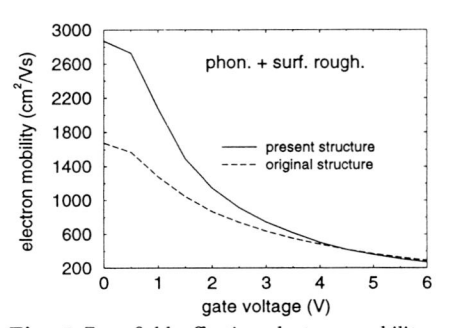

Fig. 5 Zero-field effective electron mobility as a function of the gate voltage for the proposed (solid) and original (dashed) structures.

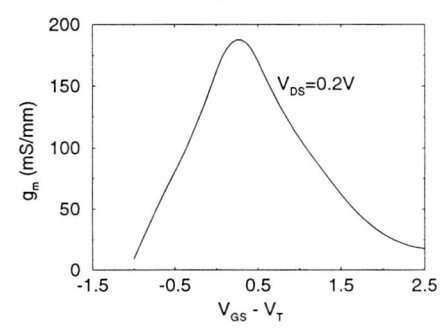

Fig. 6 Linear transconductance versus gate voltage for the proposed structure assuming $L_{eff} = 0.5\ \mu m$ and $V_{DS} = 0.2$ V.

On the Integral Representations of Electrical Characteristics in Si Devices

S. Biesemans, K. De Meyer

ASP-division, IMEC
Kapeldreef 75, B-3001 Leuven, BELGIUM

Abstract

It is shown that the 1D hydrodynamic model of the fundamental transport equations in differential form can be transformed into an equivalent integral representation. The advantage of this procedure lies in the fact that integrals are generally easier to evaluate than the corresponding differential equations. The technique of the integral representations is applied to two examples. In case of a MOSFET, a closed form analytical expression for the carrier concentration and the velocity is obtained. In case the electric field is a step-function with a strong discontinuity, the influence of the diffusion effect as well as the mobility model on the steady state velocity overshoot is analysed without the need for a dedicated numerical solver.

1. Introduction

Numerical modeling of transport in semiconductor devices plays a crucial role in their development. As MOSFET's are scaled down to the 0.1 µm range, the channel length approaches the mean free path of charge carriers and effects as non-stationary and quantum transport become apparent.

In conventional semiconductor devices, most quantum transport effects can be treated indirectly. For instance the effects of the rapidly varying crystal potential on electron transport can be modeled through the concepts of effective masses, energy gaps and the positively charged quasi-particle holes. On the macroscopic level, charge transport can then be modeled using the concepts of the semiclassical model.

In 1969, Rees [1] predicted an overshoot of the carrier mean velocity in semiconductors following rapid changes of the electric field by a self-consistent solution of the Boltzmann equation. In 1976, Shur [2] demonstrated that the full solution of the Boltzmann equation is not necessary. Using a coupled system of simplified particle-, momentum- and energy balance equations, called the hydrodynamic model, the basic physical mechanism underlying the velocity overshoot can be explained.

In principle, the Monte Carlo technique is superior to the solution of the moment equations, however, for the fundamental physical understanding of high field- and strong gradient effects, it is often satisfactory and preferable to go to analytic solutions. For instance, the saturation of carrier velocity in presence of constant high electric fields can be analytically assessed [3] using the three balance equations. It is shown that velocity saturation is caused by carrier heating (T_e>300°K). However, when strong gradients in the field are present, ensemble (diffusion) effects make the local models ($\mu(E)$) inaccurate, and non-local models ($\mu(T_e)$ or $\mu(E_n)$) prevail.

The purpose of this report is to present the efficient technique of integral representations to calculate the physical quantities from the hydrodynamic model in a closed form expression. It is shown that the 1D hydrodynamic model in differential form can be reformulated in a set of integral type equations. There is a distinct advantage in using them as an analysis tool for devices because they can easily be evaluated. The integral expressions are written in a general way independent of the particular μ-model used. As such, it is a good basis to evaluate and compare different μ-models.

2. Model description

The hydrodynamic model (HD) together with the Poisson equation are well assessed [3]. In principle, these four equations are coupled which makes it extremely difficult to solve them, even numerically. However, by assuming no generation-recombination, by simplifying correctly the energy balance equation (e.g.[4]) and by solving the Poisson equation independently of the HD-model [5,6], it is possible to calculate the carrier density (n), the velocity (v), the current density (J) and the carrier temperature (T_e). The potential (ψ) is calculated from the Poisson, 'n' from the particle balance, 'J' from the momentum balance and 'T_e' from the energy balance equation.

As a first example, the integral representation technique was applied within the drift-diffusion approach (T_e=300°K) in [7] to investigate the *short channel effect*. The details of the calculation can be found in [7] for a constant mobility (only dependent on the substrate concentration). In this report, we extend it to include high field effects ($\mu(E)$)

$$n(x) = \frac{n_i^2}{N_{sub}} exp\left(\frac{\psi(x)}{V_{th}}\right)\left[1-(1-exp(-\frac{V_{DS}}{V_{th}}))\frac{\int_0^x exp(\frac{\psi(u)}{V_{th}})\mu^{-1}(u)du}{\int_0^L exp(\frac{\psi(u)}{V_{th}})\mu^{-1}(u)du}\right] \quad (1)$$

with $\mu(x)$ any local high field mobility model (e.g. the Caughey-Thomas model [9]). The first factor gives the equilibrium concentration while the second one acts as a modulation factor. Fig.1 shows n(x) for both constant and field dependent mobilities together with MEDICI simulations. The field dependent mobility makes that n(x) increases to compensate for the loss in velocity. The velocity can be calculated by plugging (1) into the conventional Drift-Diffusion (DD) equation (i.e. $T_e=300°K$) and compare the expression with $J=-qn(x)v(x)$. This results in

$$v(x) = \frac{n_i^2}{N_{sub}} \frac{V_{th}(1-\exp(-\frac{V_{DS}}{V_{th}}))}{\int_0^L \exp(\frac{\psi(u)}{V_{th}})\mu^{-1}(u)du} \frac{1}{n(x)} \quad (2)$$

In a second example, we look at the problem of the step-like field (see insert of Fig.2) mentioned in [10] including the energy balance equation to investigate the *velocity overshoot*. At first, the Slotboom [4] approximation of the energy balance equation for T_e,

$$T_e(x) = T_l + \frac{2}{5}\frac{q}{k}\int_0^x E(u)\exp[\frac{u-x}{\lambda}]du \quad (3)$$

is used with $T_m = T_e(L)$ and $n_m = n(L)$ and $\lambda = 30$ nm. The total set of integral equations can be straight-forwardly calculated as in [7]. We give here the result for 'n'

$$n(x) = n_o \frac{T_l}{T_e(x)} \exp[\psi^*(x)][1-(1-\frac{T_m}{T_l}\frac{n_m}{n_o}\exp[-\psi^*(x)])\frac{\int_0^x \mu^{-1}(u)\exp[-\psi^*(u)]du}{\int_0^L \mu^{-1}(u)\exp[-\psi^*(u)]du}] \quad (4)$$

with $\quad \psi^*(x) = \int_0^x E(u)\frac{q}{kT_e(u)}du$

For this simplified case-study, all the integrals can be calculated analytically. The velocity is expressed in a similar way as (2). Fig.2 pictures $T_e(x)$ showing an increased temperature for higher fields. Fig.3 and 4 show the results for $n(x)$ and $v(x)$ respectively for both the present HD-model as well as for the conventional DD-model ($T_e=300°K$).

3. Discussion

The carrier density 'n' (3) and the velocity 'v' within the HD-model depend on the type of mobility model used. Fig.3 and 4 show clearly the difference between local (only function of E) and non-local (function of T_e) µ-models. In case of a T_e dependent mobility, a velocity overshoot is observed while in case of an E dependent µ the opposite is true. Another example, the calculation of a one carrier metal-semiconductor rectifier using an integral representation of J (expression (34)) was done by Stratton [8].
The evaluations of the integrals, done with the software package Mathematica running on a Macintosh Quadra 650, take about 20 s of time.

4. Conclusions

The 1D hydrodynamic model in differential form was transformed into an integral representation of the fundamental transport equations. The driving force for this procedure was the fact that integrals are generally easier to evaluate than the corresponding differential equations. The general applicability of the transport integrals to any transport problem was pointed out. Two examples were discussed. The integral representations are widely applicable to classic textbook examples (e.g.[9]) as well as to realistic devices (e.g.[7]).

References

[1] H.D.Rees, J. Phys. Chem. Solids, vol. 30, pp. 643-655, 1969
[2] M.S.Shur, Electronics Letters, vol. 12, pp. 615-616, 1976
[3] M.Lundstrom, " Mod. series Solid St. d.,vol. X, Addison-Wesley,1990
[4] A.D.Sadovnikov, D.J.Roulston, SISDEP Conference, pp153-156, 1993
[5] T. Toyabe, IEEE J. Solid-State Circ, pp. 375-382, 1979
[6] R. H. Yan, A. Ourmazd, K. F. Lee, IEEE TED, pp. 1704-1710, July 1992
[7] S. Biesemans, K. De Meyer, Jpn. J. Appl. Phys., vol.34.2B, pp. 917-920, 1995
[8] R. Stratton, Phys. Rev., vol.126, no.6, pp. 2002-2014, 1962
[9] D.M.Caughey, R.E.Thomas, IEEE Proc., pp.2192-2193, 1967
[10] G. Baccarani, M.R.Wordeman, Solid-State Elec., vol. 28, pp407-416, 1985

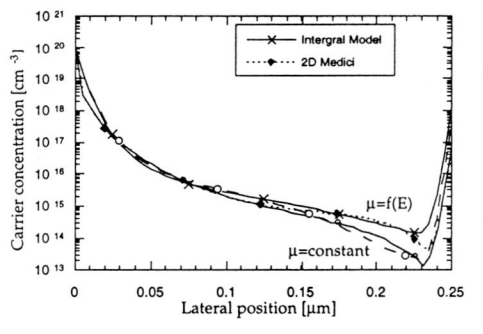

Fig.1 Comparison between MEDICI and integral model (1) for a constant and a field dependent mobility

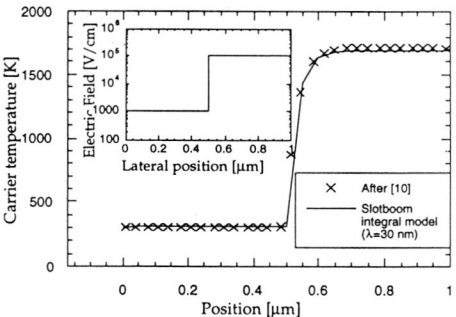

Fig.2 Carrier temperature for the given step-like electric field (insert)

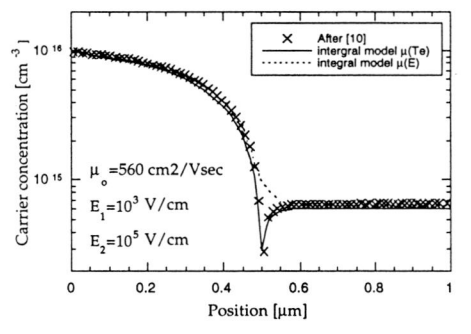

Fig.3 Comparison of the carrier concentration between the model of Baccarani [10] and the new integral model

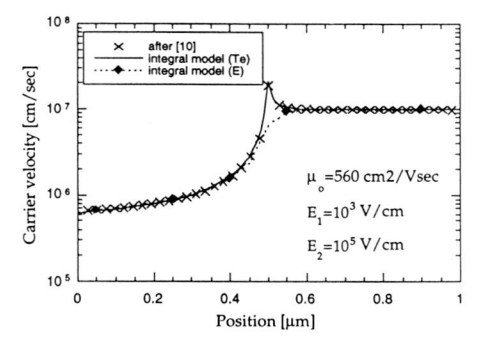

Fig.4 Comparison of the carrier velocity between the model of Baccarani and the new integral model

Large Signal Frequency Domain Device Analysis Via the Harmonic Balance Technique

B. Troyanovsky, Z. Yu, and R. W. Dutton

Center for Integrated Systems,
Stanford University,
Stanford, CA 94305, USA

Abstract

Harmonic and intermodulation distortion effects play a key role in the design and subsequent performance of analog RF/μWave systems. Due to the wide range of frequency components present in such systems, ordinary transient analysis is both extremely time-consuming and insufficiently accurate. In this paper, we present a harmonic balance version of the PISCES semiconductor device simulator. This two-dimensional device simulation tool allows for efficient, physically-based analysis of intermodulation distortion in two-dimensional device structures. Robust nonlinear relaxation methods have been developed to overcome the enormous memory and speed problems associated with fully-coupled, large-signal 2D frequency-domain analyses.

1. Introduction

Harmonic balance (HB) is a nonlinear frequency domain analysis technique which is widely used to simulate harmonic/intermodulation distortion in nonlinear high-frequency circuits [1] [2]. However, current frequency domain (HB) tools are circuit simulators which require lumped (or analytic) models for the active devices. Such models are not based on numerical solution of the semiconductor equations, and as a consequence are not predictive. They require the active semiconductor components to be characterized in advance and fit to a lumped equivalent model.

We have developed a harmonic balance version of the PISCES [3] device simulator to address the above concerns. This two-dimensional HB device simulation tool solves the drift-diffusion system of semiconductor equations in the frequency domain, using the full complement of standard PISCES physical models. A given simulation run yields the spectra corresponding to terminal current/voltage, as well as internal device potential and carrier concentrations, based on a large signal quasi-periodic input. The effects of external packaging and parasitics are handled through a mixed level circuit/device capability which supports linear elements characterized via S-parameters.

To underscore the advantages of a harmonic balance approach when the device is driven by a combination of sinusoids, we examine a typical two-tone input used for intermodulation distortion tests. The driving voltages for such tests have the form $V_{IM} = A\cos(\omega_a t + \phi_a) + B\cos(\omega_b t + \phi_b)$, where the frequencies ω_a and ω_b are closely spaced. For infinitesimally small amplitudes A and B, the device will behave linearly, and the system will respond only at the frequencies ω_a and ω_b. However, as the amplitudes increase, the nonlinearities within the system will give rise to distortion components. In general, the frequencies present in the response will then be given by the set

$$\omega_{k_1 k_2} = k_1 \omega_a + k_2 \omega_b \geq 0, \tag{1}$$

where k_1 and k_2 are integers. The close spacing of ω_a and ω_b results in some frequency components (such as $\omega_a - \omega_b$) which are many orders of magnitude smaller than those in the two-tone input. For example, if $f_a = 1GHz$ and $f_b = 1GHz - 30kHz$ (with $f_a - f_b = 30kHz$), an accurate transient analysis involves integrating over at least 10^5 periods of the high-frequency sinusoid, with spacing fine enough to resolve its harmonics. Further complications arise due to the slow decay of transients, and the relatively poor accuracy with which low-amplitude harmonics are resolved in low-distortion systems.

2. Applying Harmonic Balance to the Time-Dependent Semiconductor Equations

The harmonic balance algorithm overcomes these shortcomings by solving the problem in the frequency domain, and thus avoiding time discretization altogether. PISCES solves the drift-diffusion equations

$$\nabla \cdot (-\epsilon \psi) = q(p - n + N_D^+ - N_A^-) \tag{2}$$

$$\frac{\partial n}{\partial t} = \frac{1}{q} \nabla \cdot J_n - U \tag{3}$$

$$\frac{\partial p}{\partial t} = -\frac{1}{q} \nabla \cdot J_p - U \tag{4}$$

where in addition we have the auxiliary relations $J_n = qD_n\nabla n - q\mu_n n \nabla\psi$ and $J_p = -qD_p\nabla p - q\mu_p p \nabla\psi$. For transient analysis, these equations are discretized in both time and space, and the basic variables at each internal node k (ψ_k, n_k, p_k), along with voltages (v_q) at appropriate external nodes q, are solved for at each time step via an implicit integration scheme. The HB version of the code retains the space discretization, but assumes that the basic variables (ψ_k, n_k, p_k, v_q) have the form

$$x_n(t) = x_{n0} + \sum_{h=1}^{H} x_{nh} \cos(\omega_h t) + x_{n,h+H} \sin(\omega_h t) \tag{5}$$

where $\mathbf{x} = [\psi_1 \ n_1 \ p_1 \ \psi_2 \ n_2 \ p_2 \ \cdots \ \psi_K \ n_K \ p_K \ v_1 \ v_2 \ \cdots v_Q]^T$, K is the number of internal grid nodes, Q is the number of external circuit nodes, H is the number of harmonics used in the analysis, and $1 \leq n \leq 3K + Q$. The frequencies ω_h represent a finite subset of (1) and need not be harmonically related.

If we substitute (5) into (2) - (4) and discretize in space, we obtain a time-domain system of the form

$$F_n(t) = \delta_{cont} x'_n(t) + G_n(\mathbf{x}(t)) \rightarrow 0 \tag{6}$$

where $1 \leq n \leq 3K$, and δ_{cont} is unity for the continuity equations while being zero for the Poisson equation. The goal of the harmonic balance analysis is to determine the set of coefficients x_{nh} in (5) such that (6) is satisfied, along with the auxiliary KCL equations for the external linear elements.

Given a guess for the harmonic coefficients x_{nh}, we can sample the residual functions $F_n(t)$ at $2H + 1$ appropriate time instants to obtain the residual quasi-Fourier coefficients via a matrix transformation analogous to the DFT [1]. That is, we compute the coefficients F_{nh} such that $F_n(t) = F_{n0} + \sum_{h=1}^{H} F_{nh}\cos(\omega_h t) + F_{n,h+H}\sin(\omega_h t)$. Thus, the harmonic residual coefficients F_{nh} can be viewed as functions of the unknowns x_{nh}. If we let $\boldsymbol{\theta}$ denote the $(3K+Q)(2H+1) \times 1$ vector of coefficients x_{nh}, and if we let $\boldsymbol{\Phi}$ denote the corresponding vector of F_{nh}, we end up with a $(3K+Q)(2H+1) \times (3K+Q)(2H+1)$ nonlinear system of the form

$$\boldsymbol{\Phi}(\boldsymbol{\theta}) \to \mathbf{0} \qquad (7)$$

The Jacobian of (7) may be evaluated via straightforward application of the chain rule, and the system may then be solved using standard nonlinear solution techniques.

3. Nonlinear Relaxation Methods

Solving the harmonic balance equations with the fully-coupled direct Newton method is prohibitively expensive, as the memory needed to factor the Jacobian scales as $(2H+1)^2$ with the number of harmonics. Thus, even a modest analysis at 10 harmonics would require 441 times as much memory as a PISCES DC analysis.

Consequently, we utilize a modified block Gauss-Seidel-Newton nonlinear relaxation scheme [4] to solve (7). We define spectral error norms at each harmonic $0 \leq h \leq H$ as follows

$$e_h = \sum_{n=1}^{3K+Q} \left(|F_{nh}| + (1-\delta_h)|F_{n,h+H}| \right) \qquad (8)$$

where δ_h is unity for $h=0$, and zero otherwise. The nonlinear relaxation scheme proceeds by starting from an initial guess, which is typically a PISCES DC solution. On a given iteration, all harmonics except the one with the highest error norm are held fixed, while a single Newton step is performed on the subset corresponding to the highest e_h. The process is repeated until all error norms fall below a specified tolerance. We see that this algorithm requires only four times as much memory as a DC analysis, regardless of the number of harmonics H.

The aforementioned algorithm is guaranteed to converge at low drive levels, since in this case the harmonics decouple and the relaxation scheme reduces to a sequence of linear AC analyses. Experience has shown that the algorithm continues to exhibit extremely robust convergence in virtually all situations of practical interest when it is applied to the semiconductor equations. In addition, it offers a speed of convergence which is typically over an order of magnitude faster than that of fully-coupled Newton.

4. Examples

In this section, we present harmonic balance PISCES analyses of a high-performance silicon npn-BJT [5]. A cross-section of the device is illustrated in Fig. 1. Fig. 2 and Fig. 3 illustrate the collector current response in the frequency- and time-domain, respectively, when the device is driven by $V_{be} = 0.65 + 0.1\sin(10^{10}t)V$, $V_{ce} = 2.0V$. This harmonic balance run took slightly under an hour of simulation time on an HP-735 workstation.

Fig. 4 shows the results of a two-tone intermodulation distortion analysis carried out at 24 harmonics. The input to the BJT is $V_{be} = 0.65 + 0.02(\sin(10^8 t) + \sin(10^7 t))$, $V_{ce} = 2.0V$, with the analysis taking 3hr. 15min. High levels of intermodulation distortion are present in the collector current spectrum.

References

[1] K.S. Kundert, J.K. White, and A. Sangiovanni-Vincentelli, *Steady-State Methods for Simulating Analog and Microwave Circuits*, Kluwer Academic Publishers, 1990

[2] V. Rizzoli and A. Neri, "State of the art and present trends in nonlinear microwave CAD techniques," IEEE Trans. on Microwave Theory and Techniques, pp.343-365, Feb. 1988

[3] Z. Yu, D. Chen, L. So, and R.W. Dutton, *PISCES-2ET — Two-Dimensional Device Simulation for Silicon And Heterostructures*, Stanford University, 1994

[4] J. Ortega and W. Rheinboldt, *Iterative Solution of Nonlinear Equations in Several Variables*, Academic Press, 1970

[5] P. Vande Voorde, D. Pettengill, and S.Y. Oh, "Hybrid simulation and sensitivity analysis for advanced bipolar device design and process development," IEEE 1990 Bipolar Circuits and Technology Meeting, Minneapolis, Minnesota

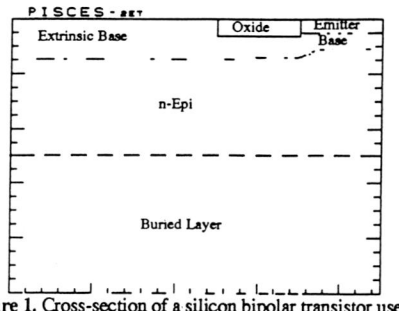

Figure 1. Cross-section of a silicon bipolar transistor used for harmonic balance PISCES analysis.

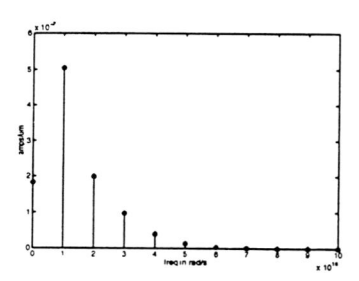

Figure 2. Collector current spectrum under a large-signal single-tone sinusoidal input.

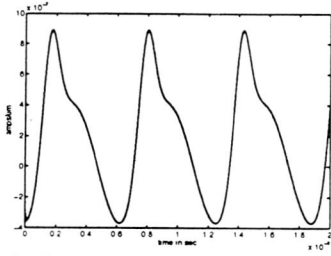

Figure 3. Time-domain representation of Figure 2. The harmonic balance run is drawn as a solid line, whereas a PISCES transient run is drawn dashed. No difference between the two can be seen.

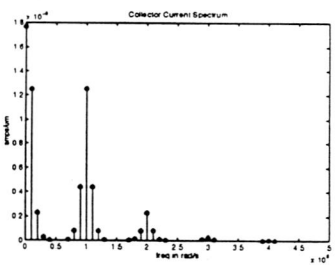

Figure 4. Collector current spectrum under a two-tone intermodulation distortion test.

A Method for Extracting the Threshold Voltage of MOSFETs Based on Current Components

K. AOYAMA

NTT LSI Laboratories,
3-1, Morinosato Wakamiya, Atsugi-shi, Kanagawa Pref., 243-01 JAPAN

Abstract

A new method for extracting the threshold voltage of MOSFETs is presented. The threshold voltage is the gate voltage at which the second difference of the logarithm of the drain current takes a minimum value. The method is applied to a 0.6-μm NMOSFET. The threshold voltage characteristics are compared with ones measured with previous methods and it is shown that the proposed method overcomes previous problems. The threshold voltage is extracted based on a physical background verified with 2D device simulation and shows a transition voltage at which drift and diffusion components in the drain current are equal.

1. Introduction

A serious problem is that threshold voltage definitions in measurements differ from the one in compact MOSFET models. The threshold voltage in compact models is defined as the gate voltage at which the surface potential ψ_s of the channel reaches the double Fermi potential in the bulk $2\phi_f$. This definition is very popular but has the disadvantages that the position where $\psi_s = 2\phi_f$ in the channel is not clear and it is difficult to measure ψ_s in MOSFETs. On the other hand, threshold voltages in measurements are extracted by the constant current (CC), the linear extrapolation (LE) and the transconductance change (TC) methods [1], [2]. With CC, the difficulties are measuring the effective channel length and width and defining a constant value of the drain current. LE should be applied only in the low drain voltage region. In the high drain voltage region, however, the square root of current extrapolation (SRE) should be applied. With extrapolation methods such as LE and SRE, the continuity in all operation voltages is lost. The threshold voltage characteristics with TC are strange for drain voltage changes, as shown in Fig. 1. A threshold voltage definition that overcomes the above disadvantages is needed. This paper presents a new method for extracting the threshold voltage: the second difference of the logarithm of the drain current minimum (SDLM).

2. Method

The diffusion component and the drift component can be respectively approximated by an exponential function and a polynomial expression, as shown in Fig. 2. The first difference of the logarithm of the drain current almost stays constant when the diffusion component is dominant and gradually approaches zero when the drift component is dominant, as shown in Fig. 3 (a). Fig. 3 (b) shows the second difference takes a minimum value at the unique transition voltage in an NMOSFET. In this work, this voltage is defined as the threshold voltage. This threshold voltage is shown in Fig. 1 for various drain voltages compared to ones extracted with previous methods. The characteristics are reasonable even for drain voltage changes.

3. Verification

With a view to certify that the dominant component in the drain current changes from diffusion to drift at the threshold voltage extracted with the above method, the rate of each component was calculated by 2D device simulation. In the simulation, an NMOSFET with $L_g = 1.0\mu m$ was used with $V_{ds} = 3.0V$ and $V_{sb} = 0V$. As a result, the minimum value is calculated with this method as shown in Fig. 4 like in measurements and the potential distributions at the interface of Si and SiO_2 obtained are those shown in Fig. 5. The drift component $[ids_{(drift)}]$ and diffusion component $[ids_{(diff)}]$ are

$$ids_{(drift)} = -q\mu n \frac{d\psi}{dx}, \quad ids_{(diff)} = q\mu n \frac{d(\psi - \xi_n)}{dx}$$

where μ is the electron mobility, n is the electron density, ψ is the conduction band potential, $(\psi - \xi_n)$ is the relative potential and ξ_n is the electron quasi-Fermi potential. At each position at the interface, the rates of drift and diffusion compete with the rates of $(d\psi/dx)$ and $[-d(\psi - \xi_n)/dx]$. Fig. 6 shows the distribution of the rate of current components through the channel. The average rate of the drift component in the channel region $[RATE_{(drift)}]$ is

$$RATE_{(drift)} = \frac{1}{L} \int_{x_s}^{x_d} rate(x) dx$$

where x_s and x_d are the source edge and the drain edge and L is the channel length. Fig. 7 shows the rate of the drift component in the channel increases as the gate voltage increases. At $V_{gs} = 0.215V$, drift equals diffusion. Therefore, the threshold voltage defined with this method shows the unique gate voltage that is the transition voltage in the dominant component in the drain current.

4. Summary

A method for extracting the threshold voltage with the second difference of the logarithm of the drain current was proposed and verified from the physical point of view using 2D device simulation. The extracted threshold voltage has the following features: It is extracted from only the drain current without both the effective channel length and width and has a physical meaning. It shows the gate voltage at which drift and diffusion in the drain current are equal each other. These features may be useful for unifying the threshold voltage in compact models and in measurement.

References

[1] M. J. Deen and Z. X. Yan, "A new method for measuring the threshold voltage of small-geometry MOSFETs from subthreshold conduction", Solid-State Electronics, vol. 33, no. 5, pp. 503-511, 1990.

[2] H. -S. Wong, M. H. White, T. J. Krutsick and R. V. Booth, "Modeling of transconductance degradation and extraction of threshold voltage in thin oxide MOSFETs", Solid-State Electronics, vol. 30, no. 9, pp. 953-968, 1987.

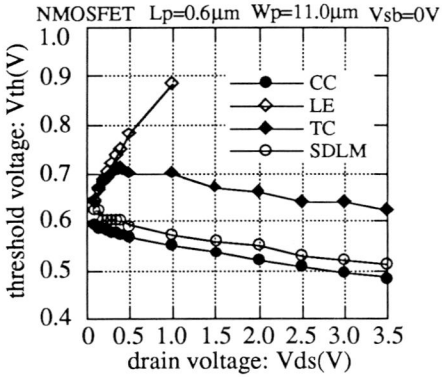

Fig. 1. Comparison of the threshold voltages extracted with previous methods and SDLM for various drain voltages using the measured drain current in the NMOSFET.

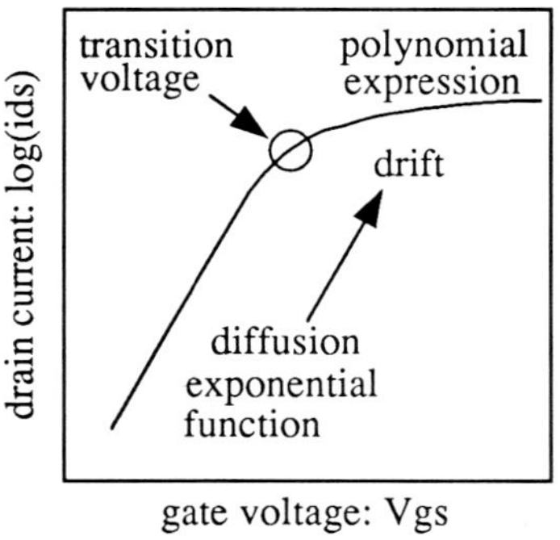

Fig. 2. The idea of extracting the threshold voltage in this work.

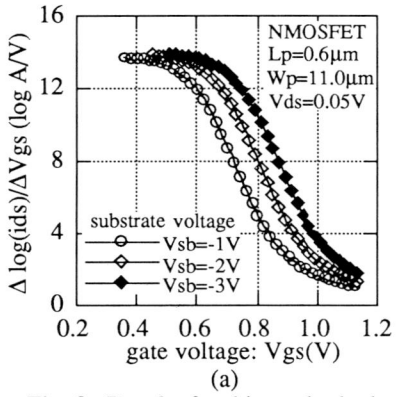

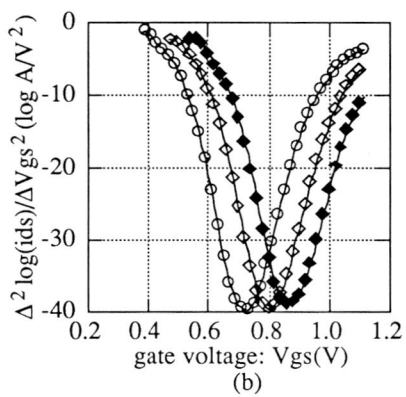

Fig. 3. Results for this method when applied to a submicron NMOSFET for the various substrate voltages. (a) The first difference of the drain current. (b) The second difference of the drain current.

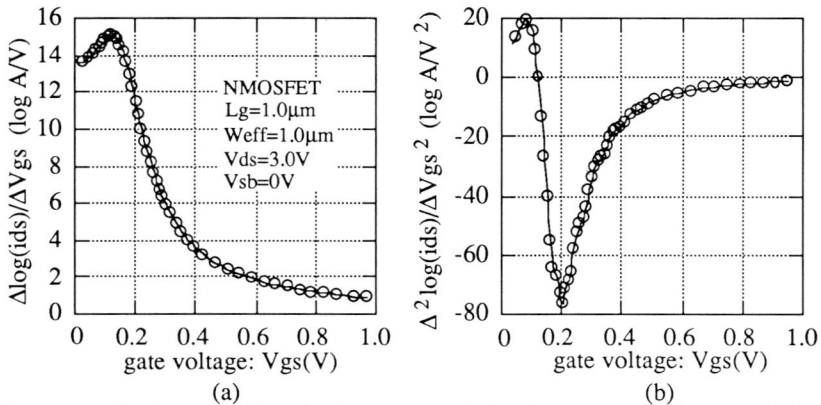

Fig. 4. The derivatives of the drain current calculated from 2D device simulation.
(a) The first difference of the drain current. (b) The second difference of the drain current.

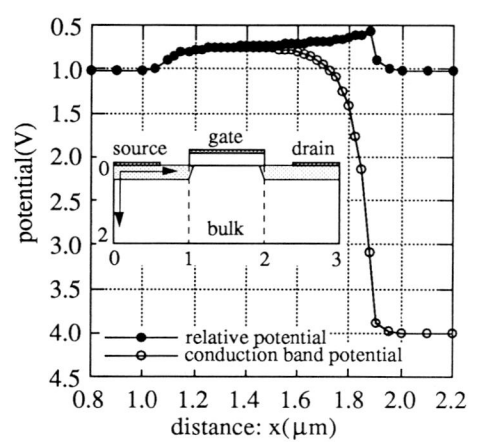

Fig. 5. The conduction band potential and the relative potential distribution at the interface at Vds=3.0V and Vgs=0.2V using 2D device simulation results. The device structure used in the simulation is also shown.

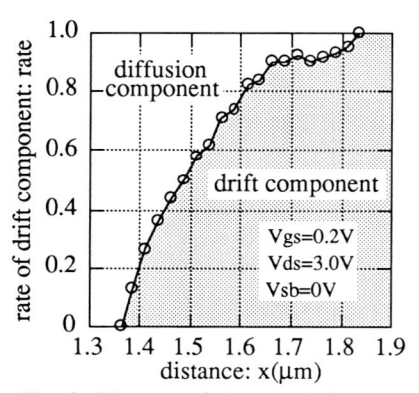

Fig. 6. The rate of the two drain current components in the channel calculated using the conduction band potential and the relative potential at the interface of Si and SiO_2.

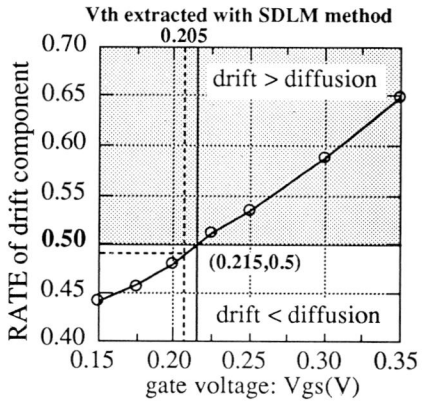

Fig. 7. The RATE of the drift current component at the interface for various gate voltages. At Vgs=0.215V, the drift component is equal to the diffusion component.

2-D MOSFET Simulation by Self-Consistent Solution of the Boltzmann and Poisson Equations Using a Generalized Spherical Harmonic Expansion

W-C Liang, Y-J. Wu, K. Hennacy, S. Singh, N. Goldsman, and I. Mayergoyz

Department of Electrical Engineering,
University of Maryland,
College Park, MD 20742 USA

Abstract

MOSFET simulation is performed by direct self-consistent solution of the Boltzmann, Poisson and Hole Continuity equations. To formulate the Boltzmann equation, a spherical harmonic approach has been developed which allows for expansion to arbitrarily high order. The self-consistent 2-dimensional MOSFET simulations incorporated four spherical harmonics. Simulation results provide the distribution function for the entire device, as well as substrate current, and average quantities including electron temperature, average velocity, and carrier concentration.

1. Introduction

The Spherical Harmonic approach to solving the Boltzmann equation is being demonstrated as a viable approach to device simulation[1, 2, 3]. We present here a new 2-D MOSFET simulation tool which employs a spherical harmonic expansion to deterministically solve the Boltzmann transport equation (BTE). The unique aspects of the approach are: (i) The spherical harmonic formulation of the BTE is performed to arbitrarily high order; (ii) A Scharfetter-Gummel type discretization has been developed for the BTE; (iii) Self-Consistent solution of the BTE, Poisson and Hole Continuity equations is achieved for the 2-D MOSFET structure; (iv) Results provide the electron distribution function, electrostatic potential and hole-concentration for the entire device; (v) From the distribution function, average quantities including electron temperature, average velocity, carrier concentration, as well as substrate current resulting from impact ionization, are obtained.

2. The Device Model

Our simulator is based on the following device model which consists of the Poisson equation, the BTE for electrons and the current-continuity equation for holes:

$$\nabla_{\mathbf{r}}^2 \phi(\mathbf{r}) = \frac{e}{\epsilon_s}[n(\mathbf{r}) - p(\mathbf{r}) + N_A(\mathbf{r}) - N_D(\mathbf{r})] \quad (1)$$

$$\frac{1}{\hbar}\nabla_{\mathbf{k}}\varepsilon \cdot \nabla_{\mathbf{r}}f(\mathbf{k},\mathbf{r}) + \frac{e}{\hbar}\nabla_{\mathbf{r}}\phi(\mathbf{r}) \cdot \nabla_{\mathbf{k}}f(\mathbf{k},\mathbf{r}) = \left[\frac{\partial f(\mathbf{k},\mathbf{r})}{\partial t}\right]_c \quad (2)$$

$$\nabla_{\mathbf{r}} \cdot [\mu_p p(\mathbf{r})\nabla_{\mathbf{r}}\phi(\mathbf{r}) + \mu_p V_t \nabla_{\mathbf{r}}p(\mathbf{r})] = R(\phi, n, p) \quad (3)$$

where: $f(\mathbf{k},\mathbf{r})$ is the distribution function; $n(\mathbf{r}) = \frac{1}{4\pi^3}\int f(\mathbf{k},\mathbf{r})d\mathbf{k}$ is the electron concentration; $p(\mathbf{r})$ is the hole concentration; $\phi(\mathbf{r})$ is the potential; We incorporate a nonparabolic band-structure, as well as acoustic, optical and intervalley phonon scattering, and impact ionization.

3. Formulation of the 2-D BTE to Arbitrarily High-Order

For steady-state 2-D MOSFET simulation, the BTE is a 5-dimensional integro-differential equation, and is therefore extremely difficult to solve. Using the SH expansion method, the BTE is reduced into a 3-dimensional system of differential-difference equations which is tractable for MOSFET simulation. We also employ the Hamiltonian transformation[3]. In contrast to other recent works, which were based on a low order SH expansion[1, 3], we have generalized the expansion approach and formulated the BTE to arbitrarily high-order SH accuracy[2]. With this approach, the momentum distribution function is expressed in terms of an infinite series of spherical harmonics: $f(\vec{r}, \vec{k}) = \sum_{l=0}^{\infty}\sum_{m=-l}^{l} f_l^m(\vec{r}, \varepsilon) Y_l^m(\theta, \phi)$. Where $f_l^m(\vec{r}, \varepsilon)$ represent the unknown expansion coefficients; and the spherical harmonics basis functions $Y_l^m(\theta, \phi)$ provide the angular dependence of the distribution function. To determine the coefficients, $f_l^m(\vec{r}, \varepsilon)$, we first substitute the above summation into the BTE. Next, we project the BTE onto each of the SH basis functions:

$$\int_0^{2\pi} d\phi \int_{-1}^{1} d(\cos\theta) Y_l^{m*}(\theta, \phi) \left\{\frac{\partial f}{\partial t} + \vec{v}\cdot\frac{\partial f}{\partial \vec{x}} + (-q)\vec{E}\cdot\frac{\partial f}{\partial \vec{p}} - \left[\frac{\partial f}{\partial t}\right]_c\right\} = 0 \quad (4)$$

By performing a similar projection onto each of the SH basis functions, an infinite system of coupled equations is generated for the unknown coefficients. We then take advantage of the SH recurrence relations to facilitate performing the projections indicated by Eqn. (4). The interesting result is that almost all of the infinite terms in each equation vanish identically due to orthogonality, and the coupling between equations is only through neighbors. Another extremely useful result is that each equation has an identical form. The system can therefore be automatically generated to arbitrarily high order and then be solved numerically. The equation for the l, m SH coefficient is given below. The equation for any of the other SH coefficients is obtained by simply appropriately changing the value of the indices l, m to other allowed integers:

$$\left\{\sum_{i=1}^{2} v(\varepsilon)\left[\frac{\partial}{\partial x_i} - eE_i\left(\frac{\partial}{\partial \varepsilon} - \frac{l-1}{2}\frac{\gamma'}{\gamma}\right)\right]\hat{a}_i^+ + v(\varepsilon)\left[\frac{\partial}{\partial x_i} - eE_i\left(\frac{\partial}{\partial \varepsilon} + \frac{l+2}{2}\frac{\gamma'}{\gamma}\right)\right]\hat{a}_i^-\right\}f_l^m$$

$$= \left[\frac{\partial f_l^m}{\partial t}\right]_c \quad (5)$$

where $v(\varepsilon) = \sqrt{2m\gamma}/m\gamma'$, $\gamma' = d\gamma/d\varepsilon$, and γ represents the dispersion relation; the sum is over the 2 directions in the $x-z$ plane; and we have constructed the operators $\hat{a}$ to relate nearest-neighbor coefficients with the following definitions:

$$\hat{a}_1^+ f_l^m \equiv \tfrac{1}{2}\left\{-\alpha_{l-1}^m \alpha_l^m f_{l-1}^{m-1} + \alpha_{l-1}^{-m}\alpha_l^{-m} f_{l-1}^{m+1}\right\}$$
$$\hat{a}_1^- f_l^m \equiv \tfrac{1}{2}\left\{\alpha_{l+1}^{-m+1}\alpha_l^{-m+1} f_{l+1}^{m-1} + \alpha_{l+1}^{m+1}\alpha_l^{m+1} f_{l+1}^{m+1}\right\}$$
$$\hat{a}_2^+ f_l^m \equiv \alpha_{l-1}^{-m+1}\alpha_l^m f_{l-1}^m; \qquad \hat{a}_2^- f_l^m \equiv \alpha_{l+1}^m \alpha_l^{-m+1} f_{l+1}^m; \qquad \alpha_l^m = \sqrt{\tfrac{l+m}{2l+1}}$$

4. Numerical Approach to 3-D Problem

To solve the SH-expanded BTE for a 2-D MOSFET we use eqn. (5) to generate equations for the first 4 spherical harmonics. Next, we reduce these 4 equations into one second order self-adjoint differential-difference equation similar to that of[3]. We then perform a Scharfetter-Gummel type discretization on the resulting equation. This yields a matrix which is well conditioned. We overcome problems typically associated with 3-dimensional calculations by using a fixed point SOR iterative solution technique. This method avoids direct solution of large matrix equations (and is easily parallelized). The Poisson equation is solved directly in 2-D using Gaussian elimination for banded matrices. The Hole-Continuity equation is solved using Slotboom variables and the fixed point iterative method. The Boltzmann-Poisson-Hole-Continuity system is solved self-consistently using a decoupled Gummel-type iterative process.

5. Simulation Results for a 2-D MOSFET

We performed example calculations on a $0.5\mu m$ channel length nMOSFET. We show example results for an applied bias of $V_{ds} = 3V$ and $V_{gs} = 3V$. Fig. 1 shows the resulting electrostatic potential which has been calculated self-consistently with the distribution function. In Fig. 2, the electron concentration within the device, which is obtained by numerical integration of the distribution function, has been plotted. Fig. 3 shows the energy distribution function along a plane at $0.001\mu m$ below the Oxide/Si interface. It can be easily seen, by the distribution function's non-Maxwellian form, that electrons were heated up when traveling toward the drain region. This quantifies the hot-electron concentration at the channel/drain edge, which is the region that generates reliability problems. Fig. 4 shows the distribution function along a plane in the substrate. Clearly, the distribution function is Maxwellian at this depth in the device since current densities and fields are very small. Fig. 5 shows the concentration of electrons and holes generated per second by impact ionization. These values are obtained by directly integrating the impact-ionization collision term in the BTE. Fig. 6 shows the substrate current as a function of applied bias. The curves were obtained by integrating the impact-ionization collision term over the entire energy and real space domain.

References

[1] H. Lin, N. Goldsman, and I. D. Mayergoyz, "Device Modeling by Deterministic Self-Consistent Solution of Poisson and Boltzmann Transport Equations," *Solid-State Electronics*, vol. 35, no. 6, pp. 769–778, 1992.

[2] K. Hennacy, *Spherical Harmonic and Effective Field Formulations of Boltzmann's Transport Equation: Case Studies in Silicon*. PhD thesis, University of Maryland, 1994.

[3] A. Gnudi, D. Ventura, G. Baccarani, and F. Odeh, "Two-Dimensional MOSFET Simulation by means of a Multidemensional Spherical Harmonics Expansion of the Boltzmann Transport Equation," *Solid-State Electronics*, vol. 36, p. 575, 1993.

fig.1: Electric potential

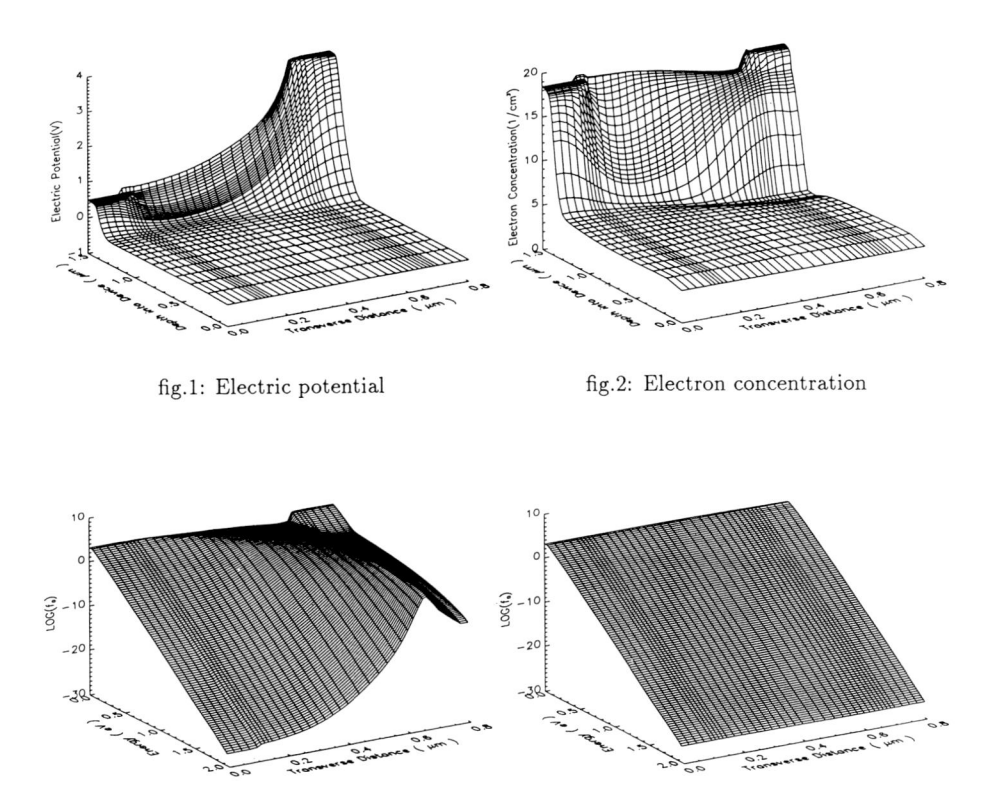

fig.2: Electron concentration

fig.3: Distribution function along the channel

fig.4: Distribution function in substrate

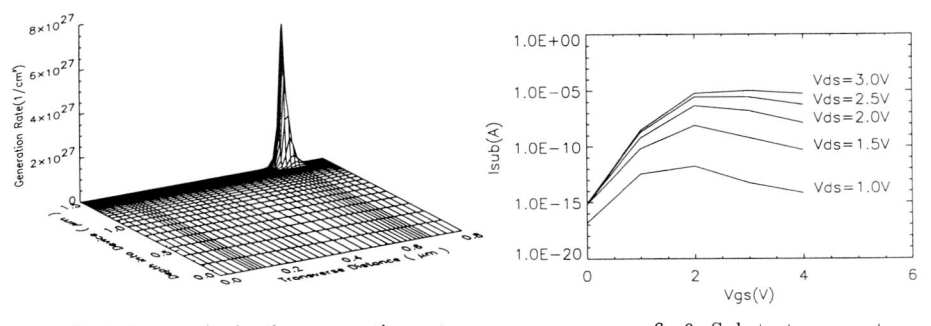

fig.5: Impact ionization generation rate

fig.6: Substrate current

Ultra High Performance, Low Power 0.2 µm CMOS Microprocessor Technology and TCAD Requirements

A. Nasr, J. Faricelli, N. Khalil, and C.-L. Huang

Ultra Large Scale Integration Operations Group
Digital Semiconductor
77 Reed Road
Hudson, MA. 01749 U.S.A
(phone: 508-568-5742)

Abstract

The process requirements for low power, ultra high performance CMOS Digital IC's are reviewed. The role, accuracy, and demands of TCAD for future process and device design are discussed. In addition, special challenges for predictive simulation of process and device behavior are highlighted. Two illustrative examples are presented. First, the interaction between the polysilicon gate process architecture and the resulting device characteristics is analyzed in details. Second, subthreshold leakage current (Id_{off}) prediction for devices with 0.2 µm length and below is addressed. A TCAD based inverse modeling strategy is proposed to enhance the predictability of sub-quarter micron CMOS device behavior.

1. Introduction

Ultra high performance, low power CMOS microprocessors are mainstream technology drivers. Operating frequencies up to 1000 Mhz and V_{dd} below 2 V will become a reality by the end of the century. The unprecedented speed of Digital's Alpha microprocessors is due in large part to the following factors: optimized CMOS transistors which balance drive current vs. reliability; process and circuit techniques which consider cost, complexity, yield and reliability; and state-of-the-art TCAD tools that reduce learning cycles.

As the cost of semiconductor fabrication facilities reaches the billion dollar mark, any tool which helps shorten the time between development and product introduction will greatly enhance return on investment. Accurate and dependable TCAD tools play a major role in achieving this goal throughout the product lifetime. During process design and layout rules definition, TCAD tools are used to investigate electrical sensitivity to various conditions. During the reliability optimization phase, simulators are critical for drain engineering and evaluation of hot-carriers reliability. TCAD tools are also used during manufacturing to enhance process robustness to fluctuations due to equipment conditions.

The progress of ULSI technologies to yield higher densities, ultra-high performance and low power chips has brought with it the need to include physical factors previously ignored in TCAD tools. For instance, the reduced channel length and gate oxide thickness have made simple threshold voltage prediction a challenge. This is mainly due to polysilicon gate depletion and dopant penetration from the gate into the silicon substrate. Furthermore, as the threshold voltage is scaled along with the power supply, off-state leakage current Id_{off} prediction and suppression become more critical. Maintaining a high Id_{on}/Id_{off} ratio as the technology and V_{dd} are scaled is of paramount importance.

One reason why the predictive capability of TCAD tools has lagged process development is that our ability to measure what we have made in the fab is limited to large structures with resolution in only one space dimension, or involves time consuming and destructive sample preparation. Without detailed characterization, there is little hope that the models in TCAD tools can be improved. This paper proposes a methodology for improving TCAD tool accuracy and predictive capabilities by a *bootstrapping* technique based on inverse modeling. This allows the extraction of physical and *inaccessible* parameters such as diffusion coefficients in multi-layer films. The same technique is also used to extract the two-dimensional (2D) doping profile which can then be used to predict short channel effects and Id_{off} currents.

2. High Performance and Low Power Requirements

Since the middle of the 1980's, Digital's CMOS technology goal has been characterized by doubling the gate density and improving the gate speed by up to 30% with each successive generation. Fig. 1 shows the maximum operating frequencies and power dissipation as a function of power supply and year of market introduction for a range of high performance Alpha microprocessors and low power portable systems. Fig. 2 shows the normalized saturation current as a function of gate oxide thickness, channel length and power supply.

The interactions between CMOS power dissipation, clock frequency, gate delay and current drive capability are well characterized. Reducing the power supply to lower power consumption will negatively impact the drive current capability and performance. In order to maintain high performance, it is necessary to carefully optimize the gate oxide thickness, threshold voltage and channel length, subject to certain reliability constraints such as hot carrier lifetime and time dependent dielectric breakdown under worse case operating conditions.

The Alpha microprocessor [1] technology uses 3.3 V with a 85 Å gate oxide and 2 V with a 65 Å oxide. Complementary n and p-type implanted amorphous silicon (α-Si) gate interconnect is used for both technologies. Deep ultra violet lithography and anti-reflective coatings are used for good control of channel length down to an L_{eff} of 0.15 μm. The lightly doped and ultra shallow phosphorous or arsenic regions provide both low Id_{off} characteristics and robust hot carrier reliability. BF_2 is used for shallow p-channel device junction and good device characteristics. Rapid thermal annealing is used for the source and drain anneal. A $CoSi_2$ salicide process with sheet resistance below $5\omega/\square$ is used for low resistance active area and gate interconnect.

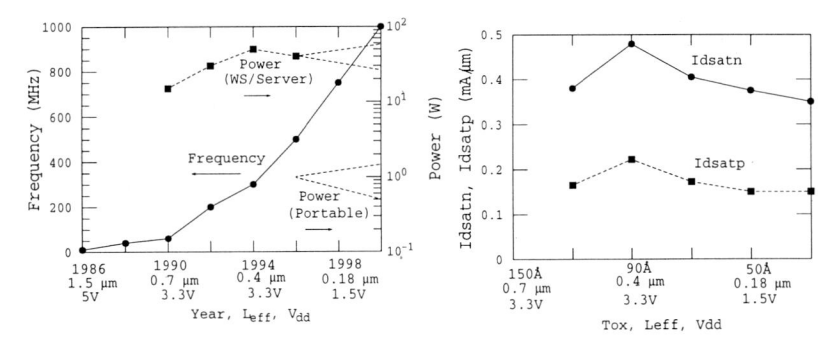

Figure 1: Operating frequency and power dissipation vs power supply and year of introduction.

Figure 2: Normalized saturation currents vs t_{ox}, L_{eff}, and V_{dd}.

3. Channel Simulation Issues and TCAD Requirements

3.1. Silicon Gate Interconnect Process Architecture

Thin gate oxides below 80 Å which are used to provide higher current gain at low power supply, are susceptible to dopant penetration during the source and drain implant. The RTA required for the formation of shallow junctions for good short channel effects, good drain-induced barrier lowering characteristics and low Id_{off} gives rise to a non-uniform dopant distribution in the polysilicon gate. The low concentration at the gate/SiO$_2$ interface accentuates poly depletion effects. This phenomenon results in higher threshold voltage, lower inversion capacitance and reduced current drive capability.

In the process of optimizing 0.2 μm gate length transistor performance for $V_{dd} = 2$ V, the p-channel device design was found to be particularly challenging for TCAD prediction. It was observed that the gate oxide thickness (t_{ox}), silicon gate interconnect deposition temperature and gate film thickness (t_{poly}) were all interrelated in determining the long channel device threshold. Conventional polysilicon gate interconnect, deposited at 620°C, reduced the grain size, as shown in the transmission electron micrograph (TEM) in Fig. 3. Alternatively, silicon gate material deposited at a temperature of 530°C increased the grain size, as shown in the TEM of Fig. 4 where the average grain size is 2000Å. This reduced the boron diffusion and eliminated boron penetration through the gate oxide.

The implanted profile for BF$_2$ (2×10^{15}, 40 keV) into the polysilicon gate at 530°C and 620°C is shown in Fig. 5. Note that while polysilicon grown at 620°C results in more uniform doping distribution, we found that it is more sensitive to penetration and V_{th} shift. On the other hand, the doping profile for the α-Si case shows a lower concentration at the Si/SiO$_2$ interface. Figure 6 shows the boron doping profile in the 2200Å and 2800Å gate interconnect material. A more uniform doping profile is observed for thinner polysilicon.

3.2. Silicon Gate Interconnect Device Analysis.

Figure 7 shows the impact of polysilicon depletion and boron penetration on the long-channel MOSFET C–V characteristics for devices with 2000Å and 2800Å t_{poly}.

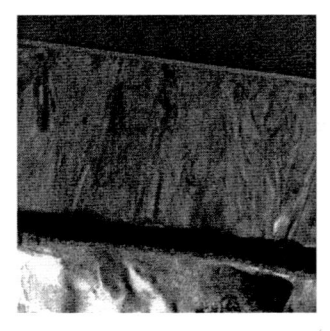

Figure 3: TEM of hi temperature (620°C) polysilicon gate interconnect (Grain size < 300 Å).

Figure 4: TEM of low temperature (530°C) α-Si gate interconnect (Grain size 2000 – 3000 Å).

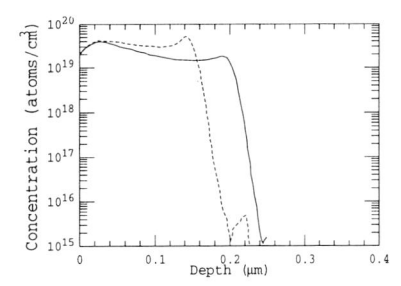

Figure 5: SIMS of BF_2 implant with 2000 Å (dashed) and 2800 Å (solid) α-Si.

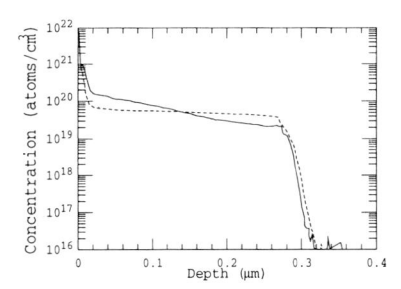

Figure 6: SIMS BF_2 comparison in α-Si (dashed) and polysilicon interconnect (solid).

The processing conditions and extracted parameter values for both Devices A (solid line) and B (dashed line) are given in Table I. Note that as t_{poly} is reduced, the polysilicon concentration, N_p, increases. The increase in N_p results in substantial improvement in the inversion capacitance for Device A (see Fig. 7). The major drawback of thinning t_{poly}, however, is the penetration of implanted specie (boron in this case) from the gate to the channel region, thus affecting the threshold voltage stability [2].

Since t_{poly} has such dramatic impact on the polysilicon depletion and boron penetration for the p-channel MOSFET with a thin gate oxide, more careful study of the source/drain implant condition is in order. Figure 8 shows comparison of MOSFET C–V degradation for three implant conditions. The process conditions for these devices are also listed in Table I as Device C (long-short dashed line), D (solid line) and E (short dashed line). It is important to note that by changing implant condition from Devices C to E, the degradation of C–V characteristics in the inversion regime increases approximately 5%. In addition, we note that, despite the increase in the N_p concentration for Device C, the boron penetration is still invisible in Fig. 8. This, however, is not the case for Device A.

We notice that by changing the t_{poly} from 2200Å (Device C) to 2000Å (Device A), both N_p and boron penetration increase. In other words, the sensitivity of

Table I: Comparison of extracted parameter values for Devices A to E.

Devices	A	B	C	D	E
t_{ox} (Å)	80.2	80.2	83.4	83.3	83.4
t_{poly} (Å)	2000	2800	2200	2200	2200
P$^+$ S/D Dose	2.5×10^{15}	2.5×10^{15}	2.5×10^{15}	2.5×10^{15}	2×10^{15}
P$^+$ S/D Energy	40	40	40	35	35
N_p ($10^{19}\times$ cm^{-3})	2.4	1.1	1.4	1	0.85

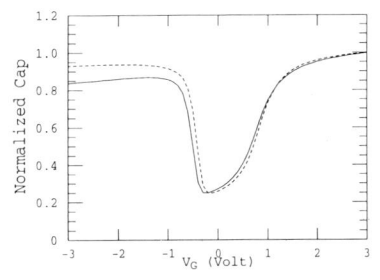

Figure 7: Gate capacitance as a function of polygate thickness (solid line – Device A and dashed line – Device B).

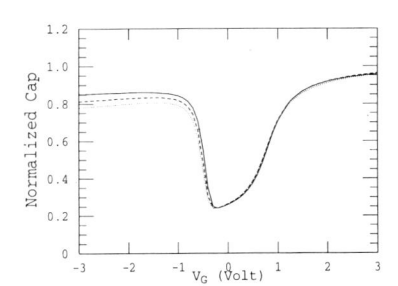

Figure 8: Gate capacitance as a function of S/D implant conditions (long-short dashed line – Device C, solid line – Device D and dashed line – Device E).

polysilicon depletion and boron penetration can be affected by a mere 200 Å t_{poly} difference. It, therefore, presents a tremendous challenge in designing p-channel MOSFET with conventional gate oxide that have sufficiently high N_p and without boron penetration.

3.3. P-Channel Inverse Modeling

One-dimensional simulation using SUPREM-III [4] of the above mentioned process conditions cannot consistently reproduce the observed electrical device behavior. On the other hand, an inverse modeling characterization methodology was successful in determining the profiles as explained in the following.

The average gate polysilicon doping concentration was first determined by matching simulated and experimental capacitance data in the inversion region. Thereafter, the one-dimensional doping profiles in the channel region were extracted using inverse modeling from deep depletion capacitance data [5]. In Fig. 9, the two extracted profiles corresponding to 2000Å and 2800Å t_{poly} are shown. The effect of boron penetration is apparent as a reduction in the net doping concentration near the Si/SiO$_2$ interface.

Using the extracted profiles, and the average polysilicon doping, the quasi-static C–V characteristics were simulated. By comparing the simulated and experimental measurements, it was found that for the device with boron penetration, the simulation cannot reproduce the experimental threshold voltage shift. Indeed, as a result

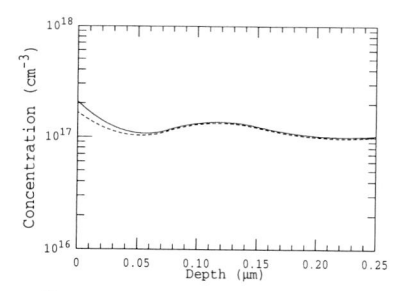

Figure 9: Comparison of channel doping profile for device with 2800Å (solid) and 2000Å (dashed) t_{poly}.

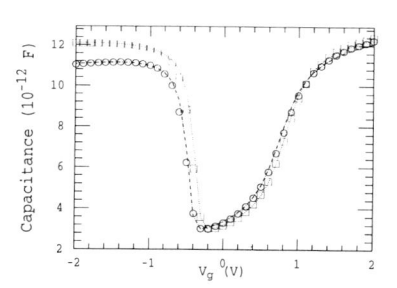

Figure 10: simulated (solid) and experimental (symbols) quasi-static C–V device with 2800Å (solid) and 2000Å (dashed).

of low, non-degenerate doping, modeling the resultant nonuniform polysilicon doping is essential for accurate device simulation. By the same token, one can use the experimental C–V data to determine the polysilicon doping profile.

The spline profile parameterization [5] was extended to represent the polysilicon doping profile. The values of the B-splines coefficients can then be extracted from experimental quasi-static data. Using this approach the polysilicon doping is determined, and an excellent fit of the experimental quasi-static C–V data is achieved as shown in Fig. 10.

It is noted that the procedure described above to determine polysilicon and channel doping profiles for various processing conditions can provide the required data for *tuning* SUPREM-III diffusion coefficients.

4. Subthreshold Simulation and TCAD Requirement

4.1. 2D physical characterization

Two of the most important quantities that must be characterized for accurate simulation of a deep submicron MOSFET are the polysilicon gate length and the dopant profile. In this discussion, we focus on 2D effects, and ignore 3D effects such as the narrow width effect on threshold voltage. In any process, but especially one under development, there can be considerable variation from the "as-drawn" poly gate length to any given poly gate on the wafer. When analyzing the characteristics of any particular transistor from a distribution of MOSFETs, it is imperative to know the actual gate length. In earlier process development efforts, we used techniques such as cross-sectional SEM to quantify the gate length. This technique, while accurate, is time consuming and can only be performed on limited numbers of devices. It also results in destroying the device under test. An alternate technique we have devised uses the measurement of the device gate capacitance in inversion to infer the gate length [6]. We first extract the gate oxide thickness t_{ox} from capacitance-voltage (C–V) measurements on a large area MOS capacitor. While extracting t_{ox}, we also extract the average polysilicon doping near the oxide interface. We then use the MINIMOS [7] device simulator to simulate the inversion capacitance of the short channel device. We vary the simulated gate length until a match is achieved with the measured capacitance. We take great lengths to include effects which would alter the inversion capacitance, such as polysilicon depletion, quantum mechanical effects important for thin t_{ox}, and various fringing capacitances [8]. This technique has been verified using cross-sectional SEM, and has shown to be effective in extracting gate lengths down to 0.1 μm.

In the area of dopant profiling, we have had considerable success in using the inverse modeling technique to extract the 2D doping profile [9]. The technique is described in the reference, but a brief description is given here for completeness. The poly gate length extraction technique described above provides the t_{ox} and polysilicon dopant concentration for the device. The channel dopant for a long channel device is extracted from deep depletion capacitance measurements. In the source/drain regions far from the gate, SIMS measurements of the source/drain dopants are used along with the area capacitance of the source/drain diode to extract the complete profile in that device region. These profiles are combined with the channel dopant profile to form an initial guess at the 2D profile. Capacitance measurements which "probe" various parts of the profile, such as gate capacitance vs. gate voltage in accumulation, are used as inputs to the 2D extraction. A device simulator (in our case, MINIMOS) is used to simulate these capacitances. A non-linear optimizer is used to determine the dopant profile which produces simulated capacitances that best fit the measured ones. A key feature of the technique is the tensor product splines representation of the 2D dopant profile which allows a completely general form for the dopant distribution, while reducing the description of the profile to a handful of coefficients. This greatly reduces the amount of work the optimizer must do compared to a straightforward mesh representation.

The results of such an extraction are shown in Fig. 11. We note that the dopant extracted in the source/drain extensions under the gate are relatively immune to small changes in the measured data or other conditions in the extraction procedure. The extracted channel dopant, on the other hand, is sensitive to assumptions about the amount of fixed charge at the Si/SiO_2 interface, or the value of the polysilicon workfunction. One strategy we have recently begun to use is to assume a small fixed interface charge and use a polysilicon workfunction that includes bandgap narrowing effects. We then let the channel dopant adjust itself to fit the measured threshold.

4.2. Id_{off} simulation issues

One important design criteria of n-channel MOSFET is the "off" current Id_{off}. Maintaining a low Id_{off} becomes increasingly difficult as device thresholds are reduced below 0.5V. Careful design of the inner source/drain junction is required to reduce short channel effects. We use two-dimensional process simulation to help design the inner junction, but many physical effects, such as boron depletion from the channel due to adjacent source/drain dopants [10], are not yet included in commercial simulators. Using the inverse modeling techniques described in the previous section, we can get some idea of what the source/drain and channel profiles look like. We can then feed the results into a device simulator and see if the correct threshold and Id_{off} behavior is predicted. This further helps validate our extraction procedure. Thus, even if we cannot yet do predictive process simulation from first principles, we can at least understand the impact of process conditions on device characteristics.

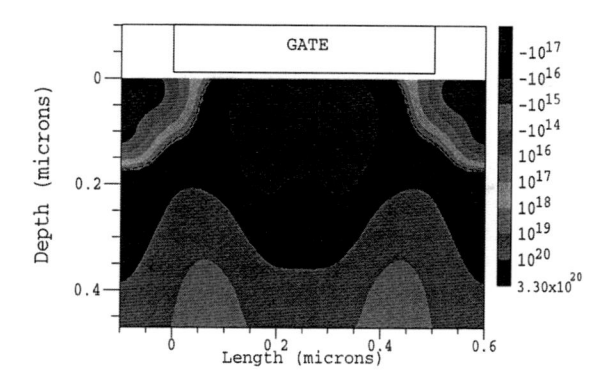

Figure 11: 2D extracted profile.

In Fig. 12, we show a plot of n-channel V_{th} vs. L_{eff} for a phosphorous moderately doped drain (MDD) structure, using a 7 degree implant in our 0.35 μm process. The figure also shows the simulated threshold voltage using a two dimensional doping profile extracted from capacitance data described in the previous section. The extraction was done on a test structure which had a gate length of 0.47 μm. We then simulated smaller channel lengths by simply scaling the 2D profile, as we had done for earlier technologies. As can be seen from the figure, scaling the profile breaks down at channel lengths below L_{eff} of 0.3 μm.

In Fig. 13, we show a plot of measured n-channel Id_{off} vs. L_{eff}, where Id_{off} is defined as the drain current at $V_{gs}=0V$, $V_{ds}=2.7V$, and 100° C. The figure also shows the simulated Id_{off} (solid line). The simulated Id_{off} is about 2.5-3 times higher than measurement down to $L_{eff}=0.3$ μm. This is reasonable, considering that the Id_{off} is very sensitive to the details of the doping profile. The simulated Id_{off} at $L_{eff}=0.22$ μm, however, is several orders of magnitude larger than measurement. This is expected, since the simulated threshold voltage is 120mV lower than measurement. An interesting observation is that not all of the increase in Id_{off} can be attributed to the inaccuracy in the threshold. Indeed, simulated Id_{off} at $L_{eff}=0.22$ μm with the threshold adjusted to match the measured data is still over an order of magnitude larger than measurement.

One can conclude that the doping profile changes considerably when the device length varies only a few tenths of microns. Thus, our initial methodology of scaling a profile extracted from one gate length is not appropriate for deep submicron MOSFET's. A better methodology is to use test structures that span the range of expected channel lengths from "short" to "long" and extract profiles for each channel length. We are in the process of designing a test chip with such features. This technique can be used to improve process simulators physical models which will then allow better prediction of subthreshold characteristics.

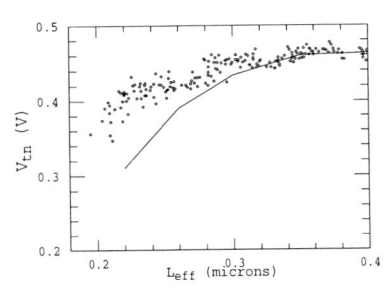

Figure 12: V_{th} v/s L_{eff} with 65 Å t_{ox}.

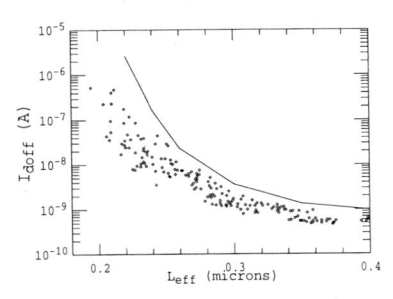

Figure 13: Id_{off} v/s L_{eff} with 65 Å t_{ox} (V_{gs}=0V, V_{ds}=2.7V, and 100° C).

5. Conclusion

We have outlined a strategy to improve TCAD prediction for sub 0.2 μm devices. Examples of polysilicon depletion, boron penetration and Id_{off} prediction requirement were addressed in details. Inverse modeling coupling with TCAD tools was proposed as a means to calibrate physical process parameters to device electrical characteristics.

References

[1] D. Dobberpuhl et al., *IEEE Int'l Solid State Circuits Conf.* p. 106, 1992.
[2] J.R. Pfiester et al., *IEEE Trans. Electron Devices* Vol. 37, p. 1842, 1990.
[3] G.Q. Lo et al., *IEEE Elect. Dev. Letters* vol. 12, p. 175, 1991.
[4] C.P. Ho and S.E. Hansen, *SUPREM-III, Tech. report 83-001*, Stanford, 1983.
[5] N. Khalil et al., *IEEE Elect. Dev. Letters*, Vol. 16, p. 17-19, 1995.
[6] C.-L. Huang et al., submitted to *IEEE Trans. Elect. Dev.*, 1995.
[7] S. Selberherr, et al., *IEEE Trans. on Elect. Dev.*, vol. 27, p. 1540, 1980.
[8] R. Rios and N. Arora, *IEDM Techn. Digest*, p.613-616, 1994.
[9] N. Khalil et al., *Accepted for presentation at ESSDERC'95*.
[10] H. Hanafi et al., *IEEE Elect. Dev. Letters*, Vol 14, No. 12, p.575-577, 1993.

Viscoelastic Modeling of Titanium Silicidation

Stephen Cea and Mark Law

Department of Electrical Engineering
University of Florida
Gaineville, FL 32611-6200
smc@tcad.ee.ufl.edu, law@tcad.ee.ufl.edu

Abstract

This paper describes titanium silicidation modeling using the process simulator FLOOPS. Erpimental work shows that viscoelastic models are required to fit the observed behavior. The capability to simulate a wide range of silicide examples are shown, but more experimental evidence is needed to calibrate the models.

Introduction

Titanium silicides are widely used in the processing of vlsi and ulsi circuits but there are still very few programs that model the growth of $TiSi_2$. In two dimensions, none of the programs to date can model both the thinning at the spacer edge and the poly smile. This abstract describes the two dimensional viscoelastic model for titanium silicide growth in the process simulator FLOOPS. The model is applied to wafer curvature data to begin to determine the mechanical properties of $TiSi_2$. The type of structures that FLOOPS can simulate are also included.

Titanium Silicide Growth Models

The model has two parts. First, we solve for the diffusion of silicon through the $TiSi_2$ to react at the titanium interface. Then we solve the flow due to the volume change of the reaction. Solving for the diffusion of silicon is done using a deal grove like reaction diffusion model[1,2,3,4] Stress dependencies in this model can be included for diffusivity, reaction rates and the mass transfer of silicon into the silicide. Which if any of these stress dependencies is correct is still under investigation.

The material deformation and flow is modeled with a viscoelastic solver which is also used for simulating oxidation. Fornara[4] used an elastic model to solve for the flow. In an earlier work[1], we used a viscous incompressible model. At the temperatures involved, 650°-800°C, oxide, polysilcon and titanium are dominated by elastic behavior. The correct model for $TiSi_2$ is still under investigation, but it is clear that there is some stress relaxation taking place. The viscoelastic solver used here is able model both elastic and viscoelastic behavior.

Viscoelastic Calibration

The Maxwell viscoelastic model consists of a viscous element and an elastic element in series. The elastic element is characterized by Young's modulus (E) and Poisson's ratio (v). Jongste et al.[4] have measured these values for C49 $TiSi_2$ using wafer curvature measurements to be E=142 Gpa and v=0.27. The viscous element relieves the stress stored the elastic element.

In two wafer curvature studies Jongste et al.[5,6] saw significant stress relaxation. The papers present data from in-situ stress measurements during RTA of titanium and silicon multilayers. In one study[5], the Si/Ti ratio of the multilayers was varied from 1.9, 2.1 and 2.4. Stress as a function of time at a constant temperature was plotted in this study. There was relaxation for all the compositions but more relaxation for Si/Ti ratios > 2. Using this data we can start to calibrate the viscosity in our model by simulating the wafer curvature. This is done by simulating the whole wafer with the viscoelastic solver. Figure 1 shows relaxation data at 525° C for a Si/Ti ratio of 2.1, along with some simulated curves. To fit the amount of relaxation with a linear viscosity requires an extremely low value of approximately 1.5e13 poises. The curve of the simulation is incorrect and the stress over relaxes. To try and better fit the data an Eyring stress dependent viscosity was used. There are two parameters needed for this model - the low stress viscosity and the nonlinear activation volume. Unfortunately neither of these parameters is known. A low stress viscosity of 1e20 poise is picked and the nonlinear activation is varied to try and fit the data. Obviously there are many choices for the low stress viscosity and more experimental evidence is needed. There are two other interesting things to note about this data. One is that the viscosity needed to fit a Si/Ti ratio of 1.9, ~1e16 at 625 C, while higher than Si/Ti ratio > 2 is still much lower than the viscosity of oxide at these temperatures. Second is that the method of relaxation seems to be the diffusion and precipitation of Si. This will probably affect the growth kinetics. One note on the applicability of this data to silicides grown from deposited Ti on substrate or poly silicon is that Norstrom et al.[7] found chains of silicon precipitates in the middle of narrow lines that were silicided with titanium. The lines exhibited the characteristic bowing seen on narrow lines. More work needs to be done but it is clear that a viscoelastic model is needed to accurately model the stress relaxation seen in $TiSi_2$ experiments.

Salicides and Local Interconnect Examples

We are currently able to simulate most of the interesting structures with titanium silicides. As our calibration of the viscoelastic model progresses we can begin to use these parameters on real world structures like salicides and local interconnects. This should allow us to begin to calibrate the stress dependencies in the diffusion reaction model. Figure 2 a shows a linear viscoelastic simulation of a salicide structure in a nitrogen ambient. The growth of titanium nitride is also modeled. Figure 2 b shows titanium silicide local interconnect over a field isolation region. Both of these figures do not yet match with experimental evidence. The purpose is to demonstrate the range of simulations FLOOPS can perform and how this can be used to simulate a wide range of silicide structures. The user is not required to provide seed or native meshes for either the $TiSi_2$ or TiN. When annealed these meshes are inserted into the structure.

Conclusion

A nonlinear viscoelastic solver has been applied to modeling titanium silicide experimental results. It is found that the stress relaxation is significant and the viscous component of $TiSi_2$ can't be ignored. The ability to simulate real world structures has also been shown.

References

1. S. Cea et al., NUPAD V Proceedings, 1994.
2. L. Borucki et al., IEDM, 1988, p 348.
3. C. M. Li et al., VPAD Conference Proceedings, 1993.
4. P. Fornara, ESSDERC Proceedings, 1994.
5. J. F. Jongste et al., Applied Surface Science, 1991, p. 212.
6. J. F. Jongste et al., J. Appl. Phys., 1993, p. 2816.
7. H. Norstrom, MRS Symp. Proc. Vol 182, 1990, p. 71.

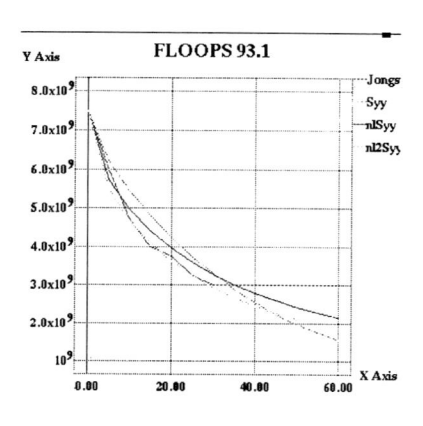

Figure 1: Stress relaxation data from Jongste[1] and simulation results: linear viscosity (S_{yy}) 1e13 poises, nonlnear viscosity (nlS_{yy}) u=1e16 poises Vd=65e4 A^3 and ($nl2S_{yy}$) u=1e20 poises Vd=1.1e10 A^3.

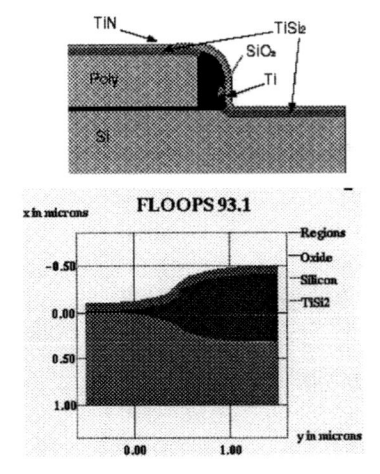

Figure 2: Silicide examples a) salicide and b) local

Multidimensional Nonlinear Viscoelastic Oxidation Modeling

Stephen Cea and Mark Law

Dept. of Electrical Engineering
University of Florida
Gainesville, Fl. 32611-2044
smc@tcad.ee.ufl.edu, law@tcad.ee.ufl.edu

Abstract

This paper describes the oxidation module for the process simulator FLOOPS. Simulation results for silicon oxidation in two and three dimensions are demonstrated.

Introduction

The accurate simulation of LOCOS and advanced LOCOS processes for submicron processes is very important. As device dimensions have decreased, the need for three dimensional simulation has become critical. Since FLOOPS is an object oriented simulator, it is easy to include and test different models. We have used two dimensional models to choose the most accurate and efficient methods. Some of these models have been implemented and tested in three dimensions.

Two Dimensional Models

The most advanced two dimensional model is a stress dependent nonlinear viscoelastic model with velocity as the unknown. It is similar to those proposed by Peng[1] and Senez[2]. Element rotation effects are taken into account in the residual stresses as they are updated at each time step. The stress dependencies on oxidant diffusion and reaction rates are from Kao[3] and Sutardja[4]. Eyring plasticity is used to model nonlinear viscosity[4] for oxide, nitride and polysilicon. The simulator uses both velocity pressure and velocity formulations. The velocity pressure technique results in smoother stress contours and better nonlinear convergence when oxide is modeled as incompressible as Peng[1] did. Senez[2] does not enforce incompressibility and gets good results. Due to the fact that this greatly eases the nonlinear convergence while giving accurate results, this is currently the recommended oxidation model in FLOOPS.

The nonlinear iterations are solved with the numerical relaxation technique[5,6]. Including the elastic effect is beneficial for this technique, in fact trying to use this on a viscous incompressible problem is extremely time consuming. Grid quality is critical for this nonlinear iteration scheme, the most important points are to avoid high aspect ratio and unevenly sized triangles. The grid in FLOOPS is not remeshed between time steps as done by Collard[6] and Senez[2]. Grid quality is maintained by splitting edges in the middle when new grid needs to be added. This helps prevent high aspect ratio triangles from forming. A second technique for helping convergence is to make each oxide grid points' position the average of it's neighbors. In brief tests this technique has greatly improved performance without loss of accuracy. For example the nonlinear stress dependent viscoelastic simulation shown in figure 1a without grid averaging took 14% more CPU time than the simulation shown in figure 1b which was run with grid averaging. When the plots of the simulations are overlaid there is no difference between the simulations with and without grid averaging. This is surprising because by changing the position of the nodes some error in the stress terms should be introduced. This error is probably no worse that interpolation error introduced by regriding and not taking into account rotation of the stresses as done by Senez[2].

Three Dimensional Models

As device dimensions decrease, modeling of LOCOS and modified LOCOS processes will have to be performed in three dimensions. The reasons for this is that three dimensional effects dominate at small dimensions. There are several numerical problems associated with three dimensional simulation of oxidation. The main problem is associated with handling the grid. At the present time linear hexahedra are supported for linear viscous and nonlinear viscoelastic simulations. This is accomplished by using velocity relaxation in the silicon and oxide to retain the grid quality and prevent nodes overrunning one another while moving the boundaries. This has allowed us to begin to investigate three dimensional LOCOS structures while a tetrahedral based moving boundary code is developed. Figure 2 shows linear viscous simulations of two LOCOS examples. The simulations are of 4um by 2um and 4um by 1um nitride lines. The oxidant contours encroach further under the narrower nitride. This causes an increased birds beak length.

Nonlinear viscoelastic simulations have been run using velocity relaxation in the oxide until the grid becomes distorted and or not fine enough. Velocity relaxation introduces some error into the residual stresses but when three dimensional cross sections of longer lines are compared to two dimensional simulations the differences are minor. Figures 3 a & b show a $1000°$ C LOCOS simulation after stripping the nitride. The nitride line was 0.6um by 2um and 1000 A thick. The final field oxide thickness is ~3500A. The birds beak length parallel to the narrow edge of the nitride is ~0.35um and along the longer edge is ~0.2um. Figure 3a demonstrates that the oxidant concentration encroaches further around the corner causing more growth under the tip of the line. Figure 3b shows the pressure contours. The higher tensile regions (white) are further under the tip indicating the increased growth and lifting of the nitride.

Conclusion

A stress dependent nonlinear viscoelastic oxidation simulator has been described. The program can simulate both two and three dimensional isolation structures.

References

1. J. P. Peng et al., COMPEL, p. 341, 1991.
2. V. Senez et al., SISDEP Tech. Digest, p. 165, 1993.
3. D.B. Kao et al., IEEE Trans. on Electron Devices, p. 25, 1988.
4. P. Sutardja et al., IEEE Trans. on Electron Devices, p. 2415, 1989.
5. H. Umimoto et al., IEEE Trans. Comp. Aided Design, p. 599, 1990.
6. D. Collard et al., NUPAD IV Proceedings, p. 21, 1992.

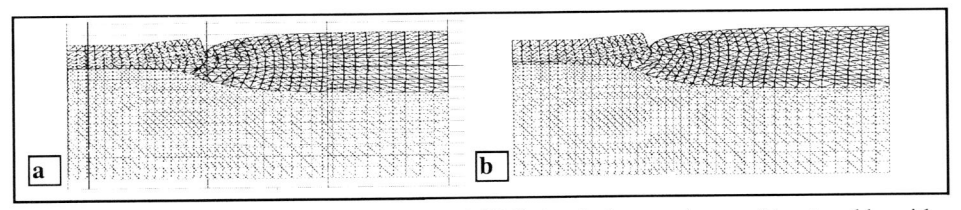

Figure 1. Comparison of 1100 C 40 min LOCOS simulation results a: without and b: with grid averaging between time steps.

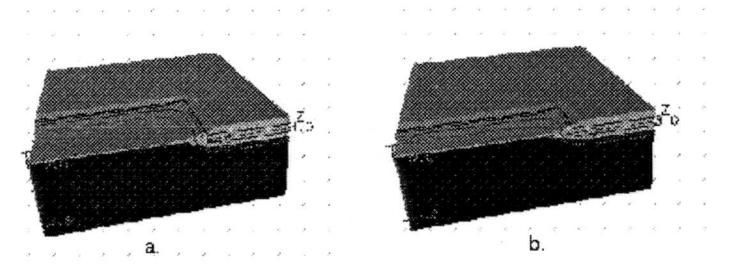

Figure 2. Linear viscous simulations of a: 4umx2.0um nitride line and b: 4umx1.0um nitride line. Oxidant contour lines are from $2.5*10^{19}$ (dark) to $5*10^{18}$ (light) cm^{-3}.

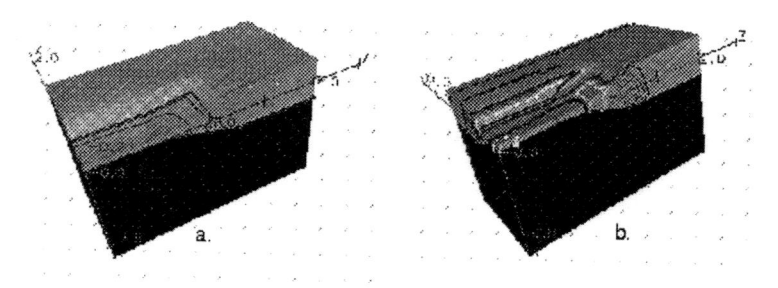

Figure 3. Nonlinear stress dependent LOCOS simulations: a: Oxidant contours (contours as above) b: Pressure contour lines from $-1*10^{10}$ (white) to $-1*10^{9}$ (black) dynes/cm^2.

Three-Dimensional Integrated Process Simulator : 3D-MIPS

M. Fujinaga, T. Kunikiyo, T. Uchida, K. Kamon, N. Kotani and T. Hirao

ULSI Laboratory, Mitsubishi Electric Corporation,
4-1 Mizuhara, Itami, Hyogo, 664 JAPAN

Abstract

We have developed a three-dimensional integrated process simulator of topography and impurity : 3D-MIPS. 3D-MIPS includes topography and impurity simulator, which can simulate deposition, etching, photolithography, BPSG flow, ion implantation, oxidation and impurity diffusion. The diffusion model, in particular, uses a novel equation which unifies diffusion and segregation. In this paper, these models and their simulation results are presented, and we demonstrate that it is possible to simulate 3D-complicated structures stably.

1. Introduction

Until now, three-dimensional (3D-) process simulators [1-3] have been developed. However, a 3D-process simulator which integrates topography and impurity has not yet been made. We have developed a new 3D-integrated process simulator(3D-MIPS: 3D-Mitsubishi Integrated Process simulator) of topography and impurity based on orthogonal mesh and mass-transport. The 3D-MIPS is a stable topography simulator with multiple-process steps using 3D-MULSS [4], and has recently implemented a impurity-diffusion model, an ion implantation model and an oxidation model. The diffusion model, in particular, has a novel equation. In this paper, these models and their simulation results will be presented, and we will demonstrate that it is possible to simulate 3D-complicated structures stably.

2. Models

2.1. Diffusion model

The diffusion model is based on the diffusion equation incorporating the electric field effect and Fair's vacancy model [5]. Neutral cluster model is used for boron, phosphorus, and antimony, and charged cluster model is used for arsenic. The used equation is described as follows:

$$\frac{\partial C_k}{\partial t} = -\nabla \cdot \boldsymbol{J}_k \tag{1}$$

$$\boldsymbol{J}_k = -\left(\frac{D_k}{m_k}\right)\nabla(C_k m_k) + Z_k a_k N_k E \tag{2}$$

$$m_k = \Pi_i m_{ki}{}^{\eta_i} \tag{3}$$

D_k : effective diffusion coefficient of impurity(No.k)
E : electric field
Z_k : the number of charges for impurity(No.k)
a_k : mobility of impurity (No.k)
N_k : concentration of active impurity (No.k)
m_k : segregation coefficient of impurity(No.k)
m_{ki} : segregation coefficient of impurity(No.k) for material(No.i)
η_i : volume rate of material(No.i) in a cell

I the right hand side of the equation (2), the first term corresponds to the new diffusion equation which is satisfied in both the inside of materials and the interface. The second term corresponds to the electric field effect.

As illustrated in Fig. 1, we assume that the impurity profile is continuous [6] and that there is a force moving the impurities in the interface of Si/SiO2. This force F can be given by the gradient of the external chemical potential difference: $\Delta\mu_{\text{ext}}$ between Si and SiO2. The flux density of diffusion can be derived from the Einstein relation and the relation between the segregation coefficient m and $\Delta\mu_{\text{ext}}$ [7] as follows:

$$\boldsymbol{J}_{\text{diff}} = -\left(\frac{D}{m}\right)\nabla(Cm) \tag{4}$$

Where the activity coefficients are assumed to be constant.

Within the Si and SiO$_2$ regions, m is constant and this equation becomes a normal diffusion one. We have, therefore, been able to obtain the unified equation of impurity diffusion which is satisfied in both the Si and SiO2 regions and the interfaces.

2.2. Implantation model

The implantation can deal with the impurities of boron, phosphorus, arsenic, antimony, and BF2. The depth profile is approximated by the Half Gauss or the PearsonIV distribution plus the exponential tail [6]. Then, by using the complementary error function, the depth profile is extended horizontally. The implantation to multiple layers is made possible by the respective stopping powers.

2.3. Oxidation model

The oxidation model was based on the 1-dimensional Deal-Grove model. The LOCOS structure is calculated by the horizontal extension using the Guillemot method [8].

2.4. Topography simulation model

We made a stable 3D-topography simulator with sequential steps by improving the algorithm of the 3D- topography simulator: 3D-MULSS [4]. The material surface is given by the equi-volume rate(0.5). The deposition model can deal with simple, isotropic, angle-dependent, Al-sputter, TiN-collimation sputter and plane depositions. The etching model can deal with simple, isotropic and anisotropic etching with mask-pattern and 2D-photo-image intensity, as calculated by Yeung method [9]. The BPSG flow is possible using the surface diffusion model [10].

3. Simulations

A simulation result of a LOCOS plus NMOS transistor structure is shown in Fig.2. The LOCOS structure was a two-dimensional simulation. 3D-ion implantations and 3D-diffusion were simulated by restarting from the 2D-LOCOS structure. The numbers of meshes are 30×32×60. The most recent calculation time using CRAY-Y/MP was 4050 sec by improving upon the DeFault Frankel method. A DRAM cell structure of Quarter micron pitch was calculated and is shown in Fig.3. The mask data was transported from the lay-out editor via the online. There were 32 process steps, the analysis region was 0.8×1.6×2.4 (μm), and the number of meshes were 20×40×60. To date, the calculation time with CRAY-Y/MP is 3297 sec.

4. Conclusions

We have developed a three-dimensional integrated process simulator of topography and impurity : 3D-MIPS. 3D-MIPS has a 3D-topography simulator which can simulate deposition , etching, photolithography and BPSG flow, which in turn can calculate scores of sequential process steps in LSI's fabrication. The 3D-MIPS includes a new model which unifies the diffusion and the segregation models, the ion implantation model and the oxidation model. We have, thus, put forward the simulation results of LOCOS plus NMOS Transistor and DRAM cell structure of quarter micron pitch. In so doing, we have demonstrated that it is possible to solve the 3D-complicated structures stably.

References

[1] S. Onga and K. Taniguchi, VLSI Symp., VI-7, 1985.
[2] K.Nishi et al, NASECODE VI,p.297, 1989.
[3] S.Odanaka et al, IEEE ICCAD-86 California, p.468, 1986.
[4] M.Fujinaga et al, TEDM Tech. Digest, p.905, 1990.
[5] D.A.Antoniadis and D.W.Dutton, IEEE Trans. ED, vol.ED-26, pp.490-500, 1979.
[6] M.Orlowski, Proceeding of VLSI Process/Device Modeling Workshop, p.1, Japan 1989.
[7] R.B.Fair and J.C.C.Tsai, J.Electrochem. Soc., p.2050, 1978.
[8] N.Guillemot at al, IEEE Trans. ED, vol.ED-34, No.5 pp.1033-1038, 1987.
[9] M.Yeung, Proc. Kodak Microelectronics Seminar INTERFACE'85, p.115, 1986.
[10] M.Fujinaga et al., Tech. Dig. of NUPAD IV, p.15, 1992.

Convensional model

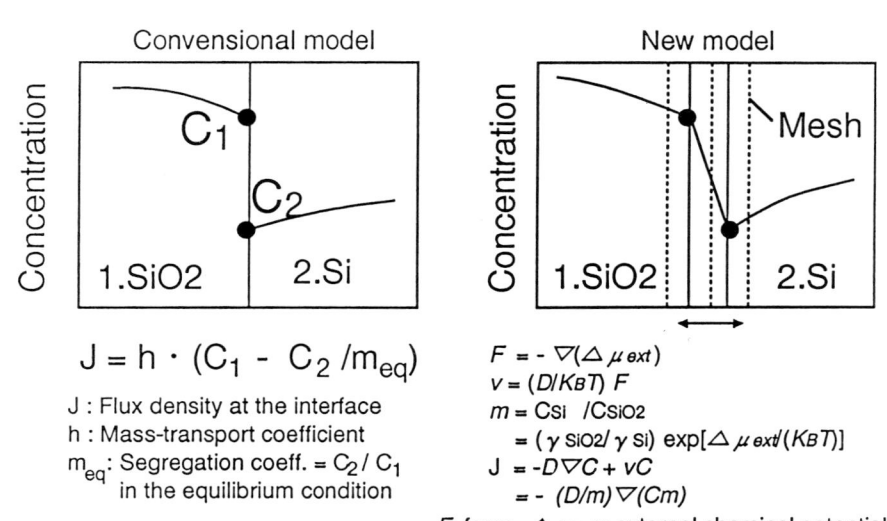

$$J = h \cdot (C_1 - C_2 / m_{eq})$$

J : Flux density at the interface
h : Mass-transport coefficient
m_{eq}: Segregation coeff. = C_2 / C_1
  in the equilibrium condition

New model

$$F = -\nabla(\Delta \mu_{ext})$$
$$v = (D/K_BT) F$$
$$m = C_{Si} / C_{SiO2}$$
$$ = (\gamma_{SiO2}/\gamma_{Si}) \exp[\Delta \mu_{ext}/(K_BT)]$$
$$J = -D\nabla C + vC$$
$$ = -(D/m)\nabla(Cm)$$

F: force, $\Delta \mu_{ext}$: external chemical potential
v: impurity average velocity, D:diffusion coeff,
m: segregation coeff., γ: activity coeff.

Fig.1 Models of Diffusion and Segregation

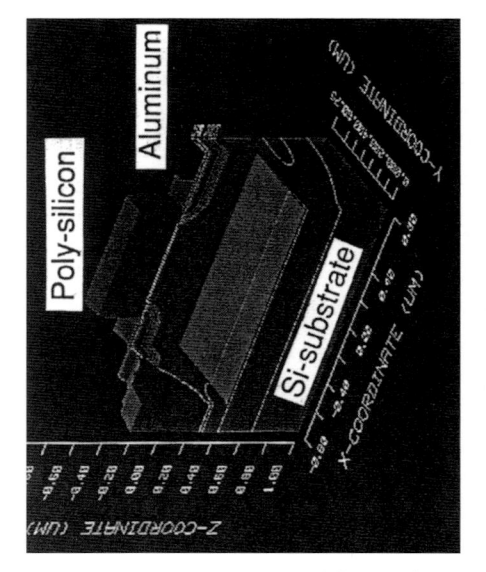

Fig.2 LOCOS + NMOS Transistor

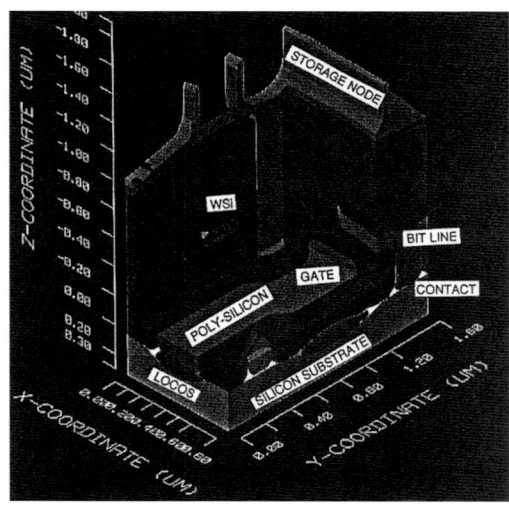

Fig.3 DRAM cell structure of quarter micron pitch

Effect of Process-Induced Mechanical Stress on Circuit Layout

Hideo Miura* and Yasunobu Tanizaki**

*Mechanical Engineering Research Laboratory, Hitachi, Ltd.
502 Kandatsu, Tsuchiura, Ibaraki 300, Japan
**Semiconductor & Integrated Circuits Div., Hitachi, Ltd.
5-20-1 Josui-honcho, Kodaira, Tokyo 187, Japan

Abstract

The effect of process-induced mechanical stress on characteristics of a simple differential amplifier circuit are analyzed using the finite element method, based on the piezoresistance effect of diffused resistors and the experimentally determined sensitivities of transistor characteristics to stress. The predicted change distribution of the resistivity of p-type diffused resistors and of the h_{fe} and V_{be} of pnp transistors agree well with the measured data.

1. Introduction

With the trend towards high integration of LSIs, device structures have been becoming increasingly complicated, and the number of thin-film materials used has been increasing. These structural changes have caused mechanical stress to increase in device structures. The stress developed is sometimes high enough to cause not only mechanical failures such as dislocations in silicon substrates and cracking or delamination of thin films, but also electronic ones. Thus, it is increasingly important to evaluate and control the mechanical stresses in the device structure in order to improve product reliability.[1]

Recently, deep isolation structures have come to be used for bipolar devices. An isolation structure is formed using the local thermal oxidation of silicon (LOCOS) process. The thickness of the newly grown oxide film reaches about 2 μm. Great stress is generated near the Si/SiO$_2$ interface during thermal oxidation, because of volumetric expansion of the newly oxidized film.[2] Passivation films such as SiO$_2$ and Si$_3$N$_4$ also have high intrinsic stress, and thus change the stress fields in silicon substrates.

Since the resistivity of diffused resistors and the electronic performance of transistors are changed by mechanical stress[3]-[6], these process-induced stresses sometimes cause serious damage to circuit performance. An example of this is a carrier-signal leakage failure that occurred in a simple differential amplifier circuit (Fig. 1) located near a thick oxide film. The failure mechanism was analyzed with respect to the change in circuit performance due to process-induced mechanical stress.

2. Stress Simulation

Mechanical stress fields in device structures were analyzed using the finite element method. Both stress-dependent oxidation and the intrinsic stress of thin films were taken into account.[7)8)] Figure 2 shows an example of the predicted residual stress distribution of normal stress σ_x after the LOCOS process. Stress concentrates at the Si/SiO$_2$ interface. A stress field develops within an area about 10 μm from the oxide edge. The stress distribution at the silicon–substrate surface is summarized in Fig. 3 The normal-stress component σ_y and shear-stress component τ_{xy} are relatively small, and exist only within 5 μm of the edge. On the other hand, the normal stress σ_x shows a highest value of about 600 MPa at 1 μm from the edge and decreases monotonically to almost zero at 15 μm from the edge.

These stress distributions are slightly changed due to passivation–film stress. Figure 4 shows the effect of the passivation film on the distribution of the residual normal stress (σ_x) at the substrate surface. When a PSG film was used for passivation, the compressive stress developed due to oxidation decreased by about 30% near the oxide edge. This was because the film had a tensile stress of about 200 MPa. On the other hand, the compressive stress increased when a silicon-nitride (P-SiN) film was used for passivation. This increase was due to the compressive stress of about 400 MPa of the nitride film. Therefore, the selection of material for the passivation film is also an important factor in mechanical stress control for silicon substrates.

3. Device–characteristic changes caused by mechanical stress

Device–characteristic changes were analyzed considering the piezoresistance–effects and experimentally determined stress sensitivities of transistor characteristics. The resistivity change of diffused resistors was calculated using the following equation:

$$dR/R = [\iint \{S_j p_{ij} \sigma_j\}^{-1} dxdy]^{1}, (j=1\cdots 6, i=1\cdots 6)$$

Where, R is resistivity, p_{ij} is the piezoresistance coefficient[6)], and σ_j is three dimensional stress component. The integration was performed on a diffused resistor area 1 μm wide and 0.1 μm deep. The calculated distribution of the resistivity change of the p-type diffused resistors is shown in Fig. 5 (measured results are also shown in this figure). The maximum change, of about 1%, was predicted at 2 μm from the oxide edge, and the measured value agreed well with this prediction. The change rate decreases monotonically to almost zero at 15 μm from the edge. The measured change rate of about 0.1% at 10 μm from the edge also shows good agreement with the predicted result.

The characteristic changes of bipolar transistors such h_{fe} and V_{be} were predicted based on the results of stress simulations and their experimentally determined stress sensitivities. The stress sensitivities were measured by applying mechanical stress to the transistors using a four-point bending test of strips cut from the device-fabricated silicon wafers. The values obtained for the pnp transistor were about 7%/100 MPa for h_{fe} and 3.5mV/MPa for V_{be}. The change distributions of h_{fe} and V_{be} predicted for the pnp transistor are shown in Fig. 6. Both of these characteristics also changed near the Si/SiO$_2$ interface. The h_{fe} and V_{be} decreased about 10% and 20 mV, respectively, at 2 μm from the edge. Both changes decreased monotonically to almost zero at 15 µm from the edge, as the resistivity change did. These predictions agreed well with the measured data.

Similar predictions were made for n-type diffused resistors and an npn transistor. The predicted resistivity change of the n-type diffused resistors is shown in Fig. 7. In this case, resistivity decreased by about 10% at 2 µm from the oxide edge. This rate of change was about ten times as high as that of p-type diffused resistors. Thus, we can see that the resistivity of n-type diffused resistors is more sensitive to mechanical stress than that of p-type diffused resistors. From this point of view, p-type diffused resistors should be used in the high stress areas to minimize the resistivity change. Figure 8 shows the predicted changes of the h_{fe} and V_{be} of the npn transistor. The h_{fe} increased by about 10% at 2 µm from the oxide edge. Though the sign of the change was the opposite, the change rate was almost same as that of a pnp transistor. The V_{be} change was about 20 mV at 2 µm from the oxide edge, which was also almost same as that of the pnp transistor. Thus, the absolute values of the stress sensitivities of h_{fe} and V_{be} for the npn transistor were almost same as those for the pnp transistor.

Some precise differential amplifiers require resistivity changes less than 0.1%. These circuits must be located more than 10 µm away from the oxide edge. This limitation can change because the stress distribution varies with the oxide film thickness and the thickness and materials of passivation films. These process parameters are important factors for mechanical stress control and thus for the improvement of device reliability. Our stress simulation is effective for the optimization of device structures and circuit layout to improve device reliability.

4. Summary

The mechanism of change of the device characteristics of a simple differential amplifier circuit located near a thick thermally oxidized film was analyzed by applying the finite element method, taking process-induced mechanical stress into account. The characteristic changes were predicted based on the piezoresistance effect of diffused resistors and the experimentally determined stress sensitivities of transistor characteristics. The predicted distribution of resistivity change of p-type diffused resistors and changes of the h_{fe} and V_{be} of a pnp transistor agreed well with the measured data. Therefore, mechanical stress simulation is clearly an effective way to design circuit layouts in consideration of performance changes due to process-induced mechanical stress.

References

1) H. Miura, et al., *Proc. of SISDEP '93*, (1993), 177.
2) H. Miura, et al., *JSME International Journal*, Vol. 36A (1993), 302.
3) C. S. Smith, *Physical Review*, Vol. 94, (1954), 42.
4) V. Zekeriya and T. P. Ma, *Appl. Phys. Lett.*, Vol. 56, (1984), 1017.
5) A. Hamada, et al., *IEEE Trans. on Electron Devices*, Vol. 38, No. 4 (1991), 895.
6) H. Miura and A. Nishimura, *Proc. of ASME WAM '94*, AMD-Vol.195, (1994), 101.
7) N. Saito, et al., *Tech. Dig. of IEDM*, (1989), 695.
8) N. Saito, et al., *Proc. of the Int. Conf. on Computational Engineering Science*, Melbourne, Australia, (1991), 880.

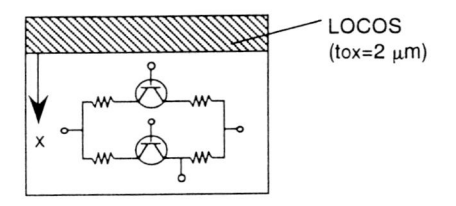

Fig. 1 A differential amplifier located near thick LOCOS

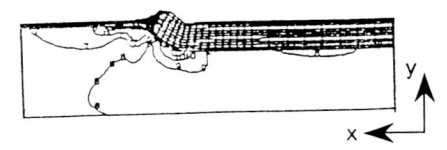

Fig. 2 Predicted stress distribution in the silicon substrate after local thermal oxidation

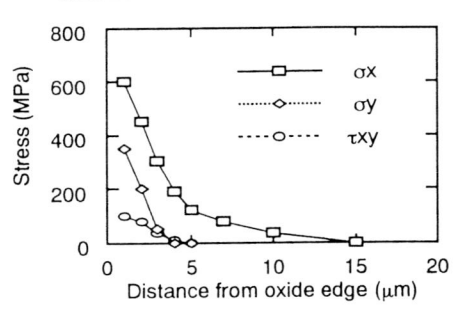

Fig. 3 Stress distribution at silicon substrate surface near thick LOCOS edge

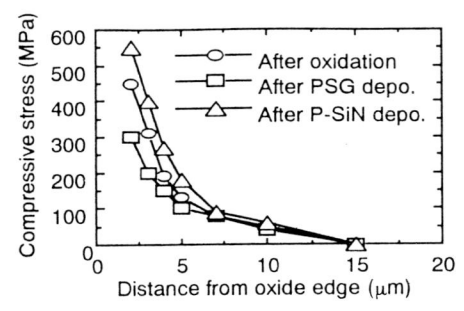

Fig. 4 Effect of passivation film on the residual stress at the silicon substrate surface

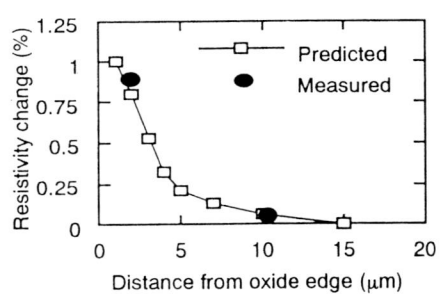

Fig. 5 Comparison of predicted resistivity changes of diffused resistors with measured changes

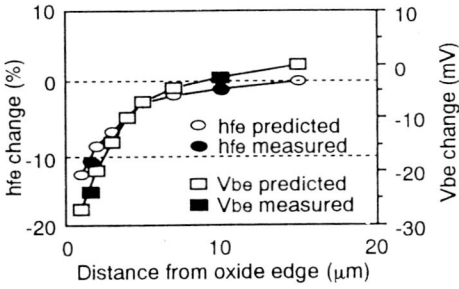

Fig. 6 Comparison of predicted changes of hfe and Vbe of a pnp transistor with measured changes

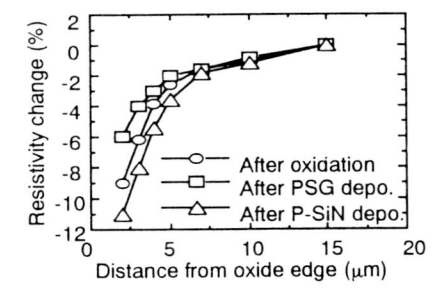

Fig. 7 Predicted resistivity changes of n-type diffused resistors

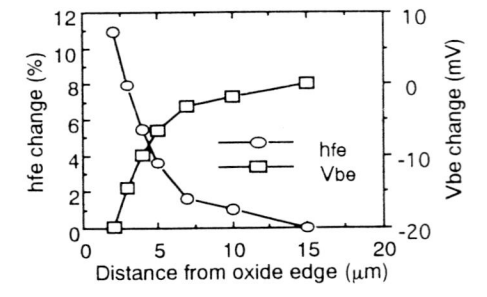

Fig. 8 Predicted changes of hfe and Vbe of an npn transistor

The Simulation System for Three-Dimensional Capacitance and Current Density Calculation with a User Friendly GUI

M. Mukai[a], T. Tatsumi[a], N. Nakauchi[a], T. Kobayashi[a], K. Koyama[a],
Y. Komatsu[a], R. Bauer[b], G. Rieger[b], S. Selberherr[b]

[a]ULSI R&D Laboratories, Semicon.Comp., SonyCorporation
4-14-1, Asahi-cho, Atsugi-shi, Kanagawa-ken, 243 JAPAN
[b]Institute for Microelectronics, TU Vienna
Gusshausstrasse 27-29, A-1040 Vienna, AUSTRIA

Abstract

For the realization of today's wiring miniaturization, it is required to accurately estimate the interconnect wiring capacitance and the current density, which has not been so far adequately accomplished. We have developed SENECA. the three-dimensional wiring capacitance and current density simulation system with graphical user interface for easy and flexible operation. The realization of this graphical input interface made it possible to reduce the man power from more than three days work to less than 30 minutes for a practical structure. The comparison between simulation and experiment proves quite good agreement. Three important applications are given to show how SENECA is utilized for practical cases.

1. Introduction

According to today's wiring miniaturization, it is required to accurately estimate the interconnect wiring capacitance and the current density, which has not been so far adequately accomplished. Firstly, correct prediction of wiring capacitance is necessary to evaluate switching speed, delay of signal and cross-talks. Wiring capacitance can be classified into wire-to-ground capacitance and wire-to-wire capacitance. For conventional processes, wire-to-wire capacitance could be ignored because the distance between wires was large and its magnitude was relatively small compared with wire-to-ground capacitance. However, as the process advances, the distance between wires becomes very small, and wire-to-wire capacitance cannot be neglected. Secondly, from the aspect of wiring reliability the concentration of the current density distribution generates heat and might cause electromigration resulting in failure there. Therefore the precise prediction of current density is also necessary to maintain reliability.

We developed SENECA, a three-dimensional simulation system capable of very easily simulating wiring capacitance and current density for practical complex structures.

2. The Simulation System SENECA

SENECA consists of three modules: the input interface SCIN, the numerical calculation part SCAP [1][2] which utilizes a three-dimensional finite element approach for the computation of wiring capacitances and current densities, and the output processor and viewer SCOUT.

The input data for SCAP consists of coordinate points and permittivity information of every hexahedron which is obtained by sectioning the object into small pieces, and the location information of electrodes. In case of practical applications, sectioning objects into hexahedrons by hand is almost impossible because the number of hexahedrons for practical objects soon reaches the order of a thousand or more. In order to resolve this problem, we developed a graphical input interface called SCIN, which enables to construct complex input structures accurately and very efficiently with very easy interactive operation. As one important feature of SCIN, rectangular parallelepiped, cylinder and spoon-cut, which are the structures used in wirings, are prepared as built-in models of input data, and the desired structure can be created by combining those built-in models and specifying parameters. Thus a defined input structure can be subsectioned into hexahedron meshes automatically and it generates thereby the input data for three-dimensional finite element numerical analysis. The realization of this graphical input interface made it possible to reduce the man power from more than three days work to less than 30 minutes for a practical structure.

SCAP can calculate three-dimensional wiring capacitances and current densities using finite element analysis. The three-dimensional Laplace equation is solved for the input configuration specified with SCIN. Tetrahedrons are used for discretization with quadratic shape functions leading to 10 parameters per element. Hexahedrons are decomposed automatically to tetrahedrons.

SCOUT, the output processor, calculates electric field, current density, generated heat from the current flow, and plots colored mapping and also vector representations. It can show any cross sections on the three-dimensional object, partial enlargement on any specified part, and print out with very simple button operation.

3. Evaluation and Practical Applications of SENECA

In order to investigate the interconnect capacitances between lines and between line and ground, SENECA was applied to the three kinds of processes A(0.7μm rule generation), B(0.5μm rule generation). C(0.35μm rule generation). Shown in Fig.1 are the structures and dimensions of process A, B, C, respectively. Firstly, the detailed comparison is made on process B between experiments and simulations. Table 1 gives some of the results showing quite good agreement of 10% difference or less. In Table 2, the ratio is shown of the capacitance between wires and between wire and ground. As the process shrinks from A to C, the role of wire to wire capacitance increases from 13.8% for process A to 65.7% for process C, resulting in the very important contribution. Fig.2 shows the twisted wiring structure and we also calculated capacitance for this structure as shown in Table 3. Usually for this kind of complex structures, actual measurement is quite difficult. Only simulation makes it possible to obtain accurate capacitances.

SENECA was applied to estimate the current density distribution in order to investigate the reliability[2] due to current localization for the layered wire with the aluminum layer crack caused by stress-induced migration. The wire is connected through barrier metal such as Ti/TiN/Ti. Fig. 3 shows the enlargement of the cracked part. High current densities are shown darker than low densities. The magnitude of the highest current density parts specified with arrows and the lowest current density part are shown in the figure. The current density ratio is around 13.6 compared with the lowest part. From this current localization we can see the importance of barrier metal structures.

4. Conclusion

We have developed SENECA, the three-dimensional wiring capacitance and current density simulation system with graphical user interface for easy and flexible operation. SENECA is now in practical use for simulating wiring capacitances of LSIs and current density estimation for achieving better reliability. The comparison between simulation and experiment proves quite good agreement. Three important applications are given to show how SENECA is utilized for practical cases.

References

[1] R. Bauer et al.,"Capacitance Calculation of VLSI Multilevel Wiring Structures," Proc.VPAD Workshop, pp.142-143, 1993.
[2] R. Bauer et al., "Calculating Coupling Capacitances of Three-Dimensional Interconnections," Proc. Solid State and Integrated Circuit Technology 92, pp.697-699, 1992.
[3] K. Hoshino et al., "Electromigration after Stress-induced Migration Test in Quarter-Micron Al Interconnects," Proc. IRPS, pp.252-255, 1994.

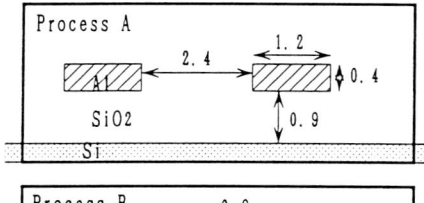

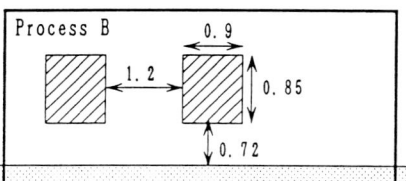

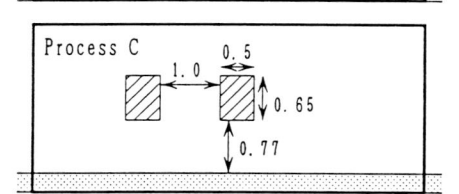

Fig. 1 Wiring size (um) of process A, B and C

Table 1 Comparison between measured and simulated data (fF/um)

	1Al (First Aluminum)		2Al (Second Aluminum)	
	wire-gnd	wire-wire	wire-gnd	wire-wire
SENECA	0.0742	0.0510	0.0485	0.0564
measured	0.0762	0.0561	0.0517	0.0607
diff	3%	10%	7%	8%

Table 2 Capacitance of A, B and C (fF/um)

	wire-gnd	wire-wire	ratio
A	0.1044	0.0145	13.8%
B	0.0871	0.0399	45.8%
C	0.0586	0.0385	65.7%

Table 3 Capacitance of crossed wires (fF/um)

wire1-ground	0.5234
wire2-ground	0.1508
wire1-wire2	0.3419

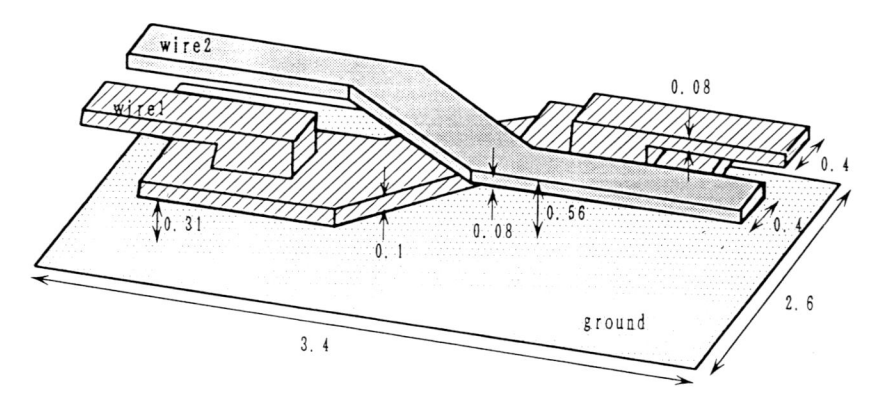

Fig. 2 Crossed wires of 256K DRAM (um)

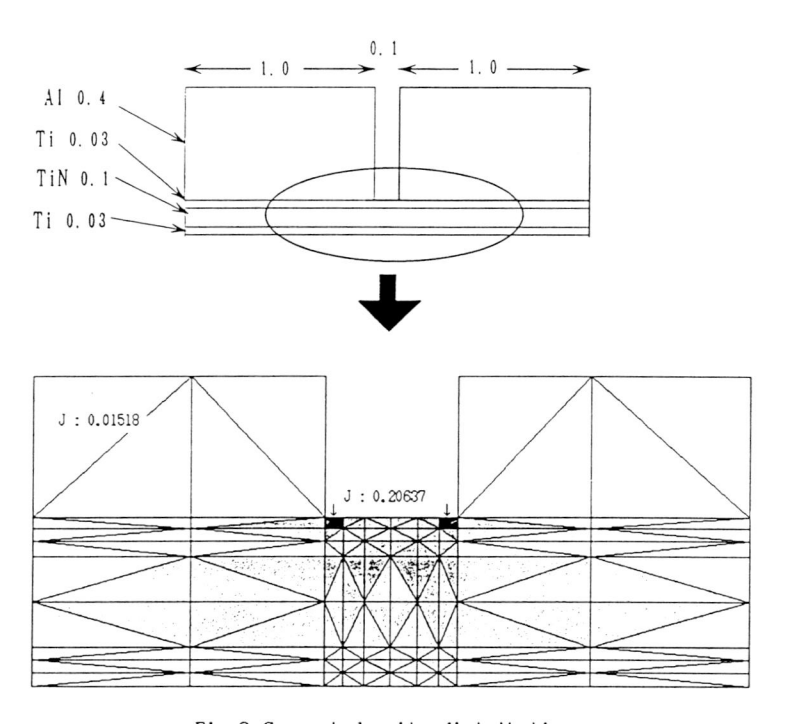

Fig.3 Current density distribution

Numerical and analytical modelling of head resistances of diffused resistors

U. Witkowski, D. Schroeder

TU Hamburg-Harburg, Techn. Electronics,
Eissendorfer Str. 38, D-21071 Hamburg, GERMANY

Abstract

Head resistances of diffused resistors are investigated by means of numerical device simulation. Outgoing from an understanding of the current flow pattern, an analytical model for the calculation of the head resistance from known geometric and technological parameters is developed. The results of the simulations and the modelling are verified by comparison with experimental data.

1. Introduction

In analog integrated circuits, diffused resistors [1] are quite common devices. They consist of a conducting strip with contact structures to the metal layer (heads) at both ends (see Fig. 1). For circuit design, precise knowledge of the resistance of the whole structure is of importance. In particular for low-ohmic resistors, the contributions of the contact heads to the total resistance become significant. In a cooperation with Philips Semiconductors in Hamburg, we investigated the head resistance by simulations and measurements of industrially fabricated devices.

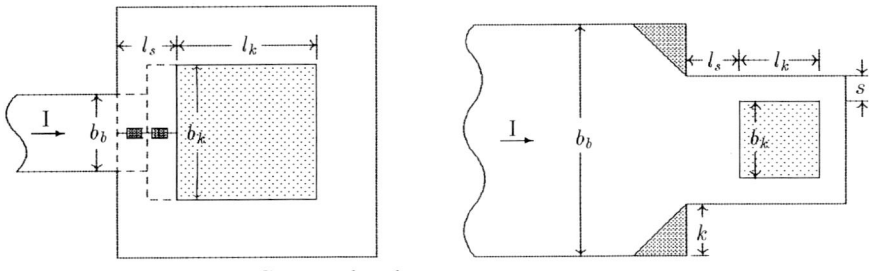

Figure 1: Contact head structures

Figure 2: Current distribution of resistor head

Table 1: Head resistances

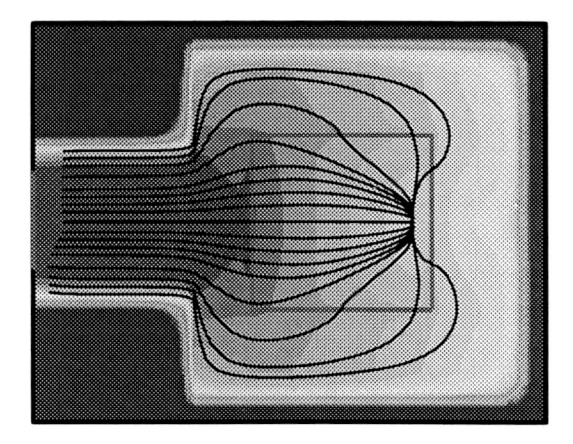

width	R_{head} [Ω]	
b_b [μm]	(simul.)	(meas.)
2.0	69.0	78
5.0	47.8	54

2. Numerical simulations

We performed numerical 3D-simulations of the head structure using the device simulator PARDESIM [2]. The finite resistance of the metal-semiconductor interface [3] has been taken into account by our model of non-ideal contacts [4, 5, 6, 7]. Figure 2 [8] shows the result of a simulation of an example structure. The resistor is made by a p-type diffusion with a sheet resistance of $125\,\Omega_\square$ into low-doped n-type silicon. The contact window is indicated by the square. The dimensions are $3.5 * 3.5\,\mu\text{m}^2$ for the contact, $1.25\,\mu$m for the collar and $2.0\,\mu$m for the width of the resistor path. The shading indicates the absolute value of the current density on the semiconductor surface. The flow lines show that the current flowing from the resistor path spreads out into the head region, where it finally enters the contact.

We simulated the two basic structures shown in Fig. 1, i.e. the two cases where the resistor path is smaller resp. wider than the contact head. The extracted head resistances for these cases are shown in Table 1, together with values from measurements done by Philips. Good agreement is obtained. (The error is about 12 %, which is within the experimental uncertainty.)

3. Analytical modelling

Based on the current flow patterns (see e.g. Fig. 2) as obtained by the simulations, we developed an analytical model for the calculation of the head resistance from known geometric and technological parameters. The results are verified by simulations and measurements.

In Fig. 2, we identify three regions in the contact head with significant flow patterns: The region between the end of the resistor path and the contact metallization, where the current spreads into the wider head diffusion; the region under the contact itself, where the current sinks into the metal; and the region on both sides of the contact, where the current flows a considerable way around the contact before entering the contact area. These three regions are modelled each by a separate resistance, i.e. the spreading resistance R_{spread}, the contact resistance R_K, and the lateral resistance

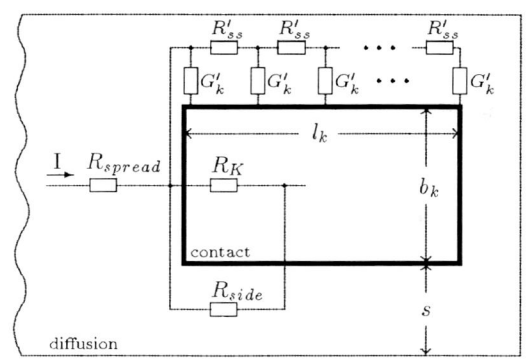

Figure 3: Modelling of lateral flow

R_{side} (cf. also Fig. 3):

$$R_{head} = R_{spread} + \left(\frac{1}{R_K} + \frac{2}{R_{side}}\right)^{-1} \tag{1}$$

The spreading resistance is composed simply by two resistors of distinct widths, as indicated in Fig. 1. The distributed inflow into the contact from underneath is modelled as a one-dimensional lossy transmission line, as shown in Fig. 4. The result is

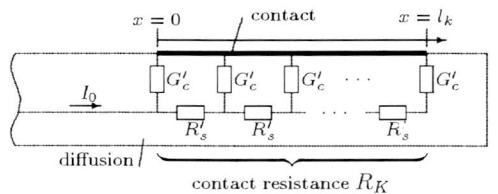

Figure 4: Modelling of contact flow

$$R_K = \frac{U(0)}{I(0)} = R_0 \coth\left(\frac{l_k}{l_t}\right), \tag{2}$$

where

$$l_t = \frac{1}{\sqrt{R'_s G'_c}} = \sqrt{\frac{\varrho_c}{R_s}}, \qquad R_0 = \sqrt{\frac{R'_s}{G'_c}} = \frac{\sqrt{R_s \varrho_c}}{b_k}. \tag{3}$$

A similar approximation is made for the lateral flow on both sides of the contact in order to obtain a model equation for R_{side}; see Fig. 3.

$$R_{side} = R_1 \coth\left(\frac{l_k}{l_1}\right), \tag{4}$$

$$l_1 = \frac{1}{\sqrt{G'_k R'_{ss}}} = \sqrt{s\, l_t \coth\left(\frac{b_k}{l_t}\right)}, \qquad R_1 = \sqrt{\frac{R'_{ss}}{G'_k}} = \sqrt{\frac{R_s}{s}\sqrt{R_s \varrho_c} \coth\left(\frac{b_k}{l_t}\right)}. \tag{5}$$

Eqns. (1-5) are the model equations for R_{head} as a function of the specific contact resistance ϱ_c, the sheet resistance R_s, and the geometric parameters.

b_k [μm]	l_k [μm]	b_b [μm]	R_{head} [Ω] (meas.)	R_{head} [Ω] (model)
3.5	3.5	2.0	78	78.9
3.5	3.5	5.0	54	55.4
6.5	3.5	8.0	33	33.2
6.5	3.5	11.0	40	41.5
12.5	3.5	17.0	24	23.4
18.5	3.5	23.0	18	16.3

Table 2: Measured and modelled head resistances

Several resistor heads with different dimensions have been investigated; the results are presented in Table 2. The table shows the measured and the modelled head resistance for various combinations of the contact size and the path width. Good agreement between the experimental results and the model is obtained.

In conclusion we note that the presented analytical model for contact head resistances can be used to accurately estimate the total resistance of diffused resistors during circuit design or process development.

Acknowledgement. The authors would like to thank R. Kuvecke (Philips Semiconductors) for helpful discussions and the provision of the experimental data.

References

[1] A.B. Grebene. *Bipolar and MOS analog integrated circuit design*. Wiley & Sons, New York 1984.

[2] O. Kalz, D. Schroeder. PARDESIM – A parallel device simulator on a transputer based MIMD-machine. In S. Selberherr, H. Stippel, E. Strasser, editors, *Proc. 5th Int. Conf. on Simulation of Semiconductor Devices and Processes (SISDEP'93), Sept. 7-9, 1993, Vienna*, Springer, Wien 1993, pp. 245-248.

[3] W. Loh et al. *Modeling and measurement of contact resistances*. IEEE Trans. Electron Devices, vol. ED-34 (1987), p. 512.

[4] D. Schroeder. An analytical model of non-ideal ohmic and Schottky contacts for device simulation. In W. Fichtner and D. Aemmer, editors, *Proc. 4th Int. Conf. on Simulation of Semiconductor Devices and Processes, Sept. 12-14, 1991, Zurich*, Hartung-Gorre, Konstanz 1991, pp. 313-319.

[5] D. Schroeder. A boundary condition for the Poisson equation at non-ideal metal-semiconductor contacts. In J.J.J. Miller, editor, *Proc. 8th Int. Conf. on the Numerical Analysis of Semiconductor Devices and Integrated Circuits (NASECODE VIII), May 19-22, 1992, Vienna*, Boole Press, Dublin 1992, pp. 105-106.

[6] D. Schroeder, T. Ostermann, O. Kalz. Nonlinear contact resistance and inhomogeneous current distribution at ohmic contacts. In S. Selberherr, H. Stippel, E. Strasser, editors, *Proc. 5th Int. Conf. on Simulation of Semiconductor Devices and Processes (SISDEP'93), Sept. 7-9, 1993, Vienna*, Springer, Wien 1993, pp. 445-448.

[7] D. Schroeder. *Modelling of interface carrier transport for device simulation*. Springer, Wien 1994.

[8] Picture created with *Picasso*, developed by ETH Zurich and ISE AG, Zurich.

New Spreading Resistance Effect For Sub-0.50 μm MOSFETs: Model and Simulation

Marius Orlowski and William J. Taylor

Advanced Products and Development Laboratory, Motorola Inc.
Austin, Texas 78721, USA

Abstract

A new major spreading resistance (SR) contribution associated with the vertical shift of the peak LDD concentration into the bulk is reported. This contribution is at least a factor of 5 larger than the SR variations of the lateral S/D profile (with peak concentration at the interface) reported so far [1]. The effects are relevant to sub-0.5 μm devices manufactured with reduced thermal budgets. A resistor network model, corroborated by 2D simulations, explains the key features of the effect, including a reversal of SR trends for shallow junctions due to the impact of accumulation resistance (AR). For the first time, ostensibly conflicting data from experiments with various sheet resistances $\varrho_\square$, junction depths χ_j, and S/D constructions can now be clearly understood.

1. Introduction

Parasitic S/D spreading resistance is known to be a limiting factor for MOSFET scalability [2]. Baccarani and Sai-Halasz [3] were first to derive an analytical expression for SR assuming an idealized uniform step p-n junction. Subsequently, Seavey [4] has shown that in practical devices Baccarani's first order derivation can underestimate SR effects up to a factor of 5. An authorative study of SR which accounts for the effects of doping gradient including accumulation resistance (AR) is in series with SR has been given by Ng and Lynch [1]. All of these studies, however, are based on the assumption that the peak concentration both of the lateral and vertical S/D profiles is at the interface. This assumption is no longer justified, especially for the S/D extension overlapping the gate electrode. First, because the nature of the lateral profile is not known precisely and is likely to have peak concentration deeper in the bulk; second, because it is advantageous to place the implantation peak somewhat deeper into the substrate to avoid otherwise sheet resistance fluctuations of up to 30% [5]. During the subsequent anneal at a reduced thermal budget the reduced diffusion might not be able to shift the concentration peak back to the interface.

2. The New Spreading Resistance Effects

The impact of concentration peak shifted into the substrate can be explained in terms of a simple resistor network, see Fig.1. The model describes the diffusion sheet resistance $R_{sh} \equiv \varrho_\square = 2 \cdot R_{s12} \cdot R_{s22}/(R_{s12} + R_{s22})$ assuming parallel current paths,

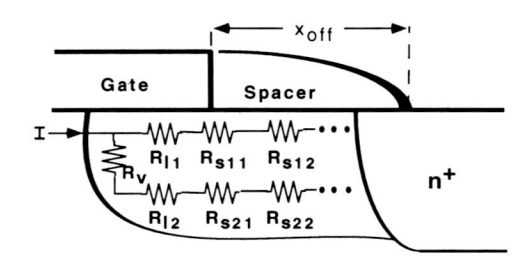

Figure 1: A resistor network describing the spreading resistance in the LDD region. Resistors R_v, R_{l1}, and R_{s11} depend upon gate bias.

and the spreading resistance by $R_{sp} = R_{tot} - R_{sh}$, where $R_{tot} = (R_{l1} + R_{s11} + R_{s12}) \cdot (R_v + R_{l2} + R_{s21} + R_{s22})/((R_{l1} + R_{s11} + R_{s12} + R_v + R_{l2} + R_{s21} + R_{s22})$. R_{tot} contains the components R_{l1} and R_{l2} which describe the increased resistance of the lateral fall-off of the LDD profile. R_v describes the resistor between the parallel resistors in the region of current spreading within the S/D regions. In other words, R_v breaks the symmetry of the resistor network with respect to the current entry from the channel. R_{l1} is the lateral resistor in the gate/drain overlap region and is a function of gate oxide field at the drain.

Of course, more involved networks can be considered, but this 'minimal' model captures the key features of the effects. To illustrate the model consider resistor values given in Table 1 in appropriate units R_o. Three vertical LDD profiles are discussed first: A) peak LDD concentrations at the interface, B) peak concentration in the bulk (close to the junction depth), and C) uniform vertical profile; see the corresponding profiles in Fig.2.

Profile	R_{l1}	R_{s11}	R_{s12}	R_v	R_{l2}	R_{s21}	R_{s22}
A	2	1	1	4	5	4	4
B	5	4	4	4	2	1	1
C	2.92	1.6	1.6	2.92	2.92	1.6	1.6
D	1	1.4	1.4	1.9	2.4	1.9	1.9
E	1	1.5	1.9	1.4	1.7	1.4	1.4

Profile	R_{sh}	R_{tot}	R_{sp}
A	1.6	3.24	1.64
B	1.6	4.95	3.35
C	1.6	3.65	2.05
D	1.6	2.75	1.15
E	1.6	2.42	0.82

Table 1: Values for resistors, in normalized units, for the model in Fig.1. Cases A, B, C correspond to the profiles in Fig.2 and neglect gate accumulation effects. Cases D and E correspond to 20 and 80 keV implants shown in Fig.5 and include V_G dependence of R_v, R_{l1}, and R_{s11}.

Table 2: Sheet resistance $R_{sh} \equiv \varrho_\square$, total resistance R_{tot}, and spreading resistance R_{sp} for resistor inputs from Table 1.

The profiles have been constructed to have identical sheet resistance $\varrho_\square$ and identical junction depth χ_j, with the same construction of lateral profile as an extension (by means of erfc(y), y being the lateral space coordinate) of the vertical LDD profile. The factor of 4 as a maximum variation of concentration values has been chosen for convenience to reflect vertical resistivity variation by one order of magnitude, according to the formula $\varrho = 1.45 \cdot 10^9 (N)^{-0.6}$ [1]. In Table 2 all the above resistances have been calculated for profile A, B, and C. It can be seen that SR for case A (peak concentration at the interface) contributes only 1.64 $R-o$ or 103%(=1.64/1.6) of $\varrho_\square$, whereas in the case B (higher concentration in the bulk) it contributes 3.35 R_o or 209% of the same $\varrho_\square$ - **a twofold increase only due to different geometry (shape) of the network (profile)!** Note that the network is composed from the same resistors; only the parallel resistor rows have been switched. Clearly, SR is a serious

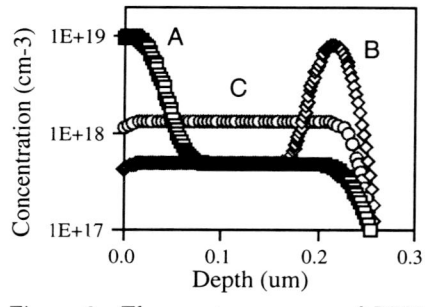

Figure 2: Three extreme cases of LDD profiles with identical sheet resistance and identical junction depth used for MINIMOS simulations of the I-V characteristics in the triode regime.

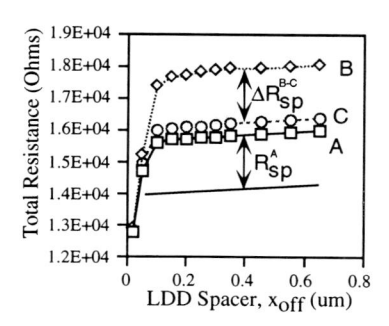

Figure 3: MINIMOS simulations of total MOSFET resistance ($V_D = 0.1V$, $V + G = V_T + 0.2V$, $L = 0.5\mu m$, $W = 1.0\mu m$) versus spacer length, x_{off}, for LDD profiles from Fig.2. Beginning at $x_{off} = 0.04\mu m$ LDD regions are completely buried under the n^+ S/D regions.

problem, when the peak concentration is shifted from the interface into the bulk. These trends, however, are reversed for shallow junctions, because of the dominant role of the accumulation resistance in case of very shallow junctions. Accumulation layer is induced in the S/D region by the gate field and extends the inversion layer of the channel into S/D regions. Simulations show that at high enough V_G the LDD accumulation layer underneath the gate in the triode region is almost independent of the original doping levels. In Table 1 resistor values are given for profiles D and E but now, in contrast to profiles A, B, and C, with R_{l1} and R_v now being modified by V_G; R_{l1} has the same value for both profiles D and E. It can be seen from Table 2 that SR for the profile D with peak concentration at the interface is larger than for profile E with concentration peak deeper in the bulk. Profiles D and E from the network model correspond to 20 and 80 keV implants shown in Fig.5.

3. SR Extraction from I-V Characteristics

SR on the source and drain side contribute in different ways to the total resistance. SR on the source side reduces not only the effective drain bias but also the effective gate field which entails higher channel resistance, whereas SR on the drain side reduces only the drain bias. In order to extract the genuine LDD SR, a transistor with n^+ source only and with LDD/n^+ drain has been constructed for various LDD profiles with exactly the same $\varrho_\square$ and χ_j shown in Fig.2. In Fig.3 the **total MOSFET resistance** is plotted for three different LDD vertical profiles, A,B,C, as a function of LDD drain offset x_{off} on the drain side. It can be seen that for large x_{off} all three curves are parallel, reflecting the same $\varrho_\square$, and merge into one point at small x_{off} at which the LDD region begins to be completely buried under the lateral profile of the n^+ region. Obviously, all other quantities being identical, the difference in the curves is due to different SR effects. As expected from the simple resistor model, profile B with a peak in the bulk displays highest SR, and profile A with the peak concentration at the interface has the smallest SR contribution. To compare the effects of lateral profile extensions versus the aforementioned effects, SR is plotted, in Fig.4 for three profiles as a function of various lateral profile extensions. It can be seen that only extreme variations of the lateral profile produce SR effects comparable with effects associated with the shift of the peak concentration into the bulk.

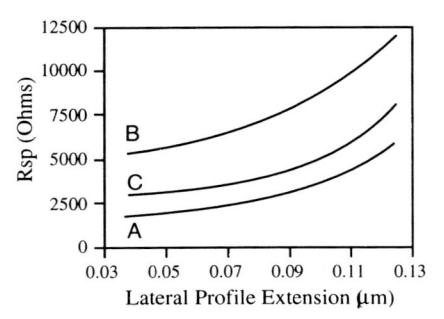

Figure 4: Spreading resistance R_{sp} for profiles A, B, C (Fig.2) as a function of lateral profile extension (slope). In agreement with ref.[1] steep lateral junctions minimize the spreading resistance. However, vertical profile variations are more significant than realistic lateral profile variations.

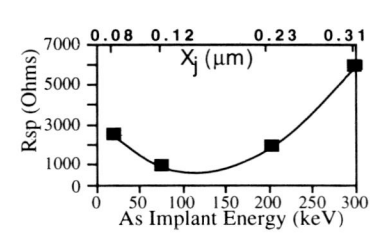

Figure 5: Spreading resistance R_{sp} for As LDD implants as a function of implant energy. (LDD dose adjusted to keep R_{sh} constant.) Reversal effect near $\chi_j = 0.15\mu m$, not predicted by previous models, is clearly seen.

4. Reverse SR Effects For Shallow Junctions

Arsenic LDD regions are investigated with implant energies, $E_{imp} = 20 - 300$ keV corresponding to $\chi_j = 0.08 - 0.32$ μm; the implant doses are adjusted to produce the same $\varrho_\square$. In Fig.5 SR is shown as a function of E_{imp}. It is seen that, in contrast to model predictions [1,3,4], SR is large not only for high E_{imp} but also **low** E_{imp}. The higher SR for very shallow junctions (low E_{imp}) is due to accumulation resistance [1] of LDD region underneath the gate, where profiles with high and low surface concentration display (in the triode region) more or less the same electron concentration. Therefore, profiles with peak shifted into the bulk take advantage of the lower resistance in the bulk, having the same high conductance at the interface by virtue of the accumulation resistance (AR). Hence, reducing the junction depth beyond 0.15 μm with a peak concentration at the interface does no longer reduce the overall SR resistance, but increases it. For deeper junctions SR effects can no longer be offset by AR effects. This happens when the surface concentration region is thicker than the accumulation region. This has important consequences for S/D definition: very shallow χ_j ($\leq 0.1\mu m$) are detrimental not only from $\varrho_\square$ fluctuations, HCI, but also from SR point of view. This study suggests that LDD profiles with concentration peak at 0.04-0.08 μm are optimal. The proposed resistor network including gate field dependent resistor $R_{l1} = R_{ac}$ modeled in ref. [1] provides an excellent basis for SPICE description of parasitic MOSFET resistances.

Acknowledgment: The authors wish to thank K. Wimmer and J. Higman for providing appropriate MINIMOS simulation capabilities and for valuable discussions.

References

[1] K. Ng and W. Lynch, **TED-33**, p.965 (1986)

[2] A. Azuma, et al, Symp. VLSI Techn., p.129 (1994)

[3] G. Baccarani and G. Sai-Halasz, EDL-4, p.27 (1983)

[4] M. Seavey, EDL-5, p.479, (1984)

[5] C. Mazuré, private communication

The Role of SEMATECH in Enabling Global TCAD Collaboration

E. M. Buturla, J. Byers, A. Husain, M. Kump, P. Lloyd, R. Manukonda, S. Runnels, D. Scharfetter

Technology Computer Aided Design
SEMATECH
2706 Montopolis Drive
Austin, Texas 78741
TCAD.DEPT@SEMATECH.ORG

Abstract

SEMATECH is a consortium of United States semiconductor manufacturers conducting precompetitive research and development for semiconductor manufacturing. This paper discusses the efforts of the Technology Computer Aided Design (TCAD) group at SEMATECH to create and foster international cooperation in a number of precompetitive TCAD-related areas. In addition, the paper addresses techniques for information exchange and the need for TCAD standards.

SEMATECH (SEmiconductor MAnufacturing TECHnology) is a consortium of United States semiconductor manufacturers working with government and academia to sponsor and conduct research aimed at assuring leadership in semiconductor manufacturing technology for the U.S. semiconductor industry. SEMATECH develops advanced semiconductor manufacturing methods, materials and equipment, and validates its development in a proving facility that simulates its members' production lines. Recently, SEMATECH has created new thrust areas, namely Design, Test, Packaging, Materials, and TCAD. This new emphasis at SEMATECH has resulted in organizational changes including the creation of a TCAD group responsible for creating a strategy and operation plan that utilizes SEMATECH resources and leverages other available external efforts.

The focus of the SEMATECH TCAD effort is to provide a precompetitive differential advantage in TCAD to the SEMATECH member companies through benchmarking, joint development, and leveraging the efforts of other research and development groups such as the Semiconductor Research Corporation and the National Labs. Transfer of resulting technology to the member companies and commercialization (if appropriate) are also part of the mission. Initially, SEMATECH TCAD efforts were kept within the United States, but recently there has been interest from the SEMATECH member companies' TCAD representatives to expand interactions with other groups. As a first step, TCAD opportunities were discussed at the recent JESSI/SEMATECH meeting in Erlangen in April.

There are a number of TCAD activities that are sufficiently important and yet pre-competitive such that they lend themselves to global cooperation. Some areas of current interest are

- Benchmarking standards
- Compact model standardization
- Interoperability standards
- Experimental impurity profiles

Benchmarks are necessary to understand how well a simulator performs. Metrics such as accuracy of the model prediction compared to experimantal data and CPU time and memory requirements can be compared between competing simulators. A good set of benchmarks can indicate to the user the ability of a particular simulator to perform for the operating range of interest. Creating a comprehensive set of benchmarks is a laborious task. It is the opinion of many TCAD practitioners that such a set would be of great value across the industry. SEMATECH recently hosted a workshop to establish a benchmarking methodology as well as to set up a suite of test cases for process, device, lithography and compact model simulators. Our plan is to define the format for test cases and then collect input files, expected results, and an explanation of the test cases. This information will be kept in an easily accesible location on the Internet and contributions to the suites will be coordinated and monitored by SEMATECH. Interactions with some European TCAD providers have occured and further are expected.

Compact model development is another prevalent but non-standardized effort among TCAD practitioners. Many compact models are currently used in the industry. However, there is no standard model which can be used for exchanging technology information in cases of joint product development and between chip manufacturers and system designers. In addition, many technology developers spend considerable effort creating accurate compact models for their technologies. To reduce the total amount of effort required, SEMATECH has sponsored workshops to study this problem and to generate a strategy for compact model evaluation and standardization. In addition, SEMATECH is supporting the very ambitious goal of creating an industry standard compact model and also is pursuing model interface development. These efforts require significant resources, thus collaboration with European developers are welcomed.

Interoperability standards is a concept that has been of interest to TCAD developers for some time. The ability to easily take some simulation code and easily integrate it with a different simulator is the objective of these standards. There was considerable interest a few years ago in "frameworks" concepts, it was felt there were technical approaches that would allow for "plug-and-play" software. Unfortunately, the implementation was very expensive, so the concept was not generally utilized. As a result, most TCAD developers have developed their own non-standard approaches. SEMA-TECH has sponsored a joint TCAD Framework integration project involving suppliers and member companies. The evaluation of that effort is currently underway. Some software developers claim that using object-oriented programming techniques will allow "plug-and-play" capability much more easily than present methods. A recent SEMATECH-sponsored workshop addressed this subject and resulted in a starting point for standards in object-oriented TCAD software.

Another important activity at SEMATECH is the creation and maintenance of a profile database. Numerous 1D impurity profiles are already in existence and others are being generated where gaps exist. Such data is costly and time consuming to create, and a reliable profile database would lead to efficiencies. Interactions between SEMATECH and IMEC for profile exchange in transient enhanced diffusion, polysilicon diffusion and silicidation have been initiated. It is expected that these efforts will result in profile exchanges.

Cooperative efforts between the US and European TCAD communities have been initiated with a number of activities now underway. Such efforts have been enhanced by the ability to communicate electronically and the use of Internet as a vehicle to provide quick and inexpensive information exchange. SEMATECH desires further global cooperative activities. Contributions and suggestions are invited. Please contact us at TCAD.DEPT@SEMATECH.ORG.

Three Dimensional Simulation for Sputter Deposition Equipment and Processes

D.S. Bang [a], Z. Krivokapic [b], M. Hohmeyer [c], J.P. McVittie [a], and K.C. Saraswat [a]

[a] Stanford University,
AEL 217, Stanford, CA 94305, USA
[b] Advanced Micro Devices,
P.O. Box 3453, Sunnyvale, CA 94088, USA
[c] ICEM CFD Engineering,
2855 Telegraph, Suite 501, Berkeley, CA 94704, USA

Abstract

A three dimensional sputter deposition simulator based upon the SPEEDIE topography simulator is presented. The simulator combines equipment models with topography evolution models in order to predict topography for VLSI metallization. Equipment scale simulation is used to determine 3D particle flux for specified wafer points. The particle flux is then used by a 3D topography simulator to determine profile evolution. The generality of the topography simulator allows deposition simulations to be performed on structures with asymmetries in the x, y, and z-directions. Examples of metal deposition simulations for contact and dual-damascene structures are presented.

1. Introduction

Sputter deposition is one of the most widely used techniques for metal deposition in VLSI fabrication. Previous approaches to modeling of sputter deposition either focused on equipment scale simulation or VLSI topography simulation [1,2]. More recent simulators combined equipment scale simulations with topography scale simulations, but continued to used symmetric VLSI topographic structures [3,4]. In this paper a 3D extension of the SPEEDIE process simulator is presented which combines equipment scale simulations with general 3D topography simulation [5].

2. Sputter Equipment Simulation

For sputter system modeling, the program uses target erosion, target emission, chamber dimensions, and analytic particle transport equations in order to determine 3D particle flux distributions for points across the wafer surface (Fig. 1). The analytic particle transport equations are generalized extensions of previous models [6-8], and are valid for chamber pressures below 2 mTorr, where the mean free paths for Al and Ti is comparable to a typical target to wafer throw distance of 5 cm [9]. Particle collisions may be considered with a Monte Carlo module which simulates collisional particle transport from target to substrate.

3. VLSI Topography Simulation

For thin film deposition process modeling, particle sticking coefficient, particle surface diffusion, and the calculated angular distribution at the substrate are applied to an initial substrate topography to determine the film's profile evolution (Fig. 2). The generality of the topography simulator allows deposition simulations to be performed on structures with asymmetries in the x, y, and z-directions (Fig. 3). Sticking coefficient is defined as the probability that a particle which strikes a surface remains at that surface. Previous experiments determined the Ti sticking coefficient to be 1.0 and surface diffusion to be negligible for sputter deposition at 250° C [9].

Film evolution is comprised of three parts: local flux calculation, surface velocity calculation, and surface regridding. Ray tracing algorithms which compares each grid's visibility versus every other grid are used to determine each surface element's visibility and direct particle flux. Surface velocity for each grid is proportional to its net flux, and the velocity of each surface node is calculated by averaging the surface velocities of each grid which surrounds the node. Initial and subsequent surface meshing is performed with ICEM [10]. The entire surface mesh is regridded after each iteration, eliminating grids which are too large, too small, or of a poor area to perimeter aspect ratio. Variable gridding size is implemented by defining subregions on the surface mesh. Surface loops are eliminated during regridding by meshing the volume above the surface topography and then keeping only those surface elements which border the meshed volume.

4. Summary

A three dimension sputter deposition simulator which combines 3D equipment scale models with 3D VLSI feature scale models to determine profile evolution for metal film deposition was presented. Examples of metal contact filling and a dual damascene liner depositions were presented with the simulator.

Acknowledgment

This research is supported by AMD, ARPA, and SRC. The authors thank IBM for use of an RS6000/590 system, and ICEM CFD Engineering for use of gridding software.

References

[1] A. Kersh, et. al., *IEDM Technical Digest,* (IEEE, Piscataway, 1992), p. 181.
[2] S. Tazawa, et. al., *IEDM Technical Digest*, (IEEE, Piscataway, 1992), p. 173.
[3] F.H. Baumann, et. al., *IEDM Technical Digest*, (IEEE, Piscataway, 1993), p. 861.
[4] D.S. Bang, et. al., *NUPAD V Technical Digest*, (IEEE, Piscataway, 1994), p. 41.
[5] SPEEDIE (Stanford Profile Emulator for Etching and Deposition in IC Engineering) is a VLSI etching and deposition simulator developed at Stanford University.
[6] I.A. Blech and H.A. Vander Plas, J. Appl. Phys., **54**, 3489, (1983).
[7] T.S. Cale, J. Vac. Sci. Technol. B **9**, 2551 (1991).
[8] M.M. Islamraja, et. al., *Proc. 7th Inter. Conf. on Numerical Analysis of Semiconductor Devices*, Copper Mt., 1991.
[9] D.S. Bang, et. al., *Proc. 10th Symp. Plasma Processing* **94-20**, (ECS, Pennington, 1994), p. 557.
[10] ICEM is a commercial gridding program developed by ICEM CFD Engineering.

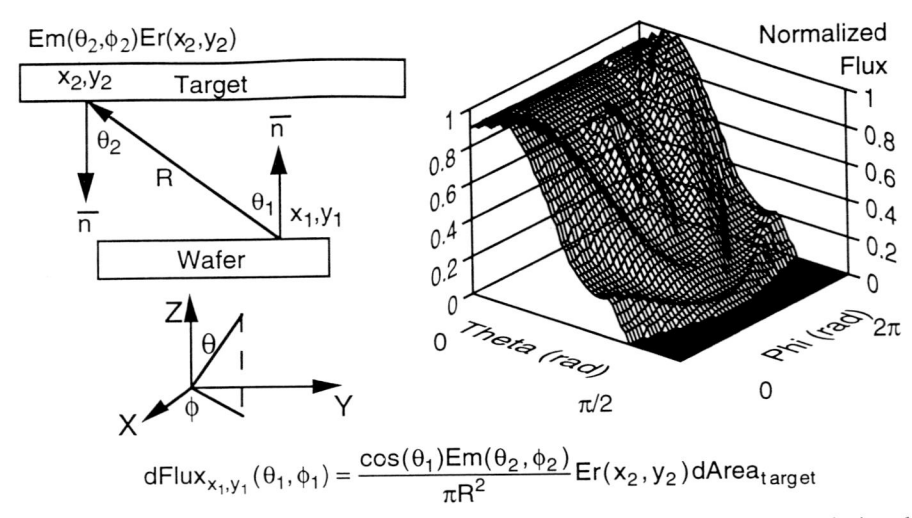

$$dFlux_{x_1,y_1}(\theta_1,\phi_1) = \frac{\cos(\theta_1)Em(\theta_2,\phi_2)}{\pi R^2} Er(x_2,y_2) dArea_{target}$$

Figure 1: The sputter system simulator uses target erosion data (Er), target emission data (Em), and chamber dimensions to determine 3D particle flux for specified wafer points. The analytic equations are valid for deposition below 2 mTorr. In this example, target erosion data and chamber dimensions for an Applied Materials system assuming cosine target emission was used to calculate the 3D flux for an off center wafer position.

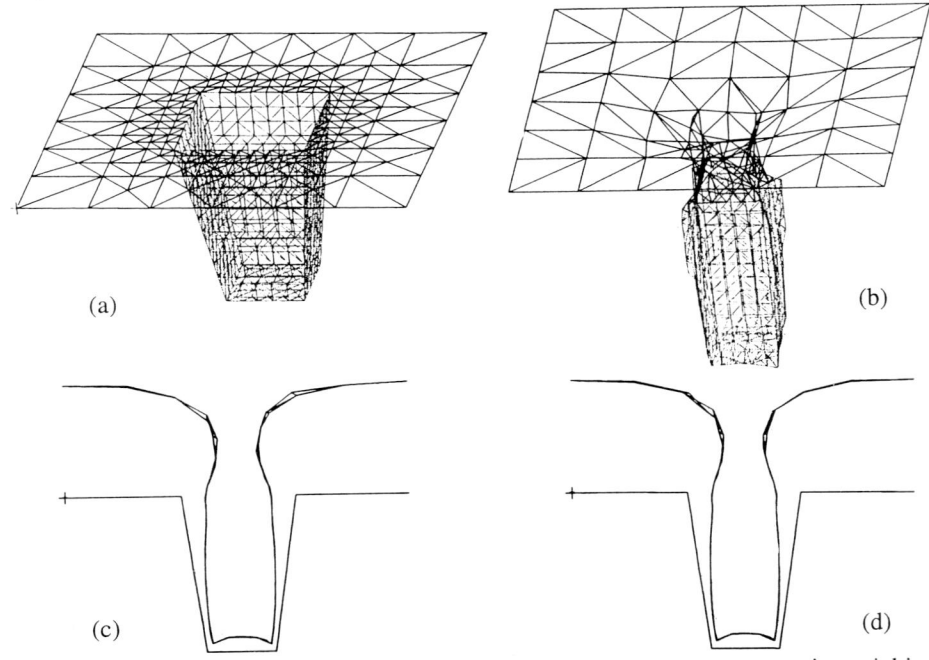

Figure 2: Simulation of 0.5 micron Ti into 0.5 micron square contact using sticking coefficient of 1.0, negligible surface diffusion, and distribution function from Fig. 1. (a) original surface mesh, (b) surface after deposition, (c) (d) two planer cross sections. Grid shading is for illustrative purposes and double grid lines in (c) and (d) represent grids which are not exactly perpendicular to the cross section plane.

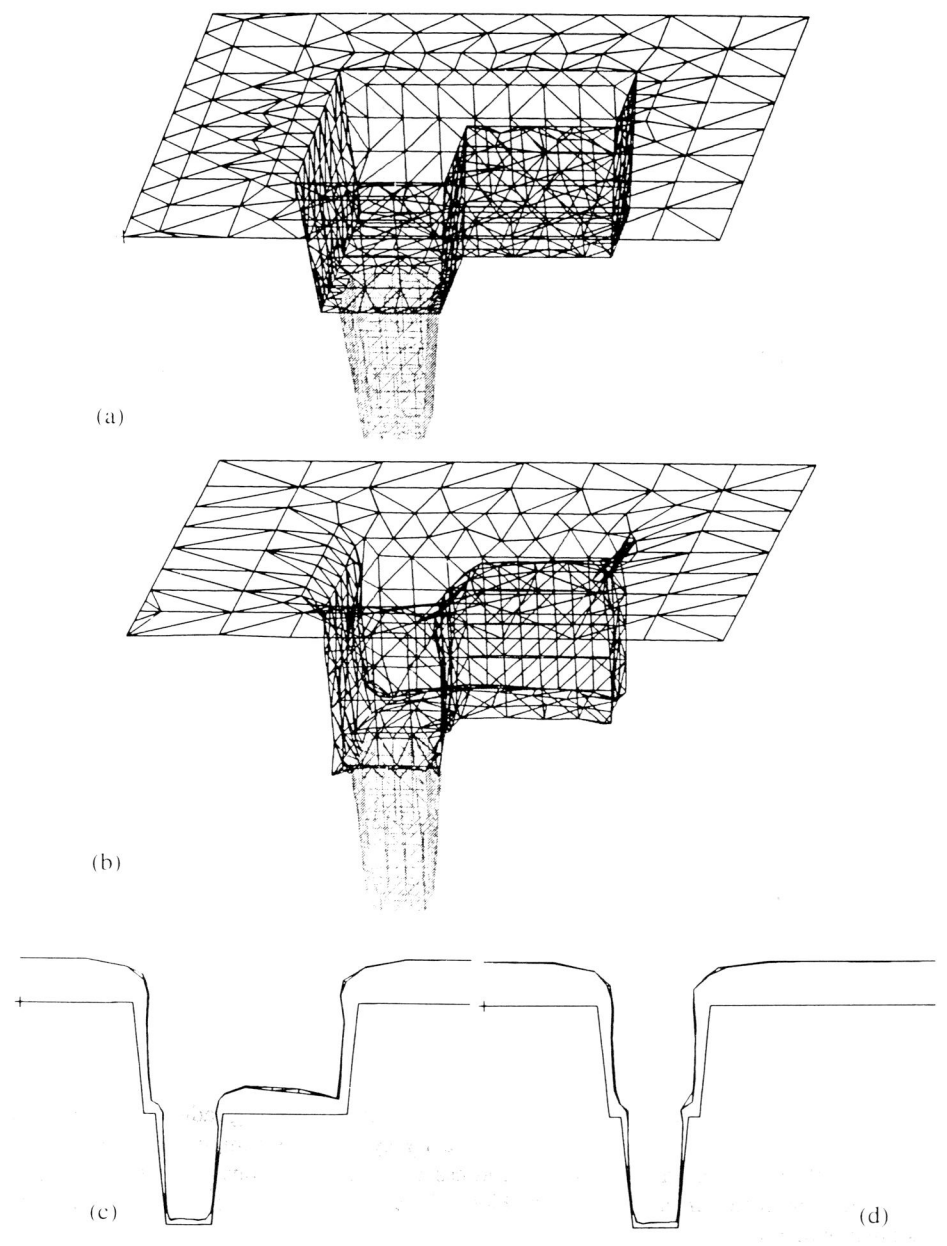

Figure 3: Simulation of 0.2 micron Ti deposition into a dual-damascene structure where the contact opening is 0.3 microns using sticking coefficient of 1.0, negligible surface diffusion, and distribution function from Fig. 1. (a) original mesh, (b) surface after deposition, (c) (d) two planer cross sections. The ability of the simulator to capture x, y, and z-axis asymmetries caused by surface asymmetries and off-center wafer position are shown in this figure. Grid shading is for illustrative purposes and double grid lines in (c) and (d) represent grids which are not exactly perpendicular to the cross section plane.

Comprehensive Reactor, Plasma, and Profile Simulator for Plasma Etch Processes

J. Zheng[a], J. P. McVittie[a], M. J. Kushner[b], and Z. Krivokapic[c]

[a]Stanford University, AEL 217, Stanford, CA 94305, USA
[b]University of Illinois, 1406 W. Green St., Urbana, IL 61801, USA
[c]Advanced Micro Devices, M/S 117, P.O. Box 3453, Sunnyvale,
CA 94088-3453, USA

Abstract

In order to achieve predictive modeling results for plasma etching processes the effects of reactor, plasma, etch chemistry, and surface kinetics have to be taken into account concurrently. The new simulator was successfully tested on a case of polysilicon etching in a He/Cl_2 chemistry.

1. Introduction

In the recent years a lot of attention has been paid to modeling of plasma etch profiles [1-5]. Etch profiles, etch rates, and their uniformities across the wafer are the main concern of process engineers. Since they depend on reactor and plasma conditions, the modeling of etch profiles has to take into account proper equipment and plasma modeling. To address this issue the profile simulator SPEEDIE (Stanford Profile Emulator for Etching and Deposition in IC Engineering) was interfaced with HPEM (Hybrid Plasma Equipment Model) from the University of Illinois. Modeling results were experimentally verified, using trenches and overhang test structures for a He/Cl_2 chemistry in a conventional commercial RIE etcher.

2. Simulator

The modules of the simulator are shown in Fig. 1. Fluxes of etching radicals and ions are predicted by using a 2-dimensional (cylindrically symmetric) computer model HPEM [6,7], which combines modules which address either different physical phenomena or different time scales in an iterative fashion. The HPEM is composed of a series of modules which are iterated to a converged solution. The electromagnetic module (EM) generates inductively coupled electric and magnetic fields in the reactor. These fields are then used in the Electron Monte Carlo Simulation (EMCS) module. In the EMCS electron trajectories are followed for many RF cycles producing the electron energy distribution as a function of position and phase. These distributions are used to produce electron impact source functions, which are transferred to the Fluid Kinetics Simulation module (FKS). In the FKS, continuity and ion momentum equations are solved for all neutral and charged particle densities, and Poisson's equation is solved for the electric potential. The FKS also imports an externally generated advective flow field produced in a hydrodynamic module (HM). The HM is a solution of the fully compressible fluid conservation equations for continuity, momentum, and energy. Slip boundary conditions may be used to extend the fluid equations to low pressure (5-20 mTorr). The plasma conductivity produced in the FKS is passed to the EM, and the species densities and time dependent electrostatic potential are passed to the EMCS. The modules are iterated until cycle averaged plasma densities converge. Acceleration algorithms are used to speed the rate of convergence of the model. For inductively coupled plasma (ICP) systems, all modules are employed. When simulating RIE systems, the EM is not used.

A circuit model is employed to self-consistently compute the DC bias of the powered electrode. Control surfaces are drawn at the boundaries of all driven or grounded metals. The conduction current (particle fluxes) and displacement currents passing through these surfaces (which may be inside dielectric materials) are computed on a cycle by cycle basis. The net currents flowing through these surfaces are then used to charge a blocking capacitor in series with the voltage generator. since the charging time for the blocking capacitor may be long compared to the simulation time, the size of the blocking capacitor is dynamically changed during the simulation.

The HPEM also includes a semianalytic model in the HPEM to address conditions where the mesh spacing, Δx, exceeds the actual sheath thickness, λ_s, at the boundaries. If the mesh is too coarse to resolve the sheath, the apparent sheath thickness is that of the mesh spacing adjacent to the wall. The sheath voltage is then dropped across the width of the numerical cell. In not resolving the sheath, the electric field in the sheath is diminished by the ratio of $\lambda_s/\Delta x$. To compensate for this effect in the HPEM, we separately compute the conduction and displacement current j_d to each location on the plasma-material boundary. With knowledge of the plasma density adjacent to the sheath we can compute the expected RF sheath amplitude thickness λ_s using the Lieberman sheath model [8]. If $\lambda_s < \Delta x$, then j_d is corrected. The revised value of j_d is then used to compute circuit parameters, among them the DC substrate bias.

The sheath potential and thickness from the HPEM are used with the Monte Carlo Sheath Transport module (MCST) of SPEEDIE to determine the ion velocity distribution (IVD). This module uses a hard ball collision model and a linear sheath field. It also considers the effect of pre-sheath ion heating on the IVD [9]. The IVD along with the fluxes of neutrals and ions from the HPEM are used in the 3-D Fluxes (3DF) module of the SPEEDIE to calculate local fluxes taking into consideration 3-D visibility along the two-dimensional cross section profiles of test structures. The surface kinetics (SK) model calculates local etch and deposition rates of the surface element and gives the etching rate of a material as a function of fluxes, ion energies, incident angles, and substrate temperature. The SK model used is a modification of the Langmuir limited adsorption model [4] to account for the effect of inhibitor deposition and its removal by ion bombardment on the effective ion flux [10,11]. The profile evolution module implements the surface movement caused by the etch and deposition rates from the SK. The surface movement algorithm uses a new segment based geometric method, in which the conservation of material, or entropy condition, is obeyed. This is done by using the under and over lapping of the cells swept out by the initial movement of the segment to determine the correct segment length and position. With this method there is no need for delooping. Re-emitted fluxes are used in the 3DF locally and in the HPEM.

3. Results

The complex simulator was tested in the case of polysilicon etching in a He/Cl_2 gas mixture. The process used in a commercially available RIE reactor was at 230 mTorr, power of 250 W, and gas flow of 75 and 150 sccm for Cl_2 and He, respectively. The following species were included in the HPEM: He, He*, He^+, Cl_2, Cl, Cl^+, Cl^-, e. The ion fluxes we quote (Fig. 3) are the sum of He^+, Cl_2^+, and Cl^+. Special overhang structures (Fig. 4) were used to extract surface parameters for the SK model and the angular spread of the IVD. The results (Fig. 5) show that the details of the experimental profile are well captured, and should be useful to predict evolution of other structures and the effects of process conditions on etch profiles.

4. Conclusions

Detailed modeling of etch profiles and studies of etch uniformities across the wafer can be obtained by using a profile simulator that takes into account reactor, plasma properties, etch chemistry, and surface kinetics effects. The feasibility of such an effort was proven in a case of a pure Cl_2/He gas mixture. Etching additives increase the complexity of computation but can be used if the proper chemical data are available.

5. Acknowledgment

Authors would like to thank Allison Holbrook for wafer preparation and etching.

4. References

[1] J. D. Bukowski, R. A. Stewart, D. B. Graves, Proceedings of the 10th Symposium on Plasma Processing, pp.87, The Electrochemical Society, 1994
[2] D. C. Gray, H. H. Sawin, J. W. Butterbaugh, J. Vac. Sci. Tech. *A9*, pp.779, 1991
[3] S. Tachi and S. Okudaira, J. Vac. Sci. Techn. *B4*, pp.459, 1986
[4] T. M. Mayer and R. A. Barker, J. Vac. Sci. Tech. *21*, pp.757, 1982
[5] H. Hübner and M. Engelhardt, J. Electrochem. Soc., *141*, pp.2453, 1994
[6] P. L. G. Ventzek, R. J. Hoekstra, and M. J. Kushner, J. Vac. Sci. Tech. *B12*, pp.461, 1994
[7] P. L. G. Ventzek, M. Grapperhaus, and M. J. Kushner, J. Vac. Sci. Tech. *B12*, pp.3118, 1994
[8] M. A. Lieberman, Trans. Plasma Sci. *17*, pp.338, 1989
[9] J. Zheng, J. P. McVittie, R. P. Brinkman, 41st AVS Symp., to be published in J. Vac. Sci. Tech., 1995
[10] G. S. Oehrlein, Y. Zhang, D. Vender, M. Haverlag, J. Vac. Sci. Tech. *A12*, pp.323, 1994
[11] J. Zheng and J. P. McVittie, Nupad V Technical Digest, pp.37, 1994

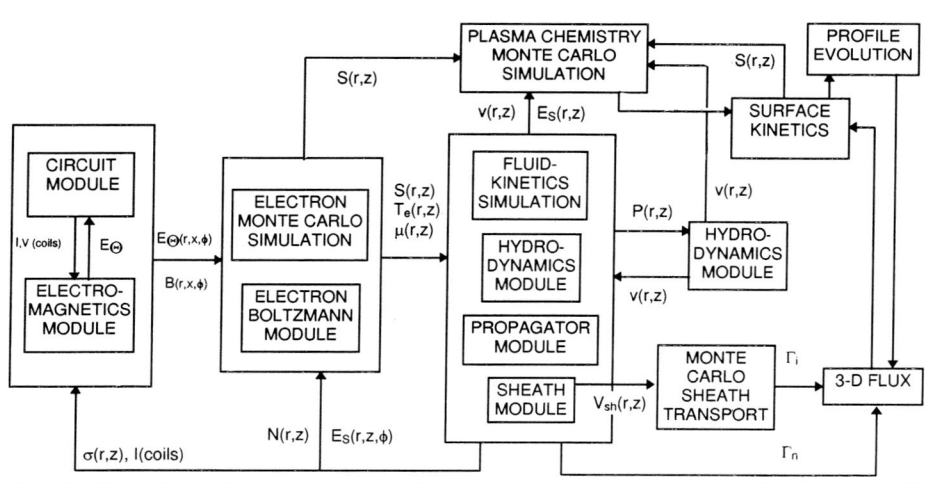

Fig. 1: Flow chart of the simulator, indicating equipment, plasma, chemistry, and profile evolution modules.

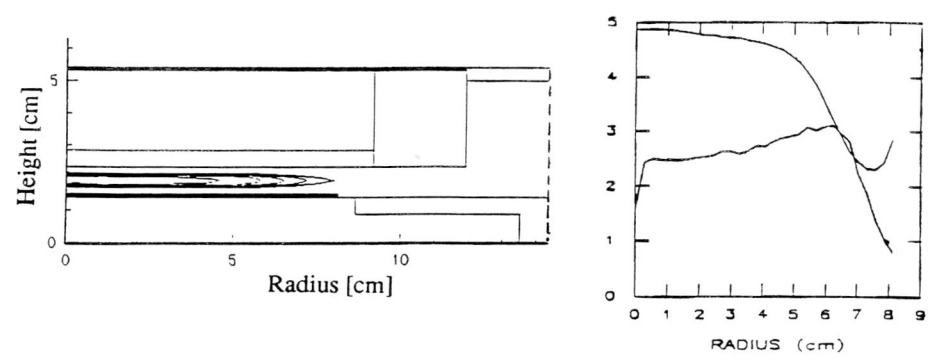

Fig. 2: Contour plot of total ion density in the reactor. maximum density is 5.5 e11 cm^{-3}.

Fig. 3: Radial distribution of ion and Cl flux in the reactor.

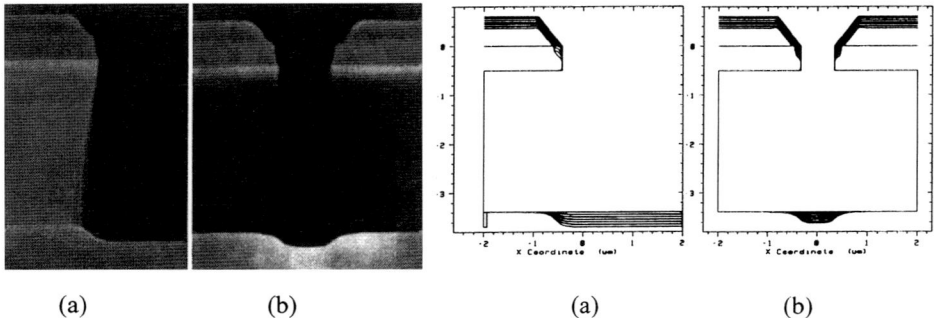

Fig. 4: SEM pictures of the overhang structures with wide opening (a) and narrow opening (b), used to determine the surface kinetics parameters.

Fig. 5: Simulated profiles of the overhang structures form Figs. 4.

Modeling the Wafer Temperature in a LPCVD Furnace

A. Kersch[a], M. Schäfer[b]

[a]SIEMENS AG, Corporate Research and Development
Otto-Hahn-Ring 6, 81730 Munich, GERMANY
[b]Fraunhofer-Institut für Integrierte Schaltungen,
Schottkystrasse 10, 91058 Erlangen, GERMANY

Abstract

In the paper the application of a newly developed radiation model to the modeling of a batch furnace is shown. The model includes multi-band spectral dependence and the treatment of semitransparent materials The results are compared with experimental data.

1. Introduction

Batch processing with large diameter wafers requires a new generation of optimized furnaces. The challenge of the equipment simulation is to provide a sufficiently accurate radiation model for the description of the heat transfer. In the past, radiation models for the complex multiwafer geometry have been developed based on the diffuse, grey approximation [1][2]. In this paper we present results obtained from first principles with a new radiation model based on the diffuse approximation with multi-band spectral dependence and allowing for semitransparent materials. The results are compared with experimental data. The radiation model is included in the CFD-simulator PHOENICS-CVD [3].

2. The Model

In a discretization of the surfaces of the enclosure into N surface elements

$$F_{ij}^l = \frac{1}{Area(j)} \int_j \int_i t_{ij}^l \frac{|n(y) \cdot r||n(x) \cdot r|}{\pi \, r^2} dy dx$$

is the generalized viewfactor in the presence of semi-transparent surfaces. t_{ij}^l is the transmittance from j to i in the $l-th$ band of the spectrum, $r = y - x$ the distance between x and y and $n(x)$ and $n(y)$ the surface normals. The radiative flux from j to i is calculated in the diffuse approximation with the help of the Gebhard factors

$$G_{ij}^l = a_i^l(T_0) \sum_{m=1,N} \left(\delta_{im} - F_{im}^l r_m^l\right)^{-1} F_{mj}^l$$

taking into account direct and multiply reflected radiation, r_m^l the reflectivity. The net flux in the $l - th$ band from j to i is then described by the matrix

$$R_{ij}^l = \left(G_{ij}^l - \delta_{ij}\right) e_j^l(T_0) Area(j) P_j^l(T_0)$$

and with

$$P_j^l(T_0) = \frac{1}{\sigma T_0^4} \int_{\nu_l}^{\nu_{l+1}} \frac{c_1 \nu^3}{e^{c_2 \nu/T} - 1},$$

is the fraction of black body radiation in the band l. The total radiation into i is then

$$Q_i = \sum_{j=1,N} \left(\sum_{l=1,L} R_{ij}^l\right) T_j^4$$

where T is the actual surface temperature during the iteration. The matrix R is calculated with an initial estimate T_0 of the surface temperature and coupled into PHOENICS-CVD.

Figure 1 shows an example of optical properties in the banded approximation.

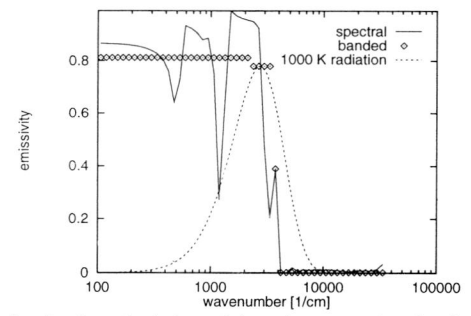

Figure 1: The hemispherical emissivity of fused quartz in the banded model in comparison with the spectral emissivity. Also shown is the black body radiation energy distribution at $1000°K$.

3. The Simulation

The geometry of the batch furnace with three heating elements in an axisymmetric simplification is shown in Figure 2.

The semitransparent quartz tube extends to the outside of the heater, the tube ends are cooled steel and constitute a radiative sink. The batch of 68 wafers was resolved into 31 wafer in a first simulation. The paddle constitutes the only deviation from axial symmetry. In one simulation the paddle was neglected, in another represented effectively as two cylindrical regions in back and front of the batch. The surfaces were discretized into about $N = 350$ radiative zones, the computation time for R was about $2hr$ on a $HP750$.

The temperature drop at the back and front end of the batch is caused on the one hand by the radiative heat loss of the first and the last wafer to the cold inlet and outlet doors. On the other hand it is caused by the view of the wafers in the batch to the reactor tube with a temperature profile. To separate these both effects, the front and the back side of the wafer batch was represented as a perfectly reflecting

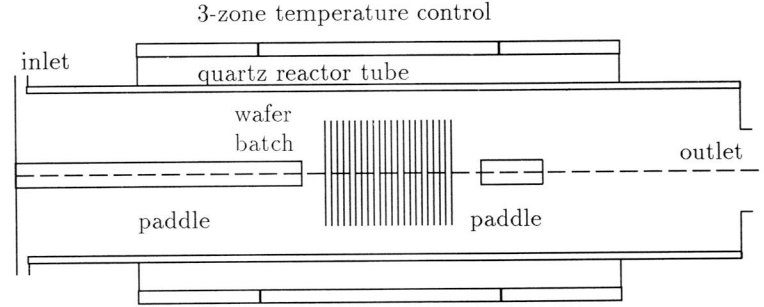

Figure 2:
Geometry of the LPCVD furnace with three heating zones. For the heating zone a constant emissivity of 0.85 was assumed; for the quartz, the silicon and the silicon carbide paddle the emissivity was taken in the banded approximation.

mirror to eliminate the heat loss to the doors, The results in Figure 3 show, that the dominant effect is the view to the doors. Furthermore, the results show a simulation with an effective representation of the paddle as cylindrical surfaces. This additional radiative interaction of the front and back side wafers diminish the temperature drop.

Figure 4 shows the comparison of the axial temperature profile with data obtained at FhG IIS-B. The quantitative agreement is good. A further comparison is made in Figure 5 with the maximal temperature difference on the wafer. The agreement is good. The experimental inhomogeneity in the center should be close to the error of the measurement.

The potential of the model to perform optimization is furthermore shown in Figure 6. The improved axial temperature profile was obtained after increasing the power of the back and front heating elements.

Figure 7 finally shows the temperature distribution in the reactor. Clearly visible is the increased temperature drop of the batch towards the inlet side with the black opening.

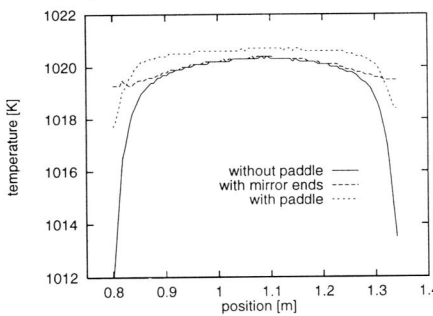

Figure 3: Temperature profile in the center of the batch (a) without paddle, (b) with mirrors at end of batch and (c) with effective representation of paddle.

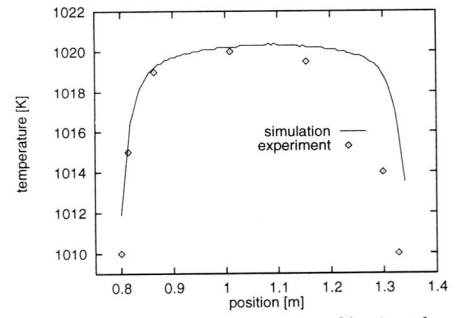

Figure 4: Temperature profile in the batch from the simulation without paddle in comparison with data.

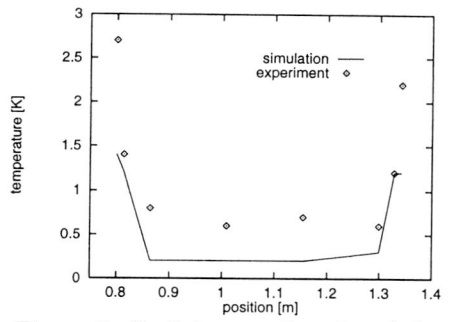

Figure 5: Radial nonuniformity of the wafers from the simulation in comparison with data.

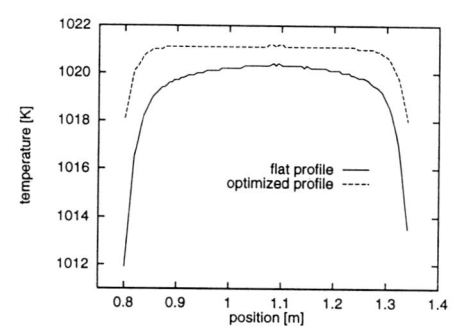

Figure 6: Optimized temperature profile in the batch with the temperatures in the heat zones set to $T_1 = 1063°K$, $T_2 = 1023°K$ and $T_3 = 1073°K$ instead of a constant $T = 1023°K$.

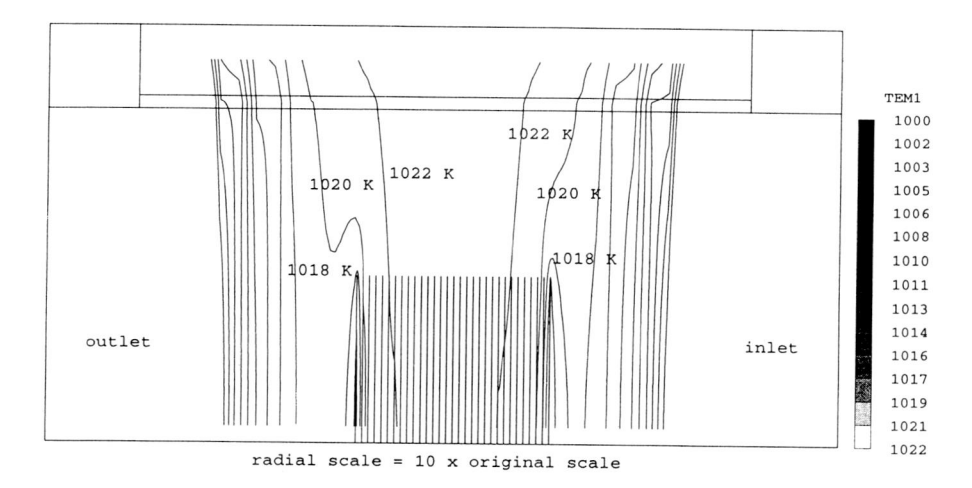

Figure 7: The temperature distribution in the reactor. Visible is the reduced temperature of the batch at the side towards the inlet. The inlet front is steel with an optical black opening of $30mm$ radius. The outlet front side is entirely steel.

References

[1] H. De Waardt and W. L. De Koning, *Automatica*, vol.28, p.243, 1982

[2] T. A. Badgwell, I. Trachtenberg and T. F. Edgar, "Modeling the Wafer Temperature Profile in a Multiwafer LPCVD Furnace", *J. Electrochem. Soc.*, vol 141, p.161, 1994

[3] PHOENICS-CVD, CHAM, Wimbledon, 1995

Determination of Electronic States in Low Dimensional Heterostructure and Quantum Wire Devices

Ali Abou-Elnour and Klaus Schünemann

Technische Universität Hamburg–Harburg
Arbeitsbereich Hochfrequenztechnik,
Wallgraben 55, D-21073 Hamburg, Germany.

Abstract

An efficient variational technique is applied to solve Schrödinger's equation in two dimensions. This model is then self-consistently used with a two-dimensional Poisson's equation solver to determine the electronic states inside low dimensional heterostructure and quantum wire devices. Finally, the advantages and limitations of the present model are discussed.

1. Introduction

The existance of a one-dimensional electron gas in low dimensional and quantum wire heterostructure devices requires an accurate and efficient model to determine the electronic states and the carrier transport properties in these devices. The self-consistent solution of Poisson's and Schrödinger's equations is belived to be one of the most accurate methods which can be used to characterize the operation and to optimize the structure of these devices.

A number of authors investigated different models which are based on the finite difference method to solve Poisson's and Schrödinger's equations self-consistently in two dimensions [1-3]. The accuracy and the computational efficiency of these models strongly depend on mesh size and discretization techniques. An alternative efficient method to solve Schrödinger's equation in one dimension by using a variational technique has been suggested [4-6]. In the present work, this method is extended to two dimensions to determine the electronic states in low dimensional and quantum wire heterostructure devices. Finally, the advantages and limitations of this method are discussed.

2. Model

The effective mass, two-dimensional Schrödinger equation is given by

$$-\frac{\hbar^2}{2m^\star}[\frac{\partial^2\psi}{\partial x^2} + \frac{\partial^2\psi}{\partial y^2}] + V(x,y)\psi = E\psi \tag{1}$$

where V(x,y) means potential energy, E eigenenergy, ψ wave function corresponding to the eigenenergy E, m^* effective mass, and $\hbar$ Planck's constant. For a semiconductor structure of length a and width b, the wave equation can be expanded as

$$\psi = \sum_{n=1}^{N}\sum_{m=1}^{M} a_{nm} \sin\frac{n\pi x}{a} \sin\frac{m\pi y}{b}. \tag{2}$$

The accuracy of this solution depends on the number of the expansion functions N and M. If N and M are infinite, the obtained wave functions are identical to the true ones. However, finite N and M still lead to very good accuracy.

The coefficients a_{nm} are obtained by means of variational integrals whose stationary values correspond to the true eigenvalues when the true eigenfunctions are inserted in the integral. The variational integral for E is given by

$$E = \frac{\frac{\hbar^2}{2m^*}\int_0^a\int_0^b [(\frac{\partial\psi}{\partial x})^2 + (\frac{\partial\psi}{\partial y})^2]dxdy + \int_0^a\int_0^b V(x,y)\psi^2 dxdy}{\int_0^a\int_0^b \psi^2(x,y)dxdy}. \tag{3}$$

The condition that (3) should be stationary is satisfied if the first-order variation in E vanishes for an arbitrary first-order variation $\delta\psi$ in ψ. Applying this condition, a matrix equation $[R]\mathbf{A} = E\mathbf{A}$ is obtained where $\mathbf{A}$ is a column vector with the elements a_{nm} and $R_{nm,n'm'} = \mathbf{I}_1 + \mathbf{I}_2$ where $\mathbf{I}_1$ and $\mathbf{I}_2$ are given by

$$\mathbf{I}_1 = \frac{\hbar^2}{2m^*}\int_0^a\int_0^b [(\frac{\partial\psi}{\partial x})^2 + (\frac{\partial\psi}{\partial y})^2]dxdy = \frac{\hbar^2}{2m^*}[(\frac{n\pi}{a})^2 + (\frac{m\pi}{b})^2]\delta_{nn'}\delta_{mm'}, \tag{4}$$

$$\mathbf{I}_2 = \int_0^a\int_0^b V(x,y)\psi^2 dxdy = \frac{4}{ab}\int_0^a\int_0^b V(x,y)\sin\frac{n\pi x}{a}\sin\frac{m\pi y}{b}\sin\frac{n'\pi x}{a}\sin\frac{m'\pi y}{b}dxdy. \tag{5}$$

Solving these equations, the subband energies and the corresponding wave functions are determined and then used to calculate the carrier distribution. Knowing the carrier distribution, the electrostatic potential is then calculated by solving Poisson's equation in two dimensions. Knowing the electrostatic potential, the new potential energy function V(x,y) is calculated and the effective V(x,y) is expressed as a linear combination of its new and old values given by

$$V_{new}(x,y) = \omega V_{old}(x,y) + (1-\omega)V_{old}(x,y) \tag{6}$$

where ω means relaxation constant which is introduced to obtain the solution safely. Schrödinger's equation is again solved to determine the new eigenenergies and the corresponding wave functions which are then used to recalculate the carrier distribution. The procedure is repeated until the initial and final values of V(x,y), within the same iteration, differ by less than a specified error.

3. Numerical results and computational performance

Schrödinger's and Poisson's equation are solved self-consistently to determine the electronic states of the structure shown in fig. 1. The electronic states and hence the device operation are strongly affected by the terminal voltages. The applied potential V_{g3} can be varied to control the barrier height and the distance between the quantum wires. The potential energy and the carrier distribution for different bias are displayed in figures 2 and 3, respectively.

From these results one can see that the present model is able to accurately determine the electronic states in two-dimensional structures. The method is able to take into account the variations in the effective mass and the boundary conditions in a more flexible way than previous models. Moreover, the accuracy does not depend on mesh size and discretization and is only affected by the number of expansion functions N and M. Using 20 expansion functions in each direction, 20-30 iterations are required to get the solution with a maximum error of 0.5 meV in V(x,y). The required CPU time for each iteration is about 80 seconds on a HP700 work station.

4. Conclusions

An efficient variational technique is investigated to solve Schrödinger's equation in 2D and is applied with a 2D Poisson solver to determine the electronic states inside low-dimensional heterostructure devices. This method overcomes the limitations of previous finite-difference methods which are arising from mesh size and discretization. Moreover, the closed form of the wave functions makes the model more tractable to determine scattering rates and transport properties inside these devices.

Acknowledgment

The authors are thankful to Prof. Dr. A. S. Omar from our institute for fruitful discussions and to the Deutsche Forschungsgemeinschaft for financial support.

References

[1] T.Kerkhoven et al, J. Appl. Phys., vol. 68, 3461 (1990).
[2] G.Snider et al, J. Appl. Phys., vol. 68, 2849 (1990).
[3] U. Ravaioli et al, Superlattice and Microstructures, vol.11, 343 (1992).
[4] A. Abou-Elnour and K. Schünemann, J. Appl. Phys., vol. 74, 3273 (1993).
[5] A. Abou-Elnour and K. Schünemann, Solid-State Elec., vol. 37, 27 (1994).
[6] A. Abou-Elnour and K. Schünemann, Solid-State Elec., vol. 37, 1817 (1994).

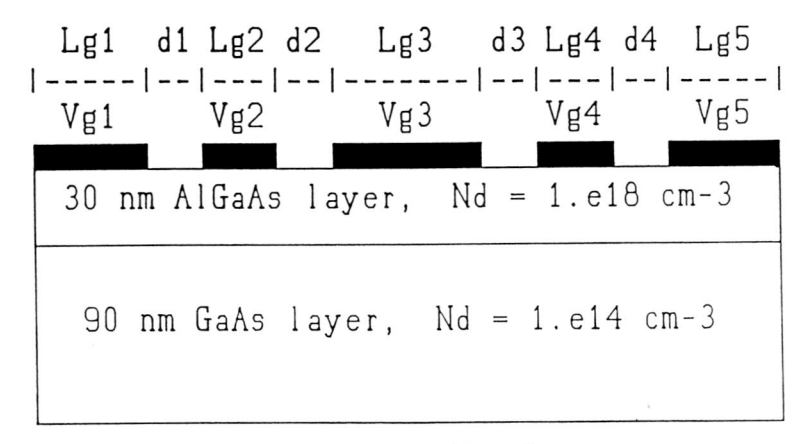

```
    Lg1   d1 Lg2 d2    Lg3      d3 Lg4 d4   Lg5
  |-----|--|---|--|--------|--|---|--|-----|
    Vg1      Vg2     Vg3         Vg4      Vg5
```

| 30 nm AlGaAs layer,   Nd = 1.e18 cm-3 |
| 90 nm GaAs layer,   Nd = 1.e14 cm-3 |

Fig. 1 The simulated AlGaAs/GaAs heterostructure.
$L_{g1}=L_{g5}=36$ nm, $L_{g2}=L_{g4}=24$ nm, $L_{g3}=48$ nm, $d_1=d_2=d_3=d_4=18$ nm

(a) (b)

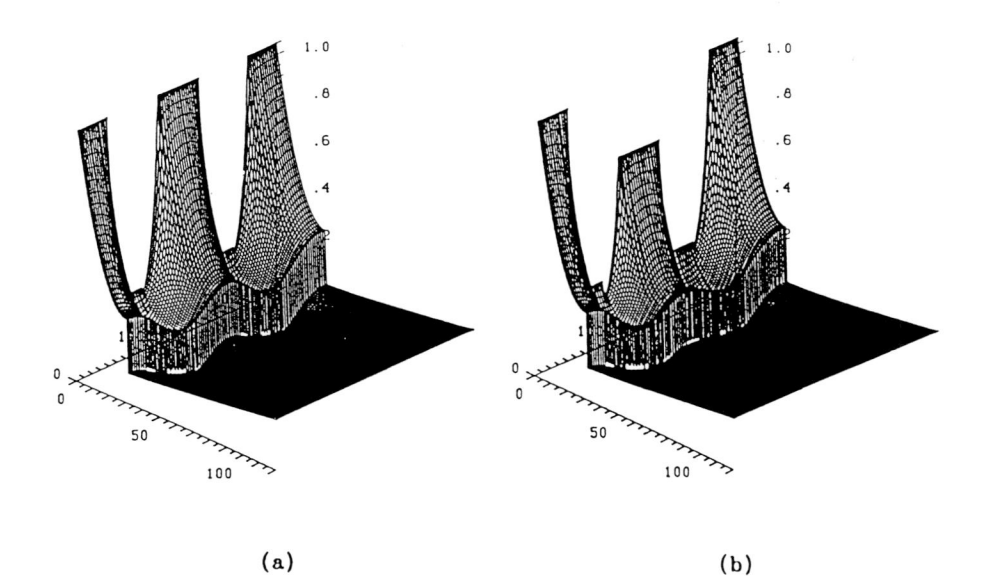

Fig. 2 The potential energy distribution [eV]
(a) $V_{g1}=V_{g5}=V_{g3}=0.0$ V, $V_{g2}=V_{g4}=0.8$ V
(b) $V_{g1}=V_{g5}=0.0$ V, $V_{g3}=0.3$ V, $V_{g2}=V_{g4}=0.8$ V

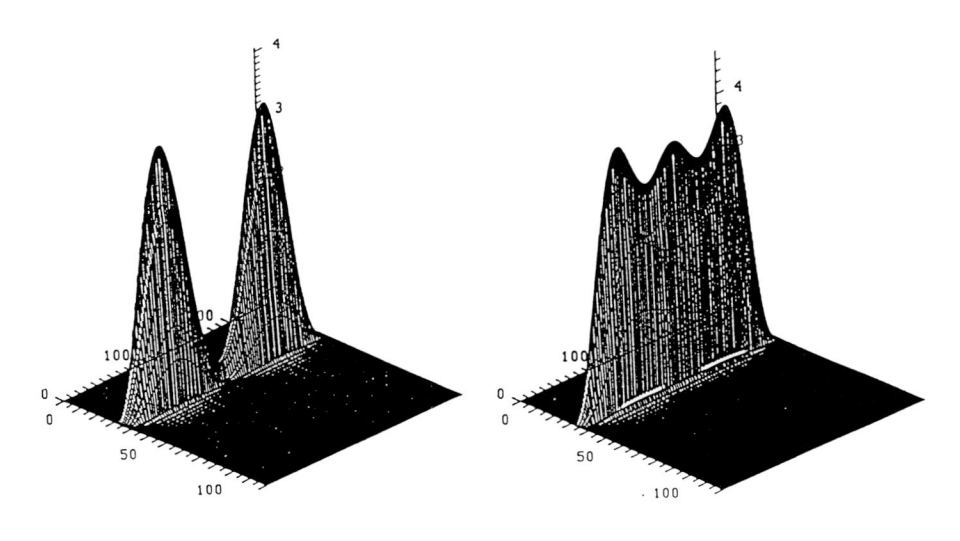

(a) (b)

Fig. 3 The carrier concentration $[10^{17}cm^{-3}]$
(a) $V_{g1}=V_{g5}=V_{g3}=0.0$ V, $V_{g2}=V_{g4}=0.8$ V
(b) $V_{g1}=V_{g5}=0.0$ V, $V_{g3}=0.3$ V, $V_{g2}=V_{g4}=0.8$ V

An Exponentially Fitted Finite Element Scheme for Diffusion Process Simulation on Coarse Grids

S. Mijalković

Faculty of Electronic Engineering, University of Niš,
Beogradska 14, 18000 Niš, YUGOSLAVIA

Abstract

A new finite element scheme for diffusion process simulation, which allows coarse grid spacings in the areas of exponentially varying concentrations and fluxes, is proposed. It employs a nonlinear test function obtained from local divergence free conditions. Two-dimensional test computations show clear superiority of the exponentially fitted finite element scheme over the standard approach, as well as its robustness regarding irregular grid geometry.

1. Introduction

The gradually increasing complexity of the multiparticle diffusion models and necessity to simulate in higher dimensions persistently challenge computational efficiency of the modern process simulators. An obvious guideline to cope with the efficiency problems in the discretization phase is to make a grid structure as coarse as possible for a given tolerance of the discrete solution accuracy. To this end, considerable effort has been directed to the development of advanced adaptive grid generation techniques for both the finite difference (FD) and finite element (FE) methods. On the other hand, the discretization fitting to the particular features of the solution is much less exploited in the diffusion process simulation as an additional grid coarsening technique. The FD scheme that exploits the exponential flux behavior to allow the coarse grid spacings in diffusion process simulation, has been proposed by Lowther [1]. The main intention of this paper is to propose a corresponding FE scheme, which could be also robust for irregular element geometry.

2. Problem formulation

The transport of the particles (impurities or point-defects) involved in the diffusion process is commonly modeled by the diffusion equations in the form:

$$\frac{\partial C}{\partial t} - \nabla \cdot (D\nabla C + Z\mu C \nabla \varphi) = r. \qquad (1)$$

C and Z are the concentration and charge state of the particle. r, D, μ and φ are concentration dependent reaction term, diffusion coefficient, mobility and built-in electric potential, which govern various interactions among particles. Introducing the normalized chemical potential $u = \log C$ and making use of the Einstein relation ($D/\mu = V_T$, with V_T representing the thermal voltage), the diffusion equation (1) can be expressed as

$$-\nabla \cdot \boldsymbol{F} + R = 0 \quad \text{with} \quad \boldsymbol{F} = De^u \nabla v. \tag{2}$$

Here $v = u + Z\varphi/V_T$ is the normalized particle electrochemical potential, while the zero-order term $R = \partial e^u / \partial t - r$ consists of the time-derivative and reaction terms. In (2), we consider both u and v (actually, u and φ) as well behaved quantities, which is consistent with the basic assumption that the components of the flux $\boldsymbol{F}$ are exponentially varying quantities.

The diffusion equation (2) is defined in a bounded domain $\Omega \subseteq \boldsymbol{R}^n$ with piecewise smooth boundary $\partial \Omega$. An initial state $u = u_0$ at $t = 0$ is defined in $\bar\Omega = \Omega \cup \partial \Omega$. Let the boundary $\partial \Omega$ consists of Dirichlet ($\partial \Omega_d$) and Neumann ($\partial \Omega_n$) segments with boundary conditions: $u = u_0$ on $\partial \Omega_d$ and $\boldsymbol{F} \cdot \boldsymbol{n} = 0$ on $\partial \Omega_n$, where $\boldsymbol{n}$ denotes the normal vector to the boundary. Without any loss of generality we consider the discretization of $\bar\Omega$ into N_e nonoverlapping elements Ω_e constructed over N_n nodes as simplexes, i.e., intervals, triangles or tetrahedra for $n = 1, 2$ or 3.

3. Exponentially fitted FE scheme

A class of generalized FE methods [2] is used to derive the FE discretization of (2). As a weak integral statement we consider

$$\int_\Omega \boldsymbol{F} \cdot \nabla \psi_i d\Omega + \int_\Omega R\psi_i dx = 0, \tag{3}$$

where ψ_i represents an exponentially fitted test function which satisfies the local divergence free problems

$$\nabla \cdot (De^u \nabla \psi_i) = 0 \quad \text{in} \quad \Omega_e \ni i \quad \text{with} \quad \psi_i(k) = \delta_{ik}. \tag{4}$$

Here i and k denote grid nodes while δ_{ik} is the Kronecker delta. Although the local divergence free problems (4) cannot be solved in a closed form, it seems appropriate to assume that $De^u \nabla \psi_i$ varies at least linearly in Ω_e to achieve first order accuracy. With this assumption, ψ_i is given by

$$\psi_i = (\phi_i - \alpha I_{\Omega_e}(e^u)) \cdot e^{u_i - u} + \alpha e^{u_i} \quad \text{and} \quad \alpha = \frac{\nabla \phi_i \cdot \nabla u}{\nabla I_{\Omega_e}(e^u) \cdot \nabla u}, \tag{5}$$

where ϕ_i is the standard linear test function and $I_{\Omega_e}(\cdot)$ represents a linear interpolation from the nodal values in the element Ω_e. In the special case of the piecewise constant u, we have $\psi_i = \phi_i$. The exponentially fitted FE scheme is obtained from (3) and (5), using ϕ_i as a finite element basis for u and v and approximating D as a piecewise constant discrete function.

The test function (5) produces an upwinding effect, that is similar to the streamline Petrov-Galerkin methods, but with no need for an external adjustment of the numerical dissipation. To avoid any occurrence of the singular and counter upwinding effects, we propose to perform $I_{\Omega_e}(\cdot)$ in obtuse elements as one-dimensional linear interpolation between nodes X_1 and X_{n+1}; here X_k, ($k = 1, \ldots, n+1$) denote coordinates of the element Ω_e nodes in ascending order along the X axis, that is aligned with ∇u. In this way, $\nabla I_{\Omega_e}(e^u) \cdot \nabla u > 0$ is guaranteed for $|\nabla u| > 0$.

4. Test computations

As a model problem we employ here the two-dimensional diffusion equation (2) in a rectangular domain ($0.4\mu m \times 0.4\mu m$), with $v = u$, $r = 0$ and assuming constant $D = 5 \cdot 10^{-15} cm^{-2}/s$. It is useful for the practical analysis of discretization schemes since the exact differential solution is available [3] for 2-D Gaussian initial state (here $R_p = 0.063\mu m$, $\Delta R_p = 0.021\mu m$, $\Delta R_{pl} = 0.018\mu m$ and $C_{max} = 10^{20} cm^{-3}$). The grid structure is selected as extremely coarse with $N_e = 106$ and $N_n = 70$. The exact solution at $t = 300s$ is shown in Figure 1. The gray and white areas denote the concentration ranges $10^i < C < 10^{i+1}$ starting from $i = 12$ at the bottom. The discrete solutions obtained with the standard FE scheme (test LIN-FE) [4] and a new exponentially fitted FE scheme (test EXP-FE) are shown in Figure 2 and Figure 3, respectively. The result of the third test (EXP-FE(o)), that employs the grid structure consisting exclusively of obtuse triangles ($N_e = 318$ and $N_n = 176$), is shown in Figure 4. For the more quantitative examination of the test computations, the relative error in the junction depth is analyzed for all above tests including also the exponentially fitted FD scheme (test EXP-FD) [1]. The progression of the junction depth error during diffusion and the dependence of the junction depth error on the substrate concentration are shown in Figure 5 and Figure 6, respectively.

It could be observed that the standard FE scheme tends to significantly overestimate the amount of diffusion and the junction depth. On the other hand, with the exponentially fitted FE scheme, besides substantial improvements of the solution accuracy, the junction depth error shows a stable accumulation during diffusion, as well as uniform distribution in the wide range of the substrate concentrations.

5. Conclusion

The efficiency of the diffusion process simulation could be significantly improved with exponentially fitted FE schemes that allow coarser grid spacings then standard approaches. A new FE scheme has combined properties of the streamline Petrov-Galerkin and the divergence free upwinding methods. The superiority of a new FE scheme over its standard counterpart is demonstrated in the test computations with the known exact analytical solution. Unlike the exponentially fitted FD scheme [1], a new exponentially fitted FE scheme appears to be robust on grid structures with obtuse triangles.

References

[1] R. E. Lowther, "A discretization scheme that allows coarse grid-spacings in finite-difference process simulation," *IEEE Trans. Computer-Aided Design*, vol. 8, pp. 837–841, Aug. 1989.

[2] I. Babuska and J. E. Osborn, "Generalized finite element methods: their performance and their relation to mixed methods," *SIAM J. Numer. Anal.*, vol. 20, No. 3, pp. 510–536, June 1983.

[3] H. Lee, R. Dutton and D. Antoniadis, "On redistribution of boron during thermal oxidation in silicon," *J. Electrochem. Soc.*, vol. 126, pp. 2001–2007, 1979.

[4] R. Ismail and G. Amaratunga, "Adaptive meshing schemes for simulating doping diffusion," *IEEE Trans. Computer-Aided Design*, vol. 9, pp. 276–289, March 1990.

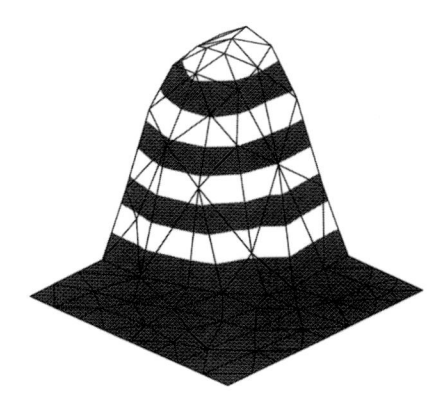

Figure 1: The exact solution.

Figure 2: The discrete solution in the test LIN-FE.

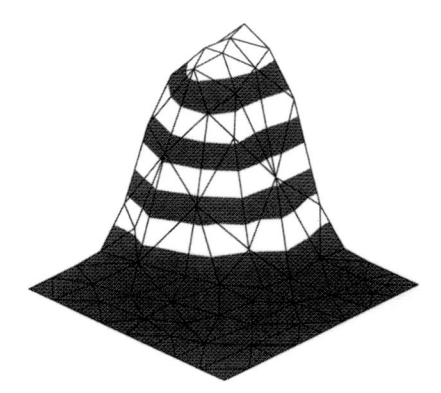

Figure 3: The discrete solution in the test EXP-FE.

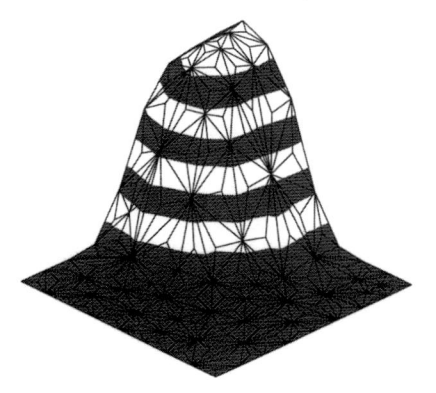

Figure 4: The discrete solution in the test EXP-FE(o).

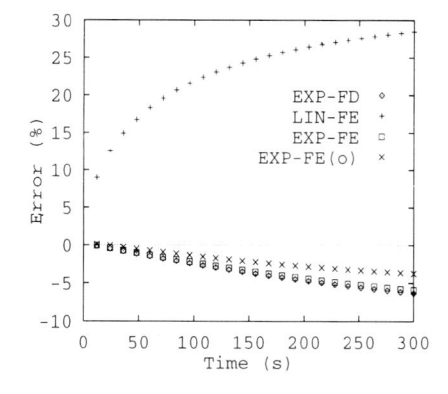

Figure 5: The progression of the junction depth error during diffusion.

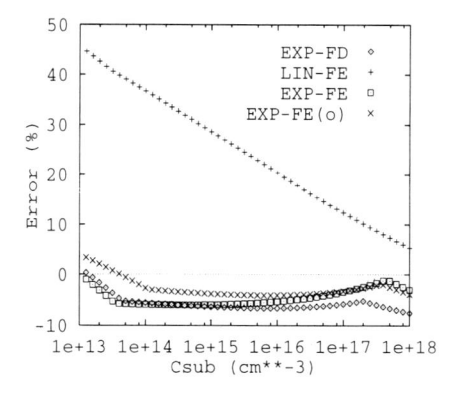

Figure 6: The dependence of the junction depth error on the substrate concentration.

Achievement of Quantitatively Accurate Simulation of Ion-Irradiated Bipolar Power Devices

P. Hazdra and J. Vobecký

Department of Microelectronics, Czech Technical University in Prague,
Technická 2, 166 27 Praha 6, Czech Republic

Abstract

Quantitatively accurate simulation of He^{2+} irradiated power diode was achieved. The results have showed that accuracy of device parameter prediction depends essentially on accurate prediction of total defect concentrations and their projected ranges while other parameters, e.g. the defect profile shape, have been proved irrelevant. The $VO^{(0/-)}$ defect level was found to be dominant for devices febricated on low-doped FZ NTD n-Si.

1. Introduction

On SISDEP'93 we presented the original way of simulating behaviour of silicon devices that were subjected to hydrogen and helium irradiation [1]. The simulation procedure comprises primary defect generation by means of Monte-Carlo simulation code TRIM-90, re-scaling of the primary defect profiles into appropriate deep-level profiles by use of experimentally oriented database, and the device simulation with full trap dynamics involved [2]. This approach provided a good qualitative agreement when applied to hydrogen, helium, and electron irradiations and their combinations [3].
The main role of the experimentally oriented database, which is a key element of the simulation, is to predict the resulting defect electronic structure and its spatial distribution from primary damage deposition and information about both the material and irradiation procedure. It works with many parameters obtained experimentally (see e.g. [4,5,7]) with big scatter and various accuracy (e.g. deep level capture and emission rates). Moreover, many re-scaling algorithms (introduction rates and distributions of secondary defects, e.g. divacancies VV, vacancy-oxygen pairs VO, vacancy-phosphorus pairs VP, etc.) are hard to verify when we proceed to higher irradiation energies. Therefore, we focused our attention in this paper on influence of particular parameters and re-scaling procedure on simulation results. The sensitivity analysis performed and careful comparison with experiment provided us with information which parameter and re-scaling factor is crucial for the proper prediction of device operation and which is irrelevant. This information is of importance for those making tedious and expensive experiments for data extraction in order to improve their models and accuracy of input simulation parameters.

2. Experiment and simulation

The device under test was P$^+$PNN$^+$ 370μm long, 16 mm diameter power diode (2.5kV/100A) fabricated on <111> FZ NTD 110 Ωcm n-type silicon. The double-diffused p-layer (8μm, $N_A = 3\times10^{19}$ cm^{-3}, 50μm, $N_A = 5\times10^{17}$ cm^{-3}) and diffused n-layer (15μm, $N_D = 10^{21}$ cm^{-3}) formed p$^+$ and n$^+$ emitter, respectively. The diode was irradiated from the anode side with defocused He^{2+} cyclotron beam with final energy 12 MeV at different doses ranging from 8x10^9 to 6x10^{10} cm^{-2}. Helium irradiation was chosen because it results only in pure damage (vacancy-related) defect levels, the distribution of which follows that one of vacancies to be accurately simulated by available means.

The magnitudes of input parameters used for the simulation procedure [2] (defect activation energies, capture and introduction rates) we chose according to our own experimental results and carefully verified results from refs. [4 - 7]. Simulated output parameters were chosen according to standard measurements available. The forward voltage drop V_f was simulated for the dc current of 100 A from which the subsequent reverse recovery process was initiated. The reverse recovery current was decreasing with the slope of 100 A/μs by use of a resistive-inductive load. The reverse recovery time t_{rr} was determined in usual way from 90% and 25% of the maximum reverse current. The soft factor was defined as a ratio of the fall and storage time [2].

3. Results and discussion

Fig.1 shows VO pairs distributions in the irradiated diode predicted by our system (simulation II $\square$) and measured by DLTS ($\blacksquare$). In order to study the influence of the defect profile shape we narrowed (sim.I $\triangle$) and widened (sim.III $\diamond$) the FWHM while the integral defect density and projected range R_p, which was predicted with good accuracy, was unchanged. The simulated trade-offs of device parameters with the above mentioned profiles are compared with measured ones on Fig.2.

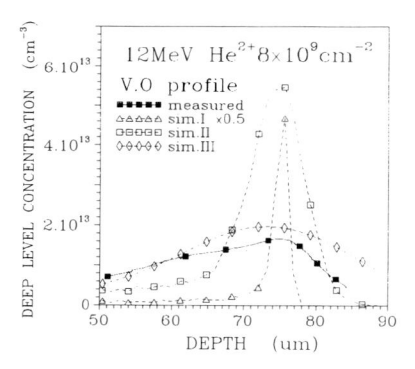

Figure 1: Predicted ($\square$) and measured ($\blacksquare$) profiles of VO$^{(0/-)}$. Simulation profiles ($\triangle$, $\diamond$) have the same R_p but different FWHM.

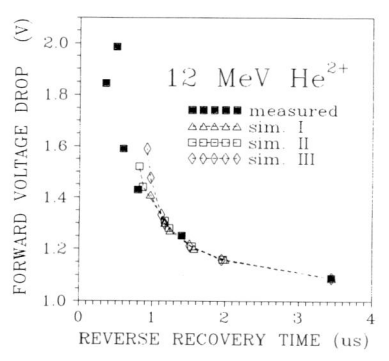

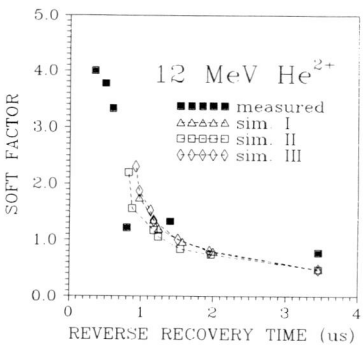

Figure 2: Measured ($\blacksquare$) and predicted ($\square$) V_f-t_{rr} (upper) and S-t_{rr} (lower) trade-off of 12 MeV He^{2+} irradiated P$^+$NN$^+$ diode. Data ($\triangle$, $\diamond$) correspond to level profiles with various FWHM (see Fig.1).

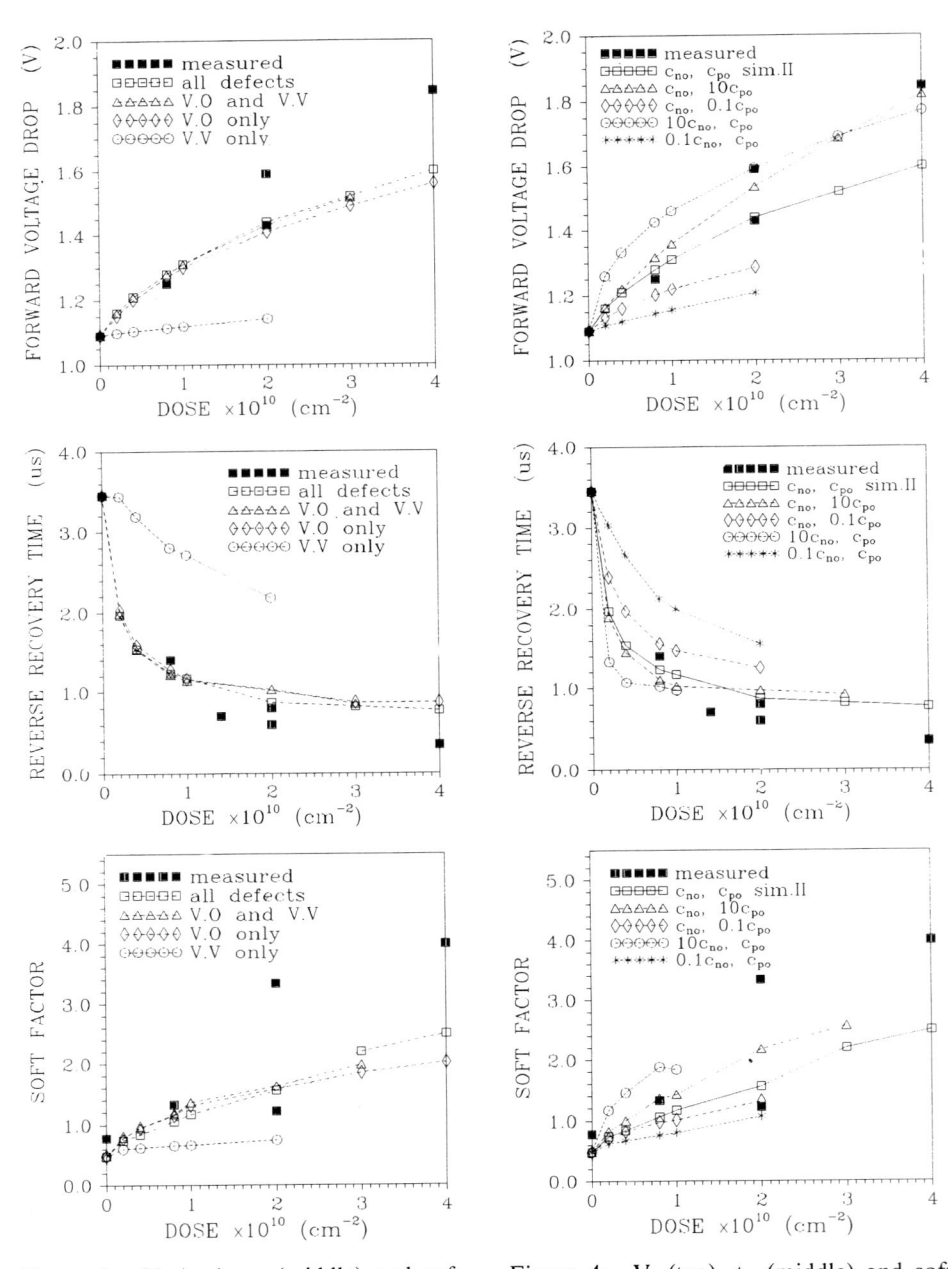

Figure 3: V_f (top), t_{rr} (middle) and soft factor (bottom) versus irradiation dose. Measured (■) and simulated data (sim. II) for all levels (□), VO$^{(0/-)}$ and VV$^{(0/-)}$ (△), VO$^{(0/-)}$ only (◇) and VV$^{(0/-)}$ only (○).

Figure 4: V_f (top), t_{rr} (middle) and soft factor (bottom) versus irradiation dose. Measured (■) and simulation (sim. II) data - capture rates c_{no}, c_{po} of level VO$^{(0/-)}$ were changed up/down by one order.

It implies that the defect profile shape is of less importance for achievement of accurate results. On the other hand, it is important to accurately predict both the projected range R_p [2], which affects the shape of the trade-off curve, and the total defect density that controls the position within a trade-off curve given by R_p. This means to know precisely an actual irradiation dose and defect introduction rate of particular defects.

The significance of individual radiation defect levels on simulation accuracy is clearly demonstrated on Fig.3. The simulations were performed for the following cases: all generally accepted defect levels resulting from helium irradiation (E1, E2, E3, H5) [2] are involved (□), two levels that are believed to be the most important ones, i.e. E1 ($VO^{(0/-)}$) and E3 ($VV^{(0/-)}$), are involved (△), and these levels are involved individually in (◇, resp. ○). These results clearly show that the VO pair is of the major importance for given starting material. It puts a clear insight into a widely discussed question concerning the defect take over from the angle of both the ON-state and reverse recovery parameters. Last but not least parameters to discuss are the capture rates the magnitudes of which are presented in publications with a rather big scatter. Fig.4 shows the agreement of measured parameters with simulated ones (sim.II) together with scatter of simulation outputs when the capture rates c_n and c_p of the VO pairs are 10 times decreased or increased. It corresponds to the range in which these rates are known from literature. It is clearly shown that c_n is important to know accurately in the first place. Further on, it is evident that capture rates are sufficient to be known with one digit accuracy.

4. Conclusions

Irradiation dose, defect introduction rate, projected range, and VO pair parameters were found relevant when quantitative agreement between measured and simulated parameters, describing both the ON-state and reverse recovery process, has to be achieved. The defect profile shape is irrelevant if both the projected range and integral defect density along the whole ion track are kept constant.

References

[1] P. Hazdra and J. Vobecký, "Modelling of Localized Lifetime Tailoring in Silicon Devices," *Simulation of Semiconductor Devices and Processes*, vol.5, Springer Verlag Wien, pp. 437-440, 1994.

[2] P. Hazdra, J. Vobecký, "Accurate Simulation of Fast Ion Irradiated Power Devices," *Solid-State Electronics*, vol. 37, No. 1, pp. 127-134, 1994.

[3] J. Vobecký, P. Hazdra, J. Voves, F. Spurný, J. Homola, "Accurate Simulation of Combined Electron and Ion Irradiated Silicon Devices for Local Lifetime Tailoring," *Proc. of the 6th ISPSD'94*, Davos, pp. 265 - 270, 1994.

[4] A. Hallén, "Lifetime Control in Silicon by Fast Ion Irradiation," PhD. Thesis, *Acta Universitatis Upsaliensis*, Uppsala, 1990.

[5] P. Hazdra, V. Hašlar, M. Bartoš, "The Influence of Implantation Temperature and Subsequent Annealing on Residual Implantation Defects in Silicon," *Nucl. Instr. Meth. Phys. Res.* B55, pp. 637 - 641, 1991.

[6] W. Wondrak, A. Boos, "Helium Implantation for Lifetime Control in Silicon Power Devices," *Proc. of ESSDERC'87*, Bologna, pp. 649 - 652, 1987.

[7] N. Keskitalo, "A Charge Carrier Lifetime Model for Proton Irradiated Silicon", *Licenciate of Technology*, Dept. of Radiation Sciences, Uppsala University, 1994.

Modeling of Substrate Bias Effect in Bulk and SOI SiGe-channel p-MOSFETs

Guo-fu Niu, Gang Ruan and Ting-ao Tang

Department of Electronic Engineering, Fudan University
Shanghai 200433, CHINA

Abstract

This paper describes the effect of substrate bias in bulk and SOI SiGe-channel p-MOSFETs. Applying a positive substrate bias to the bulk SiGe p-MOSFETs results in considerable shift of the SiGe channel threshold voltage towards more negative values, and considerable reduction of the saturated SiGe channel hole density, but has negligible effect on the surface channel threshold voltage and hole density. In SOI SiGe p-MOSFETs, the threshold voltages and hole densities are all negligibly affected by the negative substrate bias.

1. Introduction

SiGe-channel p-MOSFETs have generated substantial research work because of higher mobility of holes confined to the SiGe channel[1-6]. In earlier works[4-5], the threshold voltages and hole densities for both the SiGe channel and the surface channel were studied by assuming a zero substrate bias. However, in application circuits, the pass transistors, differential input transistors, and series load/drive transistors in CMOS gates usually have non-zero substrate bias. Thus, this paper addresses the effect of substrate bias on the threshold voltages and hole densities. The bulk n^+ gate SiGe channel modulation doped p-MOSFETs[1], and the SOI SiGe p-MOSFETs[6] will be described.

2. Analysis

2.1 Bulk modulation doped SiGe p-MOSFETs

The bulk modulation doped SiGe p-MOSFETs consists of an n^+ poly gate, a gate oxide of $t_{ox}=5$nm, an undoped Si cap of $t_{cap}=5$nm, an undoped graded SiGe channel of $t_{sige}=10$nm, an undoped Si buffer of $t_{buff}=5$nm, a modulation doped layer of 4nm with the areal boron density of Q_m, and an n-type substrate of $N_b=5 \times 10^{16}/cm^3$. Ge is graded linearly from 0.1 at the bottom SiGe/Si interface to 0.3 at the top SiGe/Si interface. Q_m is set to $1.107 \times 10^{12}/cm^2$ to adjust the SiGe channel threshold voltage at $V_b=0V$ to -0.6V which is usually required by digital circuits. At given gate-to-source bias V_g and substrate-to-source bias V_b, the Poisson's equation in SiGe/Si hetero-structure is solved iteratively[2] by assuming that the quasi-Fermi level of holes is higher than that of electrons by qV_b[7]. In implementation, the simulation depth should increase with increasing substrate bias since the depletion layer becomes wider.

The SiGe channel threshold voltage $V_{t,sige}$ is defined as the gate voltage at which the hole concentration at the top SiGe/Si interface equals the substrate doping. Similarly, the gate voltage at which the hole concentration at the surface equals the substrate doping is defined as the surface channel threshold voltage $V_{t,s}$[4]. Given V_b, $V_{t,sige}$ and $V_{t,s}$ are found using the dichotomizing search.

Fig. 1 shows the simulated SiGe channel and surface channel threshold voltages versus substrate bias. $V_{t,sige}$ increases considerably with increasing V_b from 0 to 3.0V, while $V_{t,s}$ increases slightly. As in Si MOSFETs, the substrate bias enlarges the threshold SiGe/Si interface potential by V_b, thus increasing the bulk charge and the electric field, and hence $V_{t,sige}$. At $V_{t,s}$, because the valence band at the top SiGe/Si interface is pinned at the hole quasi-Fermi level(Fig. 2), and the valence band at the surface is lower than the hole quasi-Fermi level by the following constant difference since the surface hole concentration equals N_b:

$$E_{fp} - E_v(0) = kT \ln(\frac{N_v}{N_b}) \tag{1}$$

the potential drop across the Si cap can be estimated from the valence band variation as:

$$\phi(t_{cap}) - \phi(0) \approx \frac{\Delta E_{v,t}}{q} - \frac{kT}{q} \ln(\frac{N_v}{N_b}) \tag{2}$$

where N_v is the state density in the valence band, E_{fp} is the hole quasi-Fermi level, and $\Delta E_{v,t}$ is the valence band offset at the top of the SiGe channel. The field across the Si cap according to Eq. (2) is independent of V_b, thus resulting in a nearly constant $V_{t,s}$.

Fig. 3 gives the SiGe channel and surface channel hole densities versus gate voltage at $V_b = 0$ and 1.0V. The saturated SiGe channel hole density which is a measure of the hole confinement capability is reduced by the substrate bias. This reduction also results from the increased bulk charge, which increases the electric field across the bottom SiGe/Si interface, thus reducing the SiGe channel hole density according to Gauss law. Hole confinement is thus degraded in SiGe p-MOSFETs operating at non-zero substrate bias such as pass transistors or series load transistors in CMOS gates. The saturated SiGe channel hole density versus substrate bias at t_{cap}=5 and 7nm is shown in Fig. 4, where the variation of V_b from 0 to 3.0V reduces the SiGe channel hole density by about $0.8 \times 10^{12}/cm^2$. The SiGe channel hole density increases markedly with thinning the Si cap at all V_b because of the rise of the field at the top of SiGe channel with $1/t_{cap}$, as in the case of zero V_b[5]. On the other hand, the surface channel hole density curve at $V_b = 1.0V$ almost coincides with that at $V_b = 0V$(Fig. 3), implying that the surface channel inversion hole density at a given gate voltage is negligibly affected by the substrate bias.

2.2 SOI SiGe p-MOSFETs

The channel cross section of the SOI SiGe p-MOSFETs[6] consists of an n+ poly gate, a 6.5nm gate oxide, a 7nm undoped Si cap, a 10nm undoped SiGe channel, a 5nm undoped Si buffer, a 150nm $10^{15}/cm^3$ n-type doped silicon film, a 410nm buried oxide, and a silicon substrate with the same doping as the silicon film. The calculation is the same as that for the bulk devices except that the quasi-Fermi levels for electrons and holes in the silicon film merge with one another and always equal those in the source because of the insulating buried oxide. Another difference is that the substrate-to-source bias in SOI p-MOSFETs is negative instead of being positive in the bulk case since n-channel and p-channel MOSFETs on SOI share the same substrate which is usually grounded[8].

The dependence of $V_{t,sige}$ and $V_{t,s}$ on substrate bias in SOI SiGe p-MOSFETs is given in Fig. 5. Compared to the bulk case, the SiGe channel threshold voltage is much less sensitive to the substrate bias, resulting from the thick buried oxide which causes very weak capacitive coupling between the substrate and the silicon film. For reasons similar to the bulk case, the substrate bias changes the surface channel threshold voltage negligibly. In Fig. 6, the hole densities versus gate voltage at V_b = -5.0 and 0V are plotted. It is worth noting that V_b is usually -5.0V in SOI p-MOSFETs at 5.0V supply[17]. Moving V_b from 0 to -5.0V increases the SiGe channel hole density due to increased control over the depletion charge by the substrate(back gate) which can reduce the electric field across the bottom SiGe/Si hetero-interface. The SiGe channel hole density thus increases according to the Gauss law. The increase, however, is little because of very small buried oxide capacitance. On the other hand, there is no observable difference in the surface channel hole density curve between the two substrate biases. The modeled SiGe channel hole density increases with thinning the Si cap independent of the substrate bias, as in the bulk case.

3. Conclusion

In summary, the positive substrate bias applied to the bulk devices considerably increases the SiGe channel threshold voltage and reduces SiGe channel hole density, while has little effect on the surface channel threshold voltage and hole density. In the SOI case, the threshold voltages and hole densities all show little dependence on the negative substrate bias. The SiGe channel hole density increases with thinning the Si cap independent of the substrate bias in both bulk and SOI devices.

References

[1] S. Verdonckt-Vandebroek, E.F. Crabbe, B.S. Meyerson, D.L. Harame, P.J. Restle, J.M.C. Stork, J.B. Johnson, *IEEE Trans. Electron Devices*, ED-41, 90-101(1994)
[2] P.M. Garone, V. Venkataraman, J.C. Sturm, *IEEE Electron Device Lett.*, EDL-13, 56-58(1992)
[3] S. Voinigescu, C. Salama, J. Noel, T. Kamins, *Proc. ESSDERC*, 143-147(1994)
[4] K. Iniewski, S. Voinigescu, J. Atcha, C.A.T. Salama, *Solid-State Electronics*, 36, 775-783(1993)
[5] G.F. Niu, G. Ruan, T.A. Tang, *Solid-State Electronics*, 38, 323-329, 1995
[6] D.K. Nayak, J.C.S. Woo, G.K. Yabiku, K.P. MacWilliams, J.S. Park, K.L. Wang, *IEEE Electron Device Letters*, EDL-14, 520-522(1993)
[7] S. Sze, *Physics of semiconductor devices*, John Wiley & Sons, 1981
[8] J.P. Colinge, *Silicon-On-Insulator Technology: Materials to VLSI*, Kluwer Academic, 1990

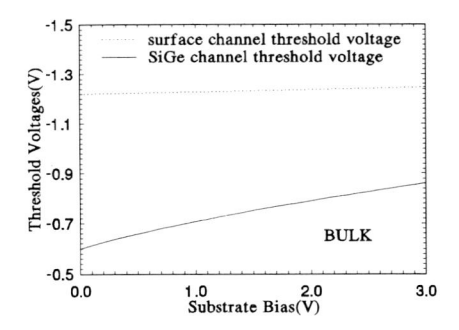

Fig. 1 Threshold voltages versus substrate bias in bulk SiGe p-MOSFETs.

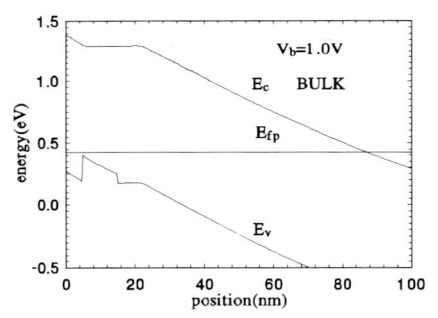

Fig. 2 Band diagram at $V_b = 1.0V$.

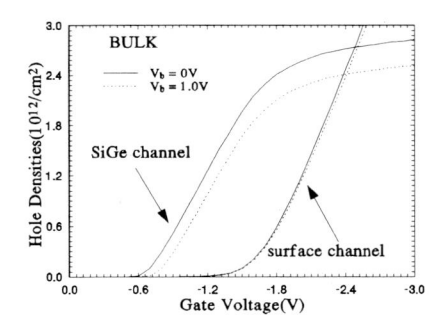

Fig. 3 Hole densities versus gate voltage at different substrate biases in bulk device.

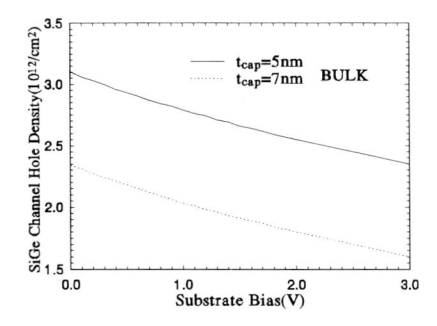

Fig. 4 Saturated SiGe channel hole density versus substrate bias for different Si cap in bulk device.

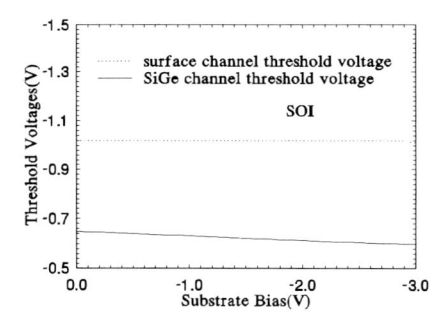

Fig. 5 Threshold voltages versus substrate bias in SOI SiGe p-MOSFETs.

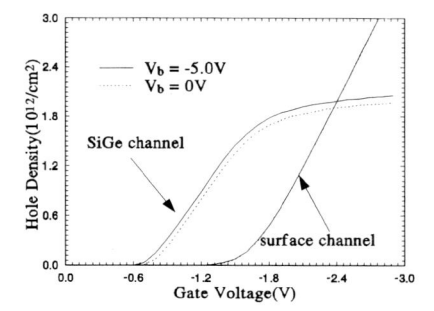

Fig. 6 Hole densities versus gate voltage at different substrate bias in SOI device.

A Very Fast Three-Dimensional Impurity Profile Simulation Incorporating An Accumulated Diffusion Length and Its Application to the Design of Power MOSFETs

Shiroo Kamohara[a], Masahiro Sugaya[a], Hitoshi Matsuo[b]

[a]Semiconductor Development Center,
Semiconductor & Integrated Circuit Div., Hitachi, Ltd.,
1-280, Higashi-koigakubo Kokubunji, Tokyo 185, Japan
[b]Central Research Laboratory, Hitachi, Ltd.,
1-280, Higashi-koigakubo Kokubunji, Tokyo 185, Japan

Abstract

In this paper, we introduce a method for the fast simulation of 3D impurity profile simulation. We analytically integrated the approximated non-linear transport equation and derived an analytical equation of 3D non-linear diffusion by introducing a new physical parameter, called the Accumulated Diffusion Length (ADL). This method allows us to simulate a 3D-profile by using a coupled 1D simulator and analytical equations, where the ADL is obtained during 1D simulation. Our new methodology is implemented in the 3D process simulator, SPIRIT-III/ADL. Our 3D-TCAD system composed of SPIRIT-III/ADL and our in-house device simulator was applied to the design of 3D-shaped power MOSFETs.

1. Introduction

Technology CAD (TCAD) systems are becoming indispensable for predicting device performance for optimize process parameters before fabrication. While two-dimensional (2D) TCAD systems are widely used [1], three-dimensional (3D) systems have not yet become practical because of the great amount of CPU time needed. Some accelerated approaches to impurity profile simulation have been proposed [2,3]. Although NIM [2] which numerically integrates the non-linear impurity transport equation gives good accuracy, this approach does not make 3D simulation practical because of its insufficient CPU-time enhancement. FABRICS [3] which uses a coupled one-dimensional (1D) simulator and an expander with analytical equations can achieve more than 100 times enhancement of CPU time, but users must define the parameter which describes lateral diffusion. This parameter must be introduced empirically because the impurity diffusion is non-linear.

In this work, we analytically integrate the approximated non-linear transport equation and derive an analytical equation of 3D non-linear diffusion by introducing a new physical parameter, called the Accumulated Diffusion Length (ADL). The ADL is obtained during 1D simulation and does not need to be defined by users. This method allows us to simulate a 3D-profile by using a coupled 1D simulator and analytical equations. Our new methodology is implemented in the 3D process simulator, SPIRIT-III/ADL. Our 3D-TCAD system composed of SPIRIT-III/ADL and our in-house device simulator CADDETH3 [4] was applied to the design of 3D-shaped power MOSFETs [5] and we confirmed that this TCAD system can be efficiently applied.

2. Methodology of 3D Process Simulation

By discretizing the time dependence, the impurity transport equation is expressed as

$$C_n(x_n,y_n,z_n) = \int_{-\infty}^{+\infty} \cdots \int_{-\infty}^{+\infty} dx_{n-1} \cdots dx_0 dy_{n-1} \cdots dy_0 dz_{n-1} \cdots dz_0$$
$$\times G^x_{n-1}(x_n-x_{n-1}) \cdots G^y_{n-1}(y_n-y_{n-1}) \cdots G^z_{n-1}(z_n-z_{n-1}) C_0(x_0,y_0,z_0) \quad ,(1)$$

where C is the impurity concentration, G is the Green's function, C_0 is the initial impurity distribution, x and y are the lateral directions, and z is the depth from the surface. Here, subscripts indicate the time step number and superscripts indicate the direction of transport. Since we are simulating power MOSFETs with a planar structure, the Green's functions in the x and y directions are expressed by Gaussian functions. To make analytical integration in the x and y directions possible, we assume the diffusion length of the mesh points other than x_i are identical to that of x_i at each time step, as shown in Fig. 1. This approximation is reasonable because that for lateral integration within a small time step only a few mesh points around x_i contribute to the concentration at x_i.

The Green's function has no explicit form in the z direction because of the structural asymmetry. However, we can replace integration in the z direction with 1D-simulation. Thus, Eq. (1) reduces to

$$C_n(x_n,y_n,z_n) = \varepsilon(x_n,y_n,z_n) f(z_n) \quad , \quad (2)$$

where

$$\varepsilon(x_n,y_n,z_n) = \sum_j \frac{1}{2}\left(\operatorname{erf}\left(\frac{b_j-x_n}{\sqrt{2}\sigma_n(x_n,y_n,z_n)}\right) - \operatorname{erf}\left(\frac{a_j-x_n}{\sqrt{2}\sigma_n(x_n,y_n,z_n)}\right) \right)$$
$$\times \frac{1}{2}\left(\operatorname{erf}\left(\frac{c_j-y_n}{\sqrt{2}\sigma_n(x_n,y_n,z_n)}\right) - \operatorname{erf}\left(\frac{d_j-y_n}{\sqrt{2}\sigma_n(x_n,y_n,z_n)}\right) \right) \quad ,(3)$$

$$\sigma_n(x_n,y_n,z_n)^2 = \sum_{k=0}^{n-1} \sigma_k(x_k,y_k,z_k)^2 + \sigma_\perp^2 \quad , \quad (4)$$

f is the result of one-dimensional (1D) simulation, the four points, (a_j,c_j), (a_j,d_j), (b_j,d_j), (b_j,c_j) shape the j-th rectangular region which is obtained by dividing the arbitrarily shaped mask patterns into rectangles. Here, σ_n is the ADL, where σ_k is the diffusion length at the k-th time step and $\sigma_\perp$ is lateral standard deviation of the ion implantation. The ADL is obtained through 1D-simulation because at every time step in the numerical calculation, the 1D-simulator calculates diffusion lengths at every mesh point. Then, the 1D distribution of the ADL is geometrically expanded to a 3D distribution. This approach is possible because the distribution of ADL values are almost symmetric below the rectangular region. In this algorithm, numerical simulations are performed in one dimensional only, and the 3D impurity distribution is calculated entirely by using the analytical form. To verify our approximation, we compare the impurity distributions for the x_n direction obtained from a numerical simulation with those obtained using our method. As shown in Fig. 2, good agreement is obtained. Table 1 shows the CPU-time of SPIRIT-III/ADL. In Table 1, we also show the CPU-time of SPIRIT-I [6], our in-house 2D process simulator. CPU time was reduced by more than a factor of 100 in comparison with numerical analysis by SPIRIT-I.

3. The 3D TCAD System and its Application to Power MOSFETs

Figure 3 shows a 3D TCAD system composed of SPIRIT-III/ADL and CADDETH3. The SPIRIT-III/ADL includes the input data processor, the 3D-mesh generator, SPIRIT-1D [7], and an expander. The input processor reads the mask layout information and the process recipe, and automatically generates sub-masks whose regions all have the same process conditions. Several sets of input data for SPIRIT-1D are generated for each sub-mask. SPIRIT-1D simulations are performed to obtain a set of 1D results. The expander generates 3D impurity profiles using the 3D mesh and the set of 1D simulation results including the 1D distribution of the impurity and the ADL. The results are transferred to the 3D device simulator CADDETH-3. Since the process simulation is much faster, process model parameter calibration time was greatly reduced, making it possible to quickly obtain accurate simulation results at the device level.

Figure 4 (a) shows the 3D impurity profile obtained with SPIRIT-III/ADL. Figure 4(b) shows a comparison of the channel implantation dose versus Vth from the experiments and from simulations with our 3D TCAD system. Good agreement is obtained not only for the channel implantation dose versus Vth, but also for the drain voltage versus the drain current and the gate voltage versus the drain current.

References

[1] M. E. Lee et. al., IEEE Trans. on Computer-Aided Design, Vol. 7, pp. 181-189, 1988.
[2] X. Tian et al., Proc. NASECODE VI, pp. 540-545, 1989.
[3] S. R. Nassif et al., IEEE Trans. on Computer-Aided Design, Vol. 3, pp. 40-46, 1984.
[4] T. Toyabe et al., IEEE Trans. Electron Devices, ED-32,pp. 2038-2044, 1985.
[5] I. Yoshida et al., IEEE J. Solid State Circuits, Vol. SC-11, pp. 472-477, 1976.
[6] M. Ohgo et al.., IEEE Trans. on Computer-Aided Design, Vol. CAD-6, pp. 439-445, 1987.
[7] S. Kamohara et al., Proceedings of BCTM, pp.126-129, 1992.

Fig. 1. Approximation scheme.

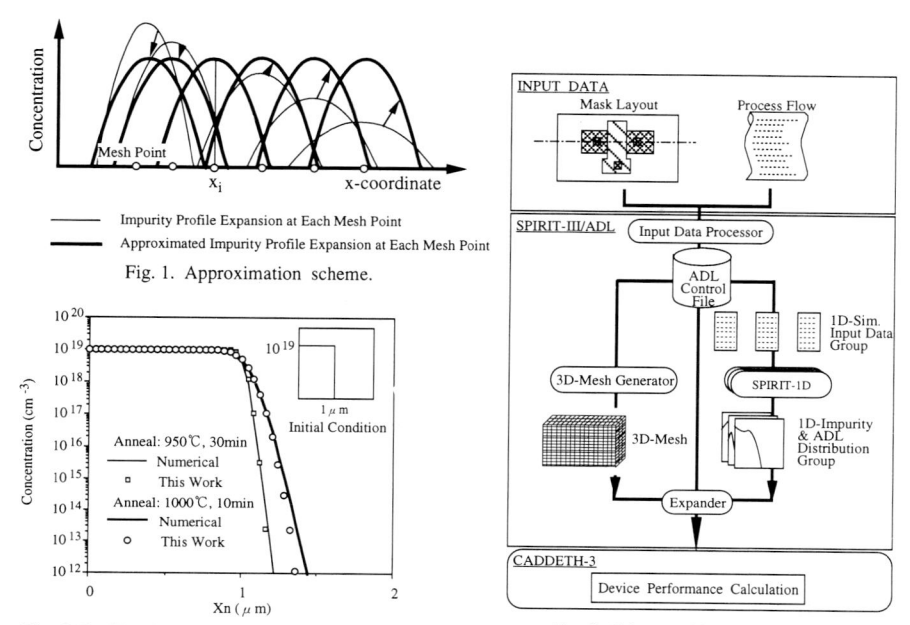

Fig. 2. Profile of x direction comparison between the numerical approach and this work.

Fig. 3. Scheme of 3D-TCAD System.

Table 1. CPU-time of SPIRIT-III/ADL for some examples.

Example	Mesh Num.	Mask Num.	Dimensions	CPU time* (sec) SPIRIT-III/ADL	SPIRIT-I [6]
NMOS	3491	6	2	23	6500
PMOS	3491	8	2	32	7000
BJT	3481	4	2	25	6300
IGBT	3481	5	2	38	7500
Power-MOS	61250	4	3	480	--

* HITACHI: M-880 Mainframe Computer

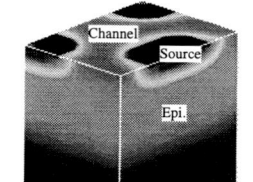

(a) Impurity Distribution of Power MOSFETs.

(b) Vth of Power MOSFETs.

Fig. 4. Application of 3D the TCAD system to Power MOSFET design.

Recovery of Vectorial Fields and Currents in Multidimensional Simulation

Daniel C. Kerr and Isaak D. Mayergoyz

Department of Electrical Engineering
University of Maryland
College Park, MD 20742 USA.

Abstract

In the context of 2-D and 3-D unstructured mixed-element meshes, a new method of recovering vectorial fields and currents in multidimensional simulation is introduced. The new method, called the method of edge elements, directly interpolates the projections of the vectors on the edges of an element into its interior. The new method is compared to two other recovery methods on the basis of resolution, consistency, and implementation ease.

1. Introduction

In the numerical simulation of semiconductor devices, the vectorial electric field and current density field (together: fields) are required throughout the domain of simulation to compute various physical models, such as mobility, impact ionization, and Joule heating. In finite-box simulations, the discretization and solution do not uniquely define the fields off of the edges joining the nodes. The fields are reconstructed from the projections of electric field or current density along the edges.

The recovery method should not only define a unique electric field and current density throughout the domain, but it should also specify how the recovered vectorial quantities should be used in the discretized equations. In general the parameter models may require the vectorial quantity along an edge, within an element, or at a node.

For accuracy, the ideal recovery method should be of high resolution and be consistent with the model and its discretization. Resolution measures the capability to distinguish between fields at adjacent locations. Inconsistency will introduce additional numerical errors. The recovery method should be consistent with the approximations of the Sharfetter-Gummel discretization, and the projections of the recovered field on the element edges should reproduce the original data.

For computer efficiency, the ideal recovery method should be easy to implement and be applicable to any 2-D or 3-D, unstructured, and mixed-element mesh. The implementation of the method should be computationally cheap, including calculation of the derivatives of the field for the Jacobian. Often, in order to apply a method only suitable for simplex meshes, nonsimplex elements are arbitrarily subdivided into simplices. However, it turns out that the numerical solutions are sensitive to the way elements are split. Splitting the elements introduces new edges ij with zero Voronoi cell areas A_{ij}, from which the field along the edge cannot be uniquely determined. Errors introduced in calculating the fields will feed back into the solution via the physical models, which may compromise the overall accuracy of the modeling effort.

2. Method of Edge Elements

The method of edge-elements (EEM) proposed below directly interpolates vectorial values defined on the edges of an element into the interior of the element. The vectorial interpolant $\mathbf{J}^k$ of the edge values F_{ij} into the interior of element k is

$$\mathbf{J}^k = \sum_{ij \in \mathcal{E}^k} F_{ij} \mathbf{e}_{ij}, \qquad (1)$$

where $\mathbf{e}_{ij}$ is the basis function of edge ij and $\mathcal{E}^k$ is the set of edges of the element. The edge basis function is defined by

$$\mathbf{e}_{ij} = d_{ij}(\lambda_i \nabla \lambda_j - \lambda_j \nabla \lambda_i), \qquad (2)$$

where d_{ij} is the length of edge ij and λ_i is the scalar finite-element shape function. The shape function λ_i for the element takes the value 1 at node i, the value 0 at all other nodes, and is linear between them. The basis function has the following characteristic properties: the tangential component of $\mathbf{e}_{ij}$ along the edge ij is equal to one, while the tangential components along all other edges are equal to zero. Reconstruction of the field using edge elements yields a non-constant vector function defined on the element which has the following properties: (1) The projection of the edge-element reconstruction on each edge of the element reproduces the original data. (2) The tangential components of the field are continuous from one element to the next. (3) The field is divergence-free. Property (2) is useful at internal interfaces between materials of different permittivities. Property (3) is consistent with the approximations made for the SG discretization of current.

To simplify the edge-element calculations, a coordinate transformation is applied to transform each element into its standard position. The coordinate transformation for an element with nodes at $\{\mathbf{r}_0, \mathbf{r}_1, \ldots\}$ into standard position is defined by $\mathbf{r}' = A^{-1}(\mathbf{r} - \mathbf{r}_0)$, where A is the Jacobian of the transformation. The edge elements transform according to

$$\mathbf{e}_{ij}(\mathbf{r}) = (A^{-1})^T \mathbf{e}_{ij}(\mathbf{r}'), \qquad (3)$$

where the primed coordinates refer to the element in standard position.

Although the definition of the edge elements can be applied to non-simplex elements, the results are inaccurate due to the quadratic component in the direction normal to the edge. To rectify this, the modified edge-elements are introduced by removing one of the factors from each squared term. In this form, edge elements are suitable for 2-D and 3-D nonsimplex elements. The standard and modified 2-D edge-elements in standard position are listed in Table 1.

The vector quantity at a location required for a parameter model may be computed by averaging the interpolant function. For example, analytically averaging over the element k results in

$$< \mathbf{J}^k > = \frac{1}{V^k} \int_{V^k} \mathbf{J}^k \, dv. \qquad (4)$$

where V^k is the volume of the element k. On the other hand, the space-varying interpolant function may be directly evaluated at the location of interest.

3. Method of Least-Squares Fitting

The method of least-squares (LSM) fitting treats the edge values in an element as measurements of the field. The recovered field is a constant field within the element

Table 1: The two-dimensional edge elements in standard position, each normalized by its length. A modified edge-element is shown in the vector plots, with node numbering indicated.

Element	Standard EE $\hat{a}_x$	$\hat{a}_y$	Modified EE $\hat{a}_x$	$\hat{a}_y$	Element averaged
Triangle					$<J_x^k> = \frac{1}{3}(2d_{01}F_{01} + d_{02}F_{02} - d_{12}F_{12})$
e_{01}	$1-y$	x	$1-y$	x	
e_{02}	y	$1-x$	y	$1-x$	$<J_y^k> = \frac{1}{3}(d_{01}F_{01} + 2d_{02}F_{02} + d_{12}F_{12})$
e_{12}	$-y$	x	$-y$	x	
Rectangle					$<J_x^k> = \frac{1}{2}(d_{01}F_{01} + d_{23}F_{23})$
e_{01}	$(1-y)^2$	0	$1-y$	0	
e_{02}	0	$(1-x)^2$	0	$1-x$	$<J_y^k> = \frac{1}{2}(d_{01}F_{01} + d_{23}F_{23})$
e_{13}	0	x^2	0	x	
e_{23}	y^2	0	y	0	

that minimizes the error along each edge, that is, the field $\mathbf{J}^k$ that minimizes $f(\mathbf{J}^k)$,

$$f(\mathbf{J}^k) = \sum_{ij \in \mathcal{E}^k} (\mathbf{J}^k \cdot \hat{\mathbf{a}}_{ij} - F_{ij})^2, \quad (5)$$

where $\hat{\mathbf{a}}_{ij}$ is a unit vector from node i to node j. The LSM can be applied uniformly to edge-, element-, or cell-field recoveries.

4. Method of Corner Averages

The method of corner averages (CAM) partitions the domain into "corners" and calculates a constant vector value in each corner. The volume V_i^k associated with the corner of element k at node i, where node i belongs to the element, is defined by the intersection of element k with the Voronoi cell at i. The corner value $\mathbf{J}_i^k$ in V_i^k is calculated by solving the system of equations

$$\mathbf{J}_i^k \cdot \hat{\mathbf{a}}_{ij} = F_{ij}, \ j \in \mathcal{N}_i^k, \quad (6)$$

where $\mathcal{N}_i^k$ is the set of nodal neighbors of i within element k.

The field at a location required for a parameter model is computed through a suitably weighted average of corner values. The original CAM [1] was proposed for triangular or prismatic elements, and it is difficult to generalize their averaging scheme to other elements. Here, a more general averaging scheme is proposed which is suitable for other elements. The generalized average is defined by

$$<\mathbf{J}> = \sum_{i,k} \mathbf{J}_i^k / V_i^k \bigg/ \sum_{i,k} 1/V_i^k, \quad (7)$$

where the indices (i, k) take on different values depending on whether the average is for a node, an element, or along an edge.

5. Discussion

The EEM was compared to the LSM and the CAM. On the basis of resolution, the EEM is clearly superior since it can distinguish any two arbitrarily close points. The other methods produce fields which are constant over various zones. On the basis of consistency, the EEM is consistent with the SG discretization. The recovered field is also consistent with the solution, since the projections of the field along element edges reproduces the original data. However, the other methods do not have this property. On the basis of implementation ease, the EEM and LSM are computationally cheaper that the CAM. The work involved in recovering the field using the EEM or LSM involves evaluating a polynomial function in the number of edges and multiplying by a matrix for the coordinate transformation.

A detailed comparison between the element-averaged EEM and the LSM was made, since it can be argued that the LSM produces fields which are optimum. It was found that the two methods yield identical results for the rectangular faces of any element, so the differences were evaluated in triangular faces. For an equilateral triangle, with nodes at $(0,0)$, $(1,0)$, and $(1/2, \sqrt{3}/2)$, the recovered fields are identical and given by

$$\mathbf{J}^k = \{0.667 F_{01} + 0.333 F_{02} - 0.333 F_{12}, 0.577 F_{02} + 0.577 F_{12}\}. \qquad (8)$$

When the upper node was moved upward, from $(1/2, \sqrt{3}/2)$ to $(1/2, 10 + \sqrt{3}/2)$, the methods produced:

$$\mathbf{J}^k_{EEM} = \{0.667 F_{01} + 3.62 F_{02} - 3.62 F_{12}, 0.501 F_{02} + 0.501 F_{12}\}, \qquad (9)$$
$$\mathbf{J}^k_{LSM} = \{0.996 F_{01} + 0.045 F_{02} - 0.045 F_{12}, 0.501 F_{02} + 0.501 F_{12}\}. \qquad (10)$$

These methods differ in the reconstruction of the x-component of the field. Which is correct? The EEM uses the geometry of the figure when averaging over the area of the element. Longer edges dominate more of the figure and therefore weigh proportionately more in the averaging. On the other hand, the LSM ignores the geometry of the figure in which the field is computed; only the relative orientation of the measurements matters. In the elongated triangle, the upper legs are rotated toward the y-direction and thus the influence of these legs on J^k_x is smaller. This means that the LSM is inaccurate for element reconstructions. The use of the LSM should be restricted to field reconstructions where all the measurements are collected at one point, for example, to reconstruct the field at the center of the Voronoi cell.

In conclusion, the newly proposed method of edge-elements is an accurate and efficient vectorial field reconstruction method. The EEM has been installed in SIMASTER, a general purpose 2-D and 3-D device simulator using unstructed meshes [2]. The EEM has proven effective in many different types of simulations.

References

[1] S. E. Laux and R. G. Byrnes, "Semiconductor device simulation using generalized mobility models," *IBM J. Res. Develop.*, vol. 29, no. 3, pp. 289–301, 1985.

[2] D. C. Kerr and I. D. Mayergoyz, "Simulation of semiconductor devices using the fixed-point iteration method on 3-D unstructured meshes," *to appear*, 1995.

An Efficient Approach to Solving The Boltzmann Transport Equation in Ultra-fast Transient Situations

Ming-C. Cheng

Department of Electrical Engineering
University of New Orleans, New Orleans, Louisiana 70148, USA

Abstract

A previously developed hydro-kinetic concept based on evolution of the distribution function is used to arrive at an efficient approach to solving the Boltzmann transport equation (BTE) in ultra-fast transient situations. The solution can properly account for effects of extreme non-equilibrium phenomena. The approach is applied to study the temporal evolution of the electron distribution in n-type Si. Results from the Monte Carlo method are also included to verify the accuracy of the proposed approach.

1. Introduction

Performance and reliability of the submicron devices have been found to be strongly influenced by the hot-electron behavior which, for example, might result in leakage gate current and device degradation [1,2]. To study these hot-electron effects, the carrier distribution function in the device needs to be determined. Therefore, in addition to non-stationary transport parameters including density n, mean energy $\bar{\epsilon}$, and average velocity $\bar{v}$, knowledge of the non-stationary and/or hot-electron distribution function also becomes crucial in small devices.

Various approaches to the distribution function have been used to study non-stationary or hot-electron phenomena in devices. Among these, the efficient approaches using the displaced Maxwellian [3], Legendre polynomial [4], and the energy-dependent distribution [5,6] are commonly used. However, these methods all have some severe limitations in highly non-stationary and/or hot-electron situations [6]. To more accurately determine the distribution function, the Monte Carlo simulation is usually used although it demands a large CPU time. In this study, an efficient approach to the hydro-kinetic distribution evolving at the velocity relaxation scale is proposed. This approach is applied to study the fast temporal evolution of the distribution function related to relaxation of the transport parameters of electrons in Si.

2. Theoretical Background - The Hydro-kinetic Transport Concept

The hydro-kinetic concept [6] is presented in Fig. 1 where the axis represents the time scale of the distribution function. The exact solution of the BTE is given by the kinetic distribution function $f(k)$ where k is the wave vector. The BTE is only valid for the time scales greater than the collision duration time, τ_c. The **hydro-kinetic** concept is based on the fact that, when use the moments (**hydrodynamic** parameters) to describe the **kinetic** distribution function, it requires an infinite set of moments, namely $f(k)$ = $f(k, n, \bar{k}, \bar{\epsilon}, \bar{k^3}, \bar{k^4}, ...)$. Therefore, relaxation times of the moments can be used to characterize $f(k)$.

In general, $\tau_n > \tau_\epsilon > \tau_m$ (the carrier density, energy, and momentum relaxation times, respectively) in semiconductors, and relaxation times of higher order moments are assumed to be less than τ_m. As illustrated in Fig. 1, after a sudden change in field, information described by the higher-order moments will vanish faster than that described by n, $\bar{\epsilon}$ and $\bar{k}$. As a consequence, after a sufficient time, $f(k)$ will evolve into a τ_m-scale hydro-kinetic distribution $f_m(k,n,\bar{\epsilon},\bar{k})$ which takes into account temporal/spatial variations through the changes in n, $\bar{\epsilon}$ and $\bar{k}$. f_m thus varies as fast as $\bar{k}$ and is valid for the scale of interest on the order of τ_m. To determine $f_m(k,n,\bar{\epsilon},\bar{k})$, n, $\bar{\epsilon}$, and $\bar{k}$ have to be solved from the hydrodynamic equations which are written as

$$\frac{\partial n}{\partial t} + \nabla \cdot (n\bar{v}) = -\frac{n}{\tau_n}, \qquad (1a)$$

$$\frac{\partial n\bar{\epsilon}}{\partial t} + \nabla \cdot \langle \epsilon v \rangle = qnE \cdot \bar{v} - \frac{n(\bar{\epsilon}-\epsilon_o)}{\tau_\epsilon}, \qquad (1b)$$

$$\frac{\partial n\bar{p}}{\partial t} + \nabla \langle p \cdot v \rangle = qnE - \frac{n\bar{p}}{\tau_m} \qquad (1c)$$

where $\bar{p}$ is the average momentum, and the relaxation times are defined in terms of integrals of the distribution function and transition rates over $\bar{k}$ space. [7] At the scale near τ_m, f_m is used to evaluate the relaxation times which therefore becomes $\bar{\epsilon}$ and $\bar{k}$ dependent, as illustrated in Fig. 1.

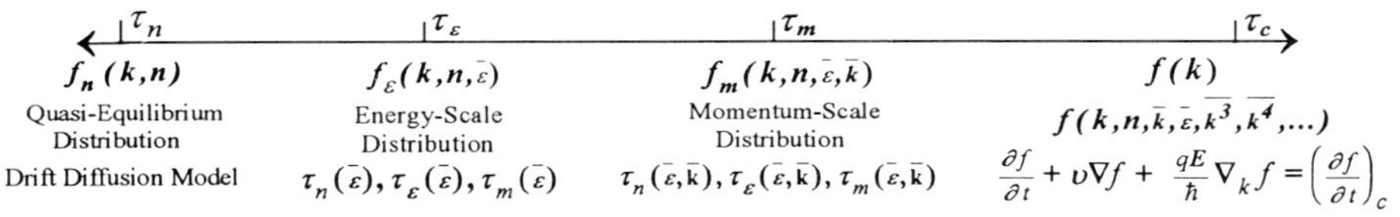

$\leftarrow \tau_n$	τ_ϵ	τ_m	$\tau_c \rightarrow$
$f_n(k,n)$	$f_\epsilon(k,n,\bar{\epsilon})$	$f_m(k,n,\bar{\epsilon},\bar{k})$	$f(k)$
Quasi-Equilibrium Distribution	Energy-Scale Distribution	Momentum-Scale Distribution	$f(k,n,\bar{k},\bar{\epsilon},k^3,k^4,...)$
Drift Diffusion Model	$\tau_n(\bar{\epsilon}), \tau_\epsilon(\bar{\epsilon}), \tau_m(\bar{\epsilon})$	$\tau_n(\bar{\epsilon},\bar{k}), \tau_\epsilon(\bar{\epsilon},\bar{k}), \tau_m(\bar{\epsilon},\bar{k})$	$\frac{\partial f}{\partial t} + v\nabla f + \frac{qE}{\hbar}\nabla_k f = \left(\frac{\partial f}{\partial t}\right)_c$

Fig. 1: Evolution of the distribution function

At a scale on the order greater than τ_m (i.e., near τ_ϵ), the $\bar{k}$ dependence becomes insignificant because generally $\tau_\epsilon > \tau_m$. The distribution therefore evolves into a τ_ϵ-scale hydro-kinetic distribution $f_\epsilon(k,n,\bar{\epsilon})$ that varies as fast as $\bar{\epsilon}$ and is only valid at scales on the order of τ_ϵ. At this scale, $f_\epsilon(k,n,\bar{\epsilon})$ is used to evaluate the relaxation times which thus becomes only energy dependent, as shown in Fig. 1. For a scale much greater than τ_ϵ, $\bar{\epsilon}$ and $\bar{k}$ are close to the steady state, and the carrier behavior can be described by the quasi-equilibrium distribution function, f_n.

3. τ_ϵ- and τ_m-Scale Hydro-kinetic Transport Models

The approach to the τ_ϵ-scale hydro-kinetic distribution f_ϵ has been introduced in a previous paper [6]. In the current study, the evolution process of the distribution from f_m into f_ϵ is presented. The evolution due to scattering is assumed to be a relaxation process influenced by $\bar{\epsilon}$ and $\bar{p}$ relaxation and the change in field. The relaxation of the hydro-kinetic distribution function can be performed numerically with given f_ϵ:

$$f_m^{\ell+1}(k) = f_\epsilon^{\ell+1}(k) + \left[f_m^\ell(k - \Delta k_m^\ell) - f_\epsilon^\ell(k)\right]\exp\left(-\Delta t^\ell/\tau_h^{\ell+1/2}\right), \qquad (2)$$

where $1/\tau_h$ is the relaxation rate for f_m evolving toward f_ϵ, and $\Delta t = t^{\ell+1} - t^\ell$. f_m^ℓ is taken as an initial distribution to evaluate the next step distribution, f_m^{l+1}. The difference between f_m and f_ϵ at each time step tends to reduce and is, for a relaxation process, proportional to $exp(-\Delta t/\tau_h)$ due to the scattering process. On the other hand, Δk_m denotes the shifted amount in k space influenced by the electric field.

f_m at each step can be solved if Δk_m and τ_h are determined. The solution of Eqs. (1a)-(1c) can be used to assist in determining Δk_m and τ_h. The relaxation of $\bar{\epsilon}$ and $\bar{k}$ at each time step can be obtained by taking the moments of the Eq. (2). The solution of these relaxation equations for $\bar{\epsilon}$ and $\bar{k}$ must be consistent with that from Eqs. (1a)-(1c). This therefore determines Δk_m and τ_h at each time step, and then f_m is determined from Eq. (2). In this study, only the energy-dependent relaxation times are used in Eqs. (1a)-(1c) to calculate the hydrodynamic parameters. To include the momentum dependence in the relaxation times, the determined f_m, as discussed in Sec. 2, needs to be used to evaluate the relaxation times that are then used again to solve Eqs. (1a)-(1c) for the hydrodynamic parameters. The iteration will provide more accurate results for hydrodynamic parameters and f_m, and will be studied in the near future.

4. Application

Using the τ_ϵ- and τ_m-scale hydro-kinetic models, the response of a homogeneous concentration of electrons in n-type Si at 300K to a rapid increase in electric field is investigated. Results including the mean energy, average velocity, and the distribution function (f_ϵ and f_m) obtained from the hydro-kinetic models are examined and compared with those determined by the Monte Carlo simulation. It is assumed that no impact ionization is involved.

A step field increasing from 5 to 30 kV/cm within 0.1 psec is applied, and $\bar{\epsilon}$ and $\bar{v}$ responding to this field are given in Figs. 2a and 2b, respectively. It is shown that $\bar{\epsilon}$ and $\bar{v}$ calculated from Eqs. (1b) and (1c) are in very good agreement with the Monte Carlo results. An evident velocity overshoot is observed due to the drastic increase in field.

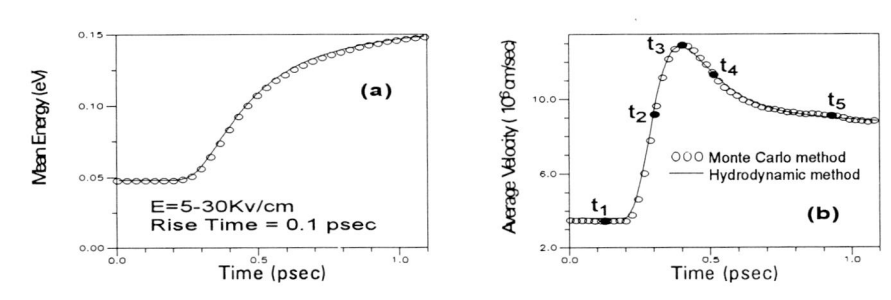

Fig. 2: Evolution of (a) mean energy (b) average velocity

The evolution of the distribution function under the influence of the step field is also shown in Figs. 3a-3e where f, f_ϵ, and f_m are illustrated at t_1-t_5. Figs. 3a-3e clearly show that the τ_ϵ-scale distribution f_ϵ evolves more slowly than f. This is because, as illustrated in Fig. 1, influenced of velocity relaxation is not properly accounted for in f_ϵ. As a result, f_ϵ starts to deviate from f when $\bar{v}$ increases rapidly due to the drastic increase in field. The discrepancy becomes significant during the overshoot interval as shown at t_2-t_4. On the contrary, f_m and f evolve closely over the simulation time since effects of velocity relaxation during the overshoot interval is properly included in f_m.

The maximum deviation between f_m and f is found at the time when $\overline{v}$ starts to decrease from the overshoot peak, as shown at t_4. However, the deviation is very small.

5. Conclusion

The study shows that the hydro-kinetic concept based on time scales of the hydrodynamic parameters can be used to characterize the evolution of the distribution function. The concept also leads to an accurate and efficient technique to solve the BTE. In the case of strong velocity overshoot, since f_m can include effects of velocity relaxation, f_m provides a much better description than f_ϵ. The CPU time required for solving the temporal evolution of the distribution and hydrodynamic parameters given in Figs. 2 and 3 is only about 10 seconds on a 486/33 PC. The approach to f_m might be an efficient method to study the phenomena associated with hot-electron effects in submicron devices.

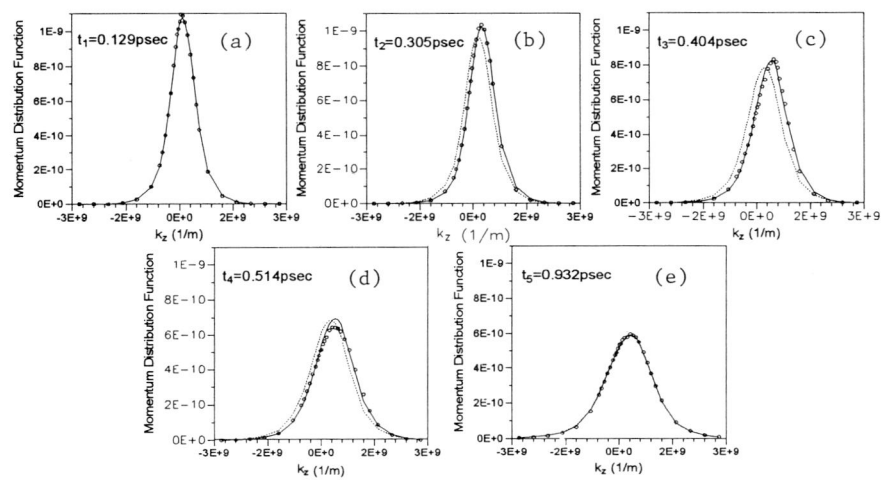

Fig. 3: Evolution of the distribution function. The times t_1-t_5 are indicated in Fig. 2. Symbols denote f calculated from the Monte Carlo method, solid lines represent f_m, and dots denote f_ϵ.

6. Acknowledgement

This research has been supported by the Nation Science Foundation under the grant number ESC-9409471.

References

[1] C. Fiegna, F. Venturi, M. Melanotte, E. Sangiorgi, and B. Ricco, IEEE Trans. Electron Dev., Vol. 38, p. 603, 1991.
[2] P. Roblin, A. Saman, and S. Bibyk, IEEE Trans. Electron. Dev., Vol. 35, p. 3229, 1988.
[3] K. F. Wu, M. P. Shaw, IEEE Trans. Electron Dev., Vol. 36, p. 603, 1989.
[4] N. Goldsman, L. Henrickson, and J. Frey, Solid-St. Electron. Vol. 34, 389 (1991).
[5] N. Goldaman and J. Frey, Solid State Electron., Vol. 31 p. 1089, 1988.
[6] Ming-C. Cheng and R. Chennupati, J. Phys. D., Vol. 28, pp. 160-173, 1995.
[7] Ming-C. Cheng and Lin Huang, J. Appl. Phys., Vol. 72, p. 3539, 1992.

Modeling of a Hot Electron Injection Laser

V. I. Tolstikhin, M. Willander

Department of Physics and Measurement Technology, Linköping University,
S-581 83 Linköping, SWEDEN

Abstract

Device concept and model of a vertically integrated transistor-laser structure for a dual gain-switching involving carrier heating are reported. Capability of a proposed three-terminal device to generate high-intensity picosecond pulses is studied using numerical simulation.

1. Introduction

Dynamic carrier heating effects in semiconductor laser diodes (LD's) attract much attention because of their potential use for a high-speed modulation [1-3]. Particularly, variation of the carrier effective temperature, properly combined with variation of the carrier concentration, has been shown as a recipe for a drastic improving the high-speed performance of gain-switched lasers [4-6]. To implement this modulation technique, an acceptable mechanism to control the carrier effective temperature should be, however, found. The possible solution is associated with injection of hot, variable energy, electrons as means to pump an LD, that can be achieved with a recently proposed three-terminal device combining the transistor and the laser features [6]. In this paper, the model of such a device, labeled as a hot electron injection laser (HEIL), is reported and the device capability of generation of good-shaped gain-switched pulses is analyzed.

2. Device concept

The schematic cross-section of a HEIL is shown in Fig. 1. The device represents a vertically integrated structure, in which an LD is inserted in the collector area of a hot electron transistor (HET). It is assumed, that electrons, injected from n$^+$- emitter, traverse the base with a transport ratio close to unity and then are accelerated into the high electric field of a collector barrier. The latter thus serves as a hot electron launcher, positioned between the base of a HET and the active region (AR) of an LD. The laser input is associated with both the carrier injection rate J, and the energy injection rate, Q, which two are connected by $Q = \varepsilon_J \times Q$, where ε_J is the energy yield per one injected electron-hole pair [4]. To vary the electron contribution to this parameter, ε_C, and inject therefore carriers at controlled energy is, after all, the idea of a HEIL. On time-scales above the time of transitional processes in electric circuit, this is implemented by two drive voltages, V_1 and V_2, applied as shown in Fig. 1. Hot electrons, injected by emitter-base biasing and then heated in electric field of a launcher, lose their energy mainly due to intercarrier scattering into an AR and so within a short thermalization time they increase the tempe-

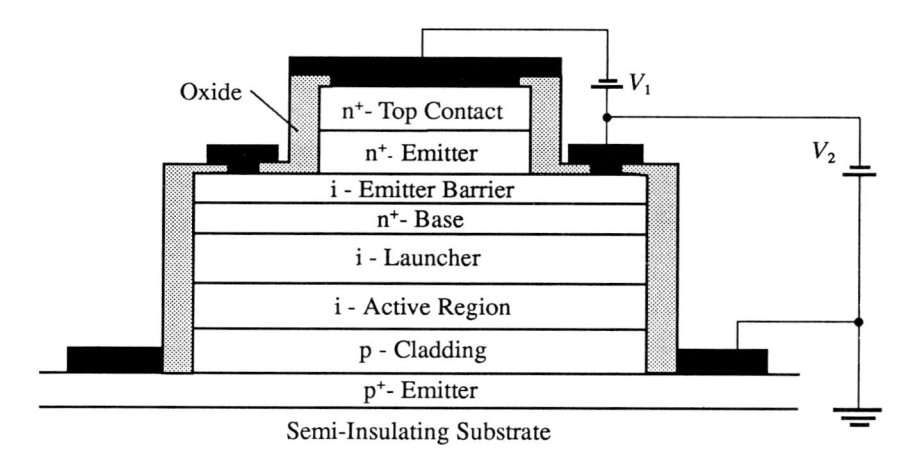

Fig. 1. Schematic structure of a hot electron injection laser.

rature of a thermalized sea along with increase of its concentration. As far as these two parameters govern the material gain, by imposing the data in signal on drive voltages one gets the possibility of a dual - concentration and temperature - gain-switching.

The device performance depends, however, on the range of change in ε_C, which is determined by the mechanism of electron transport over the launcher. If electrons traverse this region with no scattering, one benefits from injection at narrow energy distribution, that can be achieved by employing the resonant-tunneling structure as an emitter barrier [6]. In this case, hot electrons enter an AR at well-defined and variable energy, which is just ε_C. For a high scattering intensity, ε_C is determined by the level of Joule heating into the launcher and does not depend on the initial energy distribution. In that case no need for resonant-tunneling injection and a single emitter barrier of any type can be employed.

3. Model

As far as lasing is directly controlled by the carrier and the energy injection rates, not by drive voltages, the task of a HEIL modeling is divided in two following steps: 1) to find an optical output of an LD in response to $J(t)$ and $Q(t)$ as the given functions of time, and 2) to connect $J(t)$ and $Q(t)$, on the one hand, and electrical input of a device, on another.

At the first step, the earlier reported model, describing an LD as a nonequilibrium system consisting of the electrically injected carriers, the LO-phonons and the guided photons, interacting with each other [1], is employed. Homogeneity and the thermalization of the electron-hole plasma inside an AR are assumed on all actual time-scales, and the model is therefore reduced to a set of the rate equations, written for plasma concentration and effective temperature, LO-phonon occupation number and effective photon population associated with a lasing mode. By making use from the evident form of distribution functions of thermalized carriers, the interaction processes are described microscopically, the details of these calculations have been recently published elsewhere [3].

At the second step, it is assumed, that carrier injection rate, J, is a known function of emitter-base and base-collector bias voltages of a HET, V_1 and $V_2 - \Delta\Phi_{cv}$, respectively, where $\Delta\Phi_{cv}$ is the Fermi quasilevels splitting inside an AR. Thus, the problem reduces to find the energy coupled to one injected electron, ε_C. In a case that inelastic scattering into a launcher can be neglected, resonant-tunneling injection at narrow energy distribution is assumed. Then, ε_C, derived from the energy conservation condition, is

$$\varepsilon_C = \varepsilon_B(V_1) + eV_2 - \Delta\Phi_{cv} - \mu_B + \mu_C,$$

where $\varepsilon_B(V_1)$ is V_1 - dependent resonant-tunneling level, referred to the conduction band edge in a base, μ_B and μ_C are the chemical potentials of electrons in a base and an AR, respectively. In the opposite ultimate case, when phonon and intercarrier scattering are just the processes which govern the transport over a launcher, the single-temperature approximation is employed to describe the energy distribution of hot electrons in a high electric field. The Γ-L nonparabolic conduction band model is assumed, the electrons in Γ and four equivalent L valleys being described by the same effective temperature, T. The latter is determined from the balance condition, equalizing the Joule power and the energy relaxation rate due to inelastic phonon scattering, that is accounted under the homogeneous drift approximation. The relevant intra- and intervalley scattering processes are described microscopically, in a way, similar to that used in the conventional Monte Carlo simulations [7]. At the same time, the evident - Maxwellian - shape of the energy distribution is essentially used to obtain the macroscopic transport characteristics of hot electrons. The energy yield, ε_C, is then given by

$$\varepsilon_C = \frac{(1 + 3\alpha_\Gamma T)(\Delta_C + 2T) + R_{L\Gamma}(T)(1 + 3\alpha_L T)(\Delta_C + \Delta_{L\Gamma} + 2T)}{1 + 2\alpha_\Gamma T + R_{L\Gamma}(T)(1 + 2\alpha_L T)},$$

where Δ_C is the conduction band offset at the AR' edge of a launcher, $\Delta_{L\Gamma}$ is the energy gap between L- and Γ-valleys into the launcher, $R_{L\Gamma}(T) = 4(m_L/m_\Gamma)\times\exp(-\Delta_{L\Gamma}/T)$, m_i and α_i are the effective mass and the parameter of nonparabolicity in i-th valley, $i = \Gamma$, L. When deriving this equation, the transitions in real space between the equivalent valleys only are taken into account [8], and scattering by thermalized carriers is considered as a dominant energy relaxation channel for hot electrons entering an AR.

4. Modeling results

The subject of the present study is a HEIL using In(AlGa)As/InGaAsP/InP material system, operating in a wavelength 1.55 µm at room temperature. Only lasers with a bulk AR are assumed, because of the carrier bottleneck effect in a typical separate-confinement quantum-well laser makes the intended control of the enegry injection rate less efficient. A modeling example of the picosecond pulse generation from a HEIL is shown in Fig. 2. Here, the carrier injection rate remains constant (at a level $J = 3.10^{27}$ cm^{-3}s^{-1}) and the gain-switched optical pulse proceeds from modulation of only the energy injection rate, that is achieved by varying of ε_C. The stationary value is chosen high enough to suppress the generation by carrier heating coupled to injection [3, 5] and so the steady state of an LD is below the threshold. Then, switch *on* is caused by carrier cooling and switch *of* - by reverse carrier heating, produced by lowering and raising of ε_C, respectively. Note, that intensity of so switched pulse depends on the depth of the deep in a carrier temperature (which is less than 2 meV for a given numerical example) and hence it can be improved by enlarging the range of change in ε_C. To evaluate the device potential in this line, the energy yield per one injected electron, ε_C, has been computed in dependence on

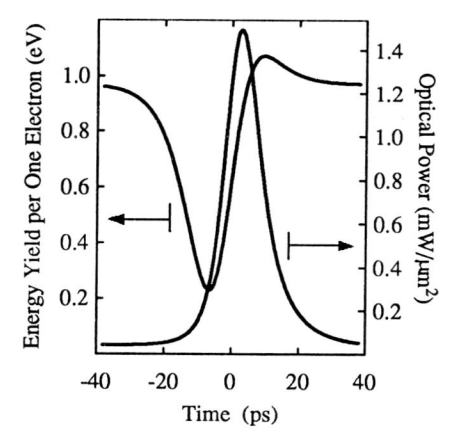

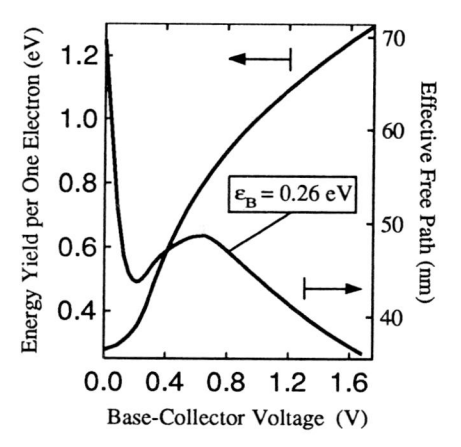

Fig. 2. Generation of picosecond optical pulses from HEIL switched by varying of energy ε_C.

Fig. 3. Energy yield per one injected electron and effective free path as functions of bias $V_2 - \Delta\Phi_{cv}$.

a base-collector bias voltage, $V_2 - \Delta\Phi_{cv}$, for two above-mentioned ultimate cases of collisionless resonant-tunneling injection and homogeneous drift over the launcher. Transition from the first to the second occures when an effective free path, λ_{eff}, defined as

$$\frac{1}{\lambda_{eff}} = \frac{1}{\varepsilon_2 - \varepsilon_1} \int_{\varepsilon_1}^{\varepsilon_2} \frac{1}{\lambda(\varepsilon)} d\varepsilon$$

where ε_1 and ε_2 are the initial and the final energies of a resonant-tunneling electron into a launcher and $\lambda(\varepsilon)$ is the enegry-dependent free path of a probe electron there, is getting less than the width of a collector barrier. Calculated value of λ_{eff} is plotted in Fig. 3 as a function of $V_2 - \Delta\Phi_{cv}$ for InP launcher. It is seen, that the probability of a collisionless transport over the collector barrier drops drastically with raising of a base-collector bias voltage in a range, needed for a high-intensity pulse generation by the above-described scheme. The drift mechanism of transport over the launcher is therefore more probable for any reasonable width of it. The associated energy yield per one injected electron is also given in Fig. 3 in dependence on $V_2 - \Delta\Phi_{cv}$ for InP launcher. By comparing these data with ones shown in Fig. 2, it is concluded that variable Joule heating in a launcher is just a mechanism insuring generation of high-intensity picosecond gain-switched pulses.

References

[1] V. I. Tolstikhin, *Techn. Phys. Lett.*, **18**, 630 (1992).
[2] S. V. Polyakov and V. I. Tolstikhin, *Techn. Phys. Lett.*, **19**, 110 (1993).
[3] V. I. Tolstikhin and M. Willander, *IEEE J. Quantum Electron.*, **QE-31**, 814 (1995).
[4] V. Gorfinkel and S. Luryi, *Appl. Phys. Lett.*, **62**, 2923 (1993).
[5] V. I. Tolstikhin and M. Willander, *J. Appl. Phys.*, **77**, 488 (1995).
[6] V. I. Tolstikhin, A. N. Mamaev, and M. Willander, *J. Appl. Phys.*, (to be published).
[7] W. Fawcett, A. Boardman, and S. Swain, *J. Phys. Chem. Solids*, **31**, 1963 (1970).
[8] I. C. Kizilyalli and K. Hess, *J. Appl. Phys.*, **65**, 2005 (1989).

Scaling Considerations of Bipolar Transistors using 3D Device Simulation

M. Schröter[1], D. J. Walkey[2]

[1]Telecom Microeletronics Centre, Northern Telecom, P.O. Box 3511, Ottawa, Ontario, Canada K1Y 4H7,
[2]Dept. of Electronics, Carleton University, Ottawa, Ontario, Canada

Abstract

The influence of 3D effects on high-speed bipolar transistors is demonstrated using a 3D mixed-mode device/circuit simulator. Basic scaling equations suitable for compact modelling are presented, and the transient behaviour predicted by a scalable bipolar compact model is compared to 3D device simulation.

1. Introduction

High-speed bipolar circuits for wireless and low-power broad-band applications often require transistors with as small an emitter area A_{E0} and, therefore ratio of emitter length to width, l_{E0}/b_{E0}, as possible. Unfortunately, the influence of the resulting 3D effects on electrical characteristics cannot be predicted by bipolar compact models presently available for circuit design. Thus, the main purpose of this paper is to identify the impact of 3D effects on high-speed bipolar transistors using 3D device simulation and to demonstrate the applicability of scaling equations for compact modelling.

2. Device simulator

The mixed-mode device/circuit simulator DEVICE described in [1] has been extended to three spatial dimensions. For present bipolar production technologies the solution of Poisson's, electron and hole continuity equation is sufficient. The program's 3D mixed-mode transient analysis capabilitiy is very useful as a vehicle for developing and verifying compact models. For spatial discretization the box integration method is used together with automatic grid generation. The linear equation system for the 3D case is solved by relaxation methods (SLOR, SOR) while for the 2D case additionally LU factorization and SIP are available.

Accurate bipolar transistor simulations are achieved by careful calibration of certain physical parameters, such as minority carrier mobility and (effective) emitter contact recombination velocity, according to literature values and to measurements of fabricated devices (cf. [2]). A special post-processing capability makes important bipolar transistor parameters, like transit frequency or sheet resistances, quickly available to the user.

DEVICE is presently being run at Northern Telecom and a couple of universities on Sun compatible workstations as well as on 386/486 UNIX PCs. A commercial package (PVWAVE from Visual Numerics) is used for graphical post-processing.

3. Investigated transistor structures and results

Fig. 1 shows a simulated 3D structure, with $l_{E0}/b_{E0} = 2$, of a self-aligned double-polysilicon bipolar transistor. For all investigations, $b_{E0} = 0.5\mu m$ and a variable emitter length l_{E0} have been assumed.

In order to obtain a meaningful comparison of electrical behaviour for transistors with different emitter size, the terminal currents have been normalized to an effective emitter area

$$A_E = b_E l_E = (b_{E0} + 2\gamma_C)(l_{E0} + 2\gamma_C) \tag{1}$$

The technology specific parameter, γ_C, takes into account emitter periphery injection and can be extracted from measurements of long (2D) transistors with different b_{E0} [3]. Using $\gamma_C = 0.051\mu m$ as obtained from 2D simulation, I_C/A_E in Fig. 2a exhibits perfect agreement at low bias indicating that (1) can be employed for 3D scaling purposes. The differences at high current densities are caused by collector current spreading which is shown in Fig. 1 for two planes. The impact of 3D emitter periphery injection and collector current spreading on the dynamic transistor behaviour is clearly demonstrated by the transit frequency f_T in Fig. 2b. The difference in peak f_T is inversely proportional to the transit time τ_f. Since the transit time τ_{fp} of electrons injected at the periphery is larger than the component τ_{fi} of electrons injected under the emitter window, the low current transit time τ_{f0} is highest and peak f_T is lowest for the 3D case with $l_{E0}/b_{E0} = 1$. Both τ_{fi} and τ_{fp} can be determined from f_T of long (2D) transistors with different emitter width. A simple derivation shows that the total (measured) low-current transit time can be "scaled" with emitter dimensions according to

$$\tau_{f0} = [(\tau_{fi} + \tau_{fp}(\gamma_C L_E/A_{E0})] / [1 + \gamma_C L_{E0}/A_{E0}] \tag{2}$$

with $L_E = L_{E0} + 4\gamma_C$, $A_{E0} = b_{E0}l_{E0}$, and $L_{E0} = 2(b_{E0}+l_{E0})$. The comparison between simulated and calculated τ_{f0} values in Table I shows good agreement. At high current densities, however, τ_f increases least for the 3D case since collector current spreading, which reduces the Kirk effect, is most pronounced there. The effect of 3D current spreading on transistor characteristics could not be predicted by presently available bipolar compact models.

In an actual process, ideal rectangular structures are usually not obtained for lithography and etching reasons. To evaluate the difference in electrical characteristics, a transistor with an emitter contact which is rounded at its corners has been simulated. The corresponding transit time τ_{f0} can still be predicted by (2) if the correct values for L_{E0}, L_E, and A_{E0} are inserted (cf. Table I). Also, I_C/A_E vs. V_{BE} of the transistor with rounded emitter contact agrees with that of the transistor with the rectangular contact.

Results for transient mixed-mode device/circuit simulation are shown in Fig. 3 for a 3D and a 2D transistor in a circuit environment. The faster switching process of the 3D structure is caused by the less pronounced high-current effects and the lower base resistance. Also shown in Fig. 3 is the switching behaviour *predicted* by a new laterally scalable version of the compact model HICUM [4].

4. Conclusions

The influence of 3D effects on electrical characteristics of small size bipolar transistors has been investigated using 3D device simulation. Collector current and (peak) transit frequency were observed to be (strongly) dependent on emitter size and geometry. The presented scaling equations for low current densities were shown to be suited for compact modelling. The scaling capabilities of a bipolar compact model were verified by transient 3D mixed-mode device/circuit simulation.

Acknowledgments

This work has been partially supported by NSERC.

References

[1] M. Schröter, "Transient and small-signal high-frequency simulation of numerical device models embedded in an external circuit", COMPEL, Vol. 10, No. 4, pp. 377-378, 1991.
[2] A. Koldehoff, M. Schröter and H.-M. Rein,"A Compact Bipolar Transistor Model for Very-High Frequency Applications with Special Regard to Narrow Emitter Stripes and High Current Densities", Solid-State Electronics, Vol. 36, 1993, pp. 1035-1048.
[3] H.-M. Rein, "A simple method for separation of the internal and external (peripheral) currents of bipolar transistors", Solid-State Electronics, Vol. 27, pp. 625-632, 1984.
[4] H.-M. Rein and M. Schröter, "A Compact Physical Large-Signal Model for High-Speed Bipolar Transistors at High Current Densities - Part II: Two-Dimensional Model and Experimental Results", IEEE Trans. Electron Dev., Vol. ED-34, 1987, pp. 1752-1761.

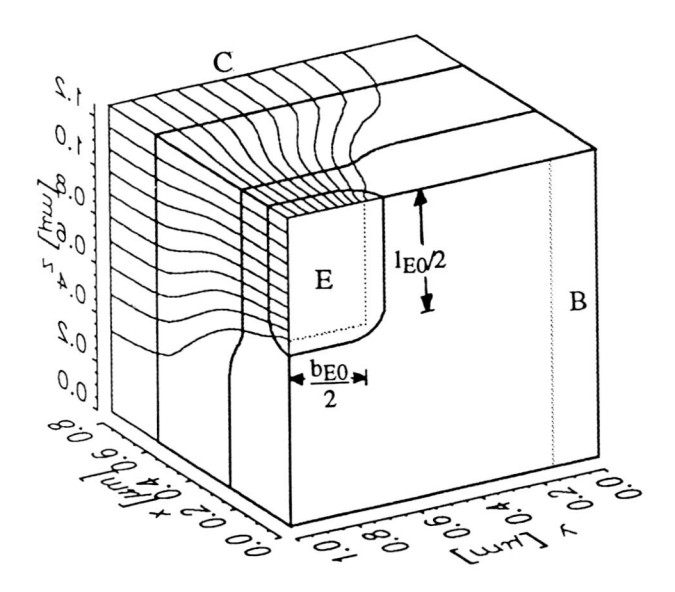

Figure 1: Simulated 3D structure ($l_{E0}/b_{E0} = 2$) with doping contours (thick lines) and transition to the buried layer. For symmetry reasons only one quarter of the relevant portion of a transistor with two base contacts and one emitter contact has been simulated. Additionally, electron current flow lines have been inserted for an operating point at high current densities with (10, 20, ...90)% of total current.

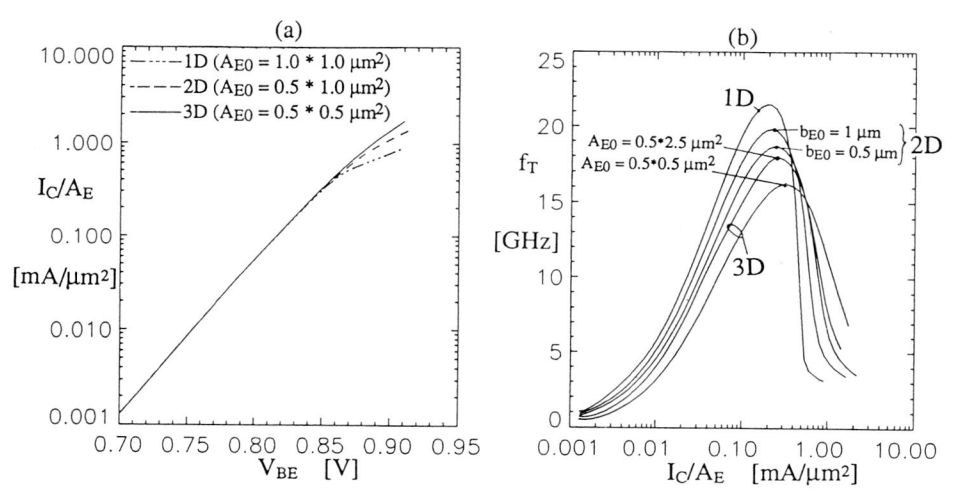

Figure 2: (a) Collector current density I_C/A_E vs. base-emitter voltage V_{BE} and (b) transit frequency f_T vs. I_C/A_E for a 3D, 2D, and 1D transistor structure. $V_{CE} = 0.8V$.

Table I: Low-current transit time τ_{f0}: comparison between simulation and the value calculated by (2) using $\tau_{fi} = 5.64ps$ and $\tau_{fp} = 11.2ps$. For the 2D (and 1D) case, $L_E = L_{E0}$.

structure	$\tau_{f0, sim}$ [ps]	$\tau_{f0, cal}$ [ps]
3D: $l_{E0}/b_{E0} = 1$	7.60	7.62
3D: $l_{E0}/b_{E0} = 2$	7.25	7.15
3D: $l_{E0}/b_{E0} = 5$	6.77	6.83
2D: $b_{E0} = 0.5\mu m$	6.60	6.60
3D: $l_{E0}/b_{E0} = 2$ (rounded emitter)	7.12	7.13

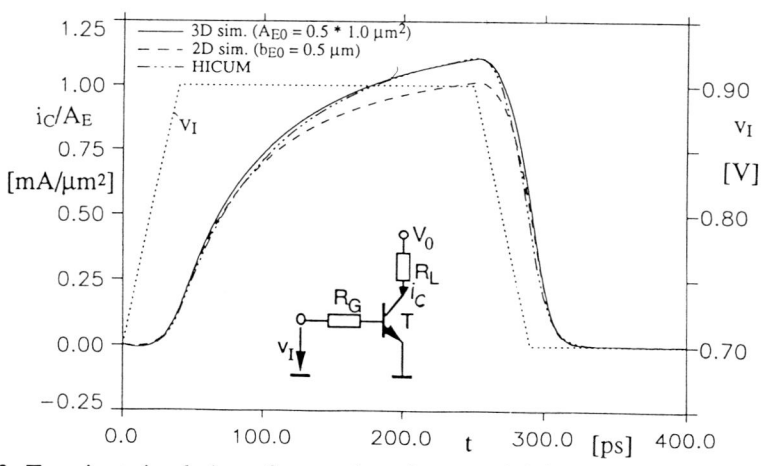

Figure 3: Transient simulation of a transistor inverter (cf. inset) using a 3D and a 2D numerical transistor as well as a compact model for T. ($R_G=R_L=400\Omega$; the internal base resistance value is in the same order as R_G.)

Three-Dimensional Monte Carlo Simulation of Boron Implantation into <100>Single-Crystal Silicon Considering Mask Structure

Myung-sik Son[a], Hwa-sik Park[b], Ho-jung Hwang[a]

[a]Semiconductor Process and Device Laboratory, Department of Electronic Engineering, Chung-Ang University, Seoul, KOREA
[b]Integrated Circuits Laboratory, Department of Electrical Engineering, Stanford University, Stanford, CA 94305, USA

Abstract

A new and computationally efficient three-dimensional Monte Carlo ion implantation simulator, TRICSI, has been developed to investigate three-dimensional mask effects for low-energy boron implantation into <100>single-crystal silicon. The simulator accurately and efficiently simulates three-dimensional implanted doping profiles under the mask structure and window. All of the typical implant parameters such as dose, tilt angle, rotation angle, in addition to energy are considered. The orientation of silicon substrate, ion beam divergence, presence of native oxide layer, wafer temperature, orientation of masking edge, masking layer thickness, and structure and size of window are also taken into account.

1. Introduction

In the crystalline silicon, one-dimensional(1D) or two-dimensional(2D) Monte Carlo(MC) ion implantation simulations have been made by many authors [1], [2], [3]. However, three-dimensional(3D) MC simulations of ion implantation in the crystalline silicon have been rarely reported and have yet to predict 3D effects depending on the mask structure and the size of the open window for ion implantation. The 3D MC simulation of ion implantation in the crystalline silicon is strongly needed to understand the 3D behaviors of implanted impurity and to compactly control the depth and lateral doping profiles of implanted impurity under smaller implant area at ever lower energy. In addition, the 3D implantation simulation based on the physical approach becomes more important under the circumstances that the 3D experiments of the implanted doping profiles are not available for the present. In this paper, we have investigated and predicted 3D mask effects by using our simulator, TRICSI (TRansport Ions into Crystal SIlicon) which is coded based on TRIM [4] and MARLOWE models [5]. A newly developed searching algorithm for a collision atom in <100>single-crystal silicon and an effective cumulative damage model for boron implantation are implemented in the simulator. In the following section, we describe the details of models employed in TRICSI.

2. Model Details

Recently, UT-MARLOWE code [1] has been developed for 1D boron implantation simulation in the crystalline silicon, and the results show a good agreement with the SIMS experiments, but it is reported that the simulation time is increased by 50% compared with that of MARLOWE because it has considered and calculated that the variation of electronic density through which the moving ion has got along its trajectory. In this paper, in order to efficiently simulate the whole volume for 3D simulation, we have newly defined and modified the value of the average electronic density for the use of ABS electronic stopping model [6] which the implanted ion has experienced along its trajectory. The value of the average electronic density used for the good agreement with the SIMS experiments [1], [7] was found 0.735 electrons/$Å^3$ over the range of the simulated energies, from 5 to 80 keV. A *priori* damage function (1) is also introduced into the cumulative damage model for boron implantation by manipulation of the maximum impact parameter P_{max} which is the same as the TRIM calculation [4]. The P_{max} is determined by the function of the ion energy, the ratio of moving ion mass to stationary silicon atom mass, and the minimum transferred energy, which is set initially to 1.5 eV for zero tilt angle and 5 eV for any tilt in our simulation. For all of collisions, the new impact parameter P'_{max} as a function of the cumulative damage probability $f(X)$ [8] is calculated by the following function, where the value of $f(X)$ is predetermined not to exceed 1 initially in the simulation.

$$P'_{max} = P_{max} \times (1 + f(X)^3) \tag{1}$$

As a result of the new P'_{max}, the collisions of the implanted ions are increased, and the otherwise channeled ions are dechanneled. The increased rate of their random smatterings decreases the depth channeling tail. The relatively simple damage model has well simulated the low-dose profiles below 1E15 ions/cm^2. To consider the lattice vibration effect, Debye model [9] has been implemented and the average displacement X_{rms} from the lattice site of the silicon is calculated by using the Debye temperature of 543K [9] for the crystalline silicon. It is assumed that the displacement due to the lattice vibration is randomly determined in 3D coordinates. We can directly calculate the impact parameter in a collision by considering the random displacement from the 3D original site of the silicon lattice atom as X_{rms} multiplied by the random number between 0 and 1. The 10 Å native oxide layer is considered as an amorphous layer on the silicon substrate and the 0.5° beam divergence is also included assuming that the incoming ion beam spreads isotropically within a cone defined by the divergence angle. Also, using the vector analysis for the direction vector of the moving ion after a collision, we can calculate two direction angles defined as azimuthal angle and the scattering angle in 3D coordinates. The nearest collisional row is first searched according to the azimuthal angle from Z axis defined in Fig. 1. Each silicon atom which is spacing by the silicon lattice constant of 5.43095 Å from the initial reference atom in the detected collisional row is examined one by one by using the scattering angle from X axis parallel to <100> direction. If a collision fails to be found in the first nearest row, the search is continued until the collision is found in the next sequential detected row or the ion energy is exhausted by the electronic collisions. Consequently, after searching for a collision in the single-crystal silicon, we can directly and accurately calculate the impact parameter and the flight-path length between collisions in 3D coordinates without the random selection of the impact parameter in the simulation of the amorphous silicon target.

3. Simulation Results

In Fig. 1, 1D simulation results obtained by using the 3D simulator have been compared with the SIMS experiments in order to demonstrate its capability and reliability. The definitions of tilt angle, rotation angle, and 3D coordinates for the simulation structure are also shown. The simulator accurately and efficiently calculates the 1D depth profiles under the area of implant window in the mask. The calculated 3D locations of implanted ions have been projected onto the desired 3D plane, so that the resulting doping profile is presented as a 2D doping profile on each projected plane. In order to investigate 3D effects due to the mask structure and the size of implant window, boron pseudoparticles were uniformly and randomly implanted into the entire area of implant window. The masking layer was assumed to be impenetrable and have the window for the implanted region with vertical edges. Fig. 2 shows the corner rounding effect that the lateral concentration profiles in the regions near the masking corners are decreased compared with those in the regions near the masking edges. The dilutes of ion concentrations around corners are due to the decrease of scattered ions into the regions near those corners, while the lateral concentration profile in the region near one specific masking corner in the structure of the window of $2500 \times 2500 \times 3/4$ is enhanced because of superposed increase of ions scattered into that region from two adjacent masking edges. In addition, in Fig. 3, the narrow window effect on the change of the lateral doping profile in a reduction of the window size has been presented: the lateral doping profile becomes circular as the size of the perfect-square window is reduced from 500×500 nm^2 to 20×20 nm^2. The tilt and the rotation angle in the simulation shown in Fig. 2, 3 are all zero. In Fig. 4, two different profiles on each projected plane in cases of 15° tilt and 0° tilt without rotation are presented to show the difference of the profiles of two different tilt cases. For the 15° tilt angle profiles, the asymmetric profile on the x-y plane is due to the 15° tilt angle, the symmetric profile on the x-z plane is due to the zero rotation angle, and the iso-concentration contours on the y-z plane are slightly shifted downwards and shrunk in the shadowed region due to the shadow effect of the masking thickness compared with 0° tilt angle profiles on the same planes.

4. Conclusions

3D low-energy boron implantation into <100>single-crystal silicon has been modeled and simulated by using the physical approach, MC method. The newly constructed MC approach assures the accuracy of the simulation as shown in the comparison of the 1D data with the SIMS experiments. The simulation results also clearly show 3D mask effects such as the corner rounding effect and the narrow window effect. The doping profile near the mask corner is enhanced or diluted due to the mask structure. In the narrow window, the impurity dilute phenomena is well presented and it finally gives circular contours of two-lateral doping profile on the <100>silicon surface. This effect arises from the fact that the lateral scattering of implanted ions is more concentrated or more diverse in the mask corner than in the mask edge. In addition, not only the planar channeling in the two-lateral doping profile, but also the axial channeling in the depth and lateral doping profile has been presented for the case of very low energy of 5 keV. It shows that the planar and axial channeling effect occurs strongly in the small implant window due to the increased critical angles for the planar and the axial channeling.

References

[1] K.M. Klein et al, IEEE Trans. on Elec. Devices, vol.39, no.7, pp.1614-1621, 1992
[2] K.M. Klein, C. Park, S.H. Yang, A.F. Tasch, IEEE IEDM Tech. Dig., 27.2.1-4, 1991
[3] G. Hobler, H. Pötzl, IEEE IEDM Tech. Dig., 27.1.1-4, pp.693-696, 1991
[4] J.F. Ziegler et al, The stopping and Range of Ions in Solids, vol. I, New York: Pergamon,1985
[5] M.T Robinson, M.Torrens, Phys. Rev. B., vol.9, no.12, pp.5008-5024, 1974
[6] C.S. Murthy, et al, IEEE Trans. on Elec. Devices, vol. 39, no.2, pp.264-273, 1992
[7] A.E.Michel et al, Appl. Phys. Lett. 44(4), pp.404-406, 1984
[8] H.J. Kang et al, J. Appl. Phys., vol.62, no.7, pp.2733-2737, 1987
[9] C. Park, K.M. Klein, A.F. Tasch, J. Electrochem. Soc., vol. 138, no.7, pp.2107-2115, 1991

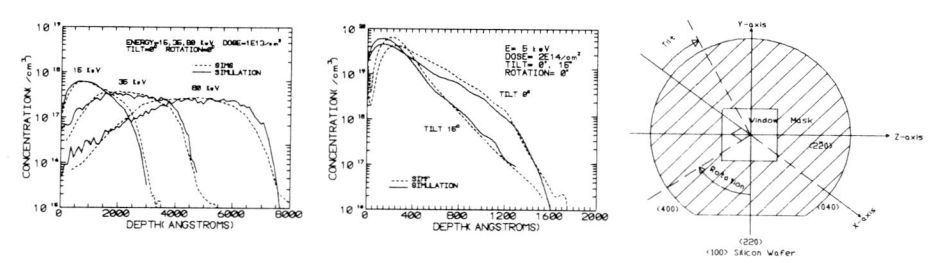

Figure 1: Comparison of 1D simulation results with the SIMS experiments [1], [7], and definitions of tilt angle, rotation angle, and 3D coordinates for the simulation structure.

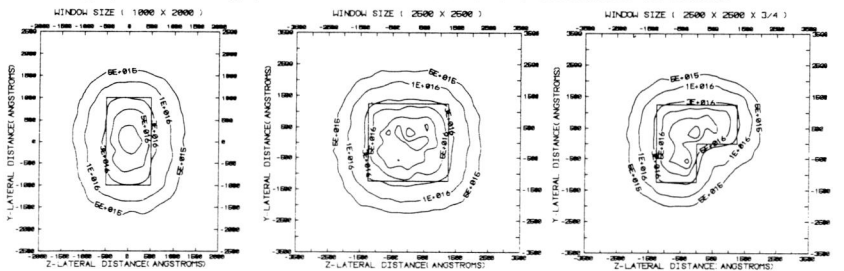

Figure 2: The corner effect of 3D ion implantation at Energy 15 keV, Dose 1E13 ions/cm^2.

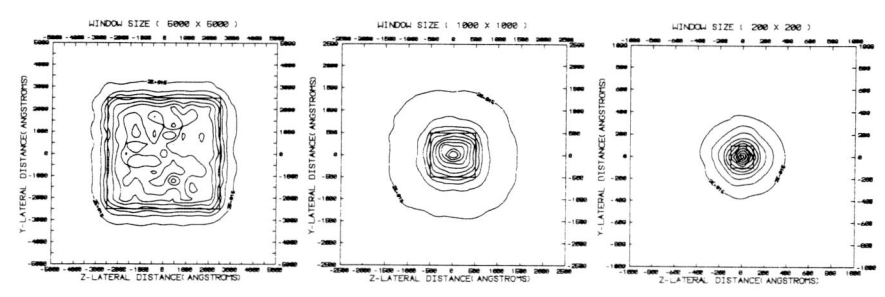

Figure 3: The narrow window effect of 3D ion implantation at Energy 15 keV, Dose 1E13 ions/cm^2.

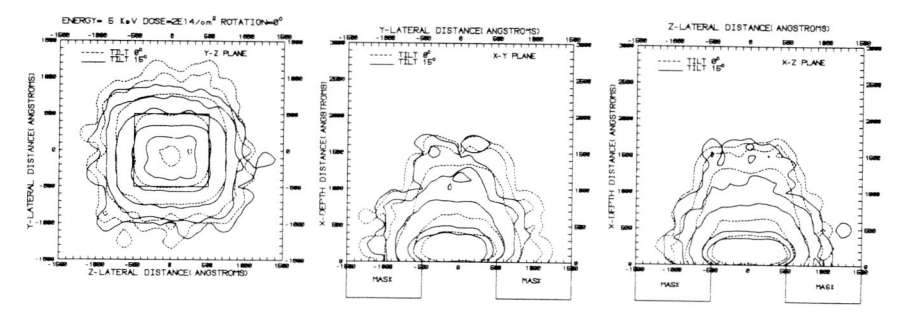

Figure 4: Comparison of the doping profiles between at $0°$ tilt and at $15°$ tilt without rotation, where the thickness of the masking layer is 500Å; for the solid or dashed contour lines, the concentration of the outermost contour is 1E15/cm^3 and from the outermost one, the concentration is increasing by one order of magnitude. The maximum concentration presented is 2E18/cm^3.

A fully 2D, Analytical Model for the Geometry and Voltage Dependence of Threshold Voltage in Submicron MOSFET's

A. Klös, A. Kostka

Solid State Electronics Laboratory, Technical University of Darmstadt
Schloßgartenstraße 8, D-64289 Darmstadt, GERMANY

Abstract

In this paper we present a physics-based, compact model for the threshold voltage shift in short-channel MOSFET's, which is based upon a new theoretical approach in MOS modeling. This method uses conformal mapping techniques to solve the 2D Poisson equation in the space charge region underneath the gate and considers inhomogeneous doping profiles therein. The derived model equations appear in closed form and require only two physical fitting parameters related to a geometry and a doping approximation. A comparison with numerical device simulations reveals a high degree of accurateness down to channel lengths of $0.2 \mu m$.

1. Introduction

For circuit development, accurate models to predict the layout dependence of the threshold voltage of short-channel devices are needed. Various quasi two-dimensional models have been published for this purpose. Most of them are based upon the unphysical charge-sharing principle [1, 2] to approximate the influence of the source/drain space charge regions on the threshold voltage. Better accuracy is achieved by more physics-based models, which solve an approximation of Poisson's equation in the device cross-section [3]. To arrive at analytical model equations suitable for circuit simulators, these models have to introduce process and layout dependent fitting parameters. Therefore, a great number of devices of different geometries are required for parameter extraction. Also, they have no forecasting ability to estimate the sensitivity of threshold voltage to process data or to investigate fictive processes at all.

The model presented in this paper follows a fully two-dimensional approach and from this, it does not need any fitting parameter to describe the two-dimensional effect of threshold voltage shift. The only fitting parameters used are one for simplification of the device structure and one for consideration of an implantation profile.

2. Solving Poisson's Equation

The device cross-section of a MOSFET with the definitions of process and geometry parameters is shown in Fig. 1a. To model the short-channel effects upon the threshold

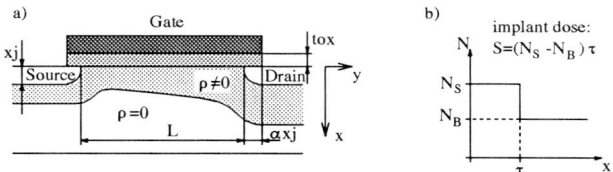

Figure 1: (a) cross-section of a MOSFET, (b) doping box approximation.

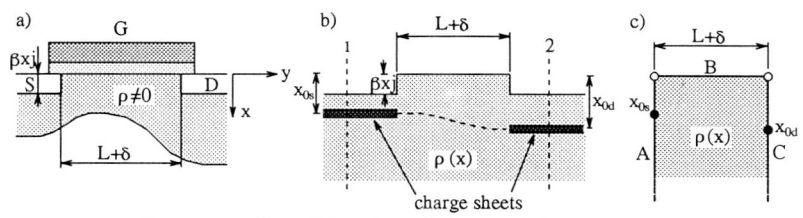

Figure 2: Simplification of the model structure.

voltage, the two-dimensional Poisson equation has to be solved. A surface potential $\phi_i + V_{sb}$ according to the onset of strong inversion is applied as boundary condition relevant for the threshold condition along the channel region. If mobile carriers are neglected, the dielectric flux density within the gate-oxide is maximal at the position y_0 where, by raising the gate voltage, the strong inversion sets on latest. Together with the work function difference Φ_{ms} and the oxide charge density q_{ox}, this flux density D_0 is used to determine the threshold voltage:

$$V_T = \Phi_{ms} - \frac{q_{ox}}{C'_{ox}} + \phi_i + \frac{D_0}{C'_{ox}} \quad \text{where} \quad C'_{ox} = \frac{\varepsilon_{si}}{t_{ox}}. \quad (1)$$

In [4, 5] conformal mapping techniques have been shown to be useful in a strictly physical, 2D analysis of planar geometries in lateral bipolar transistors, leading to closed form solutions for the Laplacian DEQ. In an attempt to apply this method in MOS modeling, the solution Φ of Poisson's equation can be found by superposition of a homogeneous, two-dimensional solution φ and a non-homogeneous, one-dimensional solution Φ_p:

$$\Delta\Phi(x,y) = -\frac{\rho}{\varepsilon_{si}} = \Delta\varphi(x,y) + \Delta\Phi_p(x) \quad \text{where} \quad \Delta\varphi(x,y) = 0, \; \Delta\Phi_p(x) = -\frac{\rho(x)}{\varepsilon_{si}}. \quad (2)$$

This is possible, as long as the space charge ρ is a function of only one coordinate. $\Phi_p(x)$ then is the one-dimensional potential solution of a long-channel MOSFET. The conformal mapping technique [4, 5] provides a method to get analytical solutions of the Laplacian DEQ $\Delta\varphi = 0$. For this, the boundary conditions of Poisson's equation have to be transformed by substracting Φ_p before solving the Laplacian DEQ.

3. Simplifying the Model Structure

In order to keep mathematics manageable, the following approximations are made. First, the doping profile of an implantation for threshold voltage shift is approximated by a dose equivalent box (Fig. 1b) to ensure closed form solutions. The depth τ of the box can be found from the best fit of the model to the body effect [6].

Secondly, to get a closed form solution for the two-dimensional potential problem outlined above, the geometry of the structure is simplified as shown in Fig. 2. In a first step the curved shape of the source/drain junctions will be replaced by a box approximation described by one shape-fitting parameter β (Fig. 2a). For homogenous substrate doping this is the only fitting parameter of the model at all. The influence of β is only of second order, and investigations using the device simulator ATLAS II for devices of different processes state the fact, that for realistic device parameters a process independent value of $\beta = 0.5$ gives a very good fit.

In a second step, a simplification is necessary for the one-dimensional description of the space charge ρ. If the space charge ρ is extended to the whole substrate (Fig. 2b), and a part of the charge underneath the source/drain regions is compressed into a charge sheet, an appropriate choice of the depths $x_{0s/d}$ of the charge sheets will result in nearly the same potential solution in the channel region as in the structure shown in Fig. 2a. A one-dimensional calculation of the potential along line 1 and 2 is used as a compatible approximation of the boundary conditions along lines A and C of Fig. 2c, respectively, which shows the shape of the area the conformal mapping is applied to at the end, and where the charge sheets do not appear. The error caused by introducing the charge sheets outside the critical area should be small due to the relatively large distance to the point at which D_0 is determined.

4. Model for NMOS-processes

Proceeding in the way described in section 3, after some mathematics, one arrives at a set of explicit, closed form equations for V_T. Due to the extent of the model equations, we concentrate on illustrative examples for NMOS-processes with $\tau > \beta x_j$, which should hold for most of them, visualizing the progress of the derivation.

(a) L, x_j, α, β, N_B, N_S, τ, V_{ds}, V_{sb} represent the input data of the model. V_{bi} is the built-in potential of the source/drain junctions, ϕ_i the surface potential according to the onset of strong inversion and δ a definition for convenience:

$$\phi_i = \frac{kT}{q} \ln \frac{N_B N_S}{n_i^2} \quad (3) \qquad V_{bi} = \frac{E_G}{2} + \frac{kT}{q} \ln \frac{N_B}{n_i} \quad (4) \qquad \delta = 2\alpha x_j (1 - \sqrt{1-\beta^2}) \quad (5)$$

(b) From one-dimensional field calculations at the surface of a long channel device and at the edges of the depletion regions underneath source/drain the corresponding electrical fields E_0 and $E_{0s/d}$, respectively, are determined in closed form. From this, the depths of the charge sheets result in:

$$x_{0s/d} = \frac{\frac{qN_S}{2\varepsilon_{si}}(\beta x_j)^2 + \beta x_j E_{0s/d} + V_{bi} + \varphi_{s/d} - \phi_i}{\frac{qN_S}{\varepsilon_{si}}\beta x_j + E_{0s/d} - E_0} \quad \text{where } \varphi_s = 0 \text{ and } \varphi_d = V_{ds} \quad (6)$$

(c) The following parameters are used for conformal mapping:

$$b = \cosh \frac{\pi \beta x_j}{L + \delta} \quad (7) \qquad c_{s/d} = \cosh \frac{\pi x_{0s/d}}{L + \delta} \quad (8) \qquad d = \cosh \frac{\pi \tau}{L + \delta} \quad (9)$$

(d) The conformally mapped position of the surface potential minimum is given by:

$$u_0 = \frac{\sqrt{V_{bi}} - \sqrt{V_{ds} + V_{bi}}}{\sqrt{V_{bi}} + \sqrt{V_{ds} + V_{bi}}} \quad (10)$$

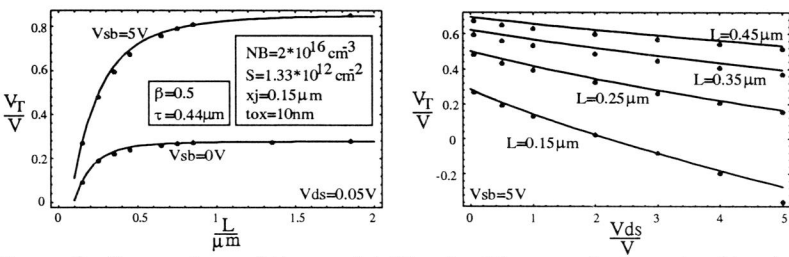

Figure 3: Comparison of the model (lines) with numerical results (dots).

(e) Solving the Laplacian DEQ results in source/drain related flux densities at u_0:

$$D_{s/d} = \frac{\varepsilon_{si} E_0}{\pi}\left(\arctan\frac{bu_0 - 1}{\sqrt{b^2-1}\sqrt{1-u_0^2}} + \frac{\pi}{2}\right) + 2\sqrt{2}\frac{L+\delta}{\pi^2}\sqrt{1-u_0}\,qN_S\,f(b,1)$$

$$+(\varepsilon_{si}(E_0 - E_{0s/d}) - qN_S\beta x_j)\frac{g(c_{s/d},b)}{\pi} + \frac{\varepsilon_{si}}{L+\delta}\sqrt{\frac{1+u_0}{1-u_0}}\,(V_{bi} + \varphi_{s/d} - \phi_i), \quad (11)$$

where f and g are abbreviations of analytical functions.

(f) Superposition of the one-dimensional solution and the results for the Laplacian DEQ gives the flux density D_0, and, hence V_T via (1):

$$D_0 = \varepsilon_{si} E_0 - D_s - D_d \quad (12)$$

Fig. 3 shows a comparison of the presented model with results of the numerical device simulator ATLAS II for a NMOS-process. Both are in very good agreement down to a channel length of about $0.2\mu m$. The only fitting parameters used are the flatband voltage and the depth τ of the implant doping box approximation.

5. Conclusion

The presented method for solving Poisson's equation in MOS devices based upon conformal mapping techniques made it possible to develop a physics-based, analytical model for the geometry and voltage dependence of the threshold voltage in short-channel devices in closed form. The small number of fitting parameters, which are all physically meaningful, qualifies the model to be useful in circuit simulators as well as in calculations of scaling behaviour. The method is a new theoretical approach in MOS modeling and not restricted to the effect of threshold voltage shift. It can be transferred on modeling other effects like the subthreshold behaviour, which is currently under work by the authors.

References

[1] L. D. Yau, Solid-State Electron., 17, pp. 1059-1063, 1974
[2] T. A. Fjeldly et al., IEEE Trans. Electron Devices, vol. 40, No. 1, p. 137, 1993
[3] Z. H. Liu et al., IEEE Trans. Electron Devices, vol. 40, No. 1, p. 86, 1993
[4] D. Freund, A. Klös, A. Kostka, Proc. ESSDERC 93, pp. 29-32, Editions Frontieres, Grenoble, 1993
[5] D. Freund, A. Klös, A. Kostka, Proc. SISDEP 93, pp. 425-428, Springer-Verlag, Wien, 1993
[6] S. Karmalkar, K. N. Bhat, Solid-State Electron., vol. 34, No. 7, pp. 681-692, 1991

On the influence of band structure and scattering rates on hot electron modeling

Chr. Jungemann[a], S. Keith[b], B. Meinerzhagen[b], and W.L. Engl[c]

[a]Fujitsu Limited 1st Process Dept.
1015, Kamikodanaka, Nakahara-ku, Kawasaki, 211 Japan
[b]Institut für Theoretische Elektrotechnik und Mikroelektronik, University of Bremen
Kufsteiner Strasse, Postfach 33 04 40, 28334 Bremen, Germany
[c]Institut für Theoretische Elektrotechnik, University of Aachen
Kopernikusstr. 16, 52074 Aachen Germany

Abstract

The relative importance of band structure and scattering rate modeling for the modeling of hot electrons is evaluated. It turns out that the influence of band structure is dominant, while the details of the scattering models are less significant. Consequently computationally rather efficient models for phonon scattering and impact ionization scattering can be applied without sacrificing accuracy as long as the full band structure is considered.

1. Introduction

For the acceptance of a Monte-Carlo model among the engineering design community a careful trade-off between computational efficiency and physical accuracy is necessary. Unfortunately there is no clear solution to this problem, so that many different Monte-Carlo models exist especially for electron transport [1]. In this paper we focus on the problem of modeling hot electrons efficiently over a broad range of energies (up to 4 eV) with sufficient accuracy. In this context it was demonstrated recently [2] that a good agreement between experimental and simulated results on several key hot electron related quantities like impact ionization coefficient and quantum yield can be achieved simultaneously by using a full band Monte-Carlo model and an impact ionization scattering rate extracted from experimental data. Since so far nothing comparable has been published for models based on a simpler description of band structure, let's assume for the moment that the adoption of the full band structure is a necessary prerequisite for reproducing these desirable results with a Monte-Carlo model. The efficiency of a Monte-Carlo model however depends not only on band structure modeling but quite heavily on the complexity of the scattering models as well and in the original paper [2] a very complex and time consuming model for phonon scattering was used. Therefore we have investigated whether the same results can be reproduced by adopting the much simpler phonon scattering model of Jacoboni et al. [5] in order to enhance the overall simulation efficiency.

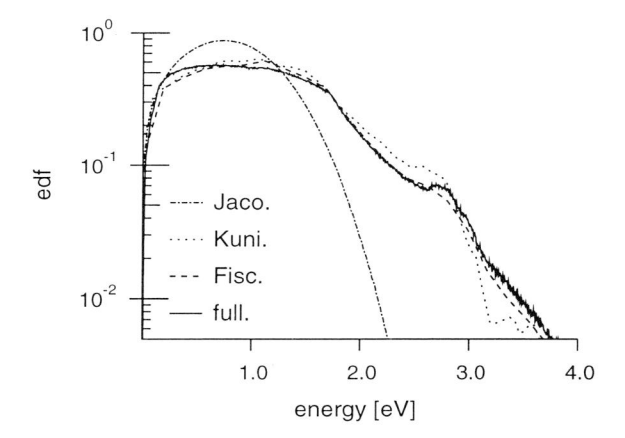

Figure 1: Energy distribution function for the new full band model and from Fischetti et al., Kunikiyo et al. [1] and the original model of Jacoboni et al. [5].

2. Description of the Model

The Monte-Carlo Model used in this paper is based on the full band structure that results from the local pseudo potential method described in [4] and on an extension of Jacoboni's model for phonon scattering to the full band case. The matrix elements of the latter model do not depend on momentum transfer so that this model is computationally very efficient. Only a slight modification of the phonon coupling constants and deformation potentials was necessary (global scaling by a factor of 0.93 was sufficient) in order to reproduce the experimentally observed velocity/field relation with the new full band model. For low energies the band structures of the original Jacoboni model [5] and our new model are in good agreement. Therefore it is a direct consequence of the nearly identical treatment of phonons in both models that all the favorable properties which Jacoboni's model has for low field mobility like the good agreement of the lattice temperature dependence with experiment are valid for our new model as well.

3. Results

In figure 1 the electron energy distribution functions for a homogeneous field of 300 kV/cm of our full band model is compared to the respective results of Fischetti et al. and Kunikiyo et al. [1] which both employed much more complex models for phonon scattering. All three results are remarkable close showing that the distribution function for such fields is insensitive to the complexity of the phonon model (see also [3]). On the other hand the distribution function is very sensitive to the band structure, which is proven by the large differences for energies larger than 2 eV between the distribution functions from Jacoboni's original model and our new model that can be observed in figure 1 as well. Please keep in mind that the two latter Monte-Carlo models employ basicly the same model for phonon scattering.

For modeling impact ionization related quantities like quantum yield and ionization coefficient we investigated the two scattering rates for impact ionization used by Cartier et al. [2] and Thoma et al. [6]. Both scattering rates include unknown parameters that have to be extracted from experiment. In order to make a consistent comparison of both rates we kept the functional form of both rates but adjusted them

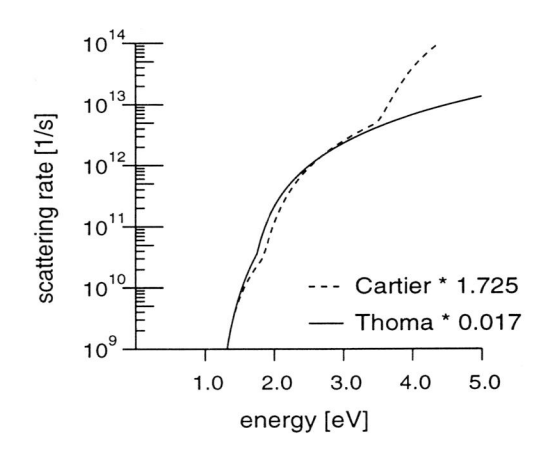

Figure 2: Impact ionization scattering rate of Thoma et al. [6] with a scaling factor of 0.017 and of Cartier et al. [2] with a scaling factor of 1.725.

by global scaling factors such that the quantum yield data of DiMaria et al. [7] at 3 eV was reproduced by our full band model. Doing so the scattering rates as a function of energy became remarkable close below 3.5 eV as can be seen in figure 2. In figures 3 and 4 the results of [7, 8] for quantum yield and impact ionization are compared to the respective results from our new full band model. It can be seen that the agreement of our simulator with the experimental data is just as good as it was reported in [2]. Moreover it becomes clear that this good agreement is insensitive to the exact functional form of the impact ionization scattering rate.

4. Summary

The above results show that even under the assumption that the adoption of the full band structure is necessary to achieve good modeling results for hot electrons over a broad range of electron energies, there are still a lot of possibilities to keep the simulator efficiency at a level that is acceptable for engineering design purposes. For MOS devices the new full band model is only about a factor of two slower than the Monte-Carlo device simulator that we have developed previously [9] and which was based on an analytical band structure description. Thus even on modern high speed workstations full band Monte-Carlo device simulation is feasible.

References

[1] A. Abramo et al., IEEE ED **4**, 1646 (1994)
[2] E. Cartier et al., Appl. Phys. Lett. **62**, 3339 (1993)
[3] J.M. Higman, IWCE 94, 7 (1994)
[4] J.R. Chelykowski et al., Phys. Rev. B **14**, 556 (1976)
[5] C. Jacoboni et al., Rev. Mod. Phys. **55**, 645 (1983)
[6] R. Thoma et al., J. Appl. Phys. bf 69, 2300 (1991)
[7] D.J. DiMaria et al., J. Appl. Phys. **57**, 1214 (1985)
[8] R. van Overstraeten et al., Solid-State Electron. **13**, 583 (1970)
[9] H.-J. Peifer et al., IEDM Tech. Dig., 131 (1991)

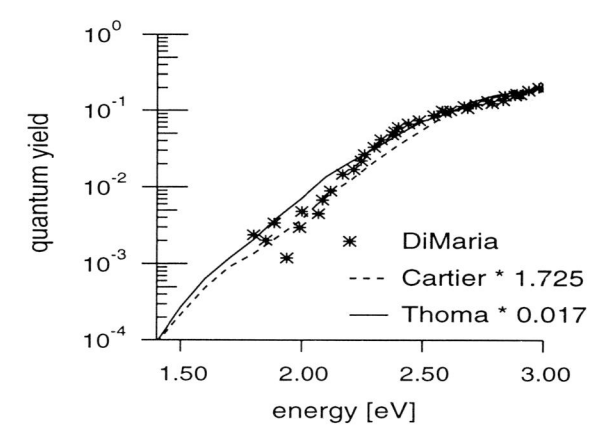

Figure 3: Quantum yield obtained with the full band model with the scaled II scattering rate of Thoma et al. [6] and Cartier et al. [2]. Experimental results: DiMaria et al. [7].

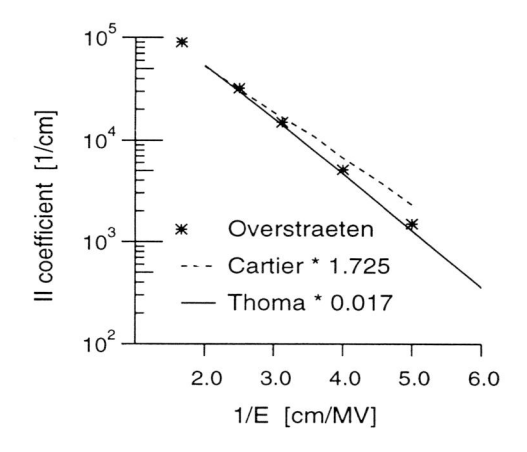

Figure 4: Impact ionization coefficient for the same models as above. Experimental results van Overstraeten et al. [8].

Finite Element Monte Carlo Simulation of Recess Gate FETs

S. Babiker, A. Asenov, J.R. Barker and S.P. Beaumont

Nanoelectronics Research Centre
Department of Electronics and Electrical Engineering
Glasgow University
Glasgow, G12 8QQ, Scotland, UK

Abstract

In this paper we report on a new Monte Carlo (MC) module incorporated in our Heterojunction 2D Finite element FET simulator H2F [1]. For the first time this module combines a precise description of the device geometry with realistic particle simulation of the non-equilibrium hot carrier transport in ultra-short recess gate compound FETs. The capabilities of the new finite element MC module are illustrated in example simulations of two compound FETs fabricated in the Nanoelectronics Research Centre of Glasgow University.

1. Introduction

The ensemble Monte Carlo approach plays an important role in the simulation of submicrometer compound FETs where both steady-state and transient device behaviour are dominated by velocity overshoot effects. A large variety of devices including MESFETs [2], HEMTs [3] and pseudomorphic HEMTs [4] have been extensively studied over the past decade using the MC approach, with emphasis placed on different aspects of the device operation and miniturisation. Almost all of those studies consider oversimplified planar or rectangular solution domains. This is usually dictated by the rectangular finite-difference grid used for the discretization of Poisson's equation. In addition, the surface potential pinning effects are usually neglected.

In contrast, almost all modern sub micrometer compound FETs are single or double recess devices [5] with complicated recess geometries. The length of the recess region is comparable to the length of the gate [5]. Device parasitics like access resistances and coupling capacitances are dependent on the shape and surface conditions of the recess [1]. These parasitics are critical to the device characteristics, limiting in many cases the device performance and restricting on the advantages of device miniturisation. Here we report on a new Monte Carlo simulation module suitable not only for qualitative research but for practical design of real sub micrometer FETs. It combines the extensive transport capabilities of the ensemble Monte Carlo with a precise description of the device geometry and proper handling of surface effects.

2. The Monte Carlo module

2. 1. Finite Element Grid Generation

Quadrilateral finite elements have been adopted to accurately model the device geometry; in particular the recess shape and gate profile. The triangulation includes not only the semiconductor regions but also the air space in the recess region and above the cap layer. This provides proper modelling of the interaction between the charge on the surface states and the electrostatic potential and gives a realistic fringing field pattern.

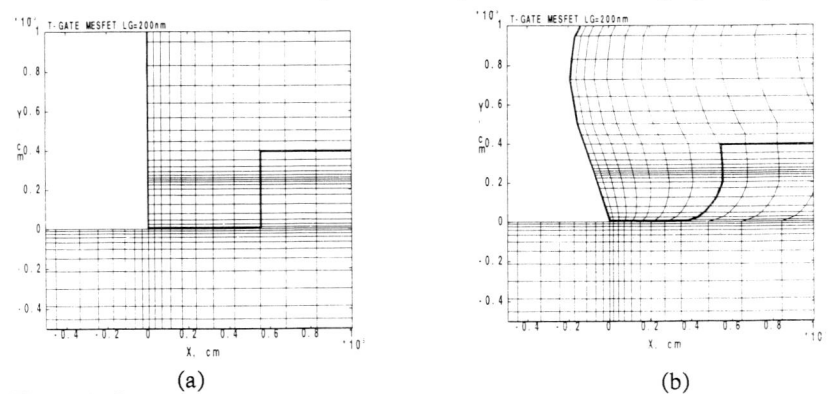

(a) (b)
Figure 1: Generation of the quadrilateral grid in the recess region: (a) the initial rectangular grid (b) the quadrilateral grid after the transformation

The grid generation proceeds in two stages. In the first stage a rectangular non-uniform grid is generated such that it conforms to the vertical layer structure and to the simplified lateral pattern of the device. In the second stage the shapes of the recess region and the gate are introduced from a data file. These are usually obtained from SEM photographs of the device cross-section. By appropriate deformation the grid is adapted to the actual geometries, Figure 1(a,b).

2.2. Solution of the Poisson Equation

The Galerkin finite element approach has been adopted to solve Poisson's equation.

$$\frac{\partial}{\partial x}\varepsilon\frac{\partial}{\partial x}\psi + \frac{\partial}{\partial y}\varepsilon\frac{\partial}{\partial y}\psi + \rho_B + \rho_I = 0 \qquad (1)$$

where $\rho_B = q(p - n + N_D^+ - N_A^-)$ is the bulk charge density and ρ_I is an interface and surface charge density, responsible for example for the surface potential pinning in recess gate FETs. The integration over quadrilateral elements during the discretization is carried out by a linear isoparametric mapping of each element into a unit square.

To provide a good initial approximation for the charge distribution in the device the nonlinear Poisson equation is solved, with electron and hole concentrations given by Boltzman or Fermi-Dirac statistics. This significantly reduces the amount of time necessary to obtain a steady-state solution.

2.3. Monte Carlo simulation in a Single Element

At the end of the finite-element triangulation process each finite element contains a single material with uniform doping concentration. Each finite element has a reference to a scattering table corresponding to the material type and doping level in the element.

A complete ensemble Monte Carlo procedure in a single quadrilateral element is the building block of the whole Monte Carlo module. During the motion of an individual particle its position with reference to the given quadrilateral finite element is tested using the linear isoparametric mapping used in the finite element formulation. If the transformed co-ordinates of the particle both lie in the range (0,1) then the particle remains within the given quadrilateral element; otherwise, the particle has crossed one of the four boundaries and is now located outside the element. Hitting a grid boundary during a free-flight-time the particle will either be transferred to the neighbouring cell, reflected, absorbed etc. depending on the type of the given element boundary.

Apart from its finite element implementation our Monte Carlo procedure is standard [6]. The time of free-flight and subsequent scattering mechanisms are chosen using the usual self-scattering scheme. The potential distribution is recalculated and the electric fields updated approximately every 5 fs depending on the minimum element size and the doping concentration in the heavily doped cap layers of our devices. The step size ensures that the carriers will transit no more than one element before the field is updated. The scattering mechanisms implemented include: ionised impurity, acoustic phonon, piezo-electric, optical phonon and polar optic scattering modes. A non-parabolic three valley (Γ,L,X) conduction band model was used for the III-V materials.

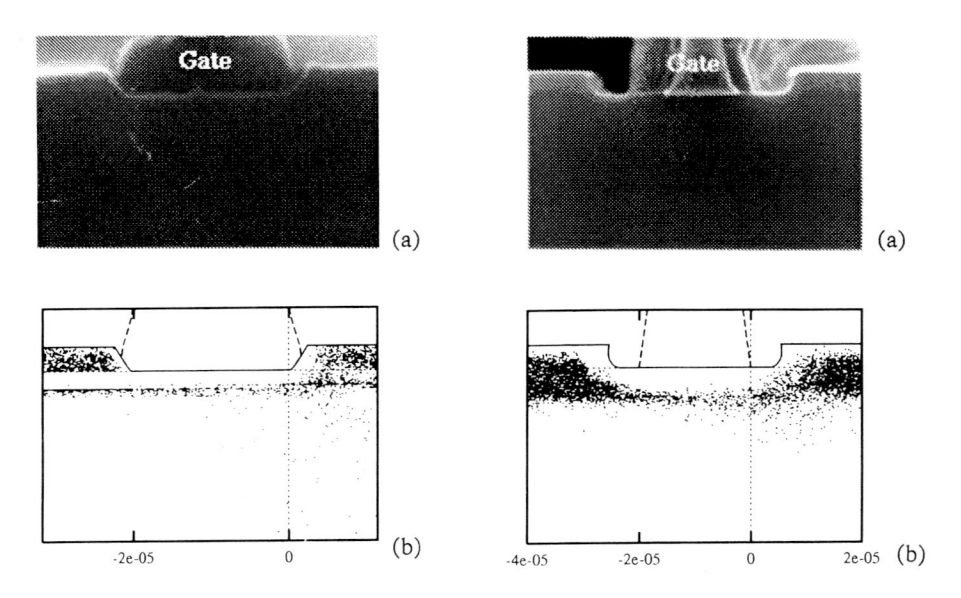

Figure 3: (a) SEM cross-section of a pseudomorphic HEMT (b) The corresponding H2F MC simulation domain

Figure 2: (a) SEM cross-sectional view of a 200 nm gate MESFET (b) The corresponding H2F MC simulation domain

3. Examples:

The ability of the quadrilateral finite element approach to precisely describe the shape of the recess region is illustrated for two short gate compound FETs with different recess shapes. Figure 2 (a) represents the SEM cross sectional view and the H2F simulation domain of a 200 nm gate length pseudomorphic HEMT fabricated virtually without offset between the gate and the recess edges. Figure 3 (a) represents the SEM cross sectional view and the H2F simulation domain of a 200 nm MESFET with 55 nm offset between the gate and recess edge. The distribution of superparticles at gate voltage $V_G = 0$ V and drain voltage $V_D = 1.5$ V are also given for both devices in Figure 2 (b) and 3 (b) respectively. In this example study we concentrate on the MESFET from Figure 3 because the effect of the recess is more pronounced in this device as a result of the relatively large gate offset. Monte Carlo simulation results regarding pseudomorphic HEMTs will be published elsewhere.

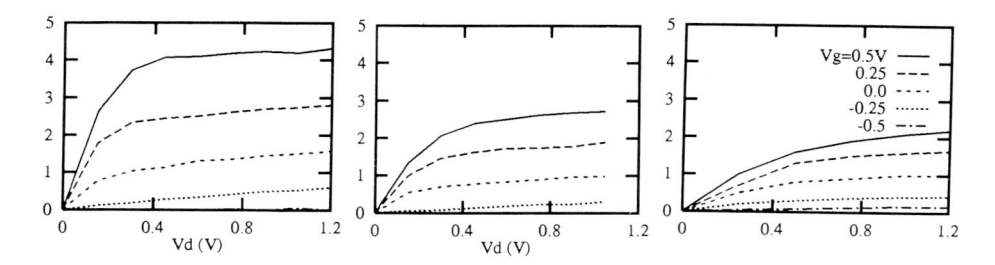

(a) (b) (c)
Figure 4: Calculated and measured I_D-V_D characteristics of a 200 nm MESFET. (a) MC simulation without surface charge (b) MC simulation with surface charge $N_s = 2.5 \times 10^{12}$ cm-2, (c) Measured characteristics

The surface conditions in the recess region critically affect the output device characteristics. The calculated and measured I_D-V_G characteristics are given in Figure 4 (a-c). The negligence of the surface potential pinning in the recess region (Figure 4 (a)) leads to unacceptable error in the calculate I_D-V_G curves. The introduction of surface charge in the recess region leads to a much better agreement between the simulations and the experiment (Figure 4 (b, c)).

4. Conclusions

In this paper we have described a new Monte Carlo module that is capable of modelling recess gate MESFETs and HEMTs. The novel feature of this module is the use of a Finite-Element discretization scheme to represent the device geometry in the best possible way. The the importance of the proper handling of the recess shape and the surface effects is demonstrated in example simulations of short MESFET and HEMTs.

References

1. A.Asenov, D. Reid, J.R. Barker, N. Cameron and S.P. Beaumont, "Finite element simulation of recess gate MESFETs and HEMTs: The simulator H2F", in *Simulation of Semiconductor Devices and Processes*, eds S.Selberherr, H.Stippel & E.Strasser, Springler Verlag, Wien, pp. 265- 268, 1993.
2. C.Moglestue, "Monte Carlo particle study of of the intrinsic noise figure in GaAs MESFETs", *IEEE Trans. Elevctron Dev.* Vol. ED-32, pp. 2092-2096, 1985.
3. G.Jensen, B.Lund and T.Fjeldly, "Monte Carlo simulation of short channel Heterostructure Field Effect Transistors", *IEEE Trans. Electron Dev.* Vol. ED-38, No. 4, pp. 840-851, 1991.
4. D.Park and K.Brennan, "Theoretical Analysis of AlGaAs/InGaAs pseudomorphic HEMTs using the ensemble Monte Carlo simulation", *IEEE Trans. Electron Dev.* Vol. ED-36, No. 3, pp. 1254-1263, 1989.
5. N.Cameron, S. Furgeson, M.R.S.Taylor, S.Beaumont, M.Holand, C. Tronche, M. Soulard and P.H. Ladbrooke, "Selectively dry gate recessed GaAs MESFETs, HEMTs and monolithic microwave integrated circuits" *J.Vac Sci Techol*, Vol. B11, No.6, pp. 2244-2248, 1993.

Coupled 2D-microscopic/macroscopic simulation of nanoelectronic heterojunction devices

Carsten Pigorsch[1], Roland Stenzel[1] and Wilfried Klix[2]

[1] Dresden University of Technology and Economics
Friedrich-List-Platz 1, D-01069 Dresden, Germany
[2] Dresden University of Technology
Mommsenstr. 13, D-01062 Dresden, Germany

Abstract

A two dimensional self consistent solution of the Schrödinger and the Poisson equation is obtained. This is coupled with a three dimensional drift-diffusion model to simulate nanometer heterojunction devices with respect to the microscopic properties of the electrons. Some examples of calculated III-V semiconductor structures are represented.

1. Introduction

The development of the semiconductor technology has made possible the production of structures and devices with small dimensions. For the simulation of these nanometer devices it is necessary to consider the wave properties of the carriers. Generally the Schrödinger equation is used to describe the microscopic energy quantisation which appears at heterojunctions [1]-[4]. The starting point of our work is the three dimensional (3D) device simulator SIMBA [5] based on a macroscopic drift-diffusion model, which is adapted to the simulation of heterostructures. The aim is the improvement of the simulation model by inserting the two dimensional (2D) Schrödinger equation. We make a point of an efficient numerical algorithm. The solution of the Schrödinger equation takes place by choice 1D or 2D depending on the structure design.

2. Simulation Model

The coupled 2D-microscopic/macroscopic model consists essentially of the microscopic Schrödinger equation and a macroscopic drift-diffusion model.
The behaviour of the electrons is described by the wave function ψ as the solution of the 2D Schrödinger equation [1], [2]

$$-\frac{\hbar^2}{2m^*}\Delta\psi(x,y) + V(x,y)\psi(x,y) = E\psi(x,y). \tag{1}$$

Here m^* is the effective mass and E are the discrete energy levels. The potential energy V is

$$V(x,y) = -q\phi(x,y) + \Delta E_C + V_{xc}(x,y) + V_{im}(x,y), \tag{2}$$

where ϕ is the electrostatic potential, ΔE_C is the conduction band discontinuity and V_{xc} is the local exchange correlation potential energy, V_{im} is the image potential [3], [6], [7]. V_{xc} and V_{im} has been neglected at first [1].
With the wave functions ψ and the corresponding energies E we obtain the microscopic electron density [2], [3], [6]

$$n(x,y) = \sum_i |\psi_i(x,y)|^2 n_i + n_{bulk} \qquad n_i = \frac{1}{\pi}\left(\frac{2m^*kT}{\hbar^2}\right)^{1/2} F_{1/2}\left(\frac{E_{Fn}-E_i}{kT}\right). \qquad (3), (4)$$

The 3D bulk electron density n_{bulk} contains the carriers in an assumed energy band above the considered discrete energies.
For the calculation of the electrostatic potential the Poisson equation is solved. The macroscopic part of the simulation model consists of the 3D continuity and transport equations for hole and electron current densities as described in [5].

3. Numerical Method

The 2D coupled algorithm is shaped as an inner iteration of the Schrödinger and the Poisson equation and an outer iteration where the transport model is solved [6].
Usually the inner iteration delivers a self consistent solution of Poisson and Schrödinger equation but for a good convergence of the whole algorithm often it is favourable to consider the transport equations after some or only one loop of the inner iteration. For a sure and successful iteration the change of the electrostatic potential is damped. The principle of the numerical algorithm is shown in Fig. 1.
The coupling of the microscopic and the macroscopic parts of the model takes place in a way described in [6]. It is based on the assumption that the sheet electron density at the heterojunction delivered by the microscopic Schrödinger equation and the density calculated by the transport equations have to be equal. For the solution of the microscopic density n by equations (3), (4) a Fermi level W_{Fn} is searched so that the integral of the electron densities in the quantum well realizes the condition above. The Fermi energy is assumed as constant in the whole area of the quantum well.
The Schrödinger equation is transformed into an eigenvalue problem by the Rayleigh-Ritz method. The wave function ψ is expressed as a sum of expansion functions in the form

$$\psi(x,y) = \sum_i \sum_j v_i(x) w_j(y).$$

In our simulations we have used Sinus functions [4] as well as B-Splines [8] for the expansion functions v_i, w_j. The advantage of Sinus functions is the independence of the number of expansion functions from the spatial discretization of the structure. The number of the expansion functions $(i \cdot j)$ is equal to the dimension of the eigenvalue problem. The resulting matrices of the eigenvalue problem are symmetric, which simplifies the solution method. We solve the problem by tridiagonalizing the symmetric matrix and than using the Sturm sequence and a bisection method to find the required eigenvalues. So the solution is less expensive than a finite difference or element method.
The Poisson equation and the transport equations are solved by the finite difference method with a non equidistant discretization.

4. Results and Discussion

The coupled microscopic/macroscopic method is applied to several heterojunction

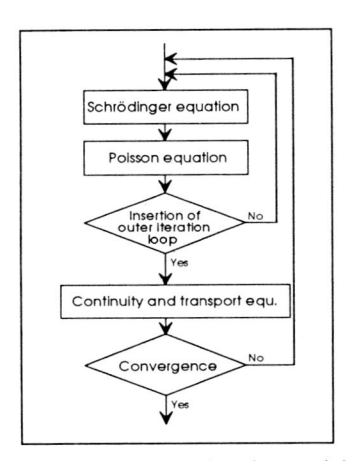

Fig. 1. Flowchart of the microscopic/macroscopic simulation (Inclusion of more model equations is possible)

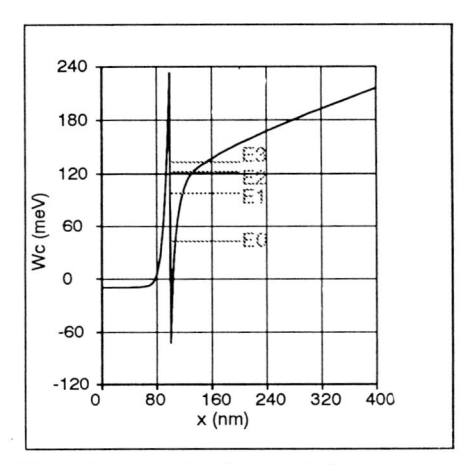

Fig. 2. Conduction band edge and discrete energy levels at a AlGaAs/GaAs heterojunction

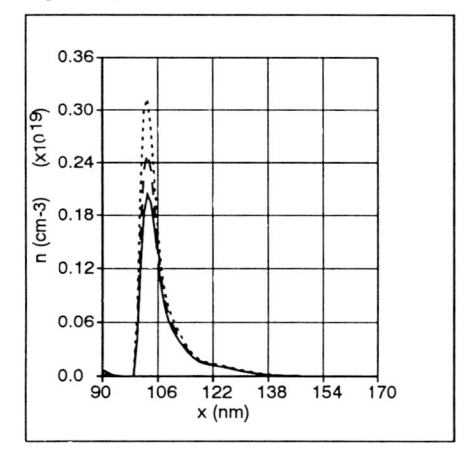

Fig. 3. Electron density at a AlGaAs/GaAs heterojunction (bias V = 0 solid line, 0.1V dashed line, 0.2V dotted line)

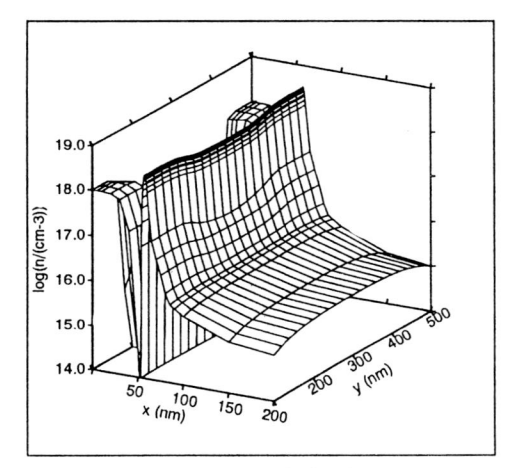

Fig. 4. Electron density distribution of a AlGaAs/GaAs-HEMT structure (L_G = 200 nm, V_{GS} = -0.5V, V_{DS} = 0)

structures. First we regard a simple abrupt GaAs/Al$_{0.3}$Ga$_{0.7}$As heterojunction. The GaAs is undoped, the Al$_{0.3}$Ga$_{0.7}$As is 10^{18} cm^{-3} n-doped. Fig. 2 is the calculated conduction band edge with the discrete energy levels at the heterojunction. With increasing bias the resulting electron density in the two dimensional electron gas increases (see Fig. 3).

We have calculate the electron density of a high electron mobility transistor (HEMT) by the 2D algorithm (Fig. 4). The HEMT consists on a 50 nm-Al$_{0.3}$Ga$_{0.7}$As-layer (10^{18} cm^{-3} n-doped) on a GaAs-substrate (undoped) with a gate length of 200 nm. Furthermore results of two quantum wire structures are shown. Fig. 5 shows a GaAs-corner in an Al$_{0.3}$Ga$_{0.7}$As-area, Fig. 6 is a square GaAs-area surrounded by Al$_{0.3}$Ga$_{0.7}$As.

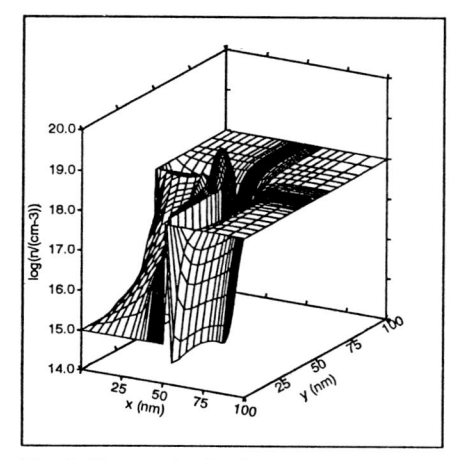

Fig. 5. Electron density distribution of a quantum wire at a AlGaAs/GaAs corner

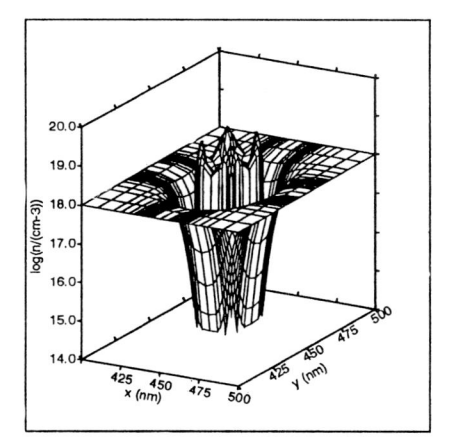

Fig. 6. Electron density distribution of a quantum wire of a (25 x 25) nm^2 GaAs area, surrounded by AlGaAs

5. Conclusion

In this paper we present a coupled microscopic/macroscopic simulation model with a complete 2D solution of the Schrödinger equation. An efficient solver for the differential equation and the resulting eigenvalue problem has been realized. The differences to the solution of a pure macroscopic model were shown by examples of several structures with III-V heterojunctions.

6. Acknowledgements

This work was supported by the German Ministry of Research and Technology under contract 01 BT 306/8.

References

[1] S. E. Laux, Numerical Methods for Calculating Self-Consistent Solutions of Electron States in Narrow Channels, Proc.of NASECODE V, pp. 270-275, 1987
[2] S. E. Laux, A. C. Warren, Self-Consistent Calculation of Electron States in Narrow Channels, IEEE IEDM 1986, pp. 567-570
[3] T. Kerkhoven et al., Efficient numerical solution of electron states in quantum wires, J. Appl. Phys., Vol. 68, No. 7, pp.3461-3469, 1990
[4] A. Abou-Elnour, K. Schuenemann, An Efficient and Accurate Self-Consistent Calculation of Electronic States in Modulation Doped Heterostructures, Solid States El., Vol. 37, No. 1, pp. 27-30, 1994
[5] R. Stenzel et al., Device Simulation of Novel In-Plane-Gated Field-Effect Transistors, Jpn. J. Appl. Phys., Vol. 33, No. 3A, pp. 1243-1247, 1994
[6] T. Wang, C. H. Hsieh, Numerical analysis of nonequilibrium electron transport in AlGaAs/InGaAs/GaAs pseudomorphic MODFET's, IEEE Trans. ED, Vol. 37, No. 9, pp. 1930-1938, 1990
[7] F. Stern, S. Das Sarma, Electron energy levels in GaAs-Ga(1-x)Al(x)As heterojunctions, Physical Review B, Vol. 30, No. 2, pp. 840-848, 1984
[8] J. Sánchez-Dehesa et al., Electronic energy of quantum-well wires, J. Appl. Phys., Vol. 73, No. 10, pp. 5027-5031, 1993

On the Discretization of van Roosbroeck's Equations with Magnetic Field

H. Gajewski[a], K. Gärtner[b]

[a]Weierstrass-Institut für Angewandte Analysis und Stochastik Berlin,
Mohrenstr. 39, D–10117 Berlin, Germany
[b]Institut für Integrierte Systeme, ETHZ, Gloriastr. 35, CH–8092 Zürich, Switzerland

Abstract

We investigate qualitative properties of the drift–diffusion model of carrier transport in semiconductors when a magnetic field is present. At first the spatially continuous problem is studied. Essentially, global stability of the thermal equilibrium is shown using the *free energy* as a Lyapunov function. This result implies exponential decay of any perturbation of the thermal equilibrium. Next, we introduce a time discretization that preserves the dissipative properties of the continuous system and assumes no more than the naturally available smoothness of the solution. Finally, we present a space discretization scheme based on weak and consistent definitions of discrete gradients and currents. Starting with a fundamental result on global stability (dissipativity) of the classical Scharfetter–Gummel scheme (without magnetic field), we adapt this scheme with respect to magnetic fields and study the M-property of the associated matrix. For two dimensional applications we formulate sufficient conditions in terms of the grid geometry and the modulus of the magnetic field such that our scheme is dissipative and yields positive solutions. These conditions cover fields up to $|\mathbf{b}|\mu_\nu \approx 0.5$ for very fine grids. This means approximately 200 Tesla for Silicon. Sufficient for some typical semiconductor sensor applications. The grid requirements might become prohibitive for large magnetic fields and complex three dimensional structures. Our techniques of defining discrete currents can be applied to similar situations, especially if projections of currents are involved in model parameters.

1. Introduction
The main issue of this paper is to provide the community, concerned with device simulations, with some of the results published in [1]. Space restrictions allow for the main results only, proofs, details and comments have to be omitted. The paper is organized to state the results mentioned in the abstract.

2. The spatially continuous problem
Let $S = (0, T)$ be a bounded time interval and let $\Omega \subset \mathbb{R}^N$, $2 \leq N \leq 3$, be a bounded Lipschitzian domain. Set $Q = S \times \Omega$. Suppose that $\partial\Omega = \Gamma_D \cup \Gamma$, $\Gamma_D \cap \Gamma \neq 0$, where Γ_D is closed and has a positive surface measure. Let us consider the following system of equations
$$-\nabla \cdot \epsilon \nabla \Psi = f + u_2 - u_1 \text{ in } Q,$$
$$\frac{\partial u_\nu}{\partial t} + q_\nu \nabla \cdot \mathbf{J}_\nu = R \text{ in } Q, \ \nu = 1, 2, \ q_1 = -1, \ q_2 = 1, \tag{1}$$
$$\mathbf{J}_\nu = \mathbf{j}_\nu - q_\nu \mathbf{b} \times \mathbf{J}_\nu, \ \mathbf{j}_\nu = -\mu_\nu u_\nu \nabla \Phi_\nu.$$

The physical meaning of the various quantities is the following: Ψ - electrostatic potential, u_ν - carrier densities, $\Phi_\nu = \Psi + q_\nu \log u_\nu$ - quasi-Fermi potentials, ϵ - dielectric permittivity, f - density of impurities, R - recombination / generation rate $R = r(x, u_1, u_2)(1 - u_1 u_2)$, $\mathbf{J}_\nu$, $\mathbf{j}_\nu$ - current densities with / without magnetic field, $\mathbf{b}$ - the magnetic field vector, μ_ν - carrier mobilities $\left(\frac{\mu_\nu}{1+|\mathbf{b}|^2}\right)(x) \geq \mu_b > 0$. We assume: $|\mathbf{b}|$, ϵ, μ_ν, $f \in L_\infty(\Omega)$, $0 < \epsilon < \epsilon_0$, $r = r(x, u_1, u_2)$ is continuous with respect to u_ν and measurable with respect to x. The growth condition $0 \leq r(x, u_1, u_2) \leq r_1(1 + |u_1| + |u_2|)$, $r_1 = \text{const} < \infty$ holds.
Using the matrix $B_\nu = \frac{1}{1+|\mathbf{b}|^2}(I + \mathbf{b}\,\mathbf{b}^T - q_\nu \mathbf{b}\times)$, the last line in (1) yields
$$\mathbf{J}_\nu = B_\nu \mathbf{j}_\nu. \qquad (2)$$
We complete the system (1) by the initial conditions $u_\nu(0, \cdot) = u_{\nu 0}(\cdot)$ in Ω and the thermal equilibrium boundary conditions
$$\Psi = \Psi_D, \ u_\nu = e^{-q_\nu \Psi_D} \text{ on } S \times \Gamma_D, \qquad (3)$$
$\mathbf{n} \cdot \epsilon \nabla \Psi + \alpha(\Psi - \Psi_\Gamma) = 0$, $\alpha \geq 0$, $\mathbf{n} \cdot \mathbf{J}_\nu = 0$ on $S \times \Gamma$, where $u_{\nu 0} \in L_\infty(\Omega)$, $\Psi_D \in L_\infty(\Gamma_D)$, Ψ_Γ in $L_\infty(\Gamma)$ are given and $\mathbf{n}$ is the outer unit normal with respect to Gamma.

2.1. Stability of the thermal equilibrium

Let $\Psi^* \in H^1(\Omega)$ be the (unique weak) solution of the boundary value problem $-\nabla \cdot \epsilon \nabla \Psi = f + e^{-\Psi} - e^\Psi$ in Ω, $\Psi = \Psi_D$ on Γ_D, $\mathbf{n} \cdot \nabla \Psi + \alpha(\Psi - \Psi_\Gamma) = 0$ on Γ.
Definition 1 The triple (Ψ^*, u_ν^*), $u_\nu^* = e^{-q_\nu \Psi^*}$, is called the thermal equilibrium solution; the functionals
$F(\Psi, u_\nu) = \int [\sum_\nu u_\nu(\log \frac{u_\nu}{u_\nu^*} - 1) + u_\nu^*] \, d\Omega + \frac{1}{2} \|\Psi - \Psi^*\|^2$,
with $\|h\|^2 = \int \epsilon |\nabla h|^2 \, d\Omega + \int \alpha h^2 \, d\Gamma$, and
$d_b(\Psi, u_\nu) = \int [\frac{1}{1+|\mathbf{b}|^2} \sum_\nu u_\nu \mu_\nu (|\nabla \Phi_\nu|^2 + (\mathbf{b}\cdot\nabla\Phi_\nu)^2) + r(x, u_1, u_2)(u_1 u_2 - 1) \log(u_1 u_2)] \, d\Omega$,
$d_b \geq 0$ are the *free energy* and the *dissipation rate*, respectively.
Proposition 1 Let (Ψ, u_ν) be a solution of (1) - (3). Then the function $L(t) = F(\Psi(t), u_1(t), u_2(t))$ satisfies: $0 \leq L(t) = L(0) - \int_0^t d_b \, ds$.
Remark 1 By proposition 1 the function L decreases strictly monotone as long as the dissipation rate is nonzero. Thus we can conclude that $L(u(t)) \to 0$ for $t \to \infty$. Moreover, L turns out to be a Lyapunov function of the system and indicates global stability of the thermal equilibrium.
The decay rate towards the thermal equilibrium can be estimated:
Corollary 1 Let (Ψ, u_ν) be a solution to (1) - (3) with $\|\Psi(t)\|_\infty \leq k < \infty \ \forall t > 0$. Then (Ψ, u_ν) tends to the thermal equilibrium solution (Ψ^*, u_ν^*). Moreover, it holds
$$\sum_\nu \|\sqrt{u_\nu(t)} - \sqrt{u_\nu^*}\|^2 + \|\Psi(t) - \Psi^*\|^2 \leq 4e^{-\lambda t} L(0) \to 0, \ t \to \infty. \qquad (4)$$
Proposition 2 Suppose $r(x, u_1, u_2) \geq r_0 = \text{const} > 0$ and $\alpha = 0$. Then the function L decreases exponentially. In particular, we have $L(t) \leq e^{-\tilde{\lambda} t} L(0)$.
The proposition shows, that the assumptions of the corollary can be cancelled in special situations even for $N = 3$, the constants λ, $\tilde{\lambda}$ are given in [1].

2.2. Time discretization

Let $S = \cup_j S_j$, $S_j = [t_{j-1}, t_j]$, $\tau_j := t_j - t_{j-1} > 0$, $t_0 = 0$. According to the backward Euler's scheme, we discretize the system as follows $-\nabla \cdot \epsilon \nabla \Psi^j = f + u_2^j - u_1^j$ in Ω, $j = 0, 1, 2, \ldots$, $\frac{u_\nu^j - u_\nu^{j-1}}{\tau_j} + q_\nu \nabla \cdot \mathbf{J}_\nu^j = R^j$ in Ω, $j = 0, 1, 2, \ldots$, $u_\nu^0 = u_{\nu 0}$.
Proposition 3 Let (Ψ^j, u_ν^j) be a solution of the time discrete system. Then the discrete function $L^j = F(\Psi^j, u_\nu^j)$ satisfies $0 \leq L^{j+1} \leq L^j \leq L^0 - \sum_{l=1}^j \tau_l d_b^l$, $d_b^l = d_b(\Psi^l, u_\nu^l) \geq 0$.

3. Space discretization

For the space discretization we use N-dimensional simplices (elements) $\mathbf{E}_l^N$ such that $\Omega = \cup_l \mathbf{E}_l^N$. A simplex $E = E_l^N$ can be represented by the $N \times (N+1)$ matrix of its vertex coordinates.

$$P = \begin{pmatrix} x_{1,1} & \cdots & x_{1,N+1} \\ \vdots & & \vdots \\ x_{N,1} & \cdots & x_{N,N+1} \end{pmatrix}, \quad \tilde{P} := \begin{pmatrix} \mathbf{1}^T \\ P \end{pmatrix}$$

where $\tilde{P}$ is an extension of P by elements of the null space of D^T (see below, $\tilde{P}$ is nonsingular for any nondegenerated simplex). $\mathbf{x}_i^T = (x_{1,i}, x_{2,i}, \ldots, x_{N,i})$ is the vector of the space coordinates of the vertex i of the simplex. The volume integral of a function is approximated by the corresponding sum over the Voronoi volume elements: $\int f \, d\Omega \approx \sum_m f_m |V_m|$, $\sum_m |V_m| = |\Omega|$.

The contribution of the recombination R to the discrete dissipation rate is given by

$$d_{rec} := \sum_\nu \sum_m |V_m| (R\Phi_\nu)(\mathbf{x}_m) = \sum_m |V_m| r (u_1 u_2 - 1) \log(u_1 u_2)(\mathbf{x}_m) \geq 0. \qquad (5)$$

3.1. Discrete Gradients

In order to discretize (1) - (3) we need projections of the currents onto the coordinate axes. Let D be a $(N+1) \times (N+1)$ matrix such that $\int \nabla u \cdot \nabla h \, dE \approx (D\mathbf{u}, \mathbf{h}) = \mathbf{h}^T D \mathbf{u}$, where $(.,.)$ is the scalar product in $\mathbb{R}^{N+1}$. By means of D we define the discrete version $\tilde{\nabla}$ of ∇ by $\tilde{\nabla} = \frac{1}{|E|} PD$. Hence the projection of $\tilde{\nabla} \mathbf{u}$ onto the x-axis is given by $(\tilde{\nabla} \mathbf{u})_x = \frac{1}{|E|}(D\mathbf{u}, \tilde{\mathbf{x}})$, $\tilde{\mathbf{x}}^T = (x_{1,1}, \ldots, x_{1,N+1})$.

We suppose the matrix D to satisfy the following conditions:
(i) D is $(N+1) \times (N+1)$ matrix, (ii) $\operatorname{rank} D = N$, $\mathbf{1}^T D = 0$, $\mathbf{1}^T = (1, \ldots, 1)$,
(iii) $(D\mathbf{u}, \mathbf{h}) = |E| (\tilde{\nabla} \mathbf{u}, \tilde{\nabla} \mathbf{h})$, $\forall \mathbf{u}, \mathbf{h}$.

Remark 2 The 'projection property' (iii) is the key for introducing consistent currents $\mathbf{j}$ on simpleces $\mathbf{E}$ in the next sections. (ii, iii) ensure the strict balance of discrete contact currents $\mathbf{j}_k$ defined by $\mathbf{j}_k = \sum_l |\mathbf{E}_l| (\mathbf{j}, \tilde{\nabla} \mathbf{h}_k)$. Where the $\mathbf{h}_k$ are suitable test functions: $h|_{\Gamma_k} = 1$, $h|_{\Gamma_j} = 0$, $j \neq k$, $h \in H^1(\Omega)$ and $\Gamma_D = \cup_i \Gamma_i$.

Lemma 1 *The matrix D is uniquely defined by (i) - (iii). Moreover, D is the finite element matrix (for piecewise linear polynomials on the simplex) of the Laplacian.*

Hence the finite element discretization satisfies: $(D\mathbf{u}, \mathbf{h}) = |E| (\tilde{\nabla} \mathbf{u}, \tilde{\nabla} \mathbf{h})$. Consistent currents can be defined via:

Lemma 2 *Let A be a $(N+1) \times (N+1)$ matrix such that $\mathbf{1}^T A = 0$ and $\operatorname{rank} A = N$. Then $\forall \mathbf{u}, \mathbf{h} \in \mathbb{R}^{N+1}$ $(A\mathbf{u}, \mathbf{h}) = |E|(\mathbf{j}_u, \tilde{\nabla} \mathbf{h})$, where $\mathbf{j}_u = \frac{1}{|E|} PA\mathbf{u}$, $\tilde{\nabla} \mathbf{h} = \frac{1}{|E|} PD\mathbf{h}$.*

3.2. Box discretization methods

The box discretization in principle approximates the Laplacian on the Voronoi volume V_m. Interpreting this with respect to the elements, one gets an discrete approximation of the Laplacian $\int \nabla u \cdot \nabla h \, dE_l \approx (A_{box} \mathbf{u}, \mathbf{h})$ on E_l, too. For $N = 2$ the matrices A_{box} and D coincide, in general we have for $N > 2$: $\frac{1}{|E|} A_{box} P \; P^T \; A_{box} \neq A_{box}$. However, because of Lemma 2 and $(A_{box} \mathbf{u}, \mathbf{1}) = 0$, we can define consistent currents and gradients on E by: $\mathbf{j}_u = \frac{1}{|E|} PA_{box} \mathbf{u}$, $\tilde{\nabla} \mathbf{h} = \frac{1}{|E|} PD\mathbf{h}$. Since time derivatives, recombination and right-hand sides can be discretized as stated above, it remains to look for discrete approximations of expressions as $\nabla e^\Psi \nabla v$, where $v = e^{-\Phi}$ is the Slotboom variable related to u_1. We have to preserve the Scharfetter–Gummel scheme - essentially an approximation of $\int (a(\Psi) \nabla v) \cdot (\nabla h) \, dE \approx (A_{SG} \mathbf{v}, \mathbf{h})$, $\mathbf{1}^T A_{SG} = 0$, $A_{SG} = A_{SG}^T$ for a strongly varying coefficient $a(\Psi)$. The index SG stands for Scharfetter–Gummel and Slotboom variables. This leads to the following representation of A_{SG} on the simplex:

$$A_{SG} \mathbf{v} = G_N^T Y G_N \mathbf{v}. \qquad (6)$$

Here Y is a weight matrix, symmetric positive definite and diagonal, G_N is a $n_e \times n_v$ matrix (n_e the number of edges, n_v the number of vertices, for details see [1]). $y_k :=$ $y_{kk} = \frac{|s_{ij}|}{|e_{ij}|}$, $\beta_s(\log a(\psi_i), \log a(\psi_j))$, $k = 1, \ldots, n_e$, $\beta_s(x,y) = e^x\beta(x-y) = \frac{x-y}{e^{-y}-e^{-x}}$, $\beta_s(x,y) > 0$, $|x|, |y| < \infty$ is related to the Bernoulli function $\beta(x) = \frac{x}{e^x-1} \geq 0$. For $a = $ const holds $A_{SG} = A_{box}$. Consistent currents on **E** are defined as before: (we still write $\mathbf{j}_u$ because $\mathbf{u} = e^\psi \mathbf{v}$ introduces a diagonal transformation of the matrix and the variable only) $\mathbf{j}_u = \frac{1}{|\mathbf{E}|}PA_{SG}\mathbf{v}$, $\tilde{\nabla}\mathbf{h} = \frac{1}{|\mathbf{E}|}PD\mathbf{h}$.

Definition 2 The discrete dissipation rate (except recombination) on a simplex **E** for one carrier density (here the electrons) is (comp. definition 1) $d_{0E} := -\mathbf{\Phi}^T A_{SG} \exp(-\mathbf{\Phi})$. (The recombination contribution $d_{recE} \geq 0$ is given by (5) with the sum restricted on **E**.)

Proposition 4 The Scharfetter-Gummel scheme (6) is dissipative, i. e. , $d_{0E} \geq 0$, for any state of the system without magnetic field.

3.3. Magnetic field

We are looking for a matrix A_{mag} approximating $\int J \cdot \nabla h \, dE \approx (A_{mag}\mathbf{v}, \mathbf{h})$, $\mathbf{1}^T A_{mag} = 0$, $A_{mag} \neq A_{mag}^T$. Here $J = J(\mathbf{b})$ denotes the continuous current with magnetic field. $J(\mathbf{b})$ is related to $j = J(\mathbf{0})$ by (see (2), the subscript $\nu = 1$ is deleted again) $J_u = B\, j_u$. We defined the discrete version of $\mathbf{j}_u$ by $\mathbf{j}_u = \frac{1}{|\mathbf{E}|}PA_{SG}\mathbf{v}$, $\mathbf{1}^T A_{SG} = \mathbf{0}$. In view of Lemma 2 we make an analogous ansatz for the discrete version of J_u $\mathbf{J}_u = \frac{1}{|\mathbf{E}|}PA_{mag}\mathbf{v}$, $\mathbf{1}^T A_{mag} = \mathbf{0}$. This yields $\tilde{P}A_{mag} = \tilde{B}\tilde{P}A_{SG}$, $\tilde{B} := \begin{pmatrix} 0 & 0 \\ 0 & B \end{pmatrix}$. Hence $A_{mag} = \tilde{P}^{-1}\tilde{B}\tilde{P}A_{SG}$. Since $\tilde{P}^{-1}$ is known, this is the desired expression for A_{mag}.

Remark 3 $\mathbf{J}_u$ has to fulfil the same boundary conditions as $\mathbf{j}_u$. The scheme is current conservative by construction.

Introducing for $N = 2$ $c_i := \cot \alpha_i$, α_i inner angles, some algebra yields: $A_{mag} = \frac{1}{1+b_z^2} A_{SG} + \delta A_{mag}$, $\delta A_{mag} = \frac{b_z}{1+b_z^2} G^T \text{diag}[\cot(\alpha_{i-1})] S_\downarrow A_{SG}$, ($S_\downarrow$ is the shift matrix). In components this reads

$$A_{SG} = \begin{pmatrix} y_1 + y_3 & -y_1 & -y_3 \\ -y_1 & y_2 + y_1 & -y_2 \\ -y_3 & -y_2 & y_3 + y_2 \end{pmatrix},$$

$$(1+b_z^2)\delta A_{mag} = b_z \begin{pmatrix} -y_3 c_3 + y_1 c_2 & -y_2(c_2+c_3) - y_1 c_2 & y_3 c_3 + y_2(c_2+c_3) \\ y_3(c_1+c_3) + y_1 c_1 & y_2 c_3 - y_1 c_1 & -y_3(c_1+c_3) - y_2 c_3 \\ -y_1(c_1+c_2) - y_3 c_1 & y_1(c_1+c_2) + y_2 c_2 & y_3 c_1 - y_2 c_2 \end{pmatrix}.$$

Proposition 5 Assume $y_i > 0$ and (i) $|b_z||\cot \alpha_i| < 1$, (ii) $-y_i(1+b_z c_{i+1}) - y_{i+1}(c_{i+1}+c_{i+2})b_z < 0$, (iii) $-y_i(1-b_z c_i) + y_{i+2}(c_i + c_{i+2})b_z < 0$. Then A_{mag} is a weak M-matrix.

Proposition 6 The change in the dissipation rate due to the magnetic field on a triangle **E** can be estimated by $|\delta d_{b\,E}| \leq \max_{i,j} \cot(\alpha_{j-1}) \sqrt{\frac{y_i}{y_j}} \sqrt{3}\, |b_z|\, d_{0\,E}$.

Remark 4 Because of $(\mathbf{a} \times \mathbf{b}) \cdot \mathbf{a} = 0$, the dissipation part δd_b of the magnetic field vanishes exactly for $N = 2$ in the continuous case. Thus $\delta d_{bE} \neq 0$ has to be considered as discretization error - for the equilateral triangle δd_{bE} is of the order $\delta \Psi^k \delta \Phi^{k'}$, $k + k' = 3$.

References

[1] H. Gajewski, K. Gärtner, On the discretization of van Roosbroeck's equations with magnetic field, Technical Report 94/14 Integrated Systems Laboratory, ETH Zurich, to appear in ZAMM.

and literature cited in [1].

Modeling of Impact Ionization in a Quasi Deterministic 3D Particle Dynamics Semiconductor Device Simulation Program[1]

K. Tarnay [a, b], F. Masszi [b], T. Kocsis [a], A. Poppe [a]

a) Technical University Budapest, Dept. Electron Devices,
H-1521 Budapest, HUNGARY, e-mail: tarnay@eet.bme.hu
b) Scanner Lab, Electronics Department, Institute of Technology, Uppsala University
P.O.Box 534, S-751 21 Uppsala, SWEDEN

Abstract

A first principle based quasi-deterministic 3D particle dynamics Monte Carlo simulation method was developed for examining mesoscopic (subhalf–micron) Si electron devices. Applying a *novel method for calculating the field and potential distributions*, the real trajectories of the carriers are exactly followed. Consequently, an important feature of this method is that *all Coulomb scattering* are inherently taken into account. A brief description of the *physical background*, the *models* and the *simulation principle* is given. A quasi deterministic model for *impact ionization* is developed and some *results* are presented.

1. Introduction

Assuming a Si device structure with a volume of $0.25 \times 0.25 \times 0.25 \ \mu m^3$ and a doping density of $10^{23} \ m^{-3}$, the number of ionized impurities is only about 10^3. The number of carriers is of the same order of magnitude.

The classical methods of semiconductor device simulation (based on the *drift-diffusion* or *hydrodynamic semiconductor equations*) use a *continuum view* and apply certain statistical considerations for the *carrier distribution functions*[2]. The potential distribution is determined by the solution of the *Poisson equation*, requiring a 3D spatial discretization resulting in 10 to 50 thousand elementary cells (ie. the average number of carriers is 1/cell or less). These methods are not valid any more if the number of carriers in the simulated structure, like in this case, lies only in the order of magnitude of a few thousands or less.

The small number of carriers suggested the development of a particle dynamics 3D Monte Carlo simulation method[3], for studying mesoscopic device behaviour, where *the trajectories of each carrier are individually and exactly followed* both in the real space and in the **k**-space (wave vector space). The simulation of 3D structures is sometimes simplified to a 2D approach. This simplification cannot be applied for the particle dynamics Monte Carlo simulation of

---

[1] The research was sponsored by
 - Digital Equipment Co. (External European Research Projects HG-001 and SW-003),
 - Swedish Board of Technical Development (NUTEK).
 - Hungarian Research Foundation (OTKA 777 and OTKA T-016748).

[2] Derived from the Boltzmann transport equation (applying the relaxation time approximation and considering the first few momenta of the electron distribution function).

[3] In particle dynamics, all charged particles - i.e. charged carriers, dopant ions, interface charges - are treated as point charges, with no charge assignment to elementary volumes. More sophisticated Monte Carlo methods (by applying charge clouds, superparticles etc.) can offer far more effective numerical solution tools, but the physics of the simulated system can easily be obscured.

charged particles[4]. However, real 3D effects, such as impact ionization or ionization by α particles can not be properly examined by 2D simulations.

2. Principle of simulation

Our 3D particle dynamics Monte Carlo semiconductor device simulator program, MicroMOS has been developed for examining sub-half micron silicon MOS transistors and is based on the *concepts of classical physics* (Newton law, Coulomb law, etc.). During development, the attention was focused to apply the deepest possible *first physical principles* inside the examined structure. The elementary particle nature of the carriers is only taken into consideration by anizotropic effective masses, and by Bragg reflection based limitations in their momenta (velocities). Thus all carriers remain in the first Brillouin zone. In this way, the bulk behaviour can be modeled quite accurately. However, near the Si-SiO$_2$ interface the *wave nature of the carriers* must be taken into consideration[5]. For a more detailed description of this method see [1], [2].

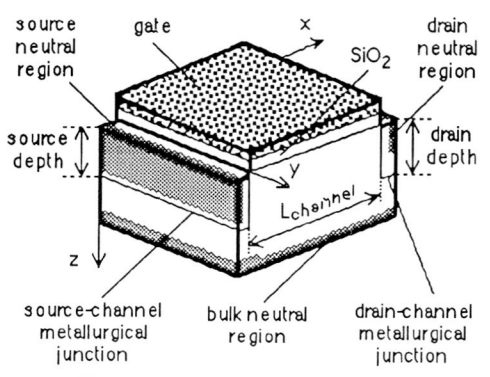

Fig. 1.: The examined MOS structure

The *state of carriers* is characterized by the space coordinate vector **r** and by the wave vector **k**.

The *carrier dynamics* is described by the Newton law

$$\mathbf{a} = m_{eff}^{-1} \cdot \mathbf{F} \tag{1}$$

where m_{eff}^{-1} is the *reciprocal effective mass tensor*.

Different transverse and longitudinal effective masses are used. Both masses are considered to be energy independent, resulting in the well known 6 ellipsoidal constant energy surfaces in the **k**-space for electrons. Different izotropic (scalar) effective masses are considered for the light and heavy holes[6], resulting in spherical constant energy surfaces in the **k**-space.

The *forces acting to the carriers* are splitted into two components:

- *the Coulomb force originating from charges inside the structure* (donors, acceptors and charged interface states), evaluated by the Coulomb law.
- *the force caused by the field of external voltages* is evaluated by a boundary value problem solution of the Laplace equation[7].

The first integral of Eq.1 yields the new carrier velocities, the second one the new real space position coordinates[8]. Initial values of the integrations are the carrier position and velocity results of the previous simulation step. To preserve neutrality, at the beginning of each simulation step electrons and holes are injected into predefined source and drain neutral regions. A similar neutral region exists deep in the bulk. Any carriers leaving the structure are annihilated at the end of each simulation step.

---

[4] The *point charges* in 2 dimensions can only be implemented as *line charges*. In this case we would get an electric field of $1/r$ dependence and a potential of logarithmic dependence, instead of the $1/r^2$ and $1/r$ behaviours, respectively. Since the force acting to the charged carriers is proportional to the field, the calculated trajectories would be sufficiently different from the real 3D ones in any 2D case.
[5] The quasi 2D nature of electrons is taken into consideration by the triangular potential well approximation. The electrons are positioned at the first momenta of their wave function. The first 3 subbands are considered.
[6] The split-off holes are not taken into consideration.
[7] With modified boundary conditions, compared to the usual solution of the Poisson equation.
[8] The lowest order numerical integration formulae are used, because the force is a strongly varying function of the space coordinates, therefore using a higher order formula will not give any accuracy improvement. On the contrary, it is possible that the result will be less accurate.

The advantage of this method is that any *Coulomb scattering process* (carrier-ionized impurity, carrier-carrier and carrier-interface state) is exactly followed in a deterministic way[9].

The *interactions between the carriers and the lattice* are described by Bragg scattering mechanisms: if a carrier leaves the first Brillouin zone, it is scattered back to the same Brillouin zone. For the shape of the Brillouin zone the Debye approximation is used[10]. Near the Si-SiO$_2$ interface f- and g- phonon *intervalley scattering processes* are taken into consideration.

Impact ionization takes place if the kinetic energy of a carrier exceeds the *impact ionization threshold energy* (for silicon W_{th}= 1.8 eV, according to set 1 of [3]). When a carrier has an energy of $W > W_{th}$, a new electron-hole pair is generated. It is assumed, that the ionizing particle looses its whole kinetic energy, and an energy equal to the band gap energy W_G is transferred to the lattice. The momentum and energy conservation conditions are fulfilled for the event. For the *Auger recombination* a phenomenological approach is used, based on a scattering cross section concept: a recombination event occurs, when the distance sinks below a critical distance between a single electron and a hole.

At the Si-SiO$_2$ interface random *elastic and specular surface scattering events* are simulated[11].

3. Results and discussion

Table 1. shows geometrical dimensions, doping concentrations, interface charge density and the operating point of the examined MOS transistor, together with physical parameters of the simulation. The simulation time step was Δt = 0.01 ps. The transistor is in an unsaturated operating point, $I_D \approx 17.3\,\mu A$. Fig. 2. shows the spatial electron distribution in the structure. It can be observed that the electrons near the silicon surface are grouped into subbands. In the followings, we summarize some results related to impact ionization during a 2 ps time interval. Table 2. shows the drain current I_D and impact ionization rates S, giving a detailed picture on the contribution of electrons corresponding to the different constant energy ellipsoid orientations. Observe, that the <100> electrons have a negative contribution to the current, because their positive x-velocity is limited by the maximum value of the corresponding wave vector (0.16 k_{Max}). Fig.3. gives the spatial distribution of the impact ionization events, caused by electrons corresponding to various constant energy ellipsoid orientations. On Fig. 4, the impact ionization rate is shown along the x-axis. Fig. 5. shows the energy distribution of the ionizing electrons.

Table 1.

Channel length	L	= 250	nm
Source length (x direction)	L_S	= 30	nm
Drain length (x direction)	L_D	= 33	nm
Source and drain depth	$D_{S,D}$	= 120	nm
Length of structure	L	= 323	nm
Depth of structure	D	= 302	nm
Width of structure	W	= 259	nm
Gate oxid thickness	d_{ox}	= 4.5	nm
Channel and bulk doping concentration	N_A	= 10^{23}	m^{-3}
Source and drain doping concentration	N_D	= 10^{24}	m^{-3}
Charged interface state density	N_{SS}	= -10^{16}	m^{-2}
Gate material	Al		
Fuchs parameter	F	= 0.5	
Impact threshold energy	W_{TH}	= 1.800	eV
Auger coefficient	B_n	= 3.5 10^{-42}	m^6/s
Drain - source voltage	U_{DS}	= 0.5	V
Gate - source voltage	U_{GS}	= 3.0	V

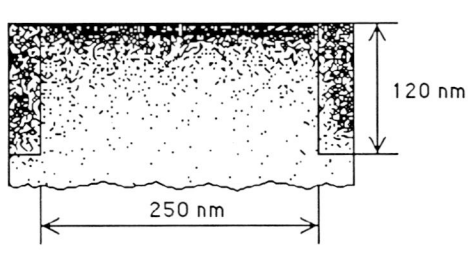

Fig. 2.: Spatial electron distribution

---
[9] Assuming that the time step Δt is small enough.
[10] The dodecaeder shaped first Brillouin zone is approximated by a sphere.
[11] The ratio of these events controlled in order to maintain an average ratio determined by the Fuchs parameter.

Table 2.

	<001>	<00$\bar{1}$>	<010>	<0$\bar{1}$0>	<100>	<$\bar{1}$00>	Σ
I$_D$ [μA]	-19.6	18.4	1.8	5.7	3.8	7.2	17,3
S [1/ps]	0.47	0.44	0.64	0.60	0.17	0.28	2.60

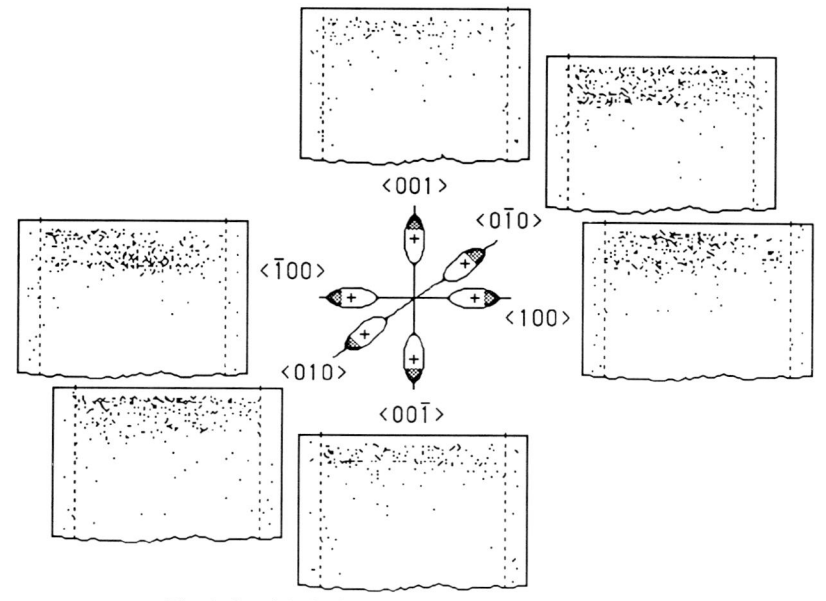

Fig. 3. Spatial distribution of impact ionization events

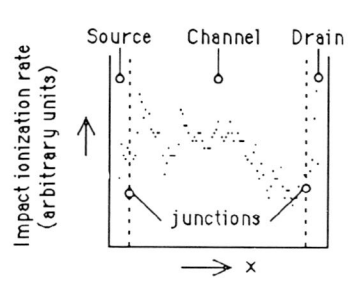

Fig. 4.: Impact ionization rate along the channel

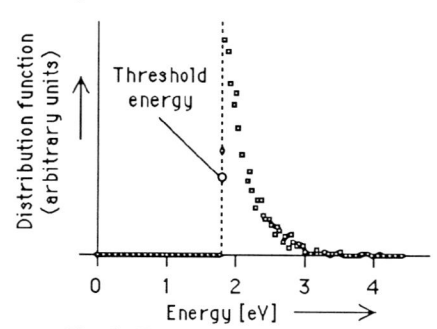

Fig. 5.: Energy distribution of ionizing electrons

References

[1] Tarnay, K.- Habermajer, I.- Poppe, A.- Kocsis, T.- Masszi, F.: Modeling the Carrier - Lattice Interactions and the Energy Transport in a 3D Particle Dynamics Monte Carlo Simulator for MOS Structures. Abstracts of NASECODE X, 21-24 June 1994, Dublin, pp. 28-29.
[2] Tarnay, K. et al..: The Impact Ionisation Process of α Particles in Mesoscopic Structures: Simulation by Monte Carlo Method. Physica Scripta **T54**, pp. 256-262, 1994
[3] Tang, J. Y.- Hess, K.: Impact Ionization of Electrons in Silicon (Steady State). J. Appl. Phys. **54**, No. 9. pp. 5130-5144, 1983

Accurate Modeling of Ti/TiN Thin Film Sputter Deposition Processes

H. Stippel and K. Reddy[*]

Corporate Technology Group, CDfM, UPD[*]
National Semiconductor Corporation
2900 Semiconductor Drive, M/S D3-677
Santa Clara, CA 95052-8090

Abstract

An accurate and user friendly tool for the simulation of Ti/TiN sputter deposition processes has been developed. Simulations have been compared to SEM measurements, which exhibit excellent agreement for both, the collimated as well as the uncollimated case. A final test showed that this simulator is capable of predicting the deposited film on the collimator sidewalls.

1. Introduction

With the growing number of metallization levels, thin film applications have become the main focus point for technologists. One of the most prominent techniques is the sputter deposition of Ti and TiN. With this method very thin layers of metallization can be produced. But, as device features shrink step coverage and film conformality become increasingly difficult. Currently, two methods are used to solve this problem: First, the deposition can be performed at high substrate temperatures; this technique is mostly applied to Al depositions. Second, a collimator can be used to decrease the spread of the incoming particle flux. In this work we focus on the effect of collimation.

A collimator is used when a temperature increase either does not improve the step coverage or is forbidden because of the impact on other properties of the device (such as concentration profiles or previous metallization levels). The collimator improves the bottom coverage of trenches or vias by removing particles with a high incoming incident angle (*beaming effect*) from the incoming particle flux [1]. However, those filtered particles are deposited on the collimator side-walls leading to a shrinking of the cell opening size. Thus, with increasing collimator lifetime, the total number of particles reaching the wafer surface decreases significantly and the beaming effect is enhanced.

For the simulations we used EVOLVE 4.0 [2] from Arizona State University, which was calibrated to the specific equipment used at National Semiconductor Corporation (M2000 from Varian). In parallel a graphical user interface (GUI) based on the VISTA-framework [3] from TU Vienna was developed.

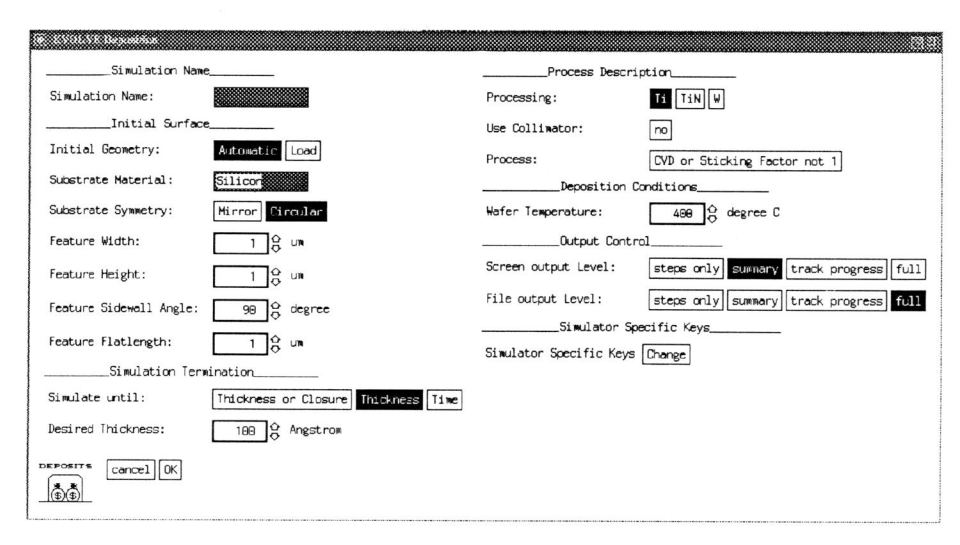

Figure 1: Graphical user interface for EVOLVE, based on VISTA.

2. Calibration of EVOLVE

For the calibration three vias with different aspect ratios were used at two different deposition temperatures for the collimated case as well as for the uncollimated case. These twelve experiments were conducted for both, Ti and TiN. It was found that for uncollimated deposition the main influencing factor were the sticking coefficients. They were determined to be 0.7 and 0.8 for Ti and TiN, respectively. Other factors such as temperature ($250° - 400°$ C), gas flow rates (Ar 20 – 80 sccm, N_2 75 – 120) and the chamber pressure (0.02 – 20 mTorr) showed no impact on deposition rates in the given ranges. The deposition power was held constant (12 kW for Ti and 20 kW for TiN). For simulation purposes the incoming particle flux distribution was approximated by a generalized cosine function.

In the case of collimated deposition the same sticking factors were applied. But the assumption of a cosine distribution for the incoming particle flux can not be sustained. This distribution depends on the collimator lifetime, because of the increasing beaming effect. Thus, the collimator lifetime becomes the key factor for collimated deposition processes. To take this fact into account the deposited film thickness on the collimator sidewalls was computed using EVOLVE. A comparison at 491 kWh showed very good agreement: The simulated thickness was 1.4 mm, which was assumed to be constant along the collimator sidewall. The measurements showed values between 1.5 mm (at the entrance of the collimator) to 1.2 mm (at the exit of the collimator).

In our experiments we used a collimator with hexagonal cells. The ratio of the cell width to the cell height was 1:1. The information about the total thickness of the collimator cell's sidewall together with the geometrical information of the collimator was then used by a Monte Carlo based simulator [4] to determine the incoming particle flux on the wafer surface. The computed flux distribution was fed into EVOLVE.

3. The Graphical User Interface

EVOLVE was integrated into VISTA for two main reasons: First, a convenient GUI (see Fig. 1), which takes care of the automatic and correct generation of the required input file, could be developed easily, based on the capabilities of the framework. Second, in this environment time-consuming and tedious tasks such as the determination of the dependence of growth rates on the collimator lifetime could be performed conveniently by executing several runs in parallel on different workstations. In the future it can also be used to substitute other less accurate deposition simulations.

The development of the GUI was necessary, because the input format of EVOLVE is too complicated and the interdependences are too complex for the every-day usage of this simulator. Additionally, the interdependences of the different parameters are crucial, but hard to obeye. All of the above is now taken care of by the GUI. The user selects the type of process to be simulated (e.g. Ti deposition) and the GUI will only allow the input of the appropriate parameters, automatically. Thus the user is reliefed from the tedious task of generating an input deck and can concentrate on the real, physical parameters.

4. Results

In Fig. 2 and Fig. 3 simulation results and comparisons to actual experiments are presented. Fig. 2 shows the result of a 1000 Å Ti deposition step without collimation in a 1.0 μm wide and 1.5 μm deep via. The input geometry was scanned in directly from the SEM photograph shown in the right half of the figure. Perfect agreement could be obtained.

In Fig. 3 results of a collimated TiN deposition are presented. The initial geometry was a 0.8 μm wide and 1.5 μm deep via on which 500 Å of TiN were deposited. As there is a notable difference in the deposition rate on the sidewall of the via compared to the deposition rate on the flat surface, the *kink* is formed. This kink is very important for the real process, because it can lead to a lack of film coverage at certain points. Thus, it's accurate modeling becomes a crucial issue for the simulation. Further investigations showed, that the formation and the appearance of the kink

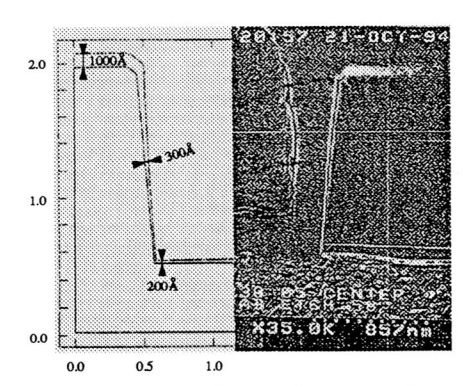

Figure 2: Uncollimated sputter deposition of Ti in a via.

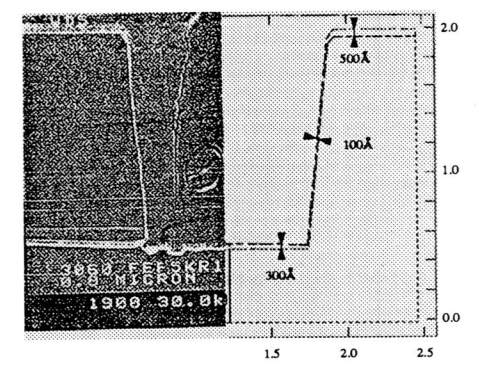

Figure 3: Collimated sputter deposition of TiN in a via.

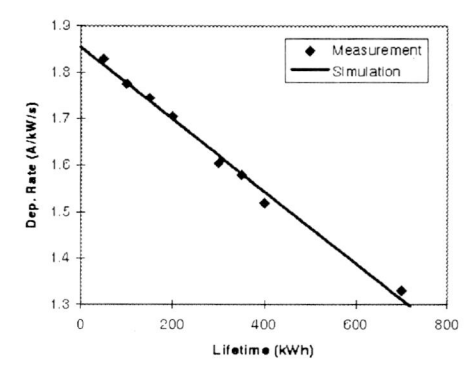

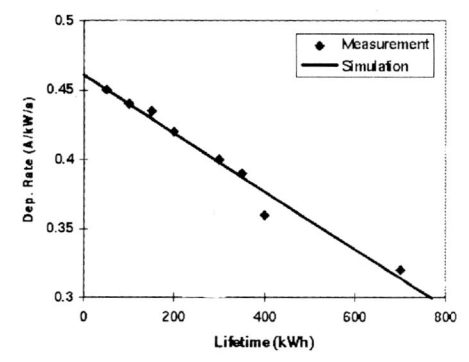

Figure 4: Deposition rate as a function of collimator lifetime for Ti.

Figure 5: Deposition rate as a function of collimator lifetime for TiN.

strongly depend on the collimator lifetime: With increasing lifetime the kink becomes larger, which can be attributed to the growing beaming effect.

Fig. 4 and Fig. 5 show the deposition rates for Ti and TiN, respectively, as a function of the collimator lifetime. The correct prediction of this dependency is critical for process engineers: When the deposition rate falls below a certain limit the film conformality can not be guaranteed anymore; voids can occur in the deposited film. Thus, the collimator needs to be changed.

5. Conclusion

We have calibrated EVOLVE for sputter Ti and TiN processes. The only parameter that had to be calibrated was the sticking factor. For collimated deposition processes the dependence of the growth rates on the collimator lifetime can be predicted for both, Ti and TiN. The deposited film thickness on the collimator side-walls could be modeled correctly.

To allow easy input definition via a GUI and parallel execution EVOLVE was integrated in VISTA. Excellent agreement with measurements was found, and the kink formation for collimated TiN depositions – crucial for this process – can be predicted accurately.

References

[1] R.V. Joshi and S. Brodsky. Collimated Sputtering of TiN/Ti Liners into Sub-Half Micron High Aspect Ratio Contacts/Lines. In *Ninth International VLSI Multilevel Interconnection Conference*, pages 253–259, June 1992.

[2] T.S. Cale. *EVOLVE – A Low Pressure Deposition Simulator, Version 4.0b*. Center for Solid State Electronics Research, ASU, August 1994.

[3] S. Selberherr, F. Fasching, C. Fischer, S. Halama, H. Pimingstorfer, H. Read, H. Stippel, P. Verhas, and K. Wimmer. The Viennese TCAD System. In *Proc. VPAD*, pages 32–35, 1991.

[4] Z. Lin. *Simulation of Flux Distributions and Flat Substrate Deposition Profiles during Collimated Sputter Deposition*. Center for Solid State Electronics Research, ASU, August 1994.

Monte Carlo Simulation of InP/InGaAs HBT with a Buried Subcollector

G. Khrenov[a], E. Kulkova[b]

[a]Computer Solid State Physics Laboratory, The University of Aizu,
Aizu-Wakamatsu City, 965-80, JAPAN
[b]Research Center "MICROEL",
Krasikov Str. 25a, Moscow, 117218, RUSSIA

Abstract

The effect of structure parameters and operation conditions on the base-collector capacitance and intrinsic transit time of HBT with buried subcollector has been investigated by using a numerical model. It is shown that non-planar geometry of HBT with buried subcollector leads to specific features of voltage dependence of the base-collector capacitance. The non-uniform collector doping profile has been proposed to reduce collector transit time and to optimize HBT structure for low-voltage operation.

1. Introduction

Heterostructure bipolar transistors (HBTs) lattice matched to InP have emerged as potential candidates for high-speed digital, microwave and long-wavelength fiber-optic communication systems because of the excellent transport property of InP and its related materials. High-speed InP/InGaAs HBT having a current gain cutoff frequency about 175 GHz has been demonstrated in [1]. The analysis of experimental data has shown that the collector capacitance charging time and intrinsic transit time dominate among the components of the total emitter-collector delay time. As a result, the main problem concerning further extension of the HBT frequency range is to reduce these times. Vertical scaling of HBT structure, resulting in the reduction of the intrinsic transit time, leads to increase of the collector capacitance and hence to increase of the collector capacitance charging time [2]. Thus, there is trade-off between the transit time and collector charging time. Effective way to reduce collector capacitance has been recently proposed in [3] and consists in reducing of the base-collector junction area in HBT with a buried subcollector. Unfortunately, fabricated HBT with buried subcollector has demonstrated large intrinsic transit time. As a result, parameters of HBT with buried subcollector must be optimized in order to realize great potential of its structure.

2. Device Structure and Model

The layout and cross-section of simulated HBTs are presented in Fig.1. As can be seen, the main feature of HBT with buried subcollector is the small area of overlap

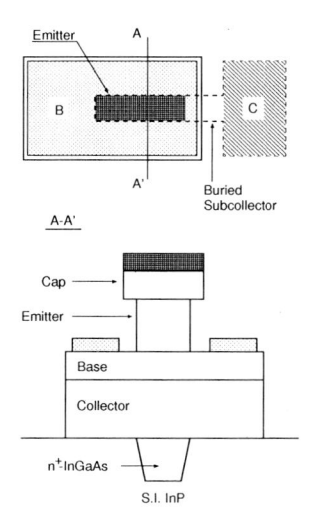

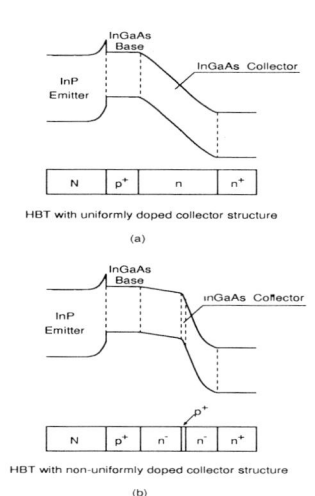

Figure 1: Schematic layout and cross-section of simulated HBT with buried subcollector.

Figure 2: Schematic band diagrams and doping of HBTs with buried subcollector: (a) uniformly doped collector and (b) non-uniformly doped collector.

between the base layer and subcollector, which leads to small total base-collector capacitance. In this study, we concentrate our attention on how base-collector capacitance and base-collector transit time are affected by applied voltages, HBT geometry and collector doping profile. The value of the total base-collector capacitance is determined from the solution of two-dimensional Poisson equation for the electrostatic potential with boundary conditions corresponding to the HBT structure presented in Fig.1. To calculate the intrinsic transit time of HBTs with buried subcollector time-dependent ensemble Monte Carlo particle simulator has been implemented. The value of the base-collector transit time is calculated directly from results of Monte Carlo simulation by using the Fourier analysis of the non-stationary collector current response [4].

3. Results and Discussion

The operation of HBTs with buried subcollector and various collector doping profiles has been investigated at temperature 300 K in the wide range of applied voltages and parameters of HBT structure. First, we have considered the HBT with conventional collector doping profile (Fig.2a). The collector capacitance C versus collector-base voltage V_{CB} is presented in Fig.3. The normalization factor C_0 corresponds to the total base-collector capacitance of the planar HBT with fully depleted collector layer. Analysis shows that the collector capacitance of HBT with buried subcollector is extremely small under the high voltages ($V_{CB} > 2.25V$), and is practically equal to the capacitance of planar HBT under the low voltages ($V_{CB} < 1.75V$). To explain this fact, a detailed analysis of the distributed base-collector capacitance $c(x)$ has been made. The distributions of $c(x)$ for voltages corresponding to different regions of C-V curve are illustrated in Fig.4. Under the low V_{CB} the value of $c(x)$ is practically constant and is the same as in the case of HBT with planar structure, because the

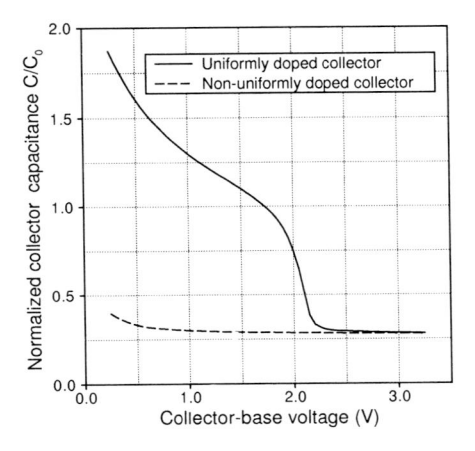

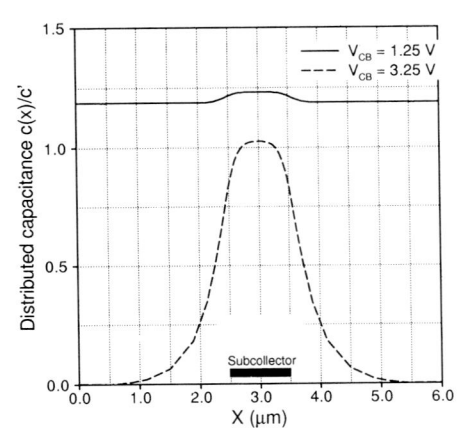

Figure 3: The normalized collector capacitance versus collector-base voltage for HBTs with buried subcollector and different collector doping profiles.

Figure 4: The distributed base-collector capacitance $c(x)$ as a function of position for HBT with uniformly doped collector structure.

lightly-doped collector layer is not fully depleted. As a result, the base-collector capacitance approximately equals the corresponding capacitance of planar HBT and varies in inverse proportion to the square root of the collector-base voltage. When the collector-base voltage increases, free carriers are driven out of the periphery part of the collector layer to center, and $c(x)$ corresponding to passive part of the base decreases. Under the high voltages (V_{CB}=3.25V), the collector layer is fully depleted and base-collector capacitance is determined as a geometrical capacitance between high conductive base layer and buried subcollector region. Thus, HBT with buried subcollector has an advantage over planar HBT only under high voltages. However, it is well known that the ultimate speed of InP/InGaAs HBT is obtained under very low collector-base voltages [5]. To overcome this contradiction we have proposed and investigated the HBT with non-uniform collector doping profile (see Fig.2b). Proposed design provides small value of the base-collector capacitance down to very low applied voltages (dashed curve in Fig.3) and has an important side benefit. The distribution of electric field corresponding to this structure is very favorable to exploit overshoot effect in collector space-charge region and hence it is possible to expect the reduction of intrinsic transit time.

The typical average electron velocity profiles for HBTs with two different collector structures are shown in Fig.5 (here the left boundary of the plot corresponds to the emitter-base interface). It is seen, that the peak values of electron velocity are practically the same for both collector structures and approximately equal to 8×10^7 cm/s. In both case the end of velocity overshoot region is associated with rapid population of the satellite valleys. The position of the peak velocity depends on the position of the high electric field domain in the collector space-charge region. For conventional uniformly doped collector structure the electric field is maximum near the base-collector interface. As a result, the electron traveling across the collector space-charge region rapidly populate the satellite valleys and lose their directional velocity. In contrast, for proposed collector structure the electric field between the base region and collector p^+ layer is low and electrons remain high velocity over a wide area, that leads to substantial reduction of the collector transit time in HBT

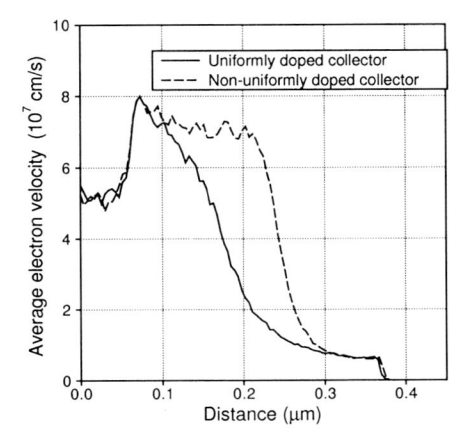

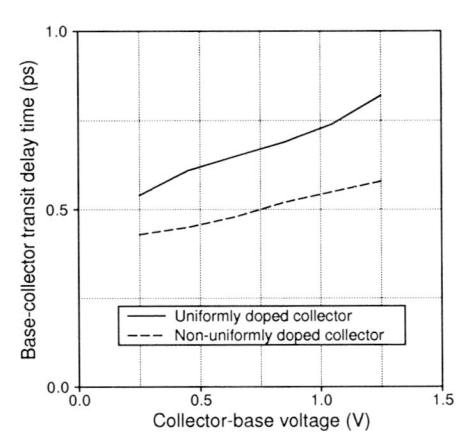

Figure 5: Average electron drift velocity corresponding to HBTs with conventional (solid curve) and proposed (dashed curve) collector structures.

Figure 6: The base-collector transit delay time versus collector-base voltage for HBTs with buried subcollector and different collector doping profiles.

with non-uniformly doped collector structure.

The base-collector transit delay times as functions of applied voltage are presented in Fig.6. It is seen, that the base-collector transit delay time of non-uniformly doped HBT is approximately 30% less than that of uniformly doped HBT, resulting from the extension of overshoot region in the collector depletion layer. This makes essential contribution to reduction of the total delay time for HBT with sub-picosecond delay. In closing we would like to point out one additional feature of non-uniformly doped collector structure. Varying sheet density of acceptors in collector p^+ layer and position of this layer it is possible to adjust HBT structure to obtain the ultimate speed at the desired range of applied voltages.

References

[1] J-I. Song, B.W-P. Hong, C.J. Palmstrom, B.P. Van der Gaag and K.B. Chough, "Ultra-High-Speed InP/InGaAs Heterojunction Bipolar Transistors," *IEEE Electron Device Lett.*, vol. 15, no. 3, pp. 94-96, 1994.

[2] G. Khrenov, V. Ryzhii and S. Kartashov, "AlGaAs/GaAs HBT Collector Optimization for High Frequency Performance," *Solid-State Electron.*, vol. 37, no. 1, pp. 213-214, 1994.

[3] J-I. Song, M.R. Frei, J.R. Hayes, R. Bhat and H.M. Cox, "Self-Aligned In-AlAs/InGaAs Heterojunction Bipolar Transistor with a Buried Subcollector Grown by Selective Epitaxy," *IEEE Electron Device Lett.*, vol. 15, no. 4, pp. 123-125, 1994.

[4] G. Khrenov, V. Ryzhii and S.Kartashov, "Fourier Analysis-Based Method for High-Frequency Performance Calculation of Heterojunction Bipolar Transistor," *Jpn. J. Appl. Phys.*, vol. 33, Pt. 1, no. 8, pp. 4550-4554, 1994.

[5] A.F.J. Levi, R.N. Nottenburg, Y.K. Chen, P.H. Beton and M.B. Panish, "Nonequilibrium Electron Dynamics in Bipolar Transistors," *Solid-State Electron.*, vol. 32, no. 12, pp. 1289-1295, 1989.

Design and Optimization of Millimeter–Wave IMPATT Oscillators Using a Consistent Model for Active and Passive Circuit Parts

Matthias Curow

Technische Universität Hamburg–Harburg
Arbeitsbereich Hochfrequenztechnik
Wallgraben 55, D–21073 Hamburg, GERMANY

Abstract

A new CAD approach for analysis, development, and optimization of millimeter–wave oscillator circuits is presented. The method has been developed with respect to universality, accuracy, and efficiency regarding the consistent modelling of active two–terminal devices operating in passive circuits. Some benefits of the approach are demonstrated by a number of results for practically realized Si and GaAs CW IMPATT oscillators at 94 and 140 GHz. It is likely that millimeter–wave IMPATT diodes operate often in a subharmonic mode in which the fundamental power may or may not be reduced compared to a single–frequency excitation.

1. Introduction

The design of millimeter–wave oscillators involves both the active device and the passive circuit. Since stable operation points are characterized by matched impedances between source and load, the main tasks are the determination of dynamic device impedances and the construction of passive circuits offering appropriate impedances at the frequencies of interest.

Usually, the active device is assumed to be driven by a sinusoidal voltage, and corresponding dynamic impedances can easily be calculated for a single–frequency excitation. However, this assumption is invalid under certain circumstances occurring in practical systems, e. g., harmonic or sub–harmonic operation (whether being intended or not). In these cases, the matching behavior of the complete circuit may significantly deviate from what is expected following the device data obtained as described before.

This problem can be overcome by a consistent treatment of the active device and the passive circuit. However, such an approach requires a substantial effort regarding the development of accurate but nevertheless numerically efficient mathematical models for semiconductor devices with short active regions (showing nonstationary carrier transport) and a suitable description of the passive network which should be applicable to different circuit types. Hence, only little work regarding this topic is available (e. g., [1], [2]).

2. Models for active device and passive load

In the present work, an extended bipolar hydrodynamic model is used that covers all effects of nonstationary transport. All relevant physical processes such as thermal generation, impact ionization, and tunneling are included, so that many important two–terminal devices such as Gunn–, PIN–, IMPATT–, BARITT–, and MITATT– devices can be examined. Special attention has been paid to the inclusion of the thermal energy and diffusive heat flux since these parameters may have a significant effect on the performance of unipolar devices. Furthermore, two additional energy balance equations for accurate modeling of impact ionization are included, allowing Monte–Carlo generated data for ionization coefficients to be used. In contrast to other models, the present one closely reproduces measured breakdown and DC operating voltages. Complete device structures (including very highly doped contact regions) are considered without any simplifications. The system of partial differential equations is solved by a full–implicit, decoupled algorithm. A special discretization of Poisson's equation causes unconditional stability of the whole solution procedure and allows large time steps compared to ordinary schemes to be used. The temperature rise across the semiconductor layers mounted on copper or diamond heat sinks is calculated separately, and appropriate sets of Monte–Carlo generated transport data are used.

The impedance calculation for waveguide mounting structures is a very complex task. The present analysis is restricted to widely used circuits consisting of either a radial line formed by a disc and a post in a rectangular waveguide of full height or a post in a reduced–height waveguide. The load characteristic as seen at the semiconductor device package plane is calculated by well–known mode–matching techniques [3], [4]. The impedance transformation caused by the device package and a series loss resistance is taken into account. It will be shown how suitable values for the parasitics can be extracted with the help of the mode–matching technique. A very good quantitative description even over broad frequency bands is then achievable. Furthermore, the influence of the bias choke can be consistently included in a modified program version [4].

The load impedance is given as a table of frequency–dependent values. A purely time–domain scheme for the inclusion of the interaction of the active device and the load impedance has been developed which is easy to implement and is well suited for highly nonlinear, broad–band circuits where harmonic balance methods might suffer from convergence problems. The impulse response of the load admittance is obtained from an inverse FFT (with typically $2^{15} - 2^{18}$ values). The convolution of this function with the time–dependent input signal gives the output response of the passive network in each time step which is used in turn to update the driving force for the active device. After a steady state is reached, the result is examined by means of Fourier analysis. Due to generally low quality factors of order 50 – 100 of the networks, stable operation points are reached in a few ns or less in most cases. Compared to the time needed for the semiconductor calculation itself, the convolution method increases the computation time only by several ten percent. Hence, overall computation times are mainly dictated by the space and time increments which have to be chosen smaller as the doping level increases. Compared to unipolar devices, computation times for bipolar devices are approx. one up to one–half order higher and vary from seconds to hours on modern RISC workstations (HP 9000/700), while typical times are some minutes for a single operation point of an IMPATT oscillator.

3. Results

Gunn and IMPATT oscillators operating in W– and D–bands have been intensively studied. A number of results for Si and GaAs IMPATT devices mounted in different circuit configurations will be presented. An important conclusion is that millimeter-wave IMPATT–diodes are usually working in a subharmonic mode with a power component at $f_0/2$. Matching at this frequency is possible because it lies typically below the avalanche frequency where the circuit impedance is capacitive. In this multifrequency operation, dynamic impedances may differ from results obtained by the usually assumed single–frequency excitation with a sinusoidal voltage. The maximum available output power at f_0 may or may not be significantly effected; no clear tendency has been found. However, compared to power losses due to subharmonic components, ohmic losses dominate at small device areas. The additional power components are absorbed by series loads such as contact or skin resistances of a few tenth of an Ohm. Since this power below the cutoff frequency does not leave the resonator and is quite small, it is usually not measured or even detected.

A particularly interesting application of the method is a sensitivity analysis around an operating point. This is useful for optimization of a circuit in order to get the maximum power from the active device. As an example, a 94 GHz IMPATT oscillator described in [5] has been modelled in detail, including package parasitics and an ohmic loss resistance of 0.35 Ω. It uses a disc–type resonator in a WR10 waveguide. Some typical results for varying mount geometries are given in Figs. 1 – 3. The operating point represents the maximum observed power output of approx. 300 – 320 mW at 95 GHz. These drawings clearly demonstrate the strong influence of disc and post diameter upon frequency and output power. However, the disc thickness mainly affects the frequency. Hence, if an oscillator has been built and optimized with respect to the output power level but its operating frequency has still to be fine tuned, another disc thickness can be chosen without affecting the power significantly.

4. Conclusion

A new simulation tool for the design and optimization of millimeter–wave oscillators is presented. A number of different semiconductor devices as well as all commonly used resonator configurations can be modeled. Consistent simulations of practically realized oscillators show a good qualitative and quantitative agreement.

Acknowledgement:
The author is indebted to the Deutsche Forschungsgemeinschaft for financial support and to the authors of [3] and [4] for supplying their programs.

References

[1] V. Stoiljkovic, M.J. Howes,, V. Postoyalko: Microwave Eng. Europe (1993) 35–40.
[2] M.F. Zybura, S.H. Jones, G.B. Tait, J.R. Jones: IEEE Guided Wave Lett. 4 (1994) 282–284.
[3] M. E. Bialkowski: AEÜ 38 (1984) 306–310.
[4] B. D. Bates, A. Ko: IEEE Trans. Microwave Theory Tech. 38 (1990) 1037–1045.
[5] H. Eisele: Ph. D. Thesis, Technical University of Munich 1989.

5. Figures

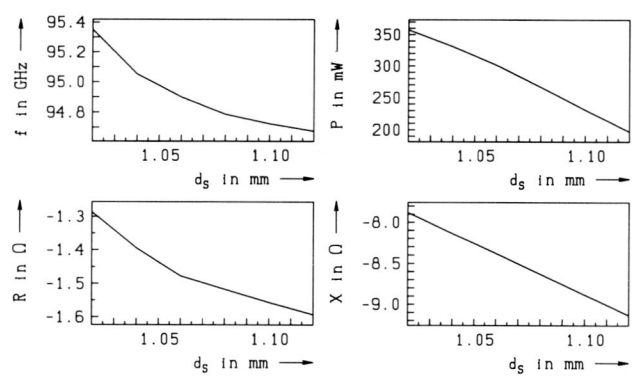

Figure 1: Operating frequency, output power, dynamic resistance and dynamic reactance vs. disc diameter.

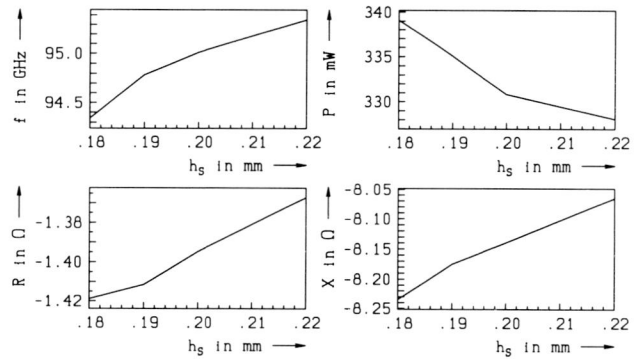

Figure 2: Operating frequency, output power, dynamic resistance and dynamic reactance vs. disc thickness.

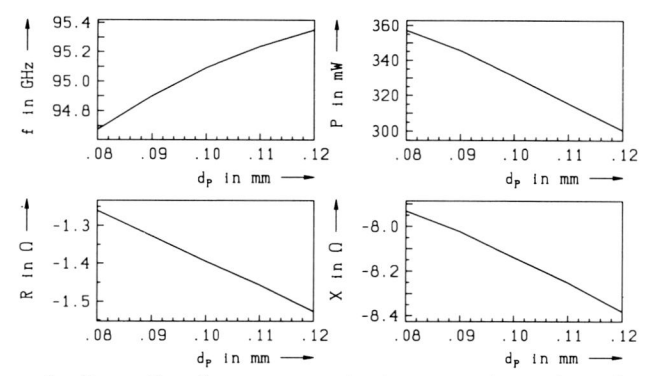

Figure 3: Operating frequency, output power, dynamic resistance and dynamic reactance vs. post diameter.

Generalised Drift-Diffusion Model of Bipolar Transport in Semiconductors

D. Reznik

Institut für Werkstoffe der Elektrotechnik, TU-Berlin
Jebensstraße 1, 10623 Berlin, GERMANY

Abstract

A generalisation of the conventional relaxation-time approximation for bipolar transport with electron-hole scattering is presented. A simple phenomenological ansatz leads to Generalised Drift-Diffusion current equations, which contain both conventional Drift-Diffusion equations and matrix-form Drift-Diffusion equations with drag currents as special cases. The effect on carrier transport in semiconductor devices under low and high injection conditions is discussed analytically and compared with simulation results.

1. Introduction

The correct description of electron-hole scattering (EHS) influence on drift- and diffusion-dominated charge carrier transport is important for the simulation of all bipolar devices, especially for bipolar power devices, as it chiefly determines the voltage drop over the device in the forward biased high injection regime. Many authors [1-6] have investigated this subject, establishing two types of current equations, which can be characterised as conventional Drift-Diffusion equations (Van Roosbroeck-type) and matrix-form Drift-Diffusion equations with drag currents (Avakyants-type). However, once fitted to achieve agreement with ohmic mobility data, both approaches cannot describe other experiments properly. Avakyants-type equations predict an injection-independent ambipolar diffusion constant, in contradiction to experimental data [9]. Both types of equations fail to explain the asymmetric steady-state carrier distributions at high current densities [10]. In this paper, the analysis of the coupled Boltzmann Equations for electrons and holes leads to a natural generalisation of the usual relaxation-time approximation for the collision terms. A simple phenomenological ansatz yields analytic generalised Drift-Diffusion current expressions (GDD), which contain both Van Roosbroeck- and Avakyants-type models as limiting cases.

2. Derivation of Model Equations

The distribution functions for electrons and holes $f_e(\mathbf{k},\mathbf{x},t), f_h(\mathbf{k},\mathbf{x},t)$ are given by the solution of the coupled Boltzmann Transport Equations (BTEs)

$$D^e f_e = \left(\frac{\partial}{\partial t} + \frac{\hbar}{m_e}\mathbf{k}\cdot\nabla_{\mathbf{x}} - \frac{q}{\hbar}\mathbf{E}\cdot\nabla_{\mathbf{k}}\right) f_e(\mathbf{k},\mathbf{x},t) = \left(\frac{\partial f_e}{\partial t}\right)_{coll},$$
$$D^h f_h = \left(\frac{\partial}{\partial t} + \frac{\hbar}{m_h}\mathbf{k}\cdot\nabla_{\mathbf{x}} + \frac{q}{\hbar}\mathbf{E}\cdot\nabla_{\mathbf{k}}\right) f_h(\mathbf{k},\mathbf{x},t) = \left(\frac{\partial f_h}{\partial t}\right)_{coll}, \quad (1)$$

with the collision terms

$$\left(\frac{\partial f_e}{\partial t}\right)_{coll} = \left(\frac{\partial f_e}{\partial t}\right)_{coll}^{latt} + \left(\frac{\partial f_e}{\partial t}\right)_{coll}^{EHS} + \left(\frac{\partial f_e}{\partial t}\right)_{coll}^{EES},$$
$$\left(\frac{\partial f_h}{\partial t}\right)_{coll} = \left(\frac{\partial f_h}{\partial t}\right)_{coll}^{latt} + \left(\frac{\partial f_h}{\partial t}\right)_{coll}^{EHS} + \left(\frac{\partial f_h}{\partial t}\right)_{coll}^{HHS},$$

which can be separated into the terms caused by lattice scattering (mainly phonon and ionized impurities scattering), electron-hole scattering (EHS) and scattering of carriers of the same type (EES and HHS). At the moment, the EES and HHS terms will be neglected, since their influence on mobility is rather small and can be included by some phenomenological corrections to the end formulae [11,12].

For the linearised BTEs, the relaxation-time approximation proves to describe the collision terms for lattice scattering reasonably in most of the cases. The collision integrals due to EHS cannot be expressed by the non equilibrium parts of the distribution function of only one carrier type, since e.g. momentum transfer from non equilibrium holes can drag the electron distribution function out of initial equilibrium. The simplest physically consistent approximation for the EHS collision terms is a modified relaxation-time ansatz, which includes additional terms caused by carrier drag:

$$\left(\frac{\partial f_e}{\partial t}\right)_{coll}^{EHS} \approx -\frac{f_e - f_e^0}{\tau_{eh}(\mathbf{k})} + \beta_e(\mathbf{k})\frac{f_h - f_h^0}{\tau_{he}(\mathbf{k})}, \quad \left(\frac{\partial f_h}{\partial t}\right)_{coll}^{EHS} \approx -\frac{f_h - f_h^0}{\tau_{he}(\mathbf{k})} + \beta_h(\mathbf{k})\frac{f_e - f_e^0}{\tau_{eh}(\mathbf{k})}. \quad (2)$$

The superscript "0" in (2) denotes equilibrium distribution functions, τ_{eh}, τ_{he} are relaxation times of electron and hole systems due to EHS, and β_e, β_h are some phenomenological coupling functions. Equation (2) is far from being mathematically exact, but it gives a qualitative description of the two main physical processes caused by EHS : the relaxation of the non equilibrium part (first terms) as well as the drag out of equilibrium due to the existence of the non equilibrium part of the scattering partner (second terms). Using the approximation (2), the relaxation-time approximation for lattice scattering and neglecting the influence of same type carrier scattering, a formal solution of the coupled BTEs (1) can be derived. A further approximation step yields the Generalised Drift-Diffusion current equations

$$j_p = q\mu_p^p(pE - \frac{kT}{q}\nabla p) - q\mu_p^n(nE + \frac{kT}{q}\nabla n)$$
$$j_n = q\mu_n^n(nE + \frac{kT}{q}\nabla n) - q\mu_n^p(pE - \frac{kT}{q}\nabla p) \quad (3)$$

with the mobilities

$$\mu_p^p = \frac{\mu_p^0(1+\mu_n^0/\mu_{np})}{\Delta}, \quad \mu_p^n = \frac{\beta\mu_n^0\mu_p^0}{\Delta\mu_{np}} \quad \mu_n^n = \frac{\mu_n^0(1+\mu_p^0/\mu_{pn})}{\Delta}, \quad \mu_n^p = \frac{\beta\mu_p^0\mu_n^0}{\Delta\mu_{pn}} \quad (4)$$

$$\Delta = 1 + \frac{\mu_n^0}{\mu_{np}} + \frac{\mu_p^0}{\mu_{pn}} + (1-\beta^2)\frac{\mu_n^0\mu_p^0}{\mu_{np}\mu_{pn}}.$$

μ_n^0, μ_p^0, are electron and hole lattice mobilities due to phonon and ionized impurity scattering, $\mu_{pn} = C/n, \mu_{np} = C/p$ are electron-hole mobilities (the constant C can weakly depend on $(n+p)$ because of screening effects) and β is a phenomenological drag parameter. A detailed motivation of (2) and (3) will be published in a longer paper [12]. It can be easily shown that for the special case $\beta=0$ the off-diagonal mobilities vanish and the equations (3,4) transform into the conventional Van Roosbroeck current equations. For the case $\beta=1$ the GDD becomes identical to the Avakyants-type equations in the form proposed by Mnatsakanov et al. [2,3,5].

3. Consequences for Bipolar Devices

For quasi neutral bulk regions, homogeneous doping implies $\nabla n \approx \nabla p$ and (3) becomes

$$j_p = qp(\mu_p^p - \mu_n^p)E - kT(\mu_p^p + \mu_p^n)\nabla p = q\mu_p^{Drift} pE - kT\mu_p^{Diff}\nabla p$$
$$j_n = qn(\mu_n^n - \mu_p^n)E + kT(\mu_n^n + \mu_n^p)\nabla n = q\mu_n^{Drift} nE + kT\mu_n^{Diff}\nabla n \quad (5)$$

Equation (5) differs from the conventional current equation due to the fact that for $\beta \neq 0$ the drift and diffusion mobilities are not equal. Using (4), one can show that for low injection conditions the *minority diffusion mobility* can be approximated as the usual Mathiessen combination of lattice and EHS mobilities and is practically independent on the choice of β. Therefore, as long as the current over a pn-junction can be described by the sum of minority diffusion currents (Shockley's approximation), the DC current-voltage characteristic should not depend on the choice of β and no difference should occur between diode characteristics calculated by Avakyants- or Van Roosbroeck current equations. Fig. 1 shows the IU-characteristics for a highly doped np-diode for $\beta = 0$ and

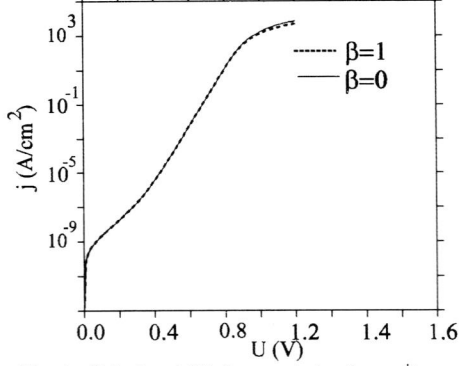

Fig. 1 : Calculated IU characteristics for a n$^+$p-diode and two values of β.

$\beta = 1$. The resulting curves coincide over a long range of current values and only differ slightly in the high current region, where deviations from the exponential behavior due to ohmic losses occur. At high currents, the voltage drop is higher for $\beta=1$, since the drift mobility decreases with rising β.

For high injection conditions, typically occurring in the low doped base of *psn*-diodes and thyristors, the voltage drop over the base region is determined by the sum of *drift mobilities*, which is therefore the quantity measured by voltage drop experiments like those performed by Dannhäuser and Krause [7,8]. Fig. 2 shows the numerically calculated current voltage characteristics of a wide base (440 μm) *psn*-diode for three different values of β. Obviously, the choice of the drag parameter β has significant influence on the forward voltage drop under high injection conditions.

The modeling of EHS also strongly affects the stationary spatial distribution of charge carriers. The simulated carrier distributions in the base are shown in Fig. 3 for a current density of 1000 A/cm^2 and three values of β: β=1 yields a distribution completely symmetrical around the minimum, while β=0 produces very strong asymmetries, moving the distribution's minimum towards the n-zone.

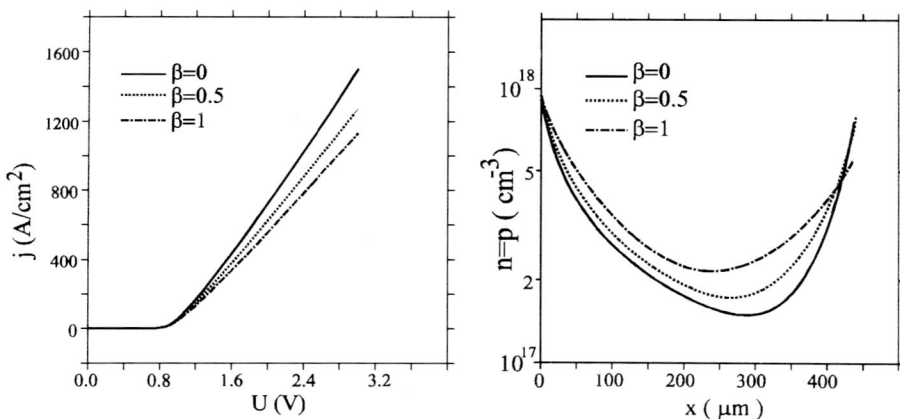

Fig. 2: Calculated current-voltage characteristics of a long base *psn*-diode

Fig. 3: Charge carrier distribution in the base of the diode from Fig. 2 at j=1000 A/cm^2

Acknowledgements

The author is indebted to R. Nürnberg and H. Gajewski, Weierstrass-Institute for Applied Analysis and Stochastics, for the implementation of the GDD into the 2D device simulator ToSCA.

References

[1] N.H. Fletcher: *Proc. IRE* **45**, 862 (1957)
[2] G.M. Avakyants: *Radiotechnika i Elektronika* **8**, 1919 (1963)
[3] T.T. Mnatsakanov, I.L. Rostovtsev, N.I. Philatov: *Solid-State-Electronics* **30**, 579 (1987)
[4] T.T. Mnatsakanov: *Phys. Stat. Sol.*, **B 143**, 225 (1987)
[5] D. Kane, R. Swanson: *Journal of Applied Physics* **72**, 5294 (1992)
[6] E. Velmre, A. Koel, F. Masszi: *SISDEP* **5**, 433 (1993)
[7] F. Dannhäuser: *Solid-State-Electronics* **15**, 1371 (1972)
[8] J. Krause: *Solid-State-Electronics* **15**, 1377 (1972)
[9] M. Rosling, H. Bleichner, M. Lundqvist, E. Nordlander: *Solid-State-Electronics* **35**, 1223 (1992)
[10] D. Reznik, W. Gerlach: *Solid-State-Electronics* **38**, 437 (1995)
[11] T. Mnatsakanov, B. Gresserov, L. Pomortseva: *Solid-State-Electronics* **38**, 225 (1995)
[12] D. Reznik, W. Gerlach: to be published in *Journal of Applied Physics*

Efficient 3D Unstructured Grid Algorithms for Modelling of Chemical Vapour Deposition in Horizontal Reactors

F. Durst[a], A.O. Galjukov[b], Yu.N. Makarov[a], M. Schäfer[a], P.A. Voinovich[b], A.I. Zhmakin[*]

[a]Lehrstuhl für Strömungsmechanik, Universität Erlangen–Nürnberg, Cauerstr. 4, D-91058 Erlangen, Germany
Tel. +49 9131 761248, Fax. +49 9131 761242; e-mail address: yuri@lstm.uni-erlangen.de

[b]Advanced Technology Center, P.O.Box 160, 198103, St.Petersburg, Russia; Tel. +7 812 251 7232; Fax: +7 812 251 6371

[*]Numerical Simulation Department, A.F.Ioffe Physical Technical Institute, Russian Academy of Sciences, St. Petersburg 194021, Russia; Tel. +7 812 2479145; e-mail address: Zhmakin@numer.ioffe.rssi.ru

Abstract

The paper deals with the use of 3D unstructured grid algorithms with adaptive local grid refinement for the computation of flow, heat and mass transfer and their application to the modelling of metalorganic chemical vapor deposition (MOCVD) of epitaxial layers of GaAs and GaN. It is shown that this approach is promising for an accurate 3D modelling of the deposition processes.

1. Introduction

At the growth conditions practically used for MOCVD of GaAs and GaN in horizontal tube reactors, high temperature and species concentration gradients are formed near the hot susceptor and the substrate on it. This is especially critical for the growth of GaN in N_2-ambient. Accurate calculation of the temperature and species concentration gradients is necessary to reproduce experimental observations and to perform optimization of the processes. Commercially available general software packages usually are not efficient enough with respect to execution time and memory requirements.

Unstructured grids were extensively used during the last decade to solve fluid flow problems, the main attention is paid to the development of 2D and 3D solvers for Euler (for example, [1]) and compressible Navier–Stokes [2] equations. The latter approach has also been implemented for the simulation of flow and deposition in CVD reactors [3]. It is also possible to use hybrid unstructured/structured grids combining the flexibility of the former with the high accuracy of the latter in computing the wall friction and heat transfer due to the possibility to align the grid lines to the solid surfaces [4]. The easiness of adaptive grid refinement is a promising feature of this approach for modelling of processes with significantly localized areas of high gradients of parameters.

2. Mathematical model

Flows in epitaxial reactors are characterized by 1) low velocities (compared to the sound speed) and 2) large temperature (and, hence, density) variations. Numerical integration of the compressible Navier–Stokes equations for very low Mach numbers presents severe problems in practice while the well-known Boussinesq approximation is invalid when the density (temperature) variations are comparable or greater than the mean values. In the present work the so-called low-Mach number Navier-Stokes equations are used [5]. The equations in dimensionless form for flows in an open system can be written as

$$\frac{\partial \varrho}{\partial t} + \nabla \cdot (\varrho \vec{V}) = 0 \qquad (1)$$

$$\frac{\partial \varrho \vec{V}}{\partial t} + \nabla \cdot (\varrho \vec{V} \vec{V}) = -\nabla P + \frac{1}{Fr} \varrho \vec{j} + \frac{1}{Re}\left[2 Div(\mu \dot{S}) - \frac{2}{3}\nabla(\mu \nabla \cdot \vec{V})\right] \qquad (2)$$

$$\varrho \frac{dT}{dt} = \frac{1}{RePr}\nabla \cdot (\lambda \nabla T) \qquad (3)$$

$$\varrho T = 1 \qquad (4)$$

where ϱ is the density, $\vec{V}$ is the velocity, P is the excess pressure, T is the temperature, μ is the dynamic viscosity, λ is the thermal conductivity, $\dot{S}$ is the deformation rate tensor, $\vec{j}$ is the unit gravitational vector, γ is the ratio of specific heats, $Re = \frac{\varrho_0 V_0 L_0}{\mu_0}$, $Pr = \frac{\mu_0 C_p}{\lambda_0}$, $Fr = \frac{V_0^2}{gL_0}$ are Reynolds, Prandtl and Froude numbers, respectively, C_p is the specific heat at constant pressure. The subindex "o" refers to the values chosen as the scales.

The growth process at practically used growth conditions is limited by the mass transport of Ga-species to the substrate, therefore, a simplified model of the deposition process similar to [5,6] is used in the present work.

$$\frac{\partial \varrho C_i}{\partial t} + \nabla \cdot (\varrho \vec{V} C_i) = \frac{1}{ReSc_i}\nabla \cdot [\varrho(\nabla C_i + C_i K_{T_i}\nabla \ln T)] + W_i \qquad (5)$$

3. Numerical method

The Watson incremental triangulation technique, generalized for the 3D case, is the basis for the grid generation method needed to produce a starting grid. The geometry is defined by a surface triangulation and the density of the grid nodes at the surface controls the grid non-uniformity in the volume of the reactor. Grid smoothing and edge swapping algorithms are employed to improve the initial grid quality.

The approximation of the basic equations is performed exploiting Ostrogradskii–Gauss and related integral theorems applied to control volumes around each vertex and to tetrahedral cells themselves to compute viscous terms. The difference equations are solved by simultaneous pointwise relaxation.

In general, refinement methods can be developed in order to minimize either the execution time or the memory requirements. Usually the memory needed for storing the grid information is larger (and sometimes substantially) than that for storing the flow variables. It may be over 50 words per tetrahedron (or over 250 per vertex) [7]. In the present approach an attempt is made to minimize the memory requirements (paying by a higher execution time for reconstruction of grid connectivity and neighbor relations). Additional improvement results from the fact that only grid refinement and no de-refinement is allowed. The refinement of the current tetrahedron results in introducing nodes in the neighbour tetrahedra having common face (they will be referred to as t–neighbours below) and common edge (e–neighbours). The algorithm is based on strict conventions for the grid elements numbering that allows for a reduction of the memory requirements to about 20 integer words per tetrahedron and to minimize the execution time for the reconstruction of the grid structure. It can be summarized as follows:

1. Mark all tetrahedra for refinement.

2. If the fraction of tetrahedra to be refined does not belong to the given range, change the refinement criterium and repeat step 1°.

3. For each marked tetrahedron loop through all edges:
    (a) introduce new vertex.
    (b) introduce new vertex into e–neighbour tetrahedra.

4. Check all tetrahedra for a number of new nodes
    (a) 6 nodes (marked tetrahedra themself)
    (b) 3 nodes belonging to one face (t–neighbours)
    (c) 1 node (e–neighbours)
    (d) other

    introduce additional nodes to (d) in order to convert them to (a) or (b) cases.

5. If new nodes have been introduced, repeat step 4°.

6. Set new connectivity relations

    (tetrahedron $\to$ 4 vertices   vertex $\to$ 1 tetrahedron).

7. Flag deleted tetrahedra and set new neghbour relations

    (tetrahedron $\to$ 4 tetrahedra).

The solution procedure is repeated after every global refinement of the grid in the whole reactor. The computations are finished when a grid-independent solution is obtained, i.e. when introduction of additional grid volumes changes the solution only within a prescribed accuracy.

4. Results

The flow regimes corresponding to GaAs and GaN MOCVD have been simulated. An example of a computational result for the modelling of MOCVD of GaAs is presented in the Figure 1. Initial (3151 vertices, 11003 tetrahedra) and final (37355 vertices, 190480 tetrahedra) grids are plotted together with the normalized growth rate distribution of the GaAs layer over the susceptor (only one half of the susceptor is shown). The clustering of the grid cells occurs in the regions of high temperature and species concentration gradients where high accuracy of the calculations is required to predict the growth rate distribution. In the parts of the flow where smooth variations of the flow parameters occur a much coarser grid is sufficient to get an accurate solution.

As a conclusion it can be stated that the proposed approach due to its high flexibility is promising for simulations of flow and deposition in CVD reactors.

Figure 1:

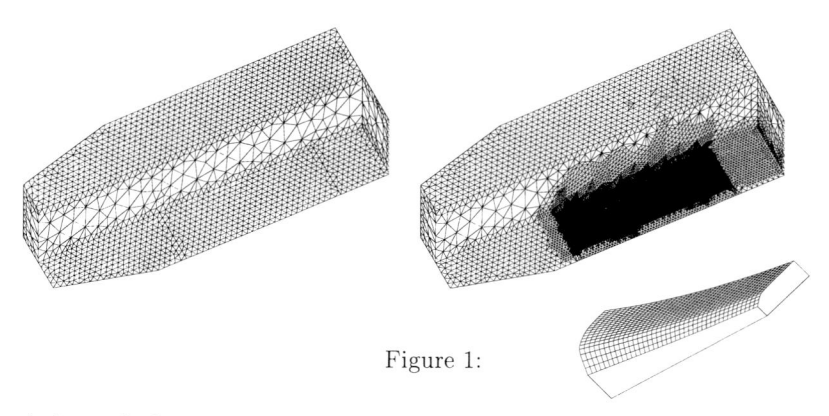

5. Acknowledgement

The works have been supported partly by BMFT project FAU4001-01IR303.

6. References

References

[1] A.A. Fursenko, D.M. Sharov, E.V. Timofeev, P.A. Voinovich. Comput.Fluids 21 (1992) 377–396.
[2] T.J. Barth. AIAA 91 (1991) 0721.
[3] A.I. Zhmakin. V Int.Congr.Comp. Appl. Math., Abstr., Leuven, (1994).
[4] D. Ofengeim, E. Timofeev, A. Fursenko, P.A. Voinovich. Proc. First Asian CFD Conference 1 (1995), 3, 383–389.
[5] Yu.N. Makarov, A.I. Zhmakin. J. Crystal Growth 94 (1989) 537–550.
[6] F.Durst, L.Kadinskii, M.Perić, M.Schäfer. J. Crystal Growth 125 (1992) 612–626.
[7] Y. Kallinderis, P. Vijayan. AIAA J. 31 (1993), 8, 1440–1447.

Preventing critical conditions in IGBT chopper circuits by a multi-step gate drive mode

W. Gerlach, U. Wiese

Inst. f. Werkstoffe der Elektrotechnik, TU Berlin, Sekr. J10
Jebensstr. 1, 10623 Berlin, GERMANY

Abstract

It will be shown, that critical conditions of the diode in a chopper circuit can be avoided using a multi-step gate drive mode. Moreover, it is demonstrated that the total losses of the chopper circuit can be reduced by an appropriate choice of the hight and length of the gate voltage steps.

INTRODUCTION

In chopper circuits and hard switching inverters, full use of the short switching times of IGBT's can only be made in combination with specially adjusted freewheeling diodes. These diodes are difficult to realize even with sophisticated technological processes and new design concepts [1, 2]. So diodes which can withstand all the stresses imposed by fast switching are scarcely available. Thus in practical circuits, the freewheeling diode may prove to be the weakest element. Especially for fast turn-on of the IGBT, the diode may be driven into an unstable state and can be destroyed [3]. To avoid this dangerous situation, it is common practice to reduce the switching speed of the IGBT current by increasing the rise time of the gate voltage via series resistors in the gate circuit. However, this increases the switching losses and may enhance the total losses appreciably at higher frequencies. In the paper it is shown, that with the multistep gate drive mode it is not only possible to prevent the dangerous situations but also to compensate the excess switching losses of the IGBT by lower on-losses.

PRINCIPLE OF THE MULTI-STEP GATE DRIVE MODE

When the IGBT is switched on fastly, a very high reverse current peak can occur in the freewheeling diode. The operating point in the IV-characteristics of the IGBT (Fig. 1) normally follows the dotted line from point A to point B. During this transition the diode takes over the full supply voltage and the excess reverse current tends to zero. But if the peak reverse current initiates the dynamic avalanche [4], the diode is no longer able to block the full supply voltage [3] and the trajectory of the operating point will end at point C. In this state the diode carries a large current at a high voltage. According to the enormous power losses it will be thermally destroyed in a short time.

Reducing the gate voltage to V_{G1} offers a simple way to avoid this critical condition effectively. V_{G1} can be made as low as the gate voltage V_{G0} which is necessary to carry the load current I_L. At $V_{G1} = V_{G0}$ the IGBT would limit the recovery current of the diode to $I_R = I_L$. Since the drain voltage V_{CE} of the IGBT remains nearly at V_{DD} up to the moment where the stored carriers in the diode are removed, the fall-off of V_{CE} starts later for lower recovery currents. Thus it is to be expected that the turn-on losses of the IGBT for switching from the operation point A to B' are increased compared to switching from A to B. But if the gate voltage is raised from V_{G1} to V_{G2} after the transition to B', the current of the IGBT will remain at the load current I_L during the transient from B' to B. During this time the recovery current from the diode is zero. Therefore it is possible to raise the gate voltage to a value, which is higher than it is allowed by the normal driving mode owing to the dynamic avalanche. As a consequence the on-voltage of the IGBT can be driven lower, reducing the on losses. This can be utilized to compensate or even to overcompensate the excess switching losses. The extent can be controlled by the hight and time delay Δt of the gate voltage step and will depend on the lengths of the on-time of the IGBT.

This approach can also be extended to a multistep waveform of the gate voltage (Fig. 2). It offers additional flexibility to adapt the drive signal to varying load conditions.

CURRENT AND VOLTAGE TRANSIENTS

The detailed operation of the chopper circuit, shown in the inlet in Fig. 3, has been investigated by isothermal numerical simulation using the device simulator MEDICI. The calculations are performed for the temperature $T = 300$ K and for the supply voltage $V_{DD} = 1600$ V. The breakdown voltage of the diode is 1650 V and that of the IGBT 1800 V. The diode is a standard PIN-diode with a homogeneous carrier lifetime of $\tau = 0.5$ μs and a forward voltage $V_F = 0.95$ V at a current density of $j_F = 200$ A/cm^2.

In Fig. 3, the current waveforms of the IGBT for different hights of a single gate voltage step are depicted. It can be recognized, that for a gate voltage of $V_G = 12$ V the dynamic avalanche in the diode will occur. Passing a sharp peak, the current steadily increases and tends to a value limited by the applied gate voltage. A similar behaviour is observed for gate voltages down to $V_{G,Cr} = 10$ V. Below this critical value, the current decays after passing a lower and broader peak to the stationary load current I_L. The quite abrupt fall off at $t = 0.25$ μs and $t = 0.37$ μs, respectively indicates a snap-off behaviour of the diode. According to the high dI/dt the voltage across the parasitic inductance L_σ drives the diode into breakdown for a short moment, as shown in Fig. 4. But carrier multiplication only produces a tail current, if the steep decay begins at a large current density, as represented by the waveform for $V_G = 9V$. For lower gate voltages, as for instance $V_G = 8V$, instead of a tail an oscillation appears in the current waveform. It is generated by the interaction of the parasitic inductance and the space charge capacitance of the diode.

From Fig. 4 it is seen, that for $U_G = 12$ V the diode cannot take over the full supply

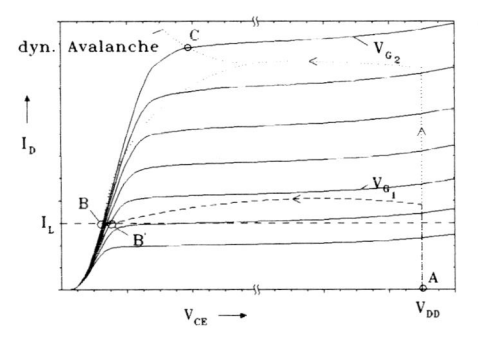

Fig. 1: Output characteristics of an IGBT and trajectories of the operating point for different gate voltages. I_L is the stationary load current.

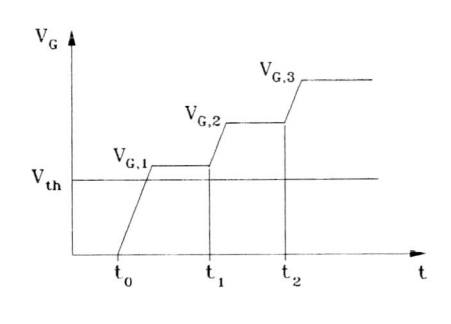

Fig. 2: Typical waveform of the multi-step drive mode

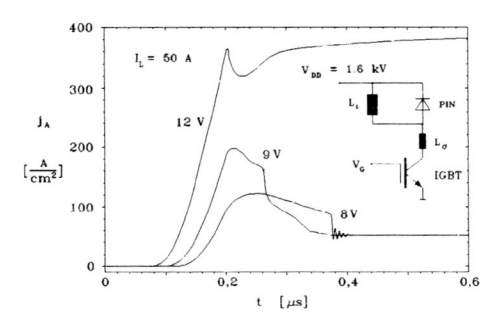

Fig. 3: Current waveforms of the IGBT for different single-stepped gate voltages (L_σ = 50 nH)

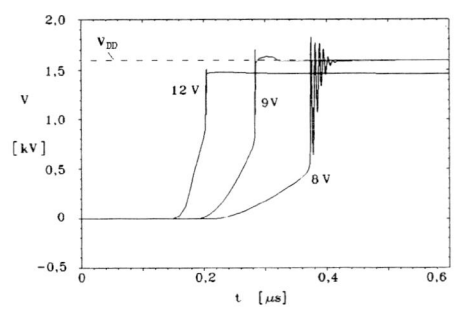

Fig. 4: Voltage transients of the PIN - diode for different gate voltages

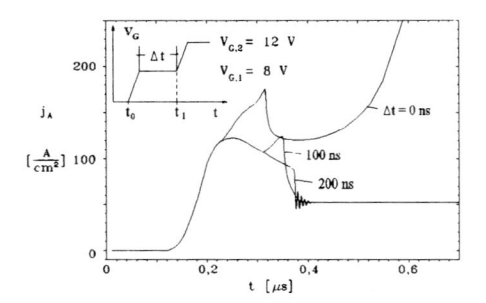

Fig. 5: Typical waveforms of the anode current in the case of a dual-step gate voltage for different time delays Δt_1 of the second step (L_σ = 50 nH)

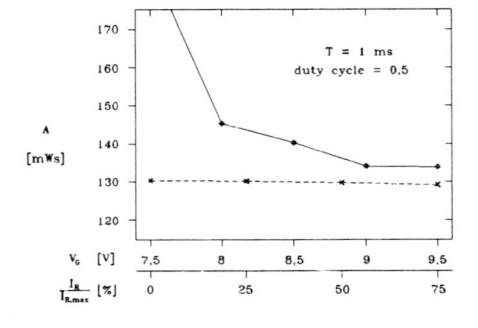

Fig. 6: Total losses of the chopper circuit
♦ : single-step drive mode
x : dual-step drive mode

voltage of 1600 V. Its blocking voltage is reduced to 1470 V so that a voltage of 130 V remains across the IGBT.

In Fig. 5 typical waveforms of the IGBT current are shown for a dual-step gate voltage. The gate voltage is ramped in the first step from 0 V to 8 V and in the second one from 8 V to 12 V within 200 ns in each case. The time delay Δt is varied from $\Delta t = 0$ to $\Delta t = 200$ ns. The case, where the second step starts after the end of the recovery process of the diode, is represented by the waveform marked by $\Delta t = 200$ ns. It is the same curve as obtained for a single-step gate voltage of 8V. Its peak current remains well below the value initiating the dynamic avalanche (peak for $\Delta t = 0$). On the other hand it could be easily lowered by a smaller V_{G1}. Moreover, if the second step starts before the recovery process is completely finished ($\Delta t = 100$ ns), the induced excess current also does not grow to a dangerous value. This demonstrates the stability of the drive mode against slight fluctuations of the recovery time.

POWER LOSSES

The total losses of the chopper circuit are calculated for a frequency of $f_o = 1$ KHz and a duty-cycle of 0.5. The results are shown in Fig. 6. For the dual-step gate drive mode V_{G1} is 8 V and $V_{G2} = 12$ V. The corresponding losses are presented for different time delays in the lower curve. Instead of expressing Δt in nanoseconds, the start, t_2, of the second step is marked by $I_R(t_2) / I_{R, max}$, where I_R is the recovery current of the diode. The total losses remain unchanged whether the second step starts after the decay of the recovery current (0 %) or about 150 ns before (75 %). This is attributed to the on losses of the IGBT, which dominate the switching losses at the frequency f_o. Although V_{G2} exceeds the critical gate voltage $V_{G, cr} = 10$ V by only 2 V, the total losses are less than those for the single-step gate drive at the maximum allowable gate voltage ($V_G = 9.5$ V).

CONCLUSION

The multi-step gate drive mode has been studied by means of two-dimensional device and circuit simulation. It is shown, that the stresses imposed on the freewheeling diode in a fast switched IGBT chopper circuit can be drastically reduced. Further, it is demonstrated that this mode can be utilized to minimize the total losses.

ACKNOWLEDGMENTS

This work has been financially supported by Siemens. We thank Dr. M. Braun, Dr. H. Mitlehner and A. Porst for helpful discussions.

REFERENCES

/ 1/ H. Schlangenotto, et.al.   IEEE Electr. Dev. Let. 10, pp. 322-324, 1989
/ 2/ M. Mehrotra, B.J. Baliga   Proc. ISPSD, pp. 190-204, 1993
/ 3/ A. Porst   Proc. ISPSD, pp. 163-170, 1994
/ 4/ H. Schlangenotto, H. Neubrand   Arch. Elektrotech. 72, pp. 113-123, 1989

Control of Plasma Dynamics within Double-Gate-Turn-Off Thyristors (D-GTO)

U. Wiesner, R. Sittig

Institut für Elektrophysik, Technische Universität Braunschweig
Hans-Sommer-Str. 66, D-38106 Braunschweig, Germany

Abstract

High voltage GTO´s require a subtle control of excess charge to exhibit low forward voltage drop as well as low turn-off loss. In this paper it is investigated whether and to what degree device behaviour could be improved if plasma dynamics is controlled with gate electrodes on cathode and anode side. Computer simulations reveal that such D-GTO´s exhibit excellent static characteristics and that with a continuous control of electron to hole current ratios on both sides a reduction of turn-off energy by a factor of about 50 seems possible.

1. Introduction

At present GTO´s are widely used in traction applications. But there still exists the desire to improve their performance. To reach this goal several authors investigated the characteristics which are achieved by applying IC-technology. The resulting IGBT´s, MCT´s and similar devices offer the advantage of considerably reduced expenses for turn-on and turn-off. On-state characteristics and switching behaviour of high voltage devices, however, depend mainly on plasma dynamics within the wide base layer. In this paper, therefore, the ultimate characteristics are investigated that could be obtained if both emitters of a GTO are completely controlled.

Following this idea the decisive compromise for switching devices [1], between high carrier concentration for low forward voltage drop and low storage charge for safe switch-off at low loss, can be avoided and charge carrier lifetime can be chosen as high as possible. Thus at turn-on and during the on-state the D-GTO may be considered as a thyristor without any emittershorts and a base width of about one carrier diffusion length. During the turn-off phase the complete control of electron- and hole-current densities on both sides allows to firstly deplete the base at a moderate voltage increase and secondly the device is definitely switched-off. At the blocking-state finally the high carrier lifetime ensures a low leakage current.

2. Device structure and static characteristics

2D-calculations were carried out using the device simulator ATLAS [2]. The assumed circuit consists of a voltage source $V_0 = 3200$ V, an ohmic load $R_L = 80\ \Omega$ and an inductive load $L_L = 400$ µH clamped to the voltage of the source by a free wheeling diode. For the interaction with these components a total device area of 1 cm² is taken into account. The structure of a segment of the investigated device is sketched in Fig. 1a. With a base doping concentration of $2.5 \cdot 10^{13}$ cm^{-3} it exhibits a blocking capability of 4500 V. An effective high-injection carrier lifetime of $\tau_{\text{eff}} = 200$ µs is assumed in the wide base layer.

The emitter control is simulated by applying the desired ratio of anode to n-gate current and cathode to p-gate current on the corresponding sides. It has to be emphasized that the following results concern only the structure given in Fig. 1a. For comparison with complete devices the voltage drop across the means for control and the corresponding losses have to be taken into account.

At 300 K a leakage current of 0.5 µA/cm² at an applied blocking voltage of 3200 V is obtained and the forward voltage drop amounts to only 1.0 V at 40 A/cm².

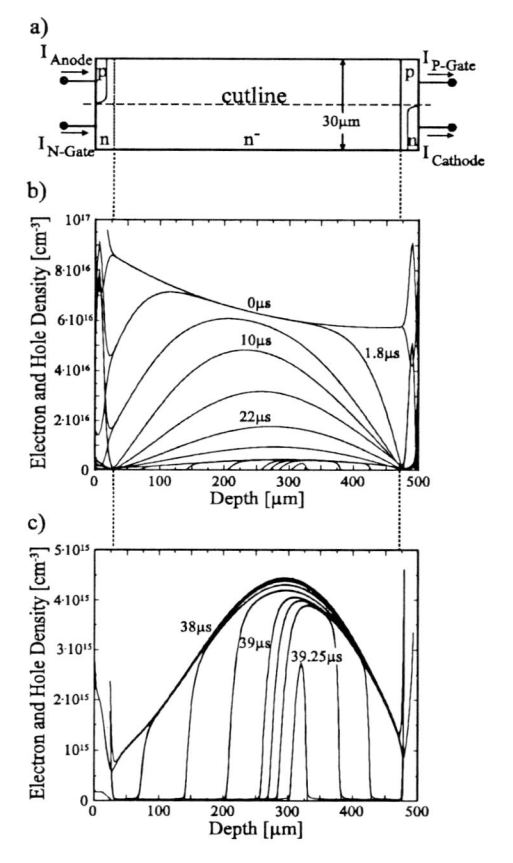

Fig. 1: a) Structure of a segment of the investigated D-GTO. Charge carrier concentration during turn-off along the cutline: b) Complete process, c) Last phase only at an enlarged scale.

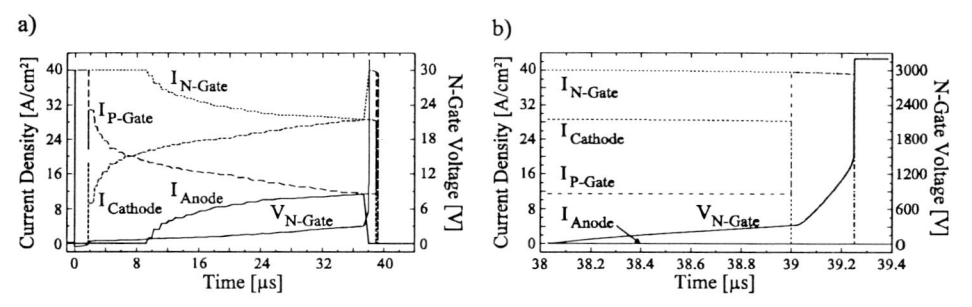

Fig. 2: Emitter and gate currents at turn-off. $V_{\text{N-Gate}}$ represents the n-gate potential when cathode or p-gate contact is grounded: a) Complete process, b) Final emitter switch-off at a spread time scale.

3. Turn-off control

As depicted in Fig. 1b the average carrier concentration in the base amounts to approximately $7 \cdot 10^{16}$ cm^{-3} during the on-state (0µs). For depletion in a first step both emitters are switched-off resulting in a pure electron current density at the anode via n-gate and a pure hole current density at the cathode side via p-gate. There the corresponding carrier concentrations drop to about $1 \cdot 10^{15}$ cm^{-3}. They have to be maintained at this level to avoid build-up of a space charge layer and a corresponding voltage increase. Therefore again rising emitter currents have to be admitted after 1.8 µs at the cathode and after 9.5 µs at the anode. These emitter current densities, however, do not add up to the applied total current density. Due to the concentration gradients carriers are removed via both gate electrodes. Finally after 38 µs the average carrier concentration is reduced to less than $3 \cdot 10^{15}$ cm^{-3}, while the voltage drop still amounts to less than 4 V.

For final switch-off a delay between turn-off of anode emitter and cathode emitter can be chosen to minimize turn-off loss and to control the position of the residual stored carriers. As soon as the anode emitter is switched-off, at t = 38 µs, holes are removed from this side and the current there is maintained by electrons being supplied from the residual reservoir. This leads to a negative space charge, an electric field peak at the nn$^-$-junction and a corresponding increase of voltage. When at t = 39 µs the cathode emitter is switched-off too, the pn-junction becomes depleted and the space charge layer extends into the carrier reservoir from the cathode side. Only 0.25 µs later all carriers are removed and voltage rises sharply up to the clamped voltage. The curves of the corresponding emitter- and gate-currents are shown in Fig. 2 and the actual power loss as well as the integrated turn-off energy are depicted in Fig. 3.

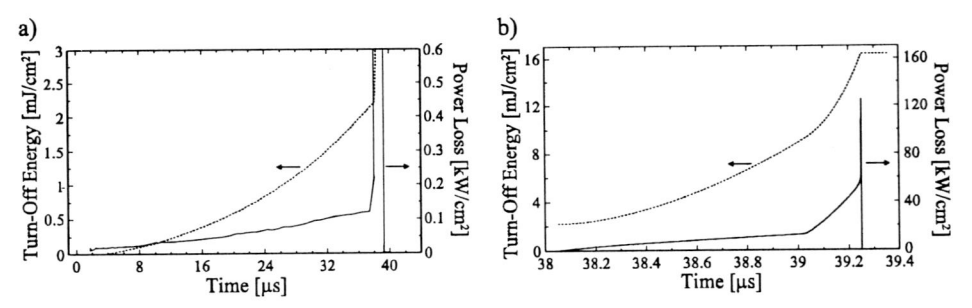

Fig. 3: Actual turn-off loss and integrated energy loss: a) Complete process, b) During the final phase at a spread time scale.

4. Comparison of turn-off losses

Experimental investigations of high voltage D-GTO's revealed already that a considerable reduction of turn-off energy loss can be obtained [3]. In these experiments, however, only delay time between turn-off of anode and cathode emitter was varied. Such a turn-off procedure was simulated, however, on a D-GTO exhibiting a blocking capability of 2000 V. A current density of 40 A/cm^2 is switched-off against an applied voltage of 1000 V. The resulting turn-off energy versus delay time is sketched in Fig. 4. By choosing a suited delay time energy loss can be reduced in this case by a factor of 9 similar to the value published in [3].

If an optimized turn-off control with a depletion phase as described in section 3 is applied, then loss for the low voltage device could be further reduced by a factor of 6 down to a turn-off energy of 0.75 mJ/cm^2 (marked by the dotted line in Fig. 4). This value can also be compared with turn-off losses reported for high voltage MCT and IGBT [4]. For these devices switched on one side only an energy of 42 mJ/cm^2 was obtained which is just the amount of the D-GTO at zero delay between anode and cathode switch-off.

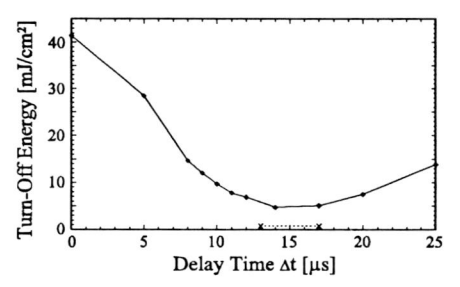

Fig. 4: Turn-off energy of a D-GTO versus delay time between switch-off of anode and cathode emitter., (I = 40A/cm^2, V = 1000 V, T = 300 K). The dotted line marks the energy loss obtained for optimized emitter control (without relevance to the delay time scale).

5. Conclusion

By computer simulations it is confirmed that D-GTO´s seem to offer excellent device characteristics for high voltage applications. Since charge carriers can be controlled from both sides a high carrier lifetime can be chosen. This way a low forward voltage drop and a low leakage current in the blocking state are reached. Applying a well adjusted emitter control during turn-off energy loss could be reduced by a factor of 50 compared to devices with one side switching.

The threshold for dynamic avalanche during the steep voltage rise is nearly doubled since electric field builds-up within two different space charge regions. Moreover even if carrier multiplication is triggered it will last only a very short time due to the reduced reservoir and it may be expected that it will not suffice to re-trigger emitter current.

However several further aspects of device behaviour have to be investigated and at present nothing can be said concerning the required expenses for the continuous control of both emitters.

References

[1] R. Sittig, "Chances, Errors and Progress - A Survey on Power Device Development", Proc. of the Conf. to the 25th anniversary of the Laboratoire d´Analyse et d´Architecture des Systemes, Toulouse, Cèpadues-Editions, Toulouse, pp 159-175, 93

[2] ATLAS 2D Device Simulation Framework, User´s Manual, Silvaco International

[3] T. Ogura, A. Nakagawa, M. Atsuta, Y. Kamei and K. Takigami, "High-Frequency 6000 V Double-Gate GTO´s", IEEE Transactions on Electron Devices, Vol. 40, No. 3, pp 628-633, March 93

[4] F. Bauer, T. Stockmeier, H. Dettmer, H. Lendenmann and W. Fichtner, "On the Suitability of BiMOS High Power Devices in intelligent, snubberless Power Conditioning Circuits", Proc. of the 6th Internat. Symposium on Power Semiconductor Devices & IC´s, Davos, Switzerland, pp 201-206, 94

A Vector Level Control Function for Generalized Octree Mesh Generation

Tao Chen, Jeffery Johnson[†], Robert W. Dutton

AEL 231D, Integrated Circuits Laboratory, Dept. of Electrical Engineering,
Stanford University, Stanford, CA 94305
[†]IBM Microelectronics Division, Essex Junction, VT 05402

Abstract

Due to the complexity of 3D geometry structure, regular gridding algorithms often generate a very large amount of mesh points which requires an enormous amount of computational power for later device and process simulations. A vector level control function and related generalized octree method is presented in this paper, and it allows mesh refinement in selected directions only. Thus the grid points can be placed more optimally and that enables efficient device and process simulations.

1. Introduction

Some quadtree/octree based mesh generation schemes have been employed for finite element methods used in device [1] and process simulations [3] because of the simplicity of the tree data structure and its advantage in faciliating adaptive mesh refinement. The quadtree mesh generation scheme recursively partitions a square into four smaller, equal size squares; the octreetakes a cube and partitions it into eight smaller cubes. This set of squares or cubes form a tree. Usually a one-level difference rule is enforced for any two neighboring cells.

While this regular quadtree method works well in 2D, in part because of the limited complexity of geometry, the analogous 3D octree method runs into problems in many cases and generating excessive grid. For example, suppose one has a BJT with base dimension of $1 \times 1 \times 0.1 \mu m^3$. Using the regular octree method, more than $50 \times 50 \times 5$ grid points are needed for merely five grid planes in the z direction for this region alone. Clearly the amount of grid points required for more complex structures can quickly become excessive. A generalized octree method which solves this problem nicely is presented in this work. The new method is able to refine in only the required directions. At each refinement stage, there are seven choices. An octant can be refined in all xyz directions, or only in xy, or xz, or yz directions, or only in x, or y, or z directions.

2. Algorithm

To achieve this generalized refinement scheme, a vector level control function is defined. Like the scalar function in [2], an integer valued 3D vector function is now defined on the solid volume. It indicates the directions for which the refinement will be performed. For example, in the previous BJT example, a vector with a large z component and moderate x, y components can be assigned to the base region. This vector level control function also can be computed based on simulation errors. For example, if the derived solution error shows that denser mesh is needed in x direction, but not in y direction, a refinement with a level control function coupled with appropriate error estimator can readily be performed.

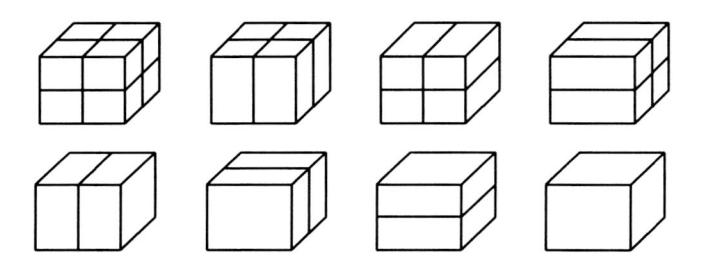

Figure 1: An octant can be refined in xyz, or xy, or xz, or yz, or only x, or y, or z directions.

Every octant also has a level vector (L_{oct}) associated with it. (The octant usually is no longer a cube.) For a given octant, to determine which directions are to be refined, the level control function (L_c) is evaluated on the center of the octant. If $L_{oct,x} < L_{c,x}$, then the octant will be refined in x direction, and similarly for the y, z directions. An octant will then be refined in one of the seven cases as shown in Fig. 1, or not at all. Each child octant will have a level vector L_{child} satisfying

$$L_{child,i} = L_{oct,i} + 1 \quad if\ i\ direction\ is\ refined, \quad (for\ i = x, y, z). \quad (1)$$

To abide the one-level difference rule, for every octant refined, its neighbors are found using the tree structure and updated. If an octant is refined in y direction, we update the four (six if refined in more than one direction) neighboring (left, right, bottom, top) octants. The level control function sets values on the centers of the neighboring octants by the following rule (this example assumes only refinement in y direction, similar rules follows easily).

$$if\ (L_{neighbor,c,y} < L_{oct,y} - 1),\ L_{neighbor,c,y} = L_{oct,y} - 1;$$
$$else\ do\ nothing.$$

After the initial tree generation, the entire octree is traversed again so that additional refinements can be performed if some values of the level control function are updated after the octants are refined. The octree generation process finishes when no more octants need to be generated. The final tree structure guarantees the one level difference in every direction and has the refinements in the directions required. Fig. 2 is a simple mesh using the generalized octree method.

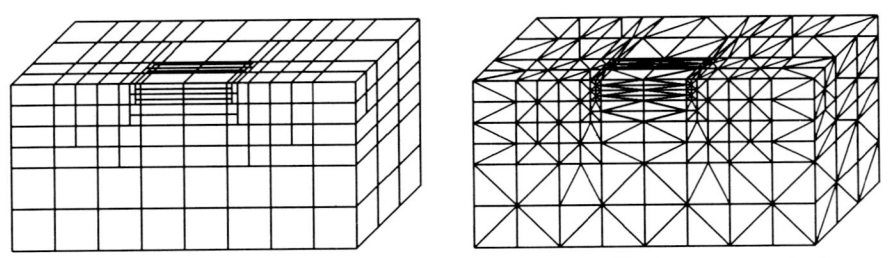

Figure 2: A simple 3D example using the generalized octree method.

Triangulation/Tetrahedralization is finally performed on the quadtree/octree. Extreme aspect ratio elements can often be generated. A local topological Delaunay transformation similar to the one proposed in [4] is then performed, where some pairs of tetrahedra are transformed into groups of three tetrahedra and the reverse process. This is analogous to the 2D case where edge swapping will produce a constrained Delaunay mesh for a convex hull. The overall mesh quality thus is improved by these topological transfromations.

3. Conclusion

This generalized octree method provides great flexibility for mesh refinement. It makes simulations with complicated structures easier due to its ability to place grids optimumly. It also provides a vehicle for the development of vector error estimators, where the refinements it performs could enhance the simulation accuracy efficiently while not adding too many points. Overall it reduces many grid points, and still generates a quality mesh.

Fig. 3 shows a mesh for a BJT used in high frequency RF circuits. It has very long but shallow base and emitter regions. The generalized octree mesh gives a very good fit according to the doping variation, and a large amount of grid points are concentrated along the junctions. Fig. 4 gives a cross section view of the mesh.

References

[1] P. Conti, N. Hitschfeld and W. Fichtner, "An Octree-Based Mixed Element Grid Allocator For Adaptive 3D Device Simulation", *NUPAD III, Technical Digest, 1991*.

[2] D. Yang, K. Law and R. Dutton, "An Automated Mesh Refinement Scheme Based On Level-Control Function", *NUPAD IV, Technical Digest, 1992*.

[3] Z. Sahul, R. Dutton and M. Noell, "Grid and Geometry Techniques for Multi-Layer Process Simulation", *Proc. SISDEP 1993*.

[4] N. Golias and T. Tsiboukis, "Three-Dimensional Automatic Adaptive Mesh Generation", *IEEE Trans. on Magnetics, Vol. 28, No. 2, March 1992*.

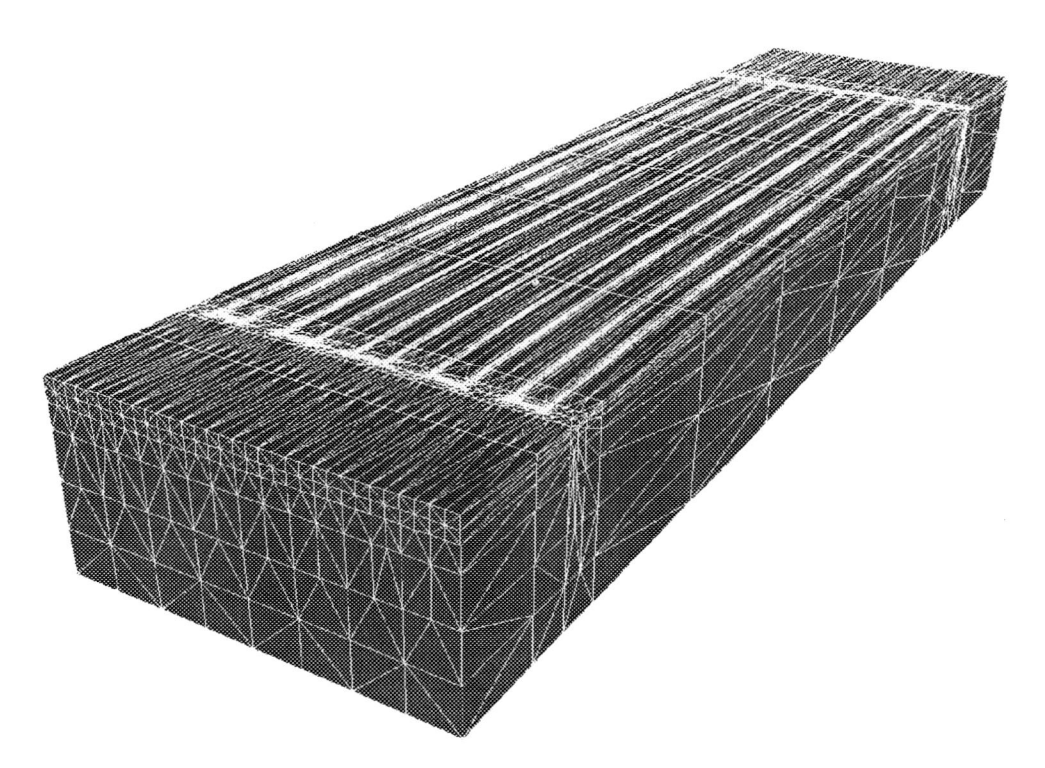

Figure 3: Mesh for a BJT used in high frequency RF circuits. The three base and two emitter regions are very long, but also very shallow. The generalized octree mesh concentrates a large amount of grid points along the junctions.

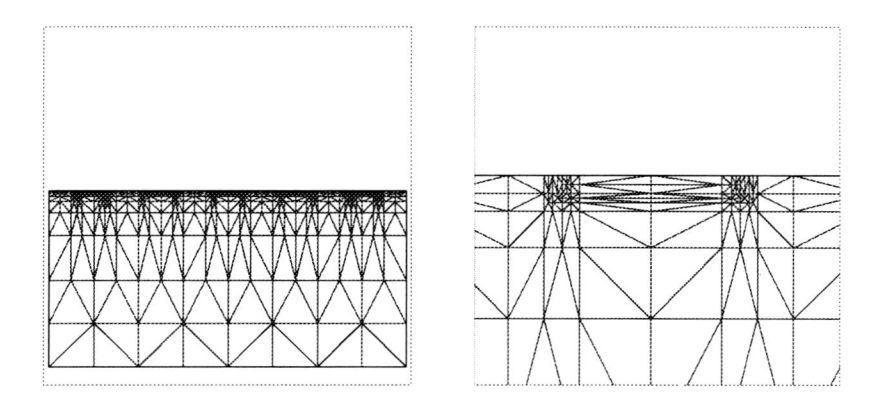

Figure 4: 2D cross section mesh of the BJT and its zoom in view of the emitter region.

Comparison of Hydrodynamic Formulations for Non-Parabolic Semiconductor Device Simulations

Arlynn W. Smith and Kevin F. Brennan

Microelectronics Research Center, Georgia Institute of Technology
Atlanta, GA USA 30332-0269

Abstract

This paper presents two non-parabolic hydrodynamic model formulations suitable for the simulation of inhomogeneous semiconductor devices. The first formulation uses the Kane dispersion relationship, $(\hbar k)^2/2m = W(1 + \alpha W)$. The second formulation makes use of a power law, $(\hbar k)^2/2m = xW^y$, for the dispersion relation. Hydrodynamic models which use the first formulation rely on the binomial expansion to obtain closed form coefficients. The power law formulation produces closed form coefficients similar to those under the parabolic band approximation.

1. Introduction

Current hydrodynamic models consist of a set of conservation equations derived by taking moments of the Boltzmann transport equation. During the derivation of the conservation equations the parabolic band approximation is used to obtain rather simple coefficients on the forcing terms in the flux equations. By relying on the parabolic band approximation higher order energy transport effects due to variations in the band structure are neglected. Accounting for band structure effects in hydrodynamic device simulation is important because parabolic models can not adequately account for high energy effects in semiconductors with non-parabolic band structures.

Non-parabolic hydrodynamic models have been reported for homogeneous material systems [1-4] using the Kane dispersion relationship [5]. The general functional form obtained is similar to parabolic hydrodynamic models with first order corrections on the diffusion term. However, the non-parabolic coefficient in the field term and the forcing terms due to non-uniform band structure are neglected in the other moment equations. Cassi and Riccò [6] introduced an alternative to the Kane relation in the form of a power law for the dispersion relationship. Instead of using a classical Kane dispersion law relating the energy and momentum, the band was fit over a specified energy range using two adjustable parameters. The approximations and assumptions implied by assuming the power law formulation were absent. It will be shown below that the power law dispersion relation leads to a more simplistic and compact formulation than the classical Kane expression.

2. Dispersion relations, concentrations, flux equations

The three dispersion relations considered in the derivation of the hydrodynamic conservation equations are; parabolic, Kane dispersion, and power law

$$\frac{(\hbar k)^2}{2m_e} = W \qquad \frac{(\hbar k)^2}{2m_e} = W(1+\alpha W) \qquad \frac{(\hbar k)^2}{2m_e} = xW^y \qquad (1)$$

where α is the non-parabolicity factor and x, y are fitting parameters over a specified energy range. If the power law is fit over the energy range $1.5 \leq W \leq 3.0$ eV as suggested in [6] the deviation in carrier concentration from the parabolic case and the Kane formulation is greater than 80% at most reduced energy values. However, when fit over the energy range $0 \leq W \leq 0.2$ eV the deviation is $\sim 2\%$, as seen in Figure 1 ($\alpha = 0.4789$). The case of the Kane dispersion relation using a binomial approximation is also included in the figure. Using the parabolic dispersion relation the particle flux equation is

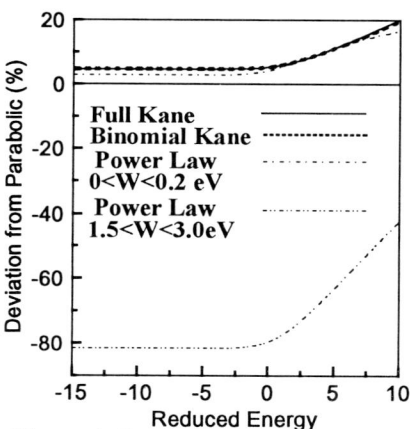

Figure 1. Deviation in carrier concentration from the parabolic case using different dispersion relations

$$-n\overline{v} = \mu KT \left[\frac{\mathcal{F}_{\frac{1}{2}}}{\mathcal{F}_{-\frac{1}{2}}} \right] \nabla n + \mu n \nabla \epsilon_c + \frac{5}{2} \mu n KT \frac{\mathcal{F}_{\frac{3}{2}}}{\mathcal{F}_{\frac{1}{2}}} \left[1 - \frac{3}{5} \frac{\mathcal{F}_{\frac{1}{2}}^2}{\mathcal{F}_{-\frac{1}{2}} \mathcal{F}_{\frac{3}{2}}} \right] \frac{\nabla T}{T} \qquad (2)$$

The flux equation for the Kane dispersion using a binomial expansion is

$$-n\overline{v} = \mu KT \left[\frac{\mathcal{F}_{\frac{1}{2}}^2 + \frac{10}{4}\alpha KT \mathcal{F}_{\frac{3}{2}} \mathcal{F}_{\frac{1}{2}}}{\mathcal{F}_{\frac{1}{2}} \mathcal{F}_{-\frac{1}{2}} + \frac{15}{4}\alpha KT \left(\mathcal{F}_{\frac{1}{2}}^2 + \frac{\mathcal{F}_{-\frac{1}{2}} \mathcal{F}_{\frac{3}{2}}}{2} \right)} \right] \nabla n + \mu \left[\frac{\mathcal{F}_{\frac{1}{2}} - \frac{21}{4}\alpha KT \mathcal{F}_{\frac{3}{2}}}{\mathcal{F}_{\frac{1}{2}} + \frac{15}{4}\alpha KT \mathcal{F}_{\frac{3}{2}}} \right] n \nabla \epsilon_c$$

$$+ \frac{\mu n KT}{\left(\mathcal{F}_{\frac{1}{2}} + \frac{15}{4}\alpha KT \mathcal{F}_{\frac{3}{2}} \right)\left(\mathcal{F}_{-\frac{1}{2}} + \frac{15}{4}\alpha KT \mathcal{F}_{\frac{1}{2}} \right)} \left[\frac{5}{2} \mathcal{F}_{-\frac{1}{2}} \mathcal{F}_{\frac{3}{2}} - \mathcal{F}_{\frac{1}{2}}^2 - \frac{\mathcal{F}_{\frac{1}{2}}^3}{2\left(\mathcal{F}_{\frac{1}{2}} + \frac{15}{4}\alpha KT \mathcal{F}_{\frac{3}{2}} \right)} \right] \frac{\nabla T}{T} \qquad (3)$$

The flux equation obtained using the power law dispersion relation is

$$-n\overline{v} = \frac{2\mu}{3xy^2}(KT)^{2-y}\frac{\left(\frac{y}{2}+1\right)\Gamma\left(\frac{y}{2}+1\right)\mathcal{F}_{\frac{y}{2}}}{\left(\frac{3}{2}y-1\right)\Gamma\left(\frac{3}{2}y-1\right)\mathcal{F}_{\frac{3}{2}y-2}}\nabla n + \frac{(2-y)\mu}{xy^2}(KT)^{1-y}\frac{\Gamma\left(\frac{y}{2}+1\right)\mathcal{F}_{\frac{y}{2}}}{\Gamma\left(\frac{3}{2}y\right)\mathcal{F}_{\frac{3}{2}y-1}}n\nabla\epsilon_c$$

$$+\frac{\mu n}{xy^2}(KT)^{2-y}\left[\frac{-y\left(\frac{y}{2}+1\right)\Gamma\left(\frac{y}{2}+1\right)\mathcal{F}_{\frac{y}{2}}}{\left(\frac{3}{2}y-1\right)\Gamma\left(\frac{3}{2}y-1\right)\mathcal{F}_{\frac{3}{2}y-2}} - \frac{(4+y)\Gamma\left(\frac{y}{2}+2\right)\mathcal{F}_{\frac{y}{2}+1}}{3\Gamma\left(\frac{3}{2}y\right)\mathcal{F}_{\frac{3}{2}y-1}}\right]\frac{\nabla T}{T} \quad (4)$$

The derivation of all these three equations is given in reference [7]. Similarly the electron energy flux equations using the three dispersion relations can also be formulated [7]. The equations were discretized by using the methods in references [8] and [9]. For the exponential terms in the discretization equation with factors composed of powers of the temperature we have made the assumption that the position dependent temperature can be replaced by the average nodal temperature.

3. Results

Figures 2 through 5 show the results of applying the model to ballistic diodes of both Si and GaAs. Figure 2 shows the current for a Si ballistic diode using the non-parabolic formulations is lower than the parabolic case, Figure 3 shows that the energy is also lower at a 1 volt bias. Figures 4 and 5 show the same trends for the GaAs ballistic diodes. One should note that at low biases the power law formulation predicts lower current than the α formulation until a certain bias voltage. At low bias, the devices are close to equilibrium and the carriers are relatively cold. Consequently, the system responds as in the drift-diffusion case resulting in a greater current for the α case than the power law case. At higher biases, the effects of carrier heating are more important, and the situation reverses, since the full hydrodynamic results dominate the 'effective' mobility.

4. References

[1] R. A. Stewart and J. N. Churchill, "A fully nonparabolic hydrodynamic model for describing hot electron transport in GaAs," Solid State Elec., Vol 33, #7, pp. 819, 1990.

[2] R. Thoma, A. Emunds, B. Meinerzhagen, H.-J. Peifer, and W. L. Engl, "Hydrodynamic equations for semiconductors with nonparabolic band structure," IEEE Trans Elec. Dev., Vol. 38, #6, pp. 1343, 1991.

[3] T. J. Bordelon, X.-L. Wang, C. M. Maziar, and A. F. Tasch, "Accounting for band structure effects in the hydrodynamic model: a first-order approach for silicon device simulation," Solid State Elec., Vol. 35, #2, pp. 131, 1992.

[4] D. Chen, E. C. Kan, U. Ravaioli, C.-W. Shu, and R. W. Dutton, "An improved energy transport model including nonparabolicity and non-maxwellian distribution

effects," IEEE Elec. Dev. Letts., Vol. 13, #1, pp. 26, 1992.
[5] E. O. Kane, "Band structure of indium antimonide," J. Phys. Chem. Solids, Vol. 1, pp. 249, 1957.
[6] D. Cassi and B. Riccò, "An analytical model of the energy distribution of hot electrons," IEEE Trans Elec. Dev., Vol. 37, #6, pp. 1514, 1990.
[7] A. W. Smith and K. F. Brennan, "Non-parabolic hydrodynamic formulations for the simulation of inhomogeneous semiconductor devices," submitted for publication in Solid State Electronics
[8] A. W. Smith, "Light confinement and hydrodynamic modelling of semiconductor structures using volumetric methods," Ph.D. Dissertation, Georgia Institute of Technology, 1992.
[9] A. W. Smith and A. Rohatgi, "Non-isothermal extension of the Scharfetter-Gummel technique for hot carrier transport in heterostructure simulations," IEEE Trans. CADICS, Vol. 12, #10, pp.1515, 1993.

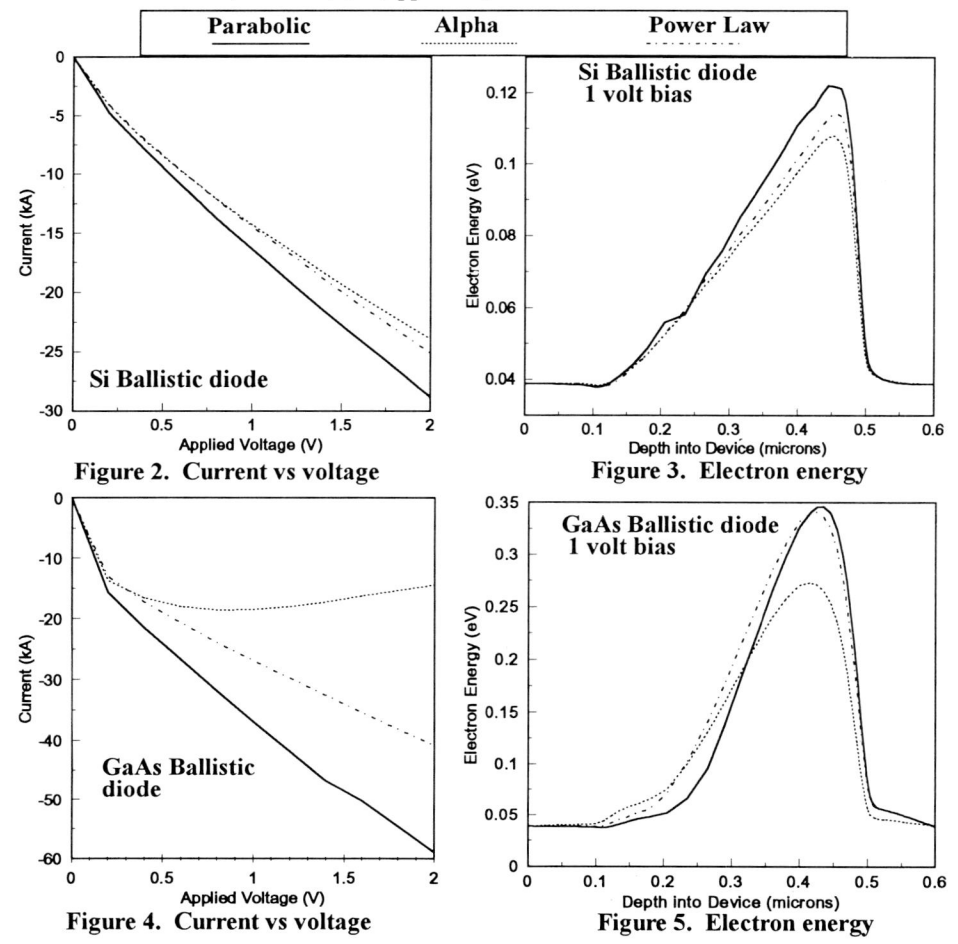

Figure 2. Current vs voltage

Figure 3. Electron energy

Figure 4. Current vs voltage

Figure 5. Electron energy

Influence of Analytical MOSFET Model Quality on Analog Circuit Simulation

M. Miura-Mattausch, A. Rahm, and O. Prigge

Corporate Research and Development, Siemens AG, 81730 Munich, Germany

Abstract

Though the importance of CAD increases, the analog circuit design is still mostly done by experience. It is known that this is because of insufficient model quality, or quality of extracted parameter values. We have developed a new MOSFET model based on the drift-diffusion approximation. By comparing a conventional piece-wise model with our precise model, critical shortage of the conventional models, which restricts the application of CAD for analog circuits, is demonstrated.

1. Introduction

Different from digital circuits, analog circuit performances are very much dependent on small signal conductances [1]. Describing these values correctly is a key in predicting the circuit performances correctly. However, these small signal values are sensitive to small deviations of applied voltages and also to small deviations of technological values. This makes prediction difficult.

Simplified piece-wise analytical MOSFET models have been developed and mostly used for circuit simulation. The main simplification of the models is the drift approximation neglecting the diffusion contribution. This simplification violates smooth transitions to different applied voltage regions. To get smooth transitions some fitting parameters are investigated [2]. To describe the saturation behavior the channel length modulation concept is introduced modeling with additional fitting parameters. Thus the piece-wise models describe transistor characteristics empirically, exactly where analog circuit performances are determined.

Circuit Performance with Our Model

Our model is based on the drift-diffusion model under the charge-sheet approximation [3]. All equations needed for circuit simulation are described analytically as functions of surface potentials at the source side and the drain side, which are solved iteratively during the circuit iteration. The key idea of our model development is to put as much as physics into the equations describing these surface potentials. Due to the inclusion of the lateral electric field in the equations, resulting surface potential values are dependent not only on applied voltages but also on channel length. It has been shown that our precise MOSFET model, in spite of the two additional iteration

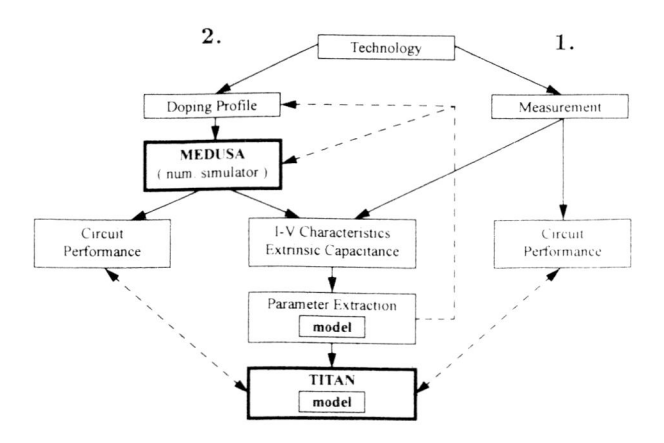

Figure 1: Flow chart of our overall simulator. TITAN is our circuit simulator.

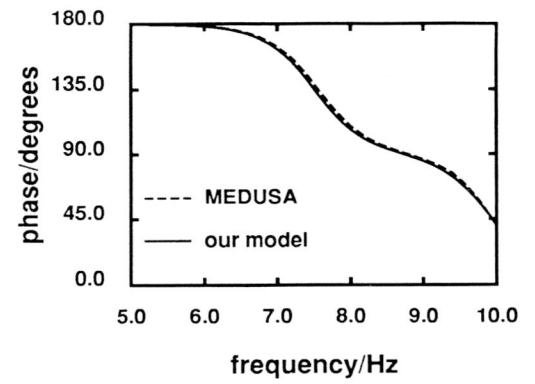

Figure 2: Comparison of the simulated phase of a linear amplifier with our model and MEDUSA.

procedures, reduces calculation time drastically in comparison with a conventional model.

As shown in Fig. 1 there are two approaches to verify the model quality. One is to produce chips with a certain technology and measure their $I-V$ characteristics in order to perform parameter extraction for circuit simulation.

The simulated circuit performances are compared with measurements. The other possibility is to use a numerical simulator in stead of measurements. We have done first the second approach. An advantage of this second approach is that we can know the transistor structure exactly even the doping profile, which can be used further for studying the influence of technological deviations on circuit performances.

Figure 2 shows a calculated phase of a linear amplifier for the channel length $L_{poly} = 1\mu m$. Figure 3 shows the influence of the oxide thickness T_{ox} variation on the circuit performance. For comparison only the T_{ox} value both in our model parameters and MEDUSA [4] input file was varied by ± 10 % from nominal thickness of $7nm$. By reducing the T_{ox} thickness the node voltage of the circuit transistor decreases, which

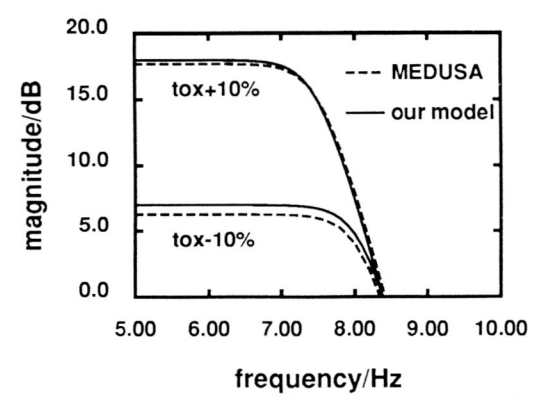

Figure 3: Influence of T_{ox} variation on the performance of a linear amplifier.

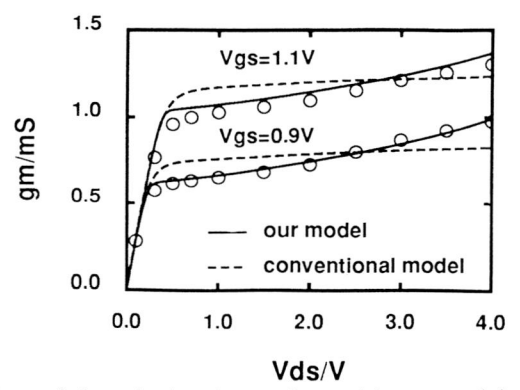

Figure 4: Comparison of the calculated g_m values with our model and a conventional model. Open circles are measurements.

causes the increase of g_{ds}. This is the reason of the amplitude decrease. Our model can predict the circuit performance not only for the nominal T_{ox} value but also the variation correctly.

2. Comparison with a piece-wise model

For the comparison with a piece-wise model we performed the first approach shown in Fig. 1. The model parameter values are extracted for a measured data set with two different models, one with our model and the other with a conventional piece-wise model based on the similar concept as BSIM3. Both models reproduce the measured $I-V$ characteristics well. Figure 4 compares the g_m values of these two models in the moderate inversion region for $L_{poly} = 0.9\mu m$. The conventional model cannot describe g_m correctly for a wide range of applied voltages. This is due to artificial fitting parameters to smooth the transition regions. These parameters can not describe complicated potential behaviors for a wide range of applied voltages. This causes inaccuracy in extracted parameter values. Figure 5 shows a comparison of a simulated amplitude of an operational amplifier. The input current is chosen so that the V_{gs} values on transistors are around 1.0V. The two models show the difference

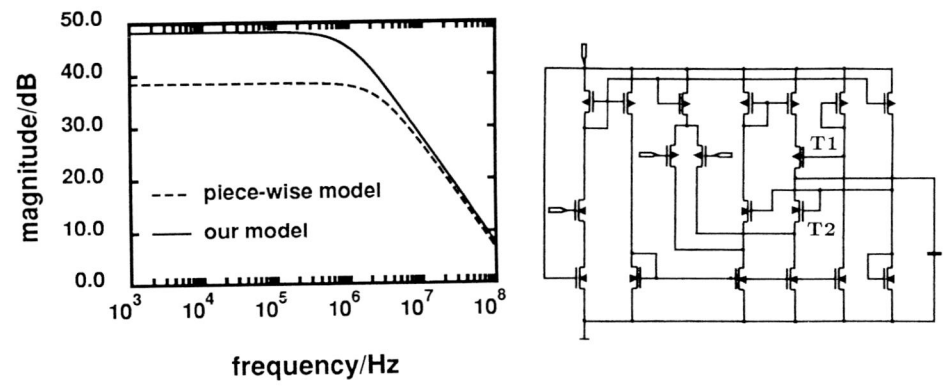

Figure 5: Simulated amplitude of an operational amplifier shown in the right.

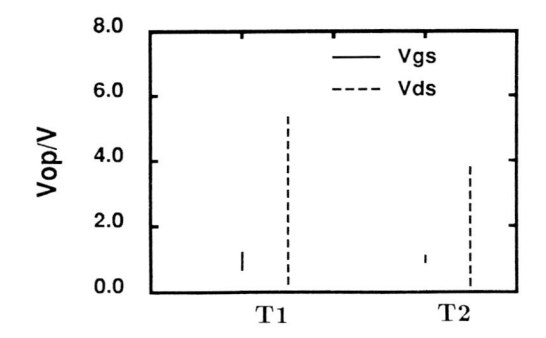

Figure 6: Voltage variation on two key transistors in the operational amplifier shown in Fig. 5 for the frequency range of $10^3 - 10^8$ Hz.

of 10dB in amplitude. Figure 6 shows the voltage variations of two key transistors in the amplifier for the frequency range of $10^3 - 10^8$ Hz. As can be seen the variation of the voltage between the source and the drain V_{ds} is large. For relative small frequency the V_{ds} values stay at their maximum, where the maximum inaccuracy of g_m occurs in the conventional model.

References

[1] Y. P. Tsividis and K. Suyama, "MOSFET modeling for analog circuit CAD: Problems and Prospects," *IEEE J. Solid-State Circuits*, vol. 29, no. 3, pp. 210-216, 1994.

[2] J. A. Power, "An enhanced SPICE MOSFET model suitable for analog applications," *IEEE Trans. Computer-Aided Design*, vol. 11, no. 11, pp. 1418-1424, 1992.

[3] M. Miura-Mattausch, U. Feldmann, A. Rahm, M. Bollu, and D. Savignac, "Unified complete MOSFET model for analysis of digital and analog circuits," *Proc. IC-CAD*, pp. 264-267, 1994.

[4] MEDUSA User's Guide, RWTH Aachen, Germany, 1989.

2-D Adaptive Simulation of Dopant Implantation and Diffusion

Chih-Chuan Lin and Mark E. Law

Department of Electrical Engineering, University of Florida
Gainesville, FL 32611
Phone: 904-392-6276 Fax: 904-392-8381
lcc@tcad.ee.ufl.edu law@tcad.ee.ufl.edu

Abstract

This paper describes the techniques for adaptive grid generation for dopant implantation and diffusion in process simulation.

1. Introduction

Mesh generation and adaption for solving dopant diffusion in process simulation is a difficult task, complicated by both the moving boundaries of oxide growth and the time dependence of the solutions. For both computational and ease of use reasons, automatic mesh generation and discretization error control is desirable. This paper describes an approach based on local error estimates [1] to refine the mesh. The results presented in this paper extend our previous work [2] to two-dimensional problems. The implementation of this approach is done in a process simulator FLorida Object-Oriented Process Simulator (FLOOPS).

2. Nodal Error Estimation

In the one dimensional part of this work, the Bank-Weiser [3] error estimator appears to be the most promising. For process simulation, the discretization error is proportional to the curvature. This is easy to compute and use. The error is used to implement the grid adaption algorithm in two dimensions and also applied to mesh refinement in self-adaptive grid generation procedure. In the benchmark, the parameter for refinement is varied from 1 to 0.1 and the coarsening parameter is one tenth of the refinement parameter. The result shows that the dopant concentration errors and profile junction depth errors are bounded and controlled by the refinement within 8% and 6% respectively when compared to solution computed on a fixed grid spacing of 10 Å. The total dose interpolation errors, critical for adaptive simulation, are within 0.2%.

3. Grid Quality Issue on Local Refinement

A smoothing technique is required to improve the grid quality after local refinement. Two approaches can be used to solve this problem, one is by Laplacian smoothing, the other is by solving an optimization problem with regard to triangulation quality[4]. The global smoothing procedure moves every grid node in the mesh to a new location, and requires the interpolation of solutions onto the new position. Two approaches for dopant interpolation have been investigated. The first is based on the dose conservation law. The second approach for solution interpolation utilizes the upwinding concept. The primarily result of utilizing smoothing and interpolation technique on the local refinement was presented for single dopant simulation[5]. In Figs 1-2, a test is carried out by simulation of a SideWAll-Masked Isolation (SWAMI) structure with a channel stop implant by Boron dose of 5×10^{15} cm^{-2}, 30 keV. The final structure is grown for 60 minutes in 1000 °C anneal under a wet oxygen ambient. The resulting structure demonstrates the capability of mesh adaption for dopant diffusion with moving boundaries. During the 60 minutes diffusion, mesh smoothing is carried out in each time step. In Fig. 1, the computed junction depth by dose normalization technique is shallower than the ones by upwinding technique. This is due the restriction of grid movement imposed upon dose normalization approach to ensure the validity of local dose conservation law. In general, the usage of the upwinding technique for interpolation is preferred since it allows faster computation. The two smoothing techniques are about the same though the Laplacian average approach shows slightly better improvement of final grid quality. The most promising is the smoothing by triangle quality optimization and the interpolation by upwinding. The smoothing is also applied to the oxide layer during oxidation. This improves the grid quality along the silicon/ oxide interfaces.

4. Applications

In order to exercise the robustness of this adaptive algorithm, the front-end to fabricate a CMOS inverter by a twin-well process is simulated with FLOOPS and the resulting grid for the isolation is shown in Fig. 3. This shows the grid adaption capability beyond the silicon layer. It allows the treatment of grid refinement or smoothing in multiple materials. The multi-layer and multi-dopant aspects therefore can be illustrated through this realistic example.

5. Conclusion

A new approach for adaptive simulation for dopant in implantation and diffusion is presented. It's suitable for multi-layer and multi-dopant in 2-D and can be extended for 3-D process simulation.

References

[1] M. Bieterman, I. Babuska, "The finite element method for parabolic equations, I. A posteriori error estimation," *Num. Math.*, vol. 40, p.339,1982.
[2] C. C. Lin, M. E. Law, and R. E. Lowther, "Automatic Grid Refinement and Higher Order Flux Discretizations of Diffusion Modeling," *IEEE Trans. Computer-Aided Design*, vol. CAD-4, No. 12, p. 436, Aug. 1993.
[3] R. E. Bank, A. Weiser, "Some a posteriori error estimators for elliptic partial differential equations," *Math. Computation*, vol. 44, p.283, 1985.
[4] C. C. Lin, M. E. Law, "Mesh Adaption and Flux Discretizations for Dopant Diffusion Modeling," *NUPAD V Proceeding*, p. 151, Honolulu, June,1995.
[5] R. E. Bank, "Moving Mesh," *NSF/SEMATECH Workshop on Gridding in VLSI TCAD*, section V, Mar. 1994.

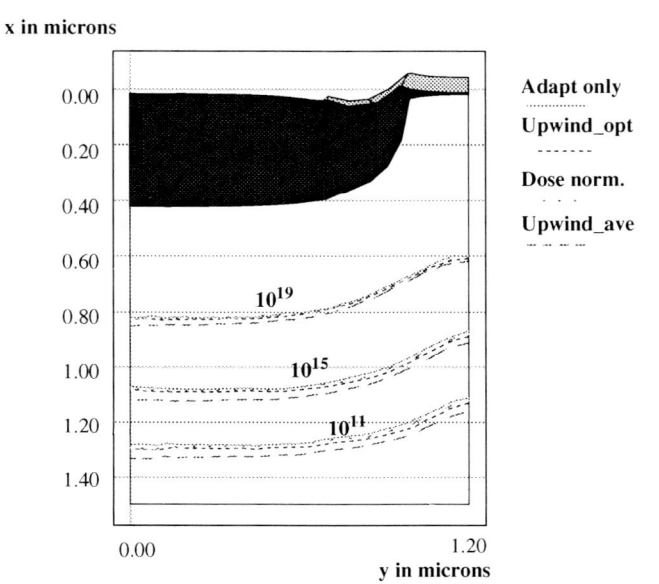

Fig. 1 The contours of dopant solution after the growth of SWAMI.

Fig. 2 The resulting smoothed mesh after the growth of SWAMI.

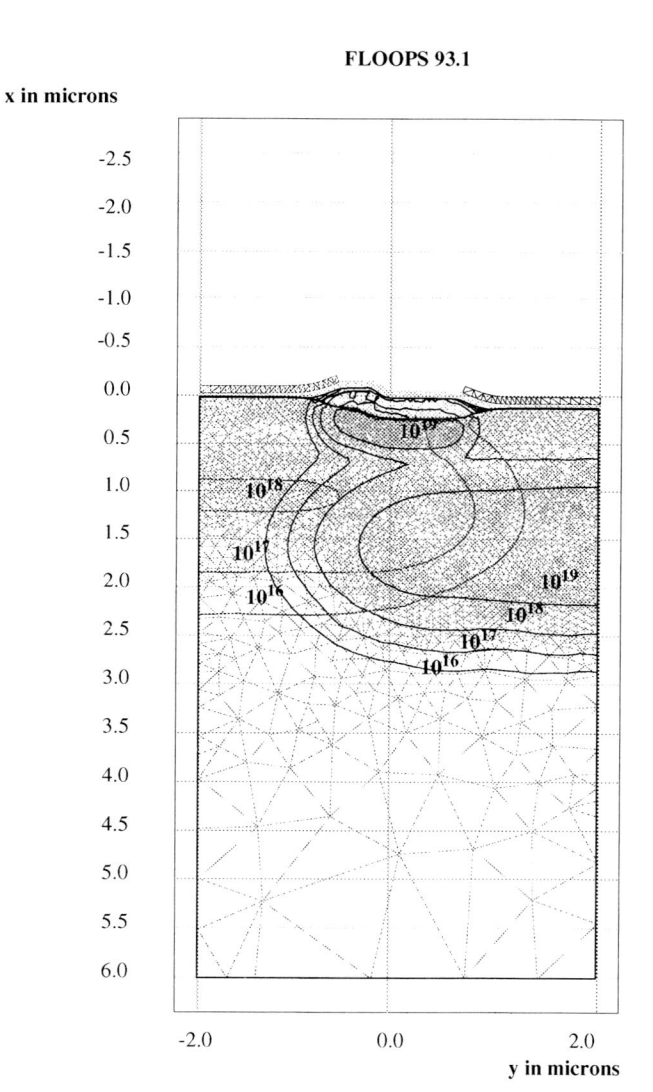

Fig. 3 The wells and isolation structure for a twin-well CMOS process.

Optimization of a Recessed LOCOS using a tuned 2-D process simulator

G.P. Carnevale[a],
P. Colpani[b], A. Marmiroli[b], A. Rebora[b],
A. Tixier[c]

[a]Università degli studi di Pavia, Pavia, Italy,
[b]SGS-THOMSON Microelectronics, Central R&D, Agrate Brianza, Italy,
[c]Université de Lille, Lille I, France

Abstract

The application of the conventional LOCOS technique (LOCal Oxidation of Silicon) to grow oxide structures for IC device isolation is no longer effective below 0.5 μm. In order to support the optimization of an alternative isolation technique, adequate simulations tools are required. In this work we describe an accurate tuning of 2-D simulation program, which uses a viscoelastic model for the oxidation of silicon, and its application to the optimization Recessed LOCOS isolation.

1. Introduction

LOCOS is the most common and simple isolation technology which has been used successfully over the last 20 years. There are several reasons for this success: the reduced number of materials involved and the use of simple and well controlled process steps (oxide growth, nitride deposition and nitride etch) are the most important ones ensuring a high manufacturability. Below the 0.5 μm limit, the lack of planarity, the lateral encroachment and the field oxide thinning prevent the use of this isolation scheme and force to the development of other structures. Due to the manufacturing process complexity and costs, simulation tools have become more and more necessary to save time and money. The aim of this work has been the calibration of a 2-D process simulation tool (*ATHENA*) [1] through the fitting of LOCOS profiles, and the application of the tuned program to the geometrical characterization and optimization of a different isolation scheme, a Recessed LOCOS structure.

2. Tuning of coefficients

Several LOCOS isolation samples have been processed by growing a 150 Å padoxide followed by a nitride layer deposition of various thicknesses (from 900 Å to 2000 Å). After etching, a wet field oxidation has been performed at different temperatures (from 920°C to 1100°C). All samples have been submitted to SEM and TEM analysis. Oxidation was simulated using the stress-dependent oxidation parameters extracted from [2]. The accurate fitting work which has been developed to produce this set is in a good agreement with previous fitting works [3], [4]. Those parameters are the activation volumes of the stress-dependent diffusivity of the oxidant in the silicon dioxide (V_d), the stress-dependent reaction rate (V_k) and the stress dependent oxide viscosity (V_c) (see table 1). Furthermore, other two important parameters are the nitride and the oxide viscosities. Their proper values have been extracted through a careful fitting of the experimental profiles (see as two examples fig. 4, 5). The fitted values are reported in table 1. As a result, the dependence of the oxide viscosity stress-independent factor shows an exponential behaviour with an activation energy of 2.3 eV. This value might be correlated with the binding energy of adjacent oxide layers. At high temperature (1000°C, 1100°C) a good agreement between the experimental photos and the simulated profiles has been achieved, while at low temperature (920°C) the result is only partially satisfactory. For this process condition the viscous equation implemented in the simulation program reaches its limit and a non-linear viscoelastic description of the nitride behaviour, similar to the oxide, is required [2].

3. Recessed LOCOS optimization

The set of coefficients defined with the previous fitting allows us to elaborate a Design Of Experiment (DOE) on the Recessed LOCOS. An Empirical Model Building program, ULYSSES [5], has been used to perform the design varying five process factors (temperature, nitride and padoxide thicknesses, etch depth and etch angle): the extracted responses (fig. 1) allow a complete morphological characterization of the isolation profile. A quadratic design has been chosen and the interpolating polynomial of the responses presents satisfactory correlation coefficients. By means of contour and pareto plots of the responses a clear understanding of the most important process factors effect has been obtained. Field oxide temperature, nitride thickness and etch depth are the most relevant ones (fig. 2). The Recessed LOCOS optimized condition has been achieved by minimizing the bird's beak under the nitride mask (l), the height of the bird's beak head ($h1$) and the difference between the bird's beak head and the field oxide surface (*diff*) (see fig. 1). The results obtained show an unexpected independence of the etch angle in the silicon on the responses (fig. 3). Besides, the optimization algorithm suggested a high oxidation temperature, two different conditions for the ratio of the nitride and padoxide thicknesses in the range $[\frac{7}{1} \div \frac{9}{1}]$, and an etch depth in the silicon substrate. One of the particular morphological conditions proposed has been processed and a satisfactory agreement with the simulation has been obtained (see fig. 6).

4. Conclusion

The physical parameters of the oxidation with a nitride mask have been successfully fitted; the tuned process simulator has been used with a DOE program which has allowed to optimize a Recessed LOCOS isolation technology, with a strong reduction of development time and costs. Ultra large scale integration requires new isolation schemes which introduce a greater complexity of development and manufacturing. Therefore, simulation tools such as EMB, RSM, DOE [5] and calibrated process simulators becomes more and more useful in development activities.

5. Acknowledgments

We are grateful for the SEM pictures of the LOCOS structures provided by *LETI* laboratory in Grenoble and for TEM analysis provided by *IMETEM* in Catania.

References

[1] Silvaco International *"ATHENA 2D Process Simulation Framework User's Guide"*
Santa Clara, California, mar 1994

[2] V. Senez, D. Collard, P. Ferreira, B. Baccus *"Simulation of advanced field isolation using calibrated viscoelastic stress analysis"*
IEDM, pp. 881-884, 1994

[3] C.S. Rafferty *"Stress effects in silicon oxidation: simulation and experiments"*
Ph. D. Thesis, pp. 184-186, Stanford University, dec 1989

[4] P. Sutardja, W.G. Oldham
IEEE Trans. Electron Devices, vol. ED-36, pp. 2415-2421, nov 1989

[5] *"ULYSSES User's Guide"*
IMEC - SGS-Thomson, jun 1993

Table 1: The equation and the parameters of the oxidation model

Fig. 1: Geometrical parameters of a Recessed LOCOS

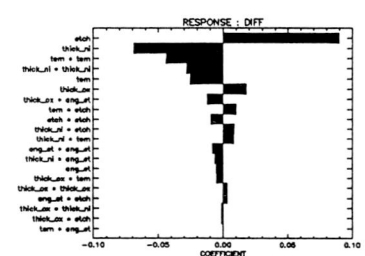

Fig. 2: Pareto plot of 'diff' response
(difference between the head and the field oxide)

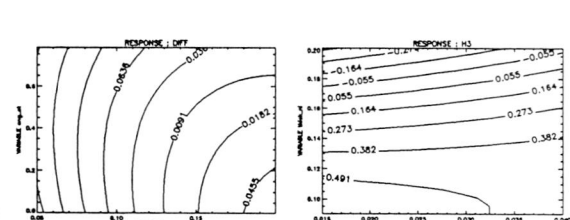

Fig. 3: Contour plot of some of the most relevant factors
(ang_et: etch angle in silicon substrate, etch: depth etch, thick_ni: nitride thickness, thick_ox: padoxide thickness)

G. P. Carnevale et al.: Optimization of a Recessed LOCOS Using a Tuned 2-D Process   289

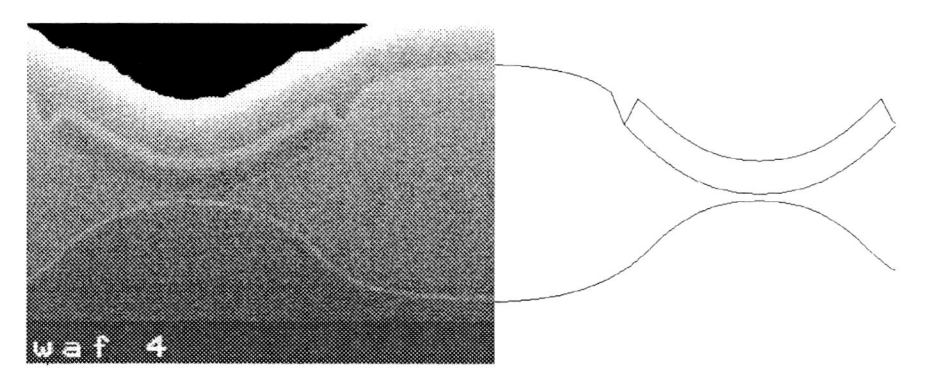

Fig. 4: SEM photo vs simulated profile (temperature 1100 °C, nitride thickness 900 Å)

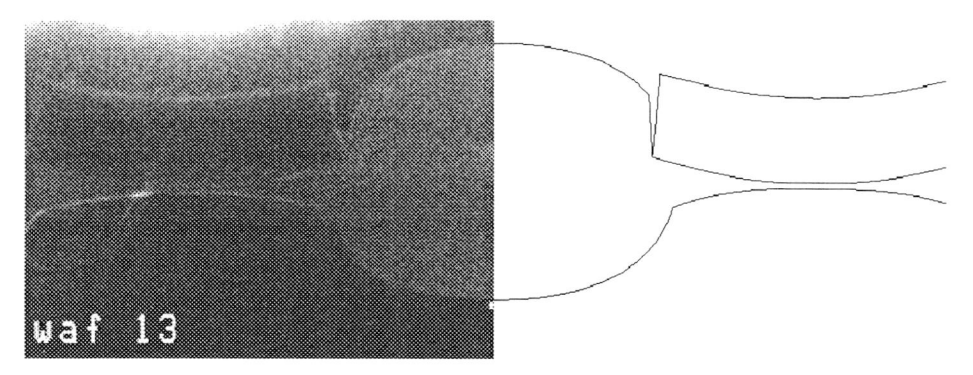

Fig. 5: SEM photo vs simulated profile (temperature 920 °C, nitride thickness 2000 Å)

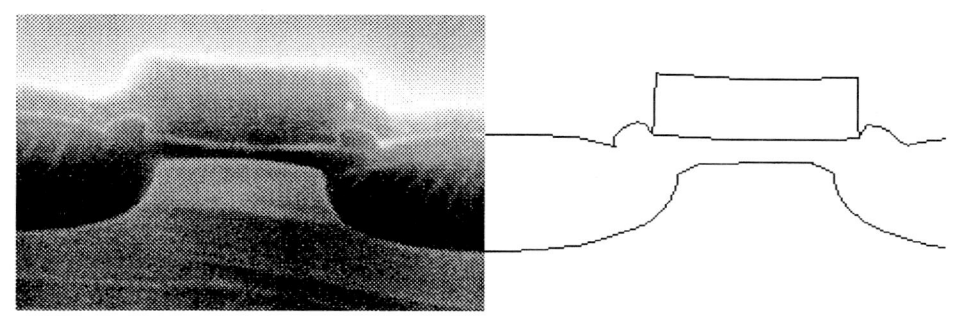

Fig. 6: Recessed SEM photo vs simulated profile

Simulation of Complex Planar Edge Termination Structures for Vertical IGBTs by Solving the Complete Semiconductor Device Equations

M. Netzel[a], R. Herzer[b]

[a]Institute for Solid State Electronics, TU Ilmenau,
PF 0565, D-98684 Ilmenau, GERMANY
[b]SEMIKRON Elektronik GmbH Nürnberg
Sigmundstrasse 200, D-90431 Nürnberg, GERMANY

Abstract

In this paper calculations of blocking characteristics for edge termination structures of 600V- and 1200V-IGBTs by means of the device-simulator ToSCA are presented. Because of the efficient grid concept and the properties of ToSCA it becomes possible to solve the complete device equations for very complex grids in relatively short time.

1. Introduction

For the extension of a medium-voltage $1.5\mu m$-MOS-process to higher voltages ($1200V$ up to $2000V$), IC-compatible edge termination structures had to be developed. The shallow pn-junctions ($3...6\mu m$) and the small total passivation layer thickness of 2.5 μm of the process makes a design of an efficient edge structure more sophisticated. But so-called OFP-FLR-structures (Offset Field Plate - Field Limiting Rings) are realizable in spite of the restrictions of the used process. For breakdown voltages up to $2000V$ a lot of field rings are necessary resulting in large and complex structures combined with small sub-elements. In the past amongst analytical methods, mainly field calculation programs solving only Poisson's equation (Depletion model) were used for the simulation of such complex edge structures [1],[2]. These tools are in most cases non-professional, a continuous support and extension of features is only restrictedly possible. Furthermore an including of the mobile carriers to detect for example channel forming or for transient calculations, is very advantageous. So it would be desirable to be able to apply the available professional and semiprofessional device simulators [3],[4],[5], to solve the complete semiconductor device equations. For the following reasons a calculation of complex edge termination structures with these tools is made more difficult:

1. the implemented mesh generators/-editors do not allow a discretization with the necessary resolution and an acceptable number of nodes

2. altogether too little number of nodes

3. convergence problems (especially if floating plates are included) and very long calculation time.

In the paper, calculations of planar edge structures by means of the device simulator ToSCA [5] and further tools of the simulation system PRODESI [6],[7],[8] are presented, for which the problems mentioned above were solved or partly overcome.

2. Mesh generation and mesh editing methods

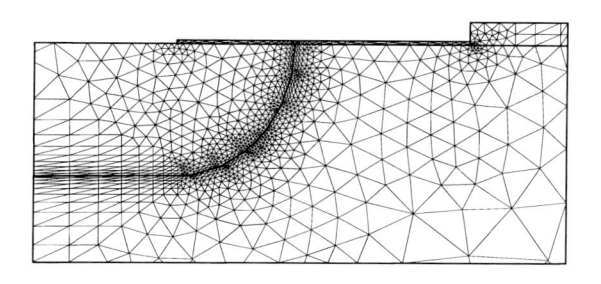

Figure 1: Grid detail of an edge structure

The triangulation was carried out with a novel mesh generator/-editor named TRIGEN [7] and TRIMAN [8], which generates automatically the complete mesh using a global description of the device. This description contains the geometry in terms of polygon curves, the doping profile, the boundary conditions and control parameters. First of all the program tries to adsorb anisotropic triangles at suitable edges or curves located in any direction. So inversion channels, pn-junctions and thin layers can be discretized with a small number of nodes. Remaining regions are covered with isotropic triangles. In Fig. 1 a detail of a termination structure illustrating the grid concept is shown. These mixed isotropic/anisotropic meshes are changeable afterwards in a wide range with the mesh editor TRIMAN. Besides the advantageous properties of isotropic and anisotropic sub-regions are maintained after all refinement procedures. Because of the connection of isotropic and anisotropic sub-meshes an optimum resolution with an acceptable number of nodes is reached. The numerical properties of the grids are not deteriorated by the anisotropic triangles. Compared to commonly used grid concepts the node saving varied from 1.5...4 depending on the structure, respectively. For the device in Fig. 2 ($1200V$ edge structure) the number of nodes would increase by a factor of 2.1 if an isotropic triangulation is performed.

3. Simulation of the electrical behaviour with ToSCA

The electrical simulation was realized by solving the complete equations of the drift-diffusion model. Additionally to the common mobility, lifetime and intrinsic conduction models the avalanche generation and the different recombination models are included in the simulation. Even with very large meshes (30.000-40.000 triangles) a calculation is possible on workstation with a medium performance in acceptable time. This is attained by the following features of ToSCA, accessory to the efficient grid concept:

1. In ToSCA the complete memory-extensive Newton Jacobian matrix is never used. The coupled equation system is solved successively by a block iteration procedure derived from Gummel's method. So considerably less memory is needed than in comparable simulators.

2. The calculation of the contact currents is done by using the Gaussian theorem and a test function. Because of this very stable contact currents and an accurate current sum are reached resulting in fast convergence behaviour.

3. Due to an automatic switching between the Newton's and Gummel's method the advantages of both methods are combined resulting in a short calculation time. Steps from 20 up to $40V$ are possible without any problems. The switching thresholds are adjustable by the user to an optimal adaption of the problem. In case of involving surfaces states for example a stress of Newton's method results in a reduced calculation time.

4. ToSCA is available in source code completely. So the maximum number of nodes/triangles can be defined by two global parameters before compiling the program. There is no limit for these parameters, so the maximum node number depends theoretically only on the memory of the workstation.

On an Alpha AXP workstation (DEC 3000/300X) with 64 MB memory devices up to 35.000 triangles were calculated with a calculation time of 30 minutes up to 6 hours and breakdown voltages ranging from 600 to $1800V$.

4. Presentation of the simulated examples

In Fig. 2 an OFP-FLR-edge structure for $1200V$-IGBTs with nine non-equidistant steps and a breakdown-voltage of $1510V$ is shown. For three versions of this edge structure the blocking capability was calculated in pseudo-3D-case by using cylindrical polar coordinates. Thereby the blocking capability of the chip corners can be calculated accurately because of their circular shape. In Fig. 3 the normalized

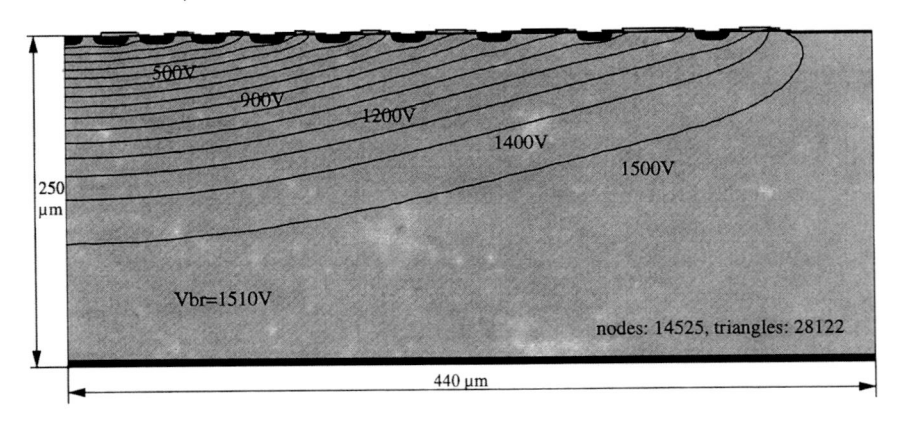

Figure 2: 1200V-OFP-FLR edge structure with non-equilateral ring-spacing

breakdown-voltages is shown for three examples with regard to their breakdown-voltages in ideal two-dimensional case. These three versions differ in the arrangement and the penetration depths of the field rings. As visible in Fig. 3 an inner radius of 150...200 microns is necessary to prevent a considerable decrease of the blocking capability in the corners. With an increase of the penetration of the field rings the breakdown-voltage goes slightly better, demonstrated by version 1 and 2.

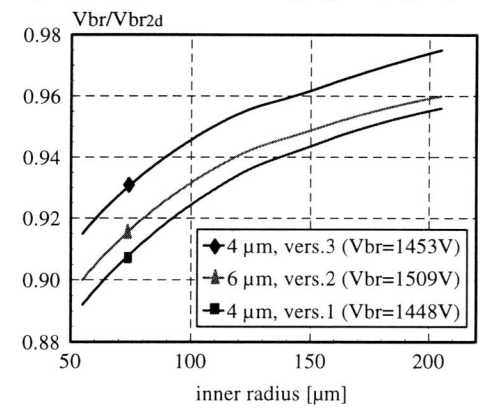

Figure 3: Blocking capability of the corners

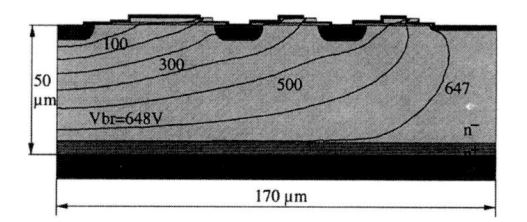

Figure 4: 600V-OFP-FLR-edge termination

But by means of an optimization of the edge structure the blocking capability of the chip corner is noticably improvable (version 3). Fig. 4 contains a $600V$-OFP-FLR edge termination structure for 600V-PT-IGBTs. Because of the less number of steps the mesh generation is distinctly easier compared to the nine-step $1200V$- edge termination. Accordingly only 6.000 triangles are sufficient to triangulate the device with a high resultion. The pn-junctions of all presented examples are dissolved with a value of $250nm$ on the field rings, on the backside emitter with $100nm$. In the substrate of the edge structures there are triangle edges up to $10\mu m$. For the triangulation of very thin layers it was taken care of the fact that all triangles have to evince at least one point inside the layer. That leads to a minimum triangle edge length lower than $50nm$ for example in the gate oxide. The required meshes for the simulation with cylindrical polar coordinates are produced by shifting the x-coordinates of all nodes or by stretching/compressing details of the complete mesh starting from a basic grid.

References

[1] BREAKDOWN (2D-Simulator), TU Berlin, Institut für Werkstoffe.
[2] BLOCKSIM (2D-Simulator), TH Ilmenau, Institut für Festkörperelektronik, 1987.
[3] SILVACO International, Santa Clara, USA, ATLAS II User Manual, 1993.
[4] TMA, Palo Alto, USA, MEDICI User Manual, 1992.
[5] H. Gajewski, H. Langmach, G. Telschow, K. Zacharias, "Der 2D-Bauelemente-simulator ToSCA," Handbuch, Berlin 1986, 1993 (unpubl.).
[6] M. Netzel, D. Schipanski, "Simulation von Bauelementen der Leistungselektronik mit dem Komplexsimulationssystem PRODESI," *21. Kolloquium Halbleiterleistungsbauelemente und Materialgüte Silizium Freiburg/Breisgau*, 1992.
[7] TRIGEN: Dreiecksgenerator für gemischte isotrope- und anisotrope Gitter, Programmbeschreibung, Institut für Halbleiterphysik, 1990,1993.
[8] TRIMAN: Gittereditor für gemischte isotrope- und anisotrope Gitter, Programmkurzbeschreibung, Technische Universität Ilmenau, 1994.

Numerical Analysis of Hot-Electron Effects in GaAs MESFETs

Y. A. Tkachenko[a], C. J. Wei[a], J. C. M. Hwang[a], D. M. Hwang[b]

[a]Lehigh University, 19 Memorial Dr W, Bethlehem PA 18015, USA
[b]Motorola, APRDL, 3501 Ed Bluestein Blvd., Austin TX 78721, USA

Abstract

For the first time, numerical simulation of GaAs metal-semiconductor field-effect transistors subjected to hot-electron-induced degradation was performed. Experimentally observed open-channel current reduction and gate-drain field relaxation were verified. Maximum electric field which governs hot-carrier phenomena was found to depend on Vdg and not on Vds, as in the case of Si MOSFET. Extended surface depletion as a result of traps observed by high-voltage electron-beam-induced current imaging agreed well with Nt of 5×10^{12} and Lt of 0.1 μ used in simulation.

1. Introduction

It is well-known that hot electrons can be trapped in the gate oxide of a Si MOSFET causing its threshold voltage to shift. For a GaAs MESFET, hot electrons can also be trapped in the silicon nitride beside the gate (Fig. 1), causing the MESFET's transconductance to decrease without affecting its threshold voltage (Fig. 2). The hot electrons are generated by biasing the MESFET with high drain-source voltage (Vds) and RF driving it beyond pinch-off. Frequently confused with other gradual degradation effects, this phenomenon has not been reported until recently [1], [2]. In [2], we observed that hot-electron effects in MESFETs are mainly dependent on the peak drain-gate voltage (Vdg). In this paper, we analyze numerically hot-electron effects in MESFETs. Similarities and differences with hot-electron effects in MOSFETs are discussed. Direct experimental verification of the spatial distribution of hot-electron-induced traps is also presented.

2. GaAs MESFET simulation

Our analysis is based on 2-dimensional numerical simulation [3] of a MESFET. The MESFET consists of a 0.5 μ gate centrally located within a 3 μ source-drain spacing. The combined build-in charge density in SiN as well as at the SiN/GaAs interface, Ni, is assumed to be $1 \times 10^{12}/cm^2$. Hot-electron-induced traps are characterized by a sheet density of Nt spread over a length of Lt from the gate toward the drain (Fig. 1). This results in an enlarged surface depletion on the drain side of the gate, as shown in Fig. 3. The amount of trap-induced channel charge depletion is proportional to Nt (Fig. 4). The MESFET channel shrinks, hence the channel current decreases, in agreement with experimental data, as shown in Fig. 5. The resulting distribution of the electric field when simulated with Vds = 0 is also nonsymmetrical. As shown in Fig. 6, the maximum field Em is reduced on the drain side by the traps. Since hot-electron effects are critically dependent on Em, it was simulated under various bias conditions and trap densities. The simulation shows that, unlike in a MOSFET, Em in a MESFET is proportional to $\sqrt{Vdg}$ instead of Vds (Fig. 7). This square root dependence can be derived by solving a 1D Poisson's equation with appropriate boundary conditions. Assuming that the surface states are uniformly distributed between gate and drain, whereas the traps are located only from 0 to Lt distance from the gate,

$$Em^2 = (q/\epsilon)Nd\{2Vdg-(q/\epsilon)[NsLdg^2-2NtLdgLt]\} \qquad (1)$$

where q is electron charge; ϵ is dielectric constant; Nd is channel doping density; Ns is surface state density; and Ldg is gate-drain spacing. Fig. 8 shows that, based on 2D simulation, Em^2 decreases linearly with increasing density of traps, Nt, in agreement with (1).

3. Electric field relaxation in the degraded MESFET.

The above simulation agrees with the experimentally observed dependence on Vdg and the tendency for hot-electron-induced degradation to gradually saturate. Figure 9 shows that the reverse gate current, which is also caused by hot electrons, is proportional to $\exp(-1/Vdg)$. Figure 10 shows that, after the MESFET undergoes hot-electron stressing, the reverse gate current decreases uniformly across the typical operating range of Vdg. Figure 11 shows that, as measured by fractional changes in the open-channel current ($Imax$), the rate of hot-electron-induced degradation is approximately proportional to the square root of time.

4. High-voltage electron-beam-induced-current imaging.

The distribution of hot-electron-induced traps was directly verified using a novel high-voltage electron-beam-induced-current imaging technique [4]. By connecting the source and drain together, a Schottky diode is built in with the gate such that the induced current is proportional to the depletion depth by the surface states. As shown in Fig. 12, due to hot-electron-induced traps, the depletion region extends farther at the drain side of the gate (0.29 ± 0.04 μ) than at the source side of the gate (0.17 ± 0.04 μ). This is in agreement with Lt = 0.1 μ used in the simulation.

Acknowledgement

This work was supported in part by the Air Force Wright Laboratory, Solid State Electronics Directorate, Microwave Division, under Contract No. F33615-90-C-1405.

References

[1] A. Watanabe, K. Fujimoto, M. Oda, T. Nakatsuka and A. Tamura, "Rapid degradation of WSi self-aligned gate GaAs MESFET by hot carrier effect", in *Proc. Int'l Reliability Physics Symp.*, 1992, pp. 127-130.
[2] Y. A. Tkachenko, Y. Lan, D. S. Whitefield, C. J. Wei, J. C. M. Hwang, L. Aucoin and S. Shanfield, "Accelerated dc screening of GaAs FETs for power slump tendency under RF overdrive," in *Dig. US Conf. GaAs Manufacturing Technology*, 1994, pp. 35-38.
[3] Silvaco International, Santa Clara, CA.
[4] D. M. Hwang, L. Dechiaro, M. C. Wang, P. S. D. Lin, C. E. Zah, S. Ovadia, T. P. Lee, D. Darby, Y. A. Tkachenko and J. C. M. Hwang, "High-voltage electron-beam-induced-current imaging of microdefects in laser diodes and MESFETs," in *Proc. Int'l Reliability Phys. Symp.*, 1994, pp. 470-477.

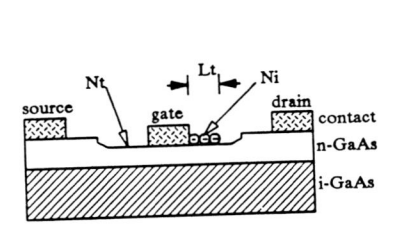

Fig. 1 MESFET structure indicating the distribution of hot-electron-induced traps.

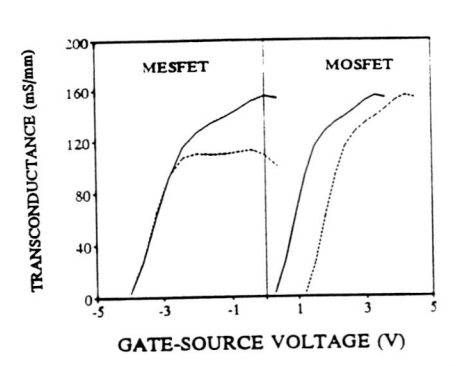

Fig. 2 Comparison of hot-electron effects of a MESFET and a MOSFET. (—) before, (---) after.

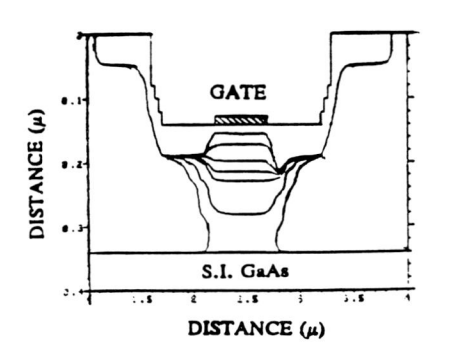

Fig. 3 Simulated depletion edge. $Nt = 5 \times 10^{12}/cm^2$. $Vds = 0$. $Vgs = 0.8, 0.6, 0.4, 0.2, 0, -1$ and -3 top down.

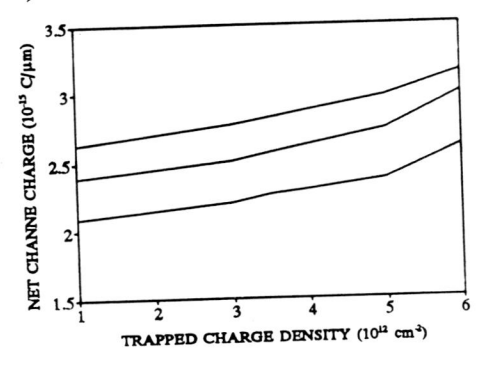

Fig. 4 Simulated net channel charge integrated on the drain side of the gate. $Vds = 1$ V. $Vgs = -0.5, 0$ and 0.5 V top down.

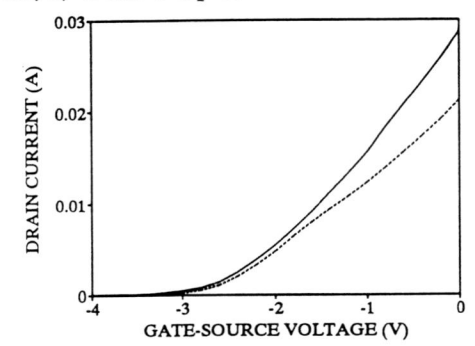

Fig. 5 Simulated transfer characteristics. (–) $Nt = 0$, (---) $Nt = 5 \times 10^{12}/cm^2$. $Vds = 0.6$ V.

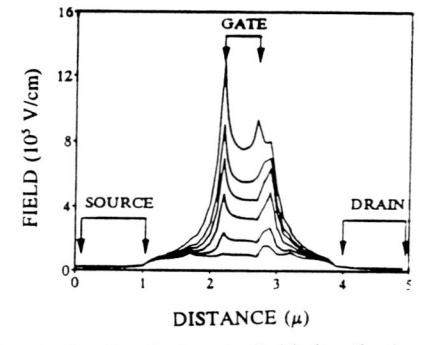

Fig. 6 Simulated electric field distribution. $Vds = 0$. $Vdg = 1, 3, 7, 11, 15$ and 23 V bottom up.

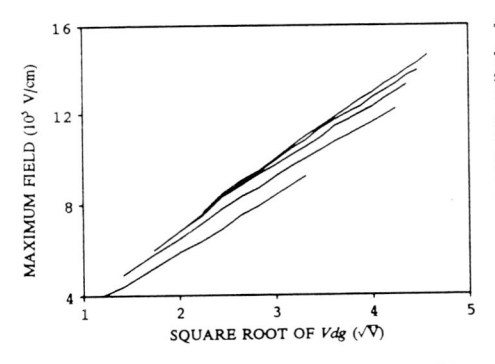

Fig. 7 Simulated maximum electric field. V_{gs} = -5, -4, -3, -2 and -1 V top down.

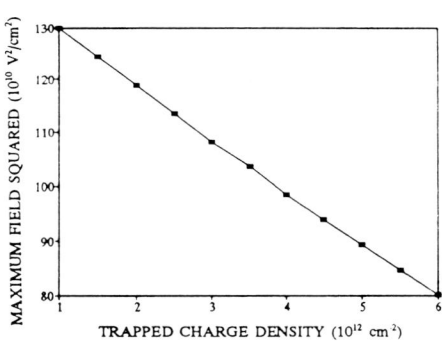

Fig. 8 Simulated maximum electric field. V_{gs} = -3 V, V_{ds} = 13 V.

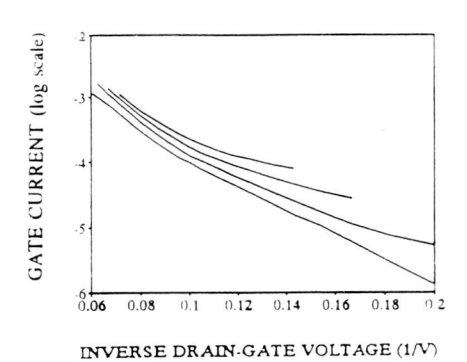

Fig. 9 Measured reverse gate current. V_{gs} = -4, -5, -6 and -7 V bottom up.

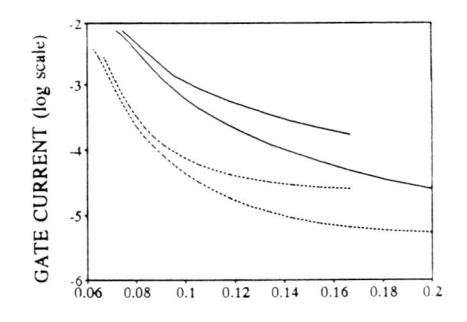

Fig. 10 Measured reverse gate current (—) before and (---) after. V_{gs} = -5 and -6 V bottom up.

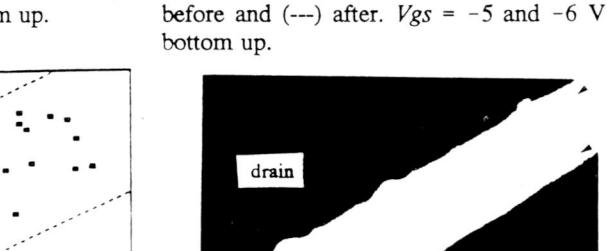

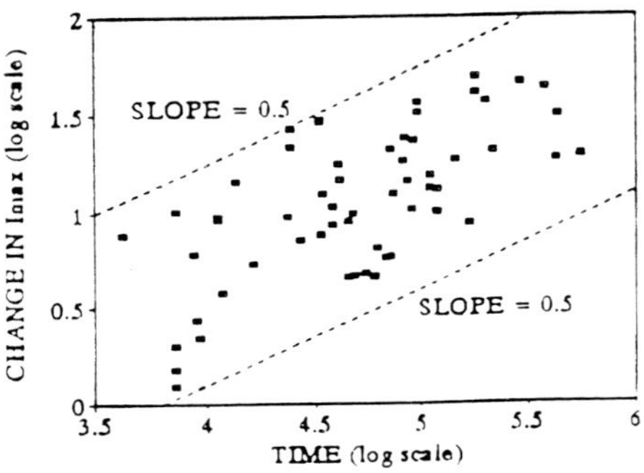

Fig. 11 Measured changes in MESFET drain current as a function of time.

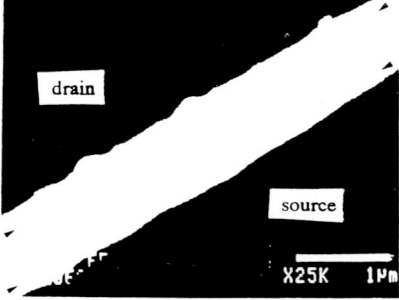

Fig. 12 High-voltage electron-beam-induced-current image of the MESFET of Fig. 1.

Capacitance Model of Microwave InP-Based Double Heterojunction Bipolar Transistors

C. J. Wei, H.-C. Chung, Y. A. Tkachenko and J. C. M. Hwang

19 Memorial Drive West, EECS Department,
Lehigh University, Bethlehem, PA 18015, USA

Abstract

A capacitance model for microwave Double Heterojunction Bipolar Transistors (DHBTs) is presented to account for their special behavior in small-signal S-parameters. A physical picture of barrier effects in DHBTs is described.

1. INTRODUCTION

InP-based double heterojunction bipolar transistors (DHBTs) are attractive for millimeter wave power applications due to their impressive power performance at high frequencies[1]. Output power of 5W/mm at 9 GHz with maximum oscillation frequency of 96 GHz has been reported[1]. The key technique is incorporation of a spacer layer between the broad-band InP sub-collector layer and InGaAs base layer to reduce the barrier which otherwise hinders collection of injected electrons. However, a barrier can not be completely eliminated resulting in a number of effects at both dc and Rf[2].
In this paper we present a capacitance model for DHBTs which accounts for the barrier effects on S-parameters. The model gives an insight into electron accumulation at interface of the barrier. It is found that fitting of the peculiar frequency behavior of the device can not be achieved without an appropriate capacitance model.

2. MODEL

Typical layer structure of the DHBT is shown in Table 1. The collector of the device consists of several layers. A spacer layer of undoped-InGaAs on the collector junction interface and a thin layer of heavily-doped InP thin layer lower the barrier height and move it towards the collector space-charge region of the reversed biased collector-base junction as shown in Figure 1. When electrons flow from the base towards the space-charge region, dynamic electron accumulation, 2D electron gas(2DEG), occurs at the trench of the barrier.
Analogous to the MOS capacitance, a collector-base junction capacitance model in presence of the barrier can be described by a depletion layer capacitance C_d in parallel with a series R_e-C_e elements as shown in Figure 2. The latter represents the 2DEG storage charge with a charging time constant, $\tau_d = R_e C_e$. The time constant τ_d represents the response time of charge build-up under voltage/current variation across the space-charge region. At lower frequency, when the

period of signal is longer than τ_d, the 2DEG screens the depletion capacitance. With frequency increasing, however, the depletion capacitance graduately dominates. The difference of this model from a MOS capacitor model is in that the capacitance is both current and voltage controlled whereas the MOS capacitance is solely voltage-controlled. The current dependence originates from the variation of 2DEG density with collector current.

All parameters of the model including the time constant τ_d are dependent on the layer structure, the barrier height and the electric field distribution. Also they are bias-dependent. At lower collector voltage, the collector region is not completely depleted as shown by the simulated electric field profile of Figure 1. The 2DEG causes a step change in the field profile. Note that for a given collector-base voltage. When 2DEG is the only type of space charge, a decrease in 2D-gas density causes a shrink the depletion region resulting in increased C_d. Therefore, the capacitance is expected to be increasing with collector current at higher frequency where C_d dominates. At lower frequency, a variance in the 2DEG can easily follow the current variance and an increasing current gives rise to an increasing capacitance C_c.

3. RESULTS

Bias-dependent S-parameters of a InAlAs-InGaAs-InP DHBT were measured with HP8510 from 0.2 to 30 GHz. The device comprises 12 emitter finger 2 by 10 um. A highly consistent equivalent circuit element extraction method[3] was used to obtain the intrinsic and extrinsic collector impedance values and their frequency dependence. Fig. 3 plots the intrinsic C-B capacitance vs frequency at $V_c = 2$ V for various injection levels.

At higher voltages, the barrier is diminished by the lowering effect of a higher field. From figure 3, there exists a turning frequency, $f_{ct} = 1.5$ GHz above which the depletion capacitance dominates. The associated time constant is $\tau_d = 1/(2\pi f_{ct}) = 105$ ps. It is also shown that the capacitance increases with collector current below f_{ct} and decreases somewhat for lower current biasing above that turning frequency, as shown in Fig.4. Using fitting technique it is straightforward to obtain the bias-dependent capacitance as well as bias-dependent time constant τ_d.

Incorporating this capacitance model into a small-signal equivalent circuit with LIBRA, a commercial software by EEsof, as shown in Figure 2, the fitting of S-parameters are considerably improved in comparison with conventional model. Fig. 5 shows compares measured S11 and S21 at $V_c = 2V$ and $I_b = 1$ mA with those obtained using conventional model. A Peculiar kink in S21 makes the accurate fitting very impossible. Fig. 6(a) and 6(b), in contrast, show excellent fitting of measured S-parameters by the modified model. The kink apparently corresponds to the turning frequency f_{ct}. The fitting parameters are listed in Table 2.

4. CONCLUSIONS

For the first time a base-collector capacitance model for DHBTs was constructed to account for the dynamic electron accumulation at InGaAs/InP collector heterojunction. This accumulation gives rise a peculiar frequency dependence of S-parameters, namely a kink in S21, which was accurately predicted by the model. As a results, valuable insight into the dynamics of electron charge build-up is obtained.

Reference

[1] M. Hafizi, P.A. Macdonald, T. Liu, D.B. Rensch and T.C. Cisco, "Microwave Power Performance of InP-based DHBTs for C- and X-band Applications," in *Digest of 1994 IEEE MTT-Symposium*, pp.671-674, San Diego, May 1994
[2] R.A. Sugeng, C.J. Wei, J.C.M. Hwang, J.I. Song, W.P. Hong and J.R. Hayes, "Junction Barrier Effects on the Microwave Power Performance of DHBTs", in *Proc. IEEE/Cornell Conference on Advanced Concepts in High-speed semiconductor Devices and Circuits*, Ithaca, Aug. 2-4, 1993, pp.52-61
[3] C.J. Wei and J.C.M. Hwang,"A new method for direct extraction of HBT equivelent circuit elements" in *IEEE MTT-S Int'l Microwave Symp. Dig.*, 1994, pp.1245-1248

Table 1 Material structure of InP power HBTs.

Name of Layer	Material	Doping(/cm³)	Thickness(nm)
Cap	n⁺⁺InGaAs	2x10¹⁹	200
Emitter	n⁺⁺InGaAs	2x10¹⁹	230
Emitter	n InAlAs	1x10¹⁷	70
Emitter	n InAlAs	1x10¹⁸	5
Spacer	i InGaAs	undoped	15
Base	p⁺⁺InGaAs	2x10¹⁹	60
Spacer	i InGaAs	undoped	10
Spacer	n⁺⁺InGaAs	2x10¹⁸	5
Collector	n⁺ InP	2x10¹⁸	15
Collector	n InP	5x10¹⁶	100
Collector	n InP	1x10¹⁶	300
Collector	n⁺⁺InP	2x10¹⁹	100
Sub-collector	n⁺⁺InGaAs	2x10¹⁹	500
Substrate	SI InP	Compensated	5 mil

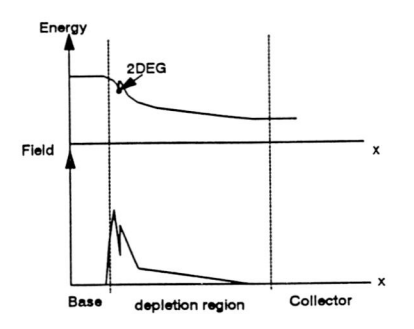

Fig. 1 Band & Field diagram in collector depletion region of DHBTs. The barrier is located in collector depletion region.

Table 2 Fitting parameters of equivalent circuit. $V_c=2$ V, $I_b=1$ mA.

$R_b(\Omega)$	0.75	$C_{ex}(pF)$	0.239
$L_b(nH)$	0.051	$C_d(pF)$	0.41
$R_e(\Omega)$	1.56	$R_e'(\Omega)$	82
$L_e(nH)$	0.018	$C'(pF)$	0.85
$R_c(\Omega)$	0.54	$C_{be}(pF)$	3.82
$L_c(nH)$	0.047	α	0.997
$R_{b2}(\Omega)$	2.75	$R_{be}(\Omega)$	0.31
		τ_c (ps)	0.13

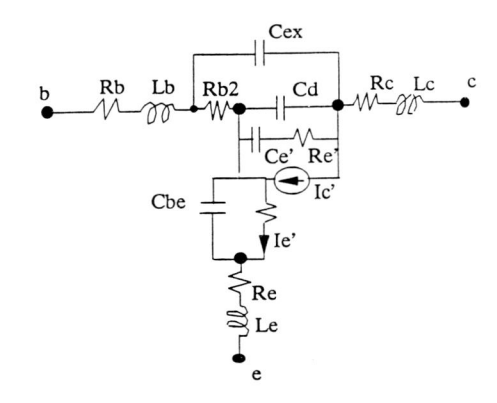

Fig. 2 Equivalent circuit of HBTs. Note the impedance between base and collctor.

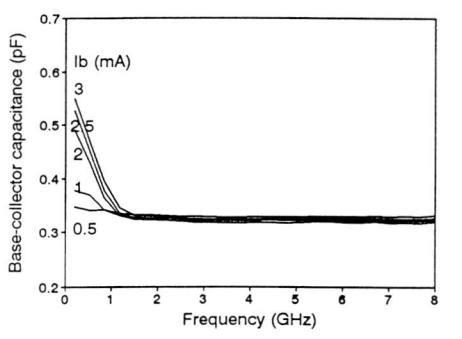

Fig 3 Total base-collactor capacitance as functions of frequency. Ib = 0.5, 1, 1.5 and 2 mA.

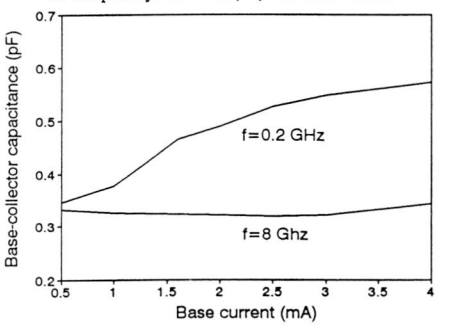

Fig. 4 Total effective Base-colletor capacitance vs. base current for f= 0.2 and 8 GHz respectively.

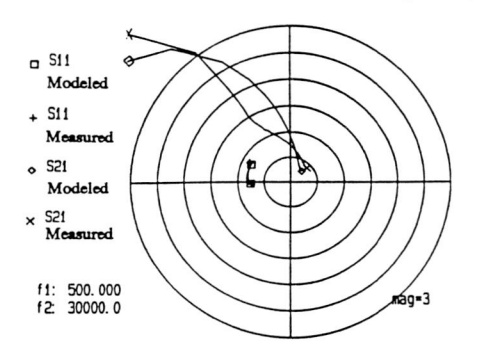

Fig. 5 S-parameter fitting using conventional model.

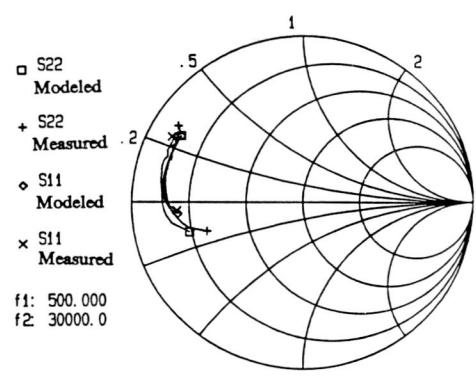

(a)

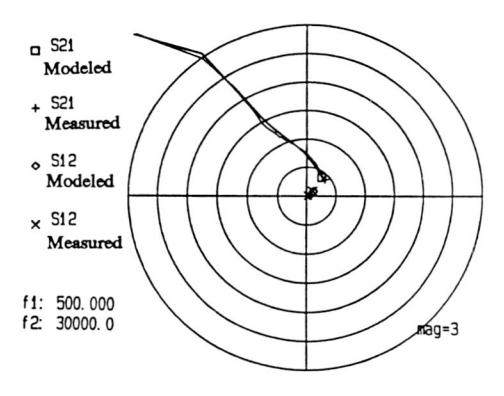

(b)

Fig. 6 S-parameter fitting using new model. (a) S11 and S22, (b) S21 and S12

Estimation of the Charge Collection for the Soft-Error Immunity by the 3D-Device Simulation and the Quantitative Investigation

Y.Ohno, T.Kishimoto*, K.Sonoda, H.Sayama, S.Komori, A.Kinomura**, Y.Horino**, K.Fujii**, T.Nishimura, N.Kotani, M.Takai* and H.Miyoshi

ULSI Laboratory, Mitsubishi Electric Corporation, 4-1, Mizuhara, Itami, Hyogo 664, Japan
* Faculty of Engineering Science and Research Center for Extreme Materials, Osaka University, Toyonaka, Osaka 560, Japan
**Osaka National Research Institute, AIST, Ikeda, Osaka 563, Japan

Abstract

The charge collection induced by incident particles was estimated by the 3-dimensional device simulation and the quantitative evaluation method using the nuclear microprobe. The role of the buried p$^+$layer was well analyzed in terms of the soft-error immunity of DRAMs. The methods developed here are applicable to optimize the well structure for the soft-error immunity of advanced DRAMs.

1, Introduction

The modification of well structure is important to get the soft-error immunity for advanced DRAMs. However, since soft-error events of DRAMs have been conventionally evaluated using energetic particles from a radioactive source such as Am which has the inherent distribution of incident energies and the amount of the incident particles, the evaluation result offers just a qualitative analysis.

In this study, the precise estimation of soft-error is demonstrated. The charge collection induced by incident particles was simulated by the 3-dimensional device simulation for the improvement of soft-error immunity. And the quantitative investigation of soft-errors was performed by using the nuclear microprobe with optional energy and controlled amount of incident particles[2]. The role of the buried p$^+$layer, which forms the retrograde well structure, against the charge collection into n$^+$ layer was evaluated by comparing the experimental results with the simulation result.

2, Estimation Methods

The charge collection efficiency was calculated using a general purpose 3-dimensional device simulator,"3D-MIDSIP (3 Dimensional-Mitsubishi Device SImulation Program)," in which a Poisson equation and continuity equations for electrons and holes were solved self-consistently. After steady state condition was calculated, electron-hole pairs created by an injected proton were distributed along the track of the particles to perform transient analysis. The depth of penetration or range of particles was determined by the initial energy E0[3]. The proton lost an average energy of 3.6eV at 300K for every electron-hole pair generation[4]. The total ionization produced by the proton was given by its initial energy E0 divided by the average energy loss. The pair generation per unit length was obtained by differentiation of the energy-range relation with the average energy loss.

Also, the new evaluation system with the proton microprobe was used for the quantitative investigation of the charge collection. Incident flux, energy and irradiated position can be easily controlled in this method. The proton energies were 1.3MeV~2.0MeV in this study. The proton with the energy of 1.3MeV almost has the same trajectory as that of the 5.0MeV alpha particle in the Si-substrate.

The measured samples had the retrograde well structure with n^+ layer and buried p^+ layer formed by the high energy ion implantation. The depth of the buried p^+ layer is 0.4~1.7μm and the peak concentration of p^+ layer is 4.2×10^{16} ~ 4.3×10^{17} cm^{-3}. The charge collection efficiency was defined as a ratio of collected charges in n^+ layers with a well structure to that without a well structure.

3, Results

Fig.1 shows the simulation result of the induced current depending on the depth of the buried p^+ layer. For the stage of funneling effect (1psec.-1nsec.), the induced current into the n^+ layer is reduced as the depth of the buried p^+ layer decreases. The shallow buried p^+ layer is so effective for the reduction of the funneling effect. However, for the stage of the diffusion effect (1nsec.-1μsec.), the induced current has no dependence on the depth of the buried p^+ layer.

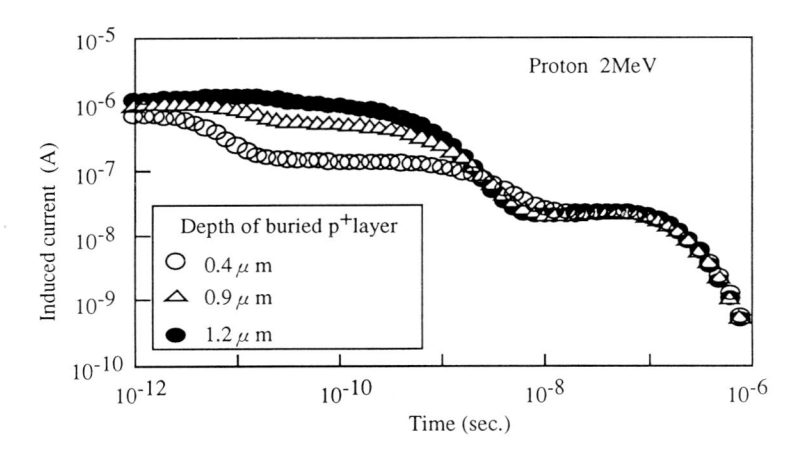

Fig.1: Dependence of the induced current on the depth of buried p^+ layer

The simulation of accumulated charge collection into the n^+ layer is shown in Fig.2. The amount of collected charges does not depend on the depth of buried p^+ layers. The most of charges are collected for the stage of diffusion effect. The quantitative investigation of charge collection was performed by using the proton microprobe with the energy of 1.3 and 2.0MeV. Fig.3 shows the experimental result of the charge collection dependence on the depth of the buried p^+ layer. The simulation result well coincides the experimental result. It is verified that the charge collection is mainly due to the diffusion of induced charges by high energy particles.

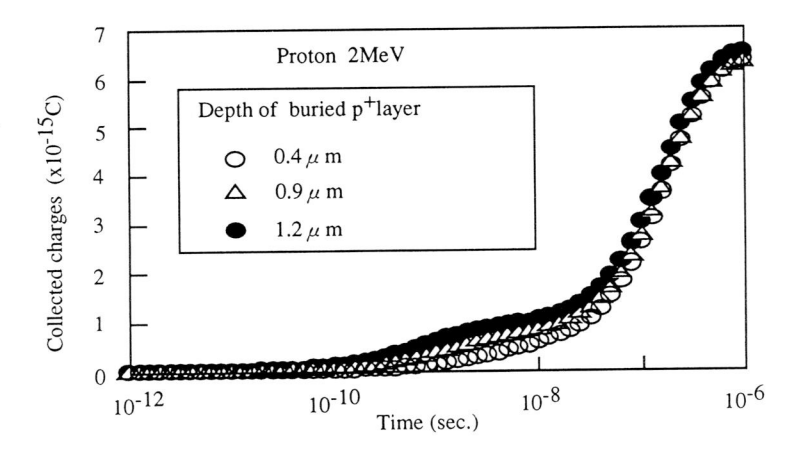

Fig.2 Dependence of the charge collection on the depth of buried p$^+$layer

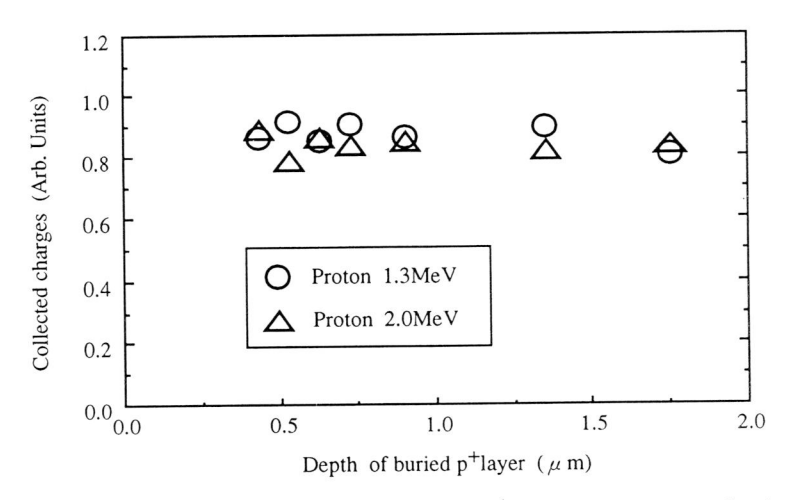

Fig.3 Relation between the depth of buried p$^+$ layer and the charge collection

Fig.4 shows the simulation of the induced current dependence on the concentration of the buried p$^+$layer. The induced current into the n$^+$layer is reduced as the concentration of the buried p$^+$layer increases for both stages of funneling and diffusion effects. Fig.5 shows the experimental result of the charge collection dependence on the concentration of buried p$^+$layer. The experimental result is well explained by the simulation. The buried p$^+$layer achieves as barrier against the charge collection caused by the diffusion and funneling effects.

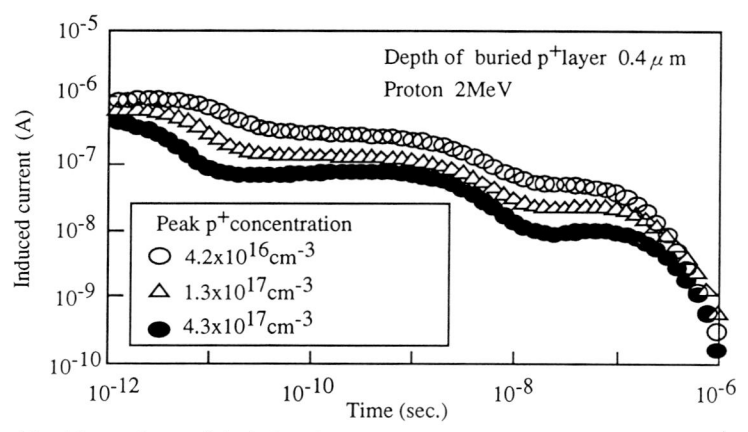

Fig.4 Dependence of the induced current on the concentration of the buried p$^+$layer

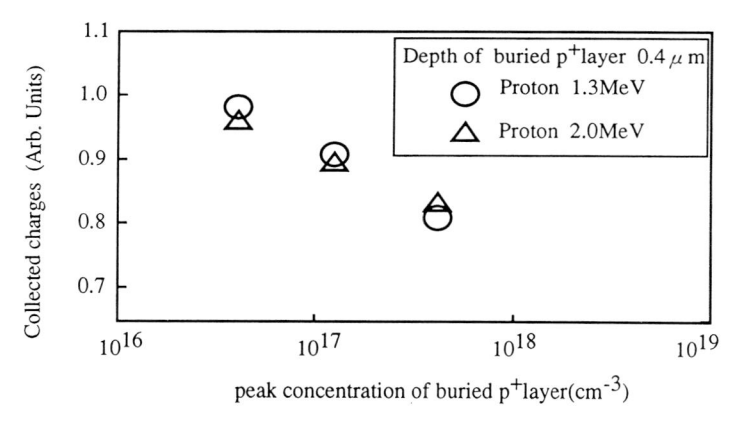

Fig.5 Relation between the concentration of the buried p$^+$layer and the charge collection

4, Summary

The 3-dimensional device simulation and the quantitative evaluation using the nuclear microprobe have been demonstrated. The quantitative study of the charge collection was realized by these investigations. The role of the buried p$^+$layer was well analyzed in terms of the soft-error immunity of DRAMs. The methods developed here are applicable to optimize the well structure for the soft-error immunity of advanced DRAMs.

References

[1] K.Tsukamoto et al.,Nucl. Instr. and Meth. B59/60, p.584 (1991)
[2] H.Sayama et al., Technical Digest of IEDM, p.815 (1992)
[3] J.F.Gibbons et al., "Projected Range Statistics [Semiconductors and Related Materials]", (1975)
[4] R.D.Ryan, IEEE Trans. Nucl. Sci., NS-20, p.473 (1973)
[5] Y.Ohno et al., IEICE Trans. Electron, vol.E77-C, no.3, p.399 (1994)

D.C. Electrothermal Hybrid BJT Model for SPICE

Janusz Zarębski, Krzysztof Górecki

Department of Radioelectronics, Gdynia Maritime Academy
Morska 81-87, 81-225 Gdynia, POLAND

Abstract

SPICE - built-in models are the isothermal models with the temperature as a parameter. Electrothermal interactions caused that the device (junction) temperature changes due to changes of dissipated power. In the paper the new BJT electrothermal model which includes the thermal effects is proposed. This model is represented by combination of SPICE-built-in isothermal model and any additionaly controlling generators modelling selfheating and some other electrical effects nonmodeled by SPICE.

1. Introduction

The strong temperature influence on the electrical characteristics of bipolar transistors is observed. On the other hand, the electrical power dissipated in BJT causes the increase of the inside temperature above the ambient one, as a result of non-ideal conditions of the heat abstraction. Therefore, the mutual interactions between the electrical and the thermal phenomena in BJT, commonly called - the electrothermal interactions, exist. Inclusion of the electrothermal interactions to modelling makes the model of a semiconductor device more adequate.
SPICE program - commonly used for circuit analysis, can not be directly applied to the electrothermal case, because there is no possibility to consider the change of device (junction) temperature caused by power dissipation. On the other hand, there were some attempts made to use SPICE to the electrothermal analysis, e.g. [1, 2]. Unfortunately, these analyses are inaccurate because of the use of too simple electrothermal device models.
In this paper an electrothermal hybrid model of BJT with much better accuracy in the whole operating range is presented. This model is represented by combination of SPICE-built-in isothermal model and additionally control generators. These generators model changes of BJT currents and voltages caused by the change of the junction temperature due to the electrothermal interactions and by other isothermal effects, like current gain factors changing and avalanche effects, which are not modelled directly by SPICE.

2. The fundamental dependences

The starting point to formulate the electrothermal hybrid model (EHM) of BJT are dependences describing GETM2 model given in [3]. Choosing the collector current i_C, the base-to-emitter voltage u_{BE} and the base-to-collector voltage u_{BC} as the independent variables, one can write for the active-normal region:

$$i_C = \beta_F \cdot i_B + I_S \cdot \left(1 + \frac{\beta_F + 1}{\beta_R}\right) \qquad (1)$$

$$u_{BE} = V_T \cdot \ln\left(\frac{i_B \cdot \beta_F}{I_S} + 1 + \frac{\beta_F}{\beta_R}\right) \qquad (2)$$

and for the active-inverse region:

$$i_C = -(\beta_R + 1) \cdot i_B - I_S \cdot \left(\frac{1}{\beta_R} + \frac{\beta_R + 1}{\beta_F}\right) \qquad (3)$$

$$u_{BC} = V_T \cdot \ln\left(\frac{i_B \cdot \beta_R}{I_S} + 1 + \frac{\beta_R}{\beta_F}\right) \qquad (4)$$

where i_B is the base current.
In Eqs (1-4) the temperature dependent parameters: the saturation current I_S, the thermal potential V_T and the current-gain factor for the active-normal region (β_F) and for the active-inverse region (β_R), respectively are given as follows:

$$I_S = I_0 \cdot \left(\frac{T_j}{T_0}\right)^2 \cdot \exp\left(\frac{-U_{go}}{V_T}\right) \qquad (5)$$

$$V_T = \frac{k}{q} \cdot T_j \qquad (6)$$

$$\beta_F = \frac{\beta_{0F} \cdot (1 + a \cdot (T_j - T_a))}{1 + b \cdot (1 + c \cdot (T_j - T_a)) \cdot |i_C|} \qquad (7)$$

$$\beta_R = \frac{\beta_{0R}}{\beta_{0F}} \cdot \beta_F \qquad (8)$$

where k, q - are the phisical constants, I_0, T_0, β_{0F}, β_{0R}, a, b, c are the model parameters and T_a is the ambient temperature. The junction temperature T_j is given by the following dependence

$$T_j = T_a + K_t \cdot (u_{CE} \cdot i_C + u_{BE} \cdot i_B) \qquad (9)$$

where the u_{CE} is the collector-to-emitter voltage, and K_t denotes the thermal resistance of the BJT. Eqs (7) and (8) denote that β_F and β_R are described by the dependences of the same form, but with different values of the coefficients β_{0F}, β_{0R}. It was assumed that parameters a, b, c are the same for β_F and β_R. So, β_F (Eq. 7) or β_R (Eq. 8) can be given as follows

$$\beta = \beta_0 \cdot f(u_{CE}, i_C) \qquad (10)$$

Now, it is easy to observe that the multiplications $\beta_F \cdot i_B$ (Eq 2) and $\beta_R \cdot i_B$ (Eq 4) can be presented as the multiplication of the isothermal collector current i_{CO} (represented by SPICE-built-in model) and the function $f(u_{CE}, i_C)$, of the form

$$\beta \cdot i_B = |i_{CO}| \cdot f(u_{CE}, i_C). \tag{11}$$

The absolute value of i_{CO} in Eq. (11) and i_C in Eq (7) is to secure the rightness of the model, both in the active-normal region ($i_C > 0$) and in the active-inverse region ($i_C < 0$).

3. The electrothermal hybrid BJT model (EHM)

The electrothermal hybrid model results directly from the transformation of the dependences from Sec. 2 to the form

$$i_C = i_{CO} + \Delta i_C \tag{12}$$
$$u_B = u_{BO} + \Delta u_B \tag{13}$$

where, i_{CO}, u_{BO} are described by the standard SPICE-built-in model with the ambient temperature as a parameter ($u_{BO} = u_{BEO}$ for the active-normal region and $u_{BO} = u_{BCO}$ for the active-inverse region). The remain part of the so transformed dependences describes the change of the collector current (Δi_C) and the base-to-emitter or the base-to-collector voltages (Δu_B), caused by the increase of the junction temperature and by the decrease of the current gain factor (both β_F and β_R) for high current density simultanously. Assuming that $\beta_F \gg \beta_R$ and $\beta_F \gg 1$, we get

$$\Delta u_B = T_j \cdot \frac{k}{q} \cdot \ln\left(\frac{|i_{CO}| \cdot f(u_{CE}, i_C)}{I_S} + 1 + \left(\frac{\beta_{0F}}{\beta_{0R}}\right)^{\varphi}\right) - T_a \cdot \frac{k}{q} \cdot \ln\left(\frac{i_{CO}}{I_{S0}} + 1 + \left(\frac{\beta_{0F}}{\beta_{0R}}\right)^{\varphi}\right) \tag{14}$$

$$\Delta i_C = -b \cdot \left(1 + c \cdot (T_j - T_a)\right) \cdot |i_C| \cdot i_C + i_{CO} \cdot a \cdot (T_j - T_a) +$$
$$+ \varphi \cdot I_S \cdot \left(\frac{1}{\beta_R} + \left(\frac{\beta_{0F}}{\beta_{0R}}\right)^{\varphi}\right) \cdot \left[1 + b \cdot \left(1 + c \cdot (T_j - T_a)\right) \cdot |i_C|\right] \tag{15}$$

where I_{S0} is the value of I_S for the ambient temperature and $\varphi = 1$ for $i_C \geq 0$ and $\varphi = -1$ for $i_C < 0$.

The finally circular form of EHM is shown in Figure 1. Additionally, in EHM the avalanche effects existing in the both BJT junctions for the appropriate high values of the junction-voltages (modelled by the additionally connected p-n diodes with satisfactory built parameters) are taken into account. The part of EHM marked as T-iso is the SPICE-built-in BJT model. The independent voltage sources (u = 0) allow the use of the values of the appropriate currents, with much better accuracy than in the case of resistors use.

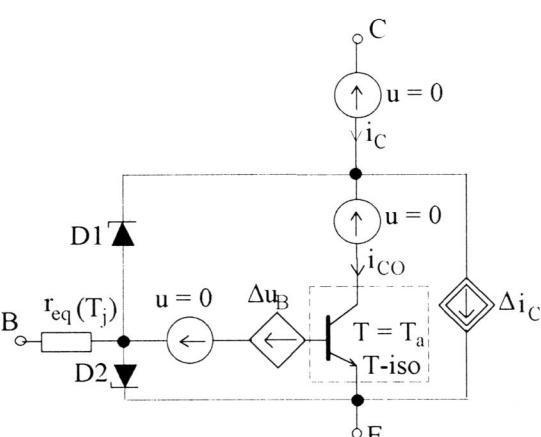

Figure 1: The circular form of EHM

To include the temperature dependence of the series BJT resistances, the linear temperature dependent equivalent resistance of the form

$$r_{eq}(T_j) = r_0 \cdot \frac{T_j}{T_a} \qquad (16)$$

where r_0 is the value of r_{eq} for $T_j = T_a$, is introduced. In this case the proper SPICE parameters represented the series BJT resistances are equal to zero.

4. Example

To illustrate the usefulness of the presented model to modelling of BJT in the electrothermal case, in Figure 2 nonisothermal $i_C(u_{CE})$ characteristics of 2N2193 transistor obtained by the use of EHM, a model proposed in [1] and by measurements, are presented. At it is seen, EHM gives a good accuracy, much better than the known model presented in [1].

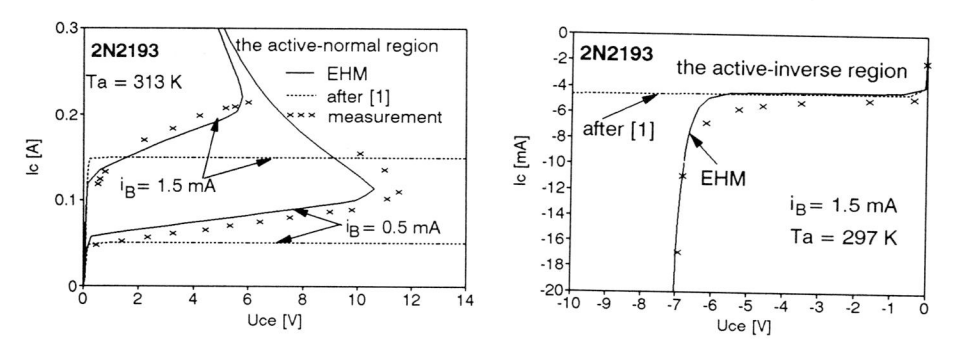

Figure 2: Nonisothermal $i_C(u_{CE})$ BJT characteristics

5. Concluding remarks

In this paper the new form of the electrothermal BJT model has been presented. This model gives a good accuracy in the whole range of the work of BJT. In EHM, apart from electrothermal interactions, also dependence of the current gain factors on the collector current and avalanche effects in the both junctions are included.

References

[1] Schurack E. at all, "Nonlinear Effects in Transistors Caused by Thermal Power Feedback Simulation and Modeling i SPICE," Proc. of IEEE ISCAS'92, San Diego 1992, pp. 879-882.
[2] Janke W., Łata Z., "SPICE Simulation of D.C. Electro-Thermal Interactions with the Use of ETM-II," Proc. of XIV Nat. Conf. CTEC, Waplewo 1991, pp. 309-314 (in English).
[3] Zarębski J., "A New Form of the BJT Model Including Electrothermal Interactions," Proc. of XI ECCTD, Davos 1993, V. 1, pp. 431-436.

Alpha–Particle Induced Soft Error Rate Evaluation Tool and User Interface

P. Oldiges

Digital Equipment Corporation
77 Reed Rd. HLO2-3/J9
Hudson, MA 01749 U.S.A.

Abstract

A simulation tool for calculating the soft error rate due to α–particle strikes in SRAM's is described. The simulator uses Monte Carlo methods to determine the initial energy and angle of α–particles, then uses layout, process and device information to determine a histogram of charge collection in a memory cell. The soft error rate is determined from a knowledge of the memory cells' critical charge. A graphical user interface for the soft error simulator is also described

1. Introduction.

A simulator was developed to aid in the understanding of soft error susceptibility of SRAMs. *SEEV* (Soft Error EValuator) uses Monte Carlo methods to determine the soft error rate (SER) of a circuit. Random numbers are used to determine initial energy, depth in the source, and incidence angle of an α–particle. An integration over all energies, incidence angles and locations of α–particle strikes in a memory cell is performed. Physical and empirical models describing the α–particle source, charge collection and interactions with materials are incorporated into *SEEV* for estimating energy and angular dependence of α–particle induced charge collection. Thicknesses and areas of overlayers for specific SRAM cells are used, and α–particle energy loss is calculated for each overlayer. Tools for SER estimation have been shown in the past [1,2]. The advantages that *SEEV* has over previous SER estimation tools are ease–of–use and speed ($10^7 \alpha$–particle trajectories can be simulated in less than 10 minutes on an Alpha workstation).

2. α–Particle Interactions with Materials.

α–particles are stopped in materials mainly by electronic interactions [3]. *SEEV* evaluates the stopping integral numerically and the resulting data is placed in a lookup table for various materials that are commonly used in the manufacture of semiconductor memories (*i.e.* Si, Al, SiO_2, and polyimide). For silicon, the knowledge of the ionization is required to determine the charge collection. Electron–hole pairs are generated by the electronic interaction between the α–particle and the silicon. For every 3.6eV loss in energy of the α–particle, one electron–hole pair is formed. *SEEV* uses an empirical formula to fit published ionization data [4].

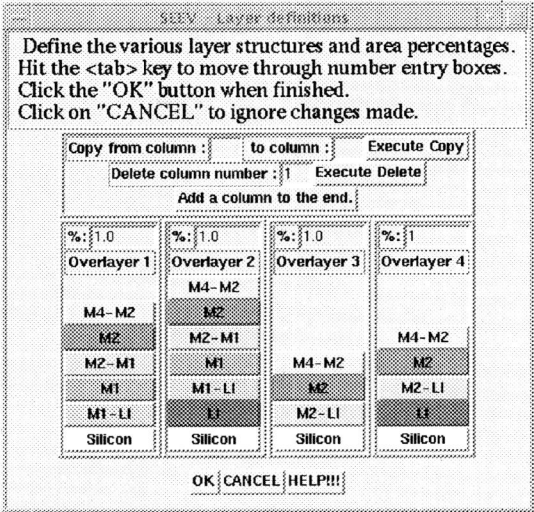

Figure 1: Widget used to define device overlayer structure in *SEEVI*.

3. Charge Collection Models in SEEV.

SEEV assumes an effective funneling length [4], derived from charge collection simulations. This effective funneling length is modified by one over the cosine of the incidence angle, although there are other methods to include angular dependence [1,5]. There is no energy dependence of the funneling length although the total charge collected does depend on the α–particle energy. The total charge collected, Q_{coll}, is calculated by integrating the ionization from the surface of the silicon to a depth along the α–particle track equal to the funneling length (modified by the angular dependence).

4. SEEV User Interface (SEEVI).

A graphical user interface was developed using the Tcl/Tk toolkit [6] to allow a user to define physical and numerical information used in the *SEEV* program. The main menu of *SEEVI* has six menu options that invoke invoke pulldown menus and are used to define:
- File Operations – Load and save an input file, set or clear input parameters or exit *SEEVI*.
- Alpha Source – Define the type of α–particle source and the parameters describing the source.
- Physical Parameters – Define effective funneling length, SRAM cell parameters, define overlayer thicknesses, and interactively add or delete various overlayer materials over the whole cell or portions of a cell. The widget used to interactively define the overlayer structure is shown in Figure 1.
- Output Parameters – Define data to be printed to an output file.
- Simulate – Perform a *SEEV* simulation.
- Title: – Text entry box for a single line title for the simulation.

5. Simulation Results.

One example of a calculation that can be performed by SEEV is the simulation of SER due to α–particles coming from ceramic package lids and from metal lines. Figure 2 shows *SEEV* calculations of the energy spectra at the silicon

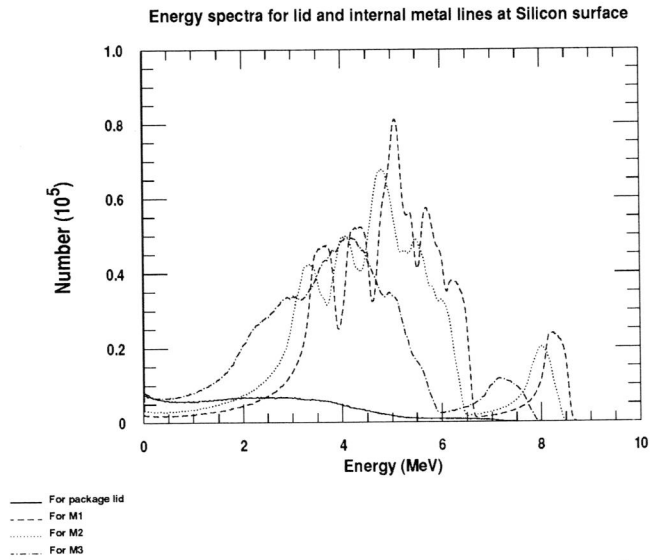

Figure 2: Energy spectra of various α–particle sources at the Silicon surface.

surface of α–particles that were generated in a package lid and in three different metal levels. Since M1 is the thinnest metal level and is closest to the silicon surface, the energy spectrum shows more discrete energy peaks. As distance away from the silicon surface increases, α–particles with varying incidence angles tend to lose more energy and the energy peaks get shifted down and more smeared out. Due to the shift in the energy and effects due to differences in incidence angle, the SER due to α–particles coming from metal lines and package lids will be different. Figure 3 shows *SEEV* calculations of relative SER for the various sources of α–particles.

Another example of a *SEEV* calculation is shown in Figure 4. The voltage dependence of SER for two different SRAM cell designs is shown. To obtain these graphs, circuit simulations of the critical charge needed to upset an SRAM cell is determined. The relative SER is then read from a graph similar to that shown in Figure 3. Measurements of a similar technology and cell to that simulated showed a SER dependence of 1.3 decades/V, in excellent agreement with simulation results of 1.29 decades/V.

References

[1] S. Satoh et al., "Cmos-sram soft error simulation system," *NUPAD–V*, pp. 181-184, 1994.

[2] G.R. Srinivasan et al., "Accurate, predictive modeling of soft error rate due to cosmic rays and chip alpha radiation," *Proc. IRPS*, pp. 12-16, Apr. 1994.

[3] W.-K. Chu et al., *Backscattering Spectrometry*, Academic Press, 1978.

[4] D.S. Yaney et al., "α–particle tracks in silicon and their effect on dynamic mos ram reliability," *IEEE TED*, vol. ED-26, no. 1, pp. 10-16, Jan. 1979.

[5] E. Takeda et al., "A cross section of α–particle–induced soft–error phenomena in vlsi's," *IEEE TED*, vol. ED-36, no. 11, pp. 2567-2575, Nov. 1989.

[6] J.K. Ousterhout. "An x11 toolkit based on the tcl language," *Proc. USENIX Conf.*, 1991.

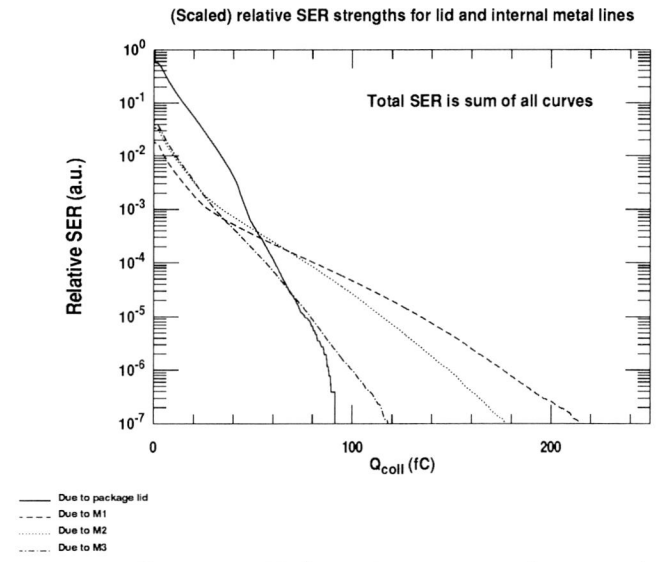

Figure 3: SER for various sources of α–particles.

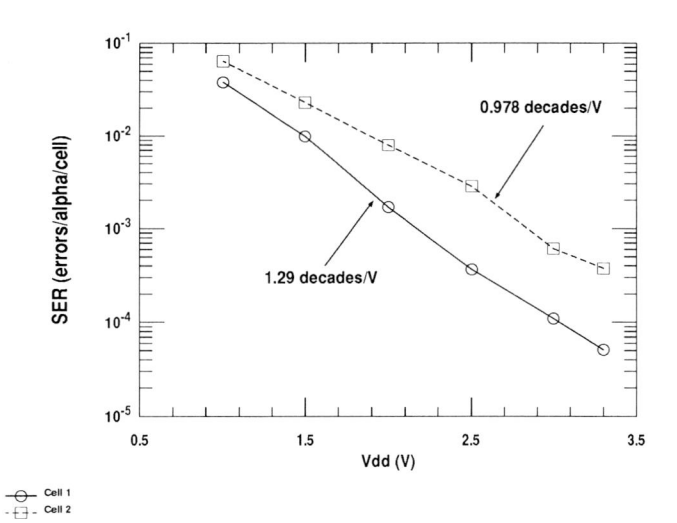

Figure 4: Calculation of SER dependence on applied external bias.

Hydrodynamic Modeling of Electronic Noise by the Transfer Impedance Method

P. Shiktorov, V. Gružinskis, E. Starikov,

L. Reggiani[†], L. Varani[‡]

Semiconductor Physics Institute,
A. Goštauto 11, 2600 Vilnius, Lithuania

[†]Istituto Nazionale di Fisica della Materia, Dipartimento di Scienza dei Materiali,
Università di Lecce, Via per Monteroni, 73100 Lecce, Italy

[‡]Centre d' Electronique de Montpellier,
Université Montpellier II, 34095 Montpellier Cedex 5, France

Abstract

The transfer impedance method in the time-domain formulation is applied to calculate the impedance field of submicron n^+nn^+ GaAs diodes in the framework of a closed hydrodynamic approach. The method enables us to determine the voltage noise-spectrum associated with velocity-fluctuations. The good agreement found with Monte Carlo simulations validates the proposed theoretical approach.

1. Introduction

The impedance-field method is widely used for noise modelling in the framework of the drift-diffusion approximation [1]. Within a one-dimensional geometry and considering single-carrier velocity-fluctuation as source of noise, the spectral density of voltage fluctuations between two terminals, $S_U(\omega)$, as measured under constant-current operation, takes the form:

$$S_U(\omega) = Ae^2 \int_0^L n(x)|\nabla Z(x;\omega)|^2 S_v(x;\omega)dx \qquad (1)$$

where ω is the cyclic frequency, e the electron charge, A the cross-section area of the device, L the lenght of the device between the probing electrodes taken along the x direction, $n(x)$ the local carrier concentration, $S_v(x,\omega)$ the local spectral density of single-carrier velocity fluctuations, $\nabla Z(x;\omega)$ the local impedance field. This latter quantity relates a perturbation of the voltage drop U with the perturbation of the conduction current-density j_d in a point x_0 through

$$\delta U(x_0;\omega) = \nabla Z(x_0;\omega)\delta j_d(x_0;\omega) \qquad (2)$$

In this work we present a procedure for the numerical calculation of the impedance field spectrum $\nabla Z(x_0;\omega)$ of a two-terminal semiconductor structure in the framework of a closed hydrodynamic approach [2,3]. The procedure is based on the modeling of a spatio-temporal response of the electrical characteristics to a local perturbation of j_d. The relevance of the approach is illustrated by calculations of $S_U(\omega)$ in submicron n^+nn^+ GaAs diodes.

2. Procedure

We describe carrier transport in the framework of the conservation equations for the carrier velocity $v(x,t)$ and mean energy $\varepsilon(x,t)$ [3]. To simulate total-current operation, we use the definition of the total current-density, J, taken to be constant both in time and space:

$$J = en(x,t)v(x,t) + \epsilon\epsilon_0 \frac{\partial E(x,t)}{\partial t} = const \qquad (3)$$

where ϵ_0 is the free-carrier permittivity, ϵ the static dielectric constant of the material and $E(x,t)$ the instantaneous local electric-field. Then the Poisson equation writes:

$$n(x,t) = N_d + \frac{\epsilon\epsilon_0}{e} \frac{\partial E(x,t)}{\partial x} \qquad (4)$$

N_d being the donor concentration. After the substitution of Eq. (4) in Eq. (3) we obtain an equation for the electric field $E(x,t)$ in the form:

$$\frac{\partial E}{\partial t} + v\frac{\partial E}{\partial x} + \frac{e}{\epsilon\epsilon_0}vN_d = \frac{1}{\epsilon\epsilon_0}J \qquad (5)$$

Equations (4) and (5) together with the velocity and energy conservation equations constitute the closed model which allows both to calculate the steady-state characteristics for a given value of J and to investigate the spatio-temporal evolution of various perturbations responsible for the electronic noise in the structure. Since J is taken to be constant, only the fluctuations of the conduction current-density $j_d = env$ are responsible for the noise. Supposing that the inititial perturbations are small enough to satisfy linearization, the difference between the time-dependent perturbed solution, $E(x,x_0,t)$, and the steady-state solution $E_s(x)$ of the unperturbed system gives the Green-function which describes the spatio-temporal evolution of the electric field perturbation caused by the local perturbation of the conduction current at the point $x = x_0$. By definition [4], the impedance field is given by

$$\nabla Z(x_0;\omega) = \int_0^L \int_0^\infty exp(-i\omega t)z(x,x_0;t)dxdt \qquad (6)$$

3. Results and discussions

Numerical simulations are performed for a n^+nn^+ GaAs diode with the following parameters: the doping levels are of $n = 10^{16}$ and $n^+ = 2 \times 10^{17}$ cm^{-3}, the cathode, n-region and anode lenghts are respectively of 0.2, 0.6 and 0.4 μm. Abrupt homojunctions are assumed. The initial perturbation is taken of the Gaussian form: $\delta(x-x_0) = exp[(x-x_0)^2/\gamma^2]/(\gamma\pi^{\frac{1}{2}})$, and $S_v(x;\omega)$ is calculated using the hydrodynamic approach developed in [2]. Typical features of the spatio-temporal evolution of the electric field perturbation initiated at $t = 0$ by the local perturbation of the conduction current in the n-region are illustrated in Figs. 1 and 2. All the results correspond to an applied voltage $U_d = 2.3$ V. Figure 1 shows the spatial distribution of $\delta E(x,t)$ at successive time moments for the initial perturbation placed in the point $x = 0.4$ μm. Figure 2 represents a time dependence of the voltage perturbation at the whole structure for the initial perturbations placed in points

$x = 0.3, 0.4, 0.5, 0.6, 0.7$ μm (respectively, curves 1 to 5). At the initial stage of the perturbation evolution, there appear two shock-waves which move along the structure toward opposite sides (see Fig. 1, solid curve). Such a behavior is typical for physical systems with nonlinear convective and diffusion processes. A quick damping of the shock-waves corresponds to an initial fast decrease of $\delta U(t)$ (see Fig. 2, $t < 0.5$ ps). In the general case, when electron heating in the n-region is small and the negative differential conductivity (NDC) is absent, the perturbation evolution ends when the shock-waves vanish entirely. If the plasma frequency is considerably higher than the damping rate of the shock-waves, the shock-wave propagation is accompanied the plasma oscillations. Such a situation is typical for the perturbations in n^+ regions. For the case considered here, the local NDC is present in the n-region and the time evolution of the perturbation is not finished with the disappearance of shock-waves. The transit through the diode of the right shock-wave creates a secondary perturbation of the electric field (caused by the dipole perturbation of carrier concentration) which grows by approaching the anode contact (see Fig. 1). The dipole-domain propagation leads to the bell-shaped evolution of the voltage perturbation (see Fig. 2, $t > 0.5$ ps). Figure 3 illustrates the spatial dependence of $|\nabla Z(x,\omega)|^2$ at the frequency $f = 10$ GHz (dashed curve). For comparison, the solid line represents the square module of the differential impedance field defined as [4]:

$$\nabla Z'(x,\omega) = \int_0^L \int_0^\infty exp(-i\omega t) z(x, x_0, t) dx_0 dt \qquad (7)$$

This quantity, which is complementary to the field impedance in Eq. (6), is used to determine the local electric field response to a perturbation of the total current flowing in the structure [3]. The spatial profile of both are reported in Fig. 3. The difference between the two shapes is due to transit-time effects associated with the formation of accumulation layers inside the n-region. Accordingly, the dashed curve, which represents the global response to a local current perturbation, exhibits its maximum around the cathode. In contrast, the continuous curve, which represents the local response to a global current-perturbation, exhibits its maximum around the anode where the accumulation-layer disappears. The spectral density of the voltage fluctuations calculated by using Eq. (1) is presented in Fig. 4 (dashed line). For comparison the solid line shows the result of the Monte Carlo simulation. The excellent agreement found between the noise spectra obtained in the framework of the hydrodynamic and Monte Carlo approaches fully supports the physical reliability of the proposed method.

This work is partially supported by the contracts ERBCHRXCT920047 and ERBCHBICT920162 from the C.E.C. and a collaborative research grant N. 931360 from NATO.

References

[1] G. Ghione and F. Filicori. *IEEE Trans. Comp. Aided Design Integrated Cir. System* **12**, 425 (1993).

[2] V. Gružinskis, E. Starikov, P. Shiktorov, L. Reggiani, L. Varani. *Appl. Phys. Lett.* **64**, 1662 (1994).

[3] V. Gružinskis, E. Starikov, P. Shiktorov, L. Reggiani, L. Varani. *J. Appl. Phys.* **76**, 5260 (1994).

[4] J.P. Nougier, J.C. Vaissiere, D. Gasquet, A. Motadid. *J. Appl. Phys.* **52**, 5683 (1981).

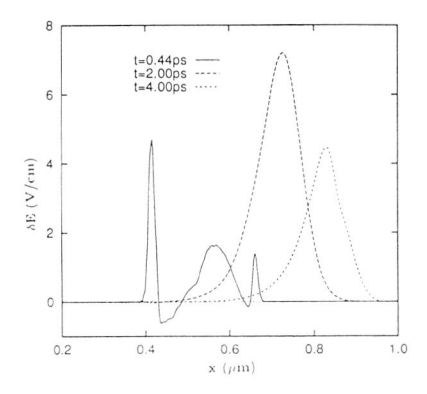

Fig. 1. Spatial profiles of the electric field perturbation at successive time moments for the initial perturbation placed at point $x = 0.4\ \mu m$.

Fig. 2. Voltage perturbation evolution for initial perturbations placed at points $x = 0.3, 0.4, 0.5, 0.6, 0.7\ \mu m$ (respectively, curves 1 to 5).

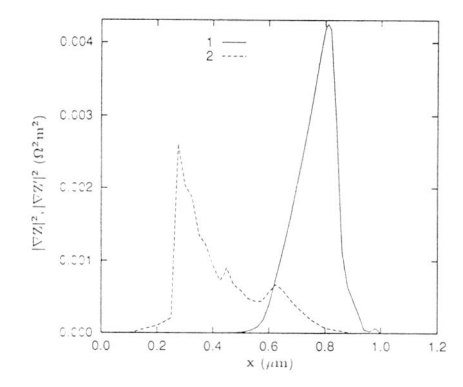

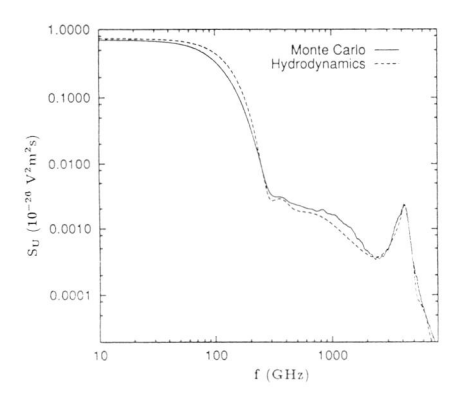

Fig. 3. Spatial profiles of the absolute value squares of the differential impedance field $|\nabla Z'(x;\omega)|^2$ (solid line) and of the impedance field $|\nabla Z(x;\omega)|^2$ (dashed line) at $f = 10\ GHz$.

Fig. 4. Frequency dependence of the spectral density of voltage fluctuations calculated by Monte Carlo and hydrodynamic approaches (solid and dashed lines, respectively)

Monte Carlo Simulation of S-Type Negative Differential Conductance in Semiconductor Heterostructures

E. Starikov, P. Shiktorov, V. Gružinskis,

L. Reggiani[†], L. Varani[‡]

Semiconductor Physics Institute,
A. Goštauto 11, 2600 Vilnius, Lithuania

[†]Istituto Nazionale di Fisica della Materia, Dipartimento di Scienza dei Materiali,
Università di Lecce, Via per Monteroni, 73100 Lecce, Italy

[‡]Centre d' Electronique de Montpellier,
Université Montpellier II, 34095 Montpellier Cedex 5, France

Abstract

Within an ensemble Monte Carlo method, we simulate a bipolar transport in a vertical layered n-GaAs/n-AlGaAs/i-GaAs heterostructure placed between n^+ and p^+ contacts. A very fast switching time and, hence, a very high frequency (up to 1 THz) of the voltage oscillations are predicted.

1. Introduction

Solid-state devices with S-type current-voltage characteristics have a wide application as fast switchers and microwave power generators. In recent years, significant attention has been payed to the heterostructure hot-electron diode (HHED) [1] based on the multilayer n^+nn^+ GaAs/AlGaAs structure. From both an experimental and theoretical analysis, this structure is found to exhibit S-type negative differential conductance (NDC) and high-frequency oscillations up to 100 GHz at temperatures below about 77 K [1,5]. To improve its high-frequency performance, we propose the replacement of the anode n^+ contact with a p^+ layer, thus introducing the possibility for an additional hole current and for a switching which occurs between the low-conductance state associated with the hot-electron current and the high-conductance state associated with the hot electron-hole current.

2. Model

We consider the vertical transport in a layered heterostructure which consists of an active region placed between heavily doped GaAs n^+ and p^+ contacts. The active region contains three layers: A, B and C. The n-GaAs layer A is placed at the n^+ contact and serves as a drift region for electrons and holes. The switching time and generation frequency depend mainly on the length of this layer. The n-AlAs layer B creates a barrier of about 0.53 eV for holes. Because of the high resistivity of AlAs, as compared with GaAs, the B layer must be as thin as possible. The undoped GaAs layer C (spacer) is introduced between the B layer and the p^+ contact to minimize the dispersion in energy of the momentum distribution of holes entering the hole barrier. As found from Monte Carlo (MC) simulations, the best device performance is obtained when the C layer thickness is of about $100 \div 300$ Å. Another important parameter is the thickness of A-B and B-C heterojunctions. Again, simulations show that abrupt junctions are undesirable for the proposed device due to their high junction-capacitance, while $20 \div 100$ Å wide junctions are more suitable. We take the junction region formed by $Al_xGa_{(1-x)}As$ where x changes monotonically from 0 to 1 with distance. All the results we shall present in the following are obtained for a structure with an active region length of 1700 Å. The length of the layer A is 1500 Å, and the doping concentration is $n = 1.5 \times 10^{17}$ cm^{-3}. Both layers B and C are 100 Å thick, with doping concentration of $n = 1.5 \times 10^{17} cm^{-3}$ and no-doping, respectively. The thickness of the A-B and B-C heterojunctions is assumed to be the same and equal to 40 Å. The doping concentration of contacts is $n^+ = p^+ = 5 \times 10^{19}$ cm^{-3}. The energy-band diagram of the proposed structure is schematically shown in Fig. 1. The graded heterojunctions enables us to smooth the discontinuity steps of the potential (or energy) at the heterojunction. Under these conditions, carrier tunneling through the potential barriers is neglected in favour of classical thermionic process. The carrier transport throughout the structure is theoretically investigated making use of a direct solution of the coupled Poisson and Boltzmann equations obtained with a MC technique. A three-valley conduction band and a single heavy-hole valence band is considered with spherically symmetric nonparabolic dispersion laws in all bands. The lattice temperature is assumed to be 300 K.

3. Results

The peculiarity of the structure is that layers A and C serve as heating regions for electrons and holes, respectively, and the intermediate layer B is the place where potential barriers prevent electrons and holes from penetrating into the diode. Figure 2 reports the current-voltage characteristics of the whole structure calculated under dynamical conditions [6]. At the lowest applied voltages both barriers act as closed gates and the system is in the lowest conductance-state. At increasing voltages, a first high conductance-state is achieved when the electron barrier acts as an open gate owing to thermionic emission of electrons via upper valleys. By further increasing the voltage, the onset of a second higher conductance-state, which corresponds to the situation where also the hole barrier acts as an open gate, takes place. The above three states (labelled respectively as I, II, and III) give rise to a conductance which is controlled, respectively, by a cold electron, a hot electron and a hot electron-hole current. The switching from one state to another is responsible for a large variation of the total current at practically the same applied voltage which, in turn, should lead to the appearance of S-type regions in the current-voltage characteristic. The simulation of the current transport throughout the given structure shows that the transition between states I and II does not lead to an S-type NDC, while the transition between states II and III is accompanied by a strong S-type behavior. This last behavior is further enphasized when the structure is connected in series with a load resistance R and in parallel with a capacitance C. Figure

3 shows the time variation of the voltage drop between the terminals of the load resistance, $U_R(t)$, when a voltage pulse $U_a(t)$ is applied to the whole circuit. The value of R is 10^{-11} Ωm^2 and $C = 200$ C_d, where C_d is the geometrical capacitance of the structure. The fast switching of $U_R(t)$ (with switching time t_{on} less than 10 ps) is independent from the sweep of $U_a(t)$. To clarify the switching mechanism, the evolution of the potential drop on the layer A, $U_A(t)$, and the average concentrations of electrons in the X-valley, $n_X(t)$, and holes in the layer A, $p(t)$, are presented in Fig. 4. Both $n_X(t)$ and $p(t)$ values are normalized to the average concentration of electrons in the layer A. The fast growth of $n_X(t)$ for $U_A(t) > 0.5$ V is associated with the onset of thermionic emission of electrons over the barrier or, in other words, to the I-II transition. The further increase of $U_A(t)$ is followed by a slow increase of the electron current which is typical in short n^+nn^+ structures. When the current becomes sufficiently large to create a voltage drop on the C region high enough for holes being able to overcome the barrier, the II-III transition starts, i.e. $p(t)$ increases from 0 to 1. The rise-time of $p(t)$ corresponds approximately to t_{on}. Obviously, for the realization of such an effect, at the hole barrier the dispersion in energy of the hole momentum-distribution in the field direction must be much less than the barrier height. The simulation shows that t_{on} of $1 \div 2$ ps can be obtained at 77 K. Under current-driven operation, the structure exhibits an oscillatory behavior in a certain range of current values. Figure 5 shows the voltage oscillations between the diode terminals when a constant total-current $j_{tot} = 3 \times 10^{10}$ A/m^2 is flowing through it. The voltage oscillates because of the periodic transitions between states II and III of the structure. The correspondig frequency of the oscilltions, 220 GHz, well correlates with a switching time of about 10 ps. Our simulations show that the terahertz frequency range can be achieved at lattice temperatures below about 77 K.

In conclusion, we have proposed a new heterostructure which, according to MC simulations, can operate as a switcher with switching times less than 10 ps. The switching mechanism is based on the fast transition from a low to a high conductance state which can occur under vertical transport. The heterostructure can be also used as a microwave generator with characteristic frequencies higher than 200 GHz. At temperatures below about 77 K the switching time can be reduced to values below 1 ps, thus implying oscillation frequencies in the THz region.

This work is partially supported by the contracts ERBCHRXCT920047 and ERBCHBICT920162 from the C.E.C. and a collaborative research grant N. 931360 from NATO.

References

[1] K. Hess, T.K. Higman, M.A. Emanuel, and J.J. Coleman, J. Appl. Phys., **60**, 3775 (1986).

[2] A.M. Belyantsev, A.A. Ignatov, V.I. Piskarev, M.A. Sinitsyn, V.I. Shashkin, B.S. Yavich, and M.L. Yakovlev, JETP Lett., **43**, 437 (1986).

[3] J. Kolodzey, J. Laskar, T.K. Higman, M.A. Emanuel, J.J. Coleman, and K. Hess, IEEE Trans. Electron Dev. Lett., **9**, 272 (1988).

[4] A. Reklaitis and G. Mykolaitis, Sol. State Electron., **37**, 147 (1993).

[5] A. Wacker and E. Schöll, Appl. Phys. Lett., **59**, 1702 (1991).

[6] V. Gruzhinsis, E. Starikov, P. Shiktorov, L. Reggiani and L. Varani, Proc. 22nd I.C.P.S., World Scientific **2**, 1628 (1995).

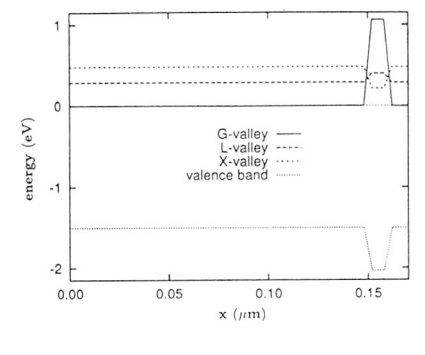

Fig. 1 - Schematic representation of the energy-band diagram of the active part of the heterostructure which consists of 0.15 μm $n - GaAs$ (layer A), 0.01 μm $n - AlAs$ (layer B) and 0.01 μm $i - GaAs$ (layer C). Solid, dashed and short-dashed lines correspond, respectively, to the Γ, L, and X-valleys of the conduction band. Dotted line shows the valence band edge.

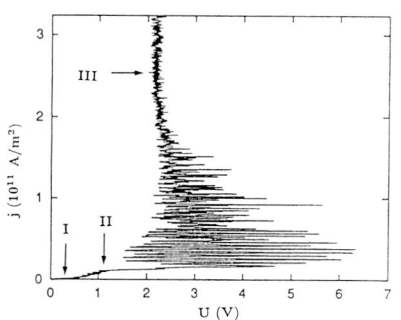

Fig. 2 - Current-voltage characteristic of the whole heterostructure obtained from a Monte Carlo simulation by taking a linearly-small increase with time of the total current j.

Fig. 3 - Time variation of the voltage applied to the whole circuit (curve 1) and of the voltage drop between the the terminals of the load resistance (curve 2).

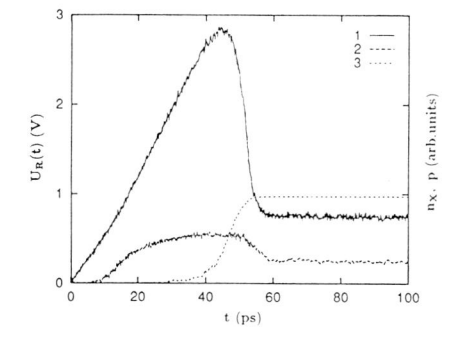

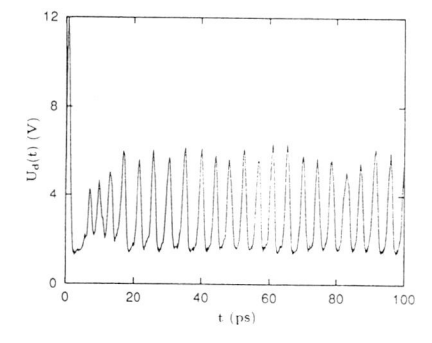

Fig. 4 - Time variation of the voltage drop at layer A, and of the average concentrations of X-valley electrons and holes in the layer A (respectively, curves 1 to 3).

Fig. 5 - Time oscillations of the voltage drop between the diode terminals calculated under constant-current operation.

Two-Barrier model for Description of Charge Carriers Transport Processes in Structures with Porous Silicon

S.P. Zimin[a], V.S. Kuznetsov[a], A.V. Prokaznikov[b]

[a]Yaroslavl State University,
Sovetskaya,14, 150000 Yaroslavl, Russia
Tel.(0852) 25-55-53, Fax(0852) 22-52-32
E-mail: phystheo@univ.uniyar.ac.ru
[b]Institute of Microelectronics RAS,
Universitetskaya,21,150007, Yaroslavl,Russia
Tel.(0852) 11-22-71, E-mail: imras@iman.yaroslavl.su

Abstract

In this work the simulation of charge carrier transport processes was carried out in the structure consisted of metal - porous silicon - mono-silicon - metal for p-type silicon on the base of hypothesis of two rectifying transitions. Approbation of this model was done and it was demonstrated a good agreement between theoretical and experimental results. The specific resistivity value was estimated for different formation regimes.

1. Introduction

Porous silicon(PS) is a new perspective material for creation different semiconductor devices, sensors, VLSI's functional elements and so on. In the present time (see [1, 2, 3] and so on) the processes of carrier drift in PS and in structures on its base are studied not complete by enough and further experimental as well as theoretical investigations are needed. The investigation of charge carrier transport in the structure metal - porous silicon - monosilicon - metal (Me-PS-MS-Me) excites special interest because the study of physical processes in this structure make it possible, on one hand, better understand PS properties and, on the other hand, to help in creation of effective electroluminescent cells in the visible region and another semiconductor devices on the base of PS. In the present work investigation of electron and hole transport was carried out in the Me-PS-MS-Me structure, which was formed on the base of p-type silicon and includes porous layer with porosity more than 50%.

2. Two-barrier model

The authors [1] have used the model, that contains one rectifying transition on the Me-PS boundary, while studying electrical properties of Me-PS-MS-Me structure in order to interpret the experimental results. The investigations during last years have showed that some experimental results can not be explained within the frameworks of this model especially for p-type silicon of high porosity. In this case two-barrier model was suggested [4], that takes into account the work of two rectifying transitions on Me-PS and PS-MS boundaries. ¿From the physical point of view the presence of these transitions is connected with dopand depletion processes in monocrystalline PS matrix at high porosity and with increasiny of forbidden band width of PS comparing to mono-silicon.

On the base of analysis of band energy diagram in the case of p-type silicon an equivalent electrical scheme was suggested. It consists of two oncoming diodes and resistor, that corresponds to a resistance of PS layer. For forward bias of voltage-current characteristic (VCC) of the structure (corresponding to the mono-Si being positive) a mathematical model was presented that described a transport of forward current through the heterojunction of PS-MS and reverse current through Schottky barrier of Me-PS. For forward biasing branch of VCC for heterojunction the expression was used that looks like this [5]:

$$I = J_s S exp(qU_1/nkT), \qquad (1)$$

for reverse biasing branch of VCC for Schottky barrier an assumption concerning generation-recombination mechanism of current transport was used [6]

$$I = S\beta^{-1/2}\sqrt{\phi_0 + U_2}, \qquad (2)$$

where I is the current through the structure, S is the contact area, J_s is saturation current density, q is the charge of an electron, n is the current transition coefficient in heterojunction, U_1 is the potential of heterojunction, U_2 is the potential of Schottky barrier, ϕ_0 is potential value of a barrier on Me-PS boundary, β is some coefficient, which depends on dopand atoms concentration, PS dielectric constant, life time of charge carriers which is considered in this work as some variation parameter. It was presupposed that PS behavior is ohmic within not large potential limits. The MS-Me was considered as ohmic one. The program "POR" was elaborated within the frameworks of these conditions taking into account charge carriers of both signs, which makes it possible to compare experimental and theoretical VCC and to calculate two very important parameters: the value of specific resistivity of PS layer and current transport coefficient for heterojunction of PS-MS.

3. An approbation of the model

An approbation of presented model was carried out for Al-PS-MS-Al and In-PS-MS-Al structures that were performed on the base of boron doped, 0.03 $Ohm \cdot cm$ silicon, boron doped, 1 $Ohm \cdot cm$ silicon an boron doped, 10 $Ohm \cdot cm$ silicon. PS was obtained by electrochemical anodization in electrolytes on the base of HF at current densities within the limits 10-60 mA/cm^2 and anodization duration time equals 5-60 min. PS layers had thickness of the order of 6-70 μm and weight porosity more than 50%. The investigations, that were carried out according to Cox-Strack method, showed that Al-PS and In-PS contacts had pronounced non-ohmic character but MS-Al transition was ohmic one. Rectifying coefficients of Me-PS-MS-Me structures within the temperature interval 120-300K were either more or less than 1. Comparison of theoretical and experimental VCC showed, that they are in good correspondence with each other and maximum differences are not exceeded 4-7%. The PS specific resistivity value, that was formed by different technological conditions, was within a large limit of values $10^4 - 10^7 Ohm \cdot cm$. It was stated that appearance of surface porous dielectric film takes place, which has specific resistivity of the order up to $10^{10} Ohm \cdot cm$ and formes by any anodization regimes.

Alongside with good conformity of theoretical and experimental characteristics, other proof of two-barrier model serviceability can serve of the experimental results, when on the same PS surface the metal was deposited in various conditions. It resulted in that, that the characteristics of Schottky barrier on various platforms were various, rectifying coefficient of Me-PS-MS-Me structures and kind of VCC is sharp changed. However use of program "POR" resulted in that, that the size determined from model of PS specific resistance under each contact was in enough narrow interval of significances, explainable by natural spread of PS specific resistance on area of film.

The usage of two-barrier model to Me-PS-MS-Me structure has allowed to describe the processes of injection hole from silicon substrate to PS, to find out the phenomenon of modulation of electrical resistance of PS layer at submission on structure of large potential and to put on record the availability of currents of outflow on perimeter of contacts. These experimental results were described by us in [4, 7]. The presence of two rectifying transitions in Me-PS-MS-Me structure manifest itself also by analysis of voltage - capacitance characteristics which will be reported in the future.

Acknowledgements

The authors are very thankful to N.V. Perch for his help in work, A.L. Vinke - for Al-PS-MS-Al samples, N.E. Mokrousov - for In-PS-MS-Al structures. This work was supported in part by the Russian of Fundamental Research Foundation under the Grant No 94-02-05460-a.

References

[1] R.S. Anderson, R.S. Muller and C.W. Tobias, "Investigation of the electrical properties of porous silicon", J. Electrochem. Soc.,vol.138, pp.3406-3411, 1991.

[2] L.A. Balagurov, N.B. Smirnov, E.A. Kozhukhova, A.F. Orlov, E.A. Petrova and A.Ya. Polyakov, "Characteristics of the contact metal/porous silicon", Izv. Acad. Nauk, vol.58, no.7, pp. 78-83, 1994.

[3] S.P. Zimin, "Hall effect in low-resistivity porous silicon", Pis'ma v JTP, vol.20, no.7, pp.55-59,1994.

[4] S.P. Zimin, V.S. Kuznetsov and A.V. Prokaznikov, "Physical peculiarities of charge carriers transport in porous silicon structures", Technical Digest Inter. Conf. ICVC'93, Taejon, pp. 179-182, 1993.

[5] S.M.Sze, "Physics of Semiconductor Devices", 2nd ed. John Willey and Sons, New York, 1981.

[6] V.I. Striha, "Theoretical foundations of functioning of metal-semiconductor contact", Kiev, 1974, 264 p.

[7] S.P. Zimin, V.S. Kuznetsov, N.V. Perch and A.V. Prokaznikov, "To the question of current transport mechanism in the structures with porous silicon", Pis'ma v JTP, vol. 20, no.22, pp. 22-26, 1994.

Monte-Carlo simulation of inverted hot carrier distribution under strong carrier-optical phonon scattering

I. Nefedov, A. Andronov

Institute for Physics of Microstructures,
Russian Academy of Science,
46 Ulyanov str., Nyzhny Novgorod,
603600, RUSSIA

Abstract

Results of Monte-Carlo simulation of hot carrier transport in semiconductors with strong carrier-optical phonon interactions under electric and magnetic fields are presented. It is shown that inverted carrier distributions and negative differential conductivity in Teraherts range can be observed in $p - Ge$ under crossed fields, in low valley of $n - GaAs$ under crossed fields and small parallel electric field and in $n - GaAs$ with weak periodic superlattice potential and narrow minigaps under strong electric field.

1. Introduction

Search for new mechanisms of inverted carrier distribution and negative differential conductivity (NDC) in semiconductors is important for development of active semiconductor devices in submillimeter range and beyond. In this report results of Monte-Carlo (MC) simulation of hot carriers transport in low doped semiconductors at low temperature concerning such mechanisms are presented. Three situations are considered: 1) the inverted population of direct optical transitions between subbands of light and heavy holes in $p - Ge$ under crossed electric and magnetic fields; 2) inversion of electron distribution over energy of cyclotron rotation in low valley of $n - GaAs$ under crossed fields with small parallel electric field; 3) rod-like carrier distribution and NDC in $n - GaAs$ with weak periodic superlattice potential and narrow minigaps under strong electric field. In all these cases highly non-equilibrium carrier distribution take place and systems can demonstrate NDC in Teraherts range.

In all these situations emission of optical phonons plays dominating role for formation of carrier distribution. In "passive" region of the momentum space (energy below an optical phonon energy $\hbar\omega_0$) scattering rate of carriers is mainly due to carrier interactions with acoustic phonons and ionized impurities and is much less than the rate in "active" region (energy above $\hbar\omega_0$) where optical phonon emission is involved. For the electric field studied carriers "flight" through the passive region almost without scattering and quickly come back after emission of optical phonon. In crossed fields carriers move along closed trajectoris with rotation centre p_c. If p_c is less than p_0 (the momentum corresponding to $\hbar\omega_0$) there are trajectories entirely lying in the passive region. So, the carriers moving along such trajectories can be accumulated. In the case of $p_c > p_0$ the motion of carriers is similar to their motion in just electric field.

The simulations presented are based on standart one-particle MC algorithm and its modifications necessary for specific problems. The modified self-scattering procedure was included for describing acoustic phonon scattering in $p - Ge$. This procedure contains additional randomization of MC algorithm and is useful when explicit form of scattering probability is absent [1]. The direct transitions for $p - Ge$ and Bragg scattering and tunneling for the superlattices were also included in MC algorithm. Every time a carrier during free flight crosses the energy corresponding to the energy involved in intersubband transition or Brillouin zone boundary a random number is chosen to find which process takes place [2, 3].

2. Laser on hot holes of $p - Ge$

In $p - Ge$ one can choose such values of crossed fields that rotation centre in light hole subband p_c lies below optical phonon while for heavy holes $p_c > p_0$. This is possible due to difference between masses of light and heavy holes. So, light holes are accumulated on the closed trajectories in the passive region and over-population of light hole subband and inversion on direct intersubband transitions may take place (see for review [4]). Detailed MC simulations of hot holes in $p - Ge$ in crossed fields at low temperature have shown that this system demonstrates negative conductivity at the frequencies above 3 Teraherts and stimulated emission at these frequencies can be observed. Influence of the stimulated emission on distribution functions of holes and optimum orientation of the fields relative to crystal axis are also estimated by MC simulation [5, 2].

3. Cyclotron rotation inversion in $n - GaAs$

Carriers get into neighbourhood of point $p = 0$ after emission of optical phonon. So, in crossed fields carriers are accumulated around "main" trajectory crossing the point $p = 0$. If $p_c \leq p_0/2$ the main trajectory is closed and corresponds to large energy of cyclotron rotation (high Landau levels) and inversion over cyclotron rotation energy can take place. MC simulation of electron transport in low valley of $n - GaAs$ has shown that in crossed fields this inversion doesn't occur. Nevertheless adding a small electric field parallel to the magnetic results in higher population of main trajectory as compared with trajectories with small cyclotron energy. MC calculations have shown that in this case the inversion over energy of cyclorton rotation is observed. However, this inversion strongly depends on system parameters and so it will be difficult to obtain this inversion in real samples.

4. Rod-like electron distribution and NDC in superlattices with weak periodic potential and narrow minibands

In superlattices with narrow minigaps at low temperature we consider the situation when top of the first miniband is just below $\hbar\omega_0$. In the electric field a carrier performs Bloch oscillations before it is scattered in the passive region or reaches the active region due to interminiband tunneling. Under these conditions narrow stretched along the field almost symmetric relative to $p = 0$ carrier distribution ("roding") and dynamic NDC at the frequencies higher than Bloch frequency occur. The NDC is a result of carrier bunching in Brillouin zone in the lowest miniband produced by interminiband tunneling [3]. The results of MC simulation demonstrate the roding in carrier distribution and dynamic NDC in Teraherts range.

References

[1] V. A. Kozlov, I. M. Nefedov, Phys. Stat. Sol. (B), vol. 109, p. 393, 1982.

[2] A. V. Muravjev, I. M. Nefedov, S. G. Pavlov, V. N. Shastin, Kvantovaia Elektronika, vol. 20, p. 142, 1993 (russian).

[3] A. A. Andronov, I. M. Nefedov, Abstract 7th Int. Conf. on Superlattices Microstructures and Microdevices, Banff, Alberia, Canada, p. 2.85, 1994.

[4] A. A. Andronov, in Spectroscopy of Nonequilibrium Electrons and Phonos (C. V. Shank and B. P. Zacharchenya eds), Elsevier Science Publishers B. V., p. 169, 1993.

[5] A. V. Muravjev, I. M. Nefedov, Yu. N. Nozdrin, V. N. Shastin, Fizika i Technika Poluprovodnikov, vol. 23, p. 1739, 1989 (russian).

Algorithms and Models for Simulation of MOCVD of III-V Layers in the Planetary Reactor

T. Bergunde[a], M. Dauelsberg[b], Yu. Egorov[**], L. Kadinski[b], Yu. N. Makarov[b], M. Schäfer[b], G. Strauch[*], M. Weyers[a]

[a]Ferdinand–Braun–Institut für Höchstfrequenztechnik,
Rudower Chaussee 5, D-12489 Berlin, Germany
[b]Lehrstuhl für Strömungsmechanik, Universität Erlangen–Nürnberg,
Cauerstr. 4, D-91058 Erlangen, Germany
[*]AIXTRON GmbH,
Kackerstr. 15–17, D-5100 Aachen, Germany
[**]Advanced Technology Center,
P.O. Box 160, 198103 St. Petersburg, Russia

Abstract

Advances in development of mathematical models and numerial techniques for modelling of MOCVD in the Planetary Reactor^{TM} are presented. Importance of coupled flow and mass transport calculations, accurate modelling of radiative heat transfer and complex chemical interactions is discussed. Advantages and disadvantages of block–structured and unstructured grid algorithms are considered.

1. Introduction

In recent years mathematical modelling and numerical simulation of metalorganic chemical vapour deposition (MOCVD) has become useful to find the optimal set of process parameters and to investigate growth mechanisms. This work is concerned with modelling of growth of III-V semiconductors in the multiwafer Planetary Reactor^{TM}, which is especially suitable for large scale production of heterostructures for various kinds of semiconductor devices due to the high degree of growth rate and compositional uniformity across the wafer and good utilization of precursor materials [1, 2]. In the reactor to be modelled, several wafers are placed on rotating satellites which in turn rotate around the central axis of the susceptor plate. Group III metalorganics and group V hydrides, both mixed with the H_2 carrier, are introduced through separate inlet channels at the center of the reactor flowing radially outwards along the growing layers. The ceiling plate is thermally coupled to the water cooled reactor top by a cooling gas mixture of Ar and H_2 that allows for controlling of the ceiling temperature through its composition.

Simplified modelling aproaches have been applied to predict growth rate profiles and to optimize technical design in this type of CVD reactor [3, 2]. However, mass transport calculations have been done in a separate step after flow and temperature prediction and radiation heat transfer has been considered in a simplified manner.

An advanced approach was proposed in [4, 5] taking into account coupled flow, heat transfer and mass transport, including chemical reactions and formation of deposits on the reactor ceiling. The aim of this work is to develop further the mathematical models and algorithms used for modelling of growth in the Planetary Reactor^{TM}.

2. Mathematical models

The mathematical model for MOCVD growth consists in the solution of coupled partial differential equations which describe conservation of total mass and momentum, heat transfer and the chemical species' mass transport in the reactor, including multicomponent diffusion and chemical reactions. Flow is coupled with mass transport of the predominant gas phase species H_2, AsH_3, TMGa and MMGa. Coupling of flow and mass transport is crucial in modelling of growth in the Planetary Reactor^{TM}, since the molar weight of the gas mixture is not uniform, especially near the entrance region where strong intermixing occurs between the flows coming from two inlet channels.

Thermal radiative heat transfer is modelled by assuming a non-participating gas mixture and semi-transparent grey-diffusive quartz walls. Radiative heat transfer is coupled with heat conduction in the quartz wall and solid parts of the reactor, including thermal solid/fluid interaction, and conductive heat transfer in the cooling gas above the ceiling plate. Detailed modelling of radiative heat transfer is crucial for the accurate determination of the temperature distribution at the ceiling plate. The ceiling temperature influences sensitively the kind and thickness of deposits. If the ceiling temperature is too high - deposition of polycristalline GaAs can take place, whereas if it is too cold - condensation of As can occur. Therefore, accurate calculation of the temperature distribution on the ceiling is important to reduce the deposits. Wave length dependencies of radiative properties of the quartz wall and the effect of deposits on radiative properties and, therefore, on the ceiling temperature are shown to be important to predict temperature distribution on the ceiling [6, 7]. In Fig 1 isotherms (a) and streamlines (b) of the flow in the reactor are shown. Heating of the nozzle and ceiling due to radiative heat transport can be seen.

By modelling of the species' mass transport homogeneous decomposition of TMGa to MMGa and deposition of polycrystalline GaAs on the ceiling are taken into account. Growth of III-V heterostructure layers on the wafer is performed at mass transport limited growth conditions. The growth rate is determined by mass transport of the group III species only, because arsine (AsH_3) is introduced into the reactor at high

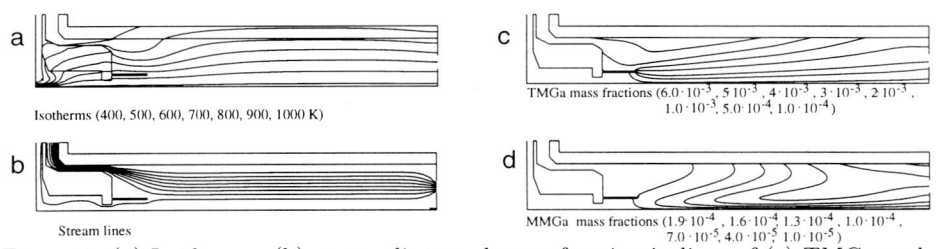

Figure 1: (a) Isotherms, (b) stream lines and mass fraction isolines of (c) TMGa and (d) MMGa in the reaction chamber. Growth conditions: inlet flow ratio between upper and lower inlet R= 85, total flow rate F= $21.5\frac{1}{min}$, $F_{AsH_3} = 100\frac{1}{min}$, growth temperature $T_g = 750°$ und system pressure $P_0 = 200$ mbar.

excess. TMGa (c) and MMGa (d) mass fraction distributions are shown in Fig. 1. Homogeneous decomposition of TMGa occurs actively in the heated gas near the susceptor. Verification of the models is performed by comparison of calculated and measured growth rate distributions (Fig. 2). Polycrystalline deposit grows on the quartz ceiling at kinetically limited conditions due to the lower temperature on the cooled ceiling compared to the substrate. Therefore, a heterogeneous rate law is assumed as mass transport boundary condition on the ceiling. Thickness of the deposited GaAs film on the ceiling can be affected by the cooling gas composition (Ar/H$_2$ ratio), this can be seen in Fig. 3. The model predicts reasonably the thickness of the deposits and effect of the cooling gas composition, as shown in Fig. 3, and modelling can help to adjust the optimal Ar/H$_2$ ratio to reduce the formation of the polycrystalline deposit on the ceiling.

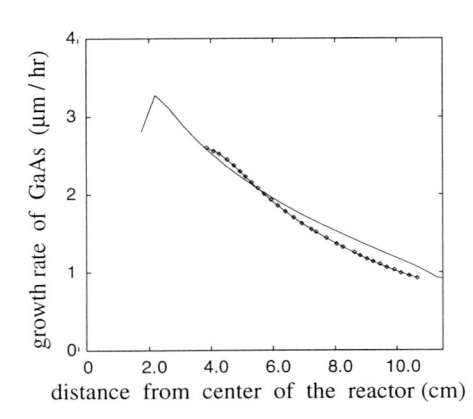

Figure 2: Calculated (—) and measured (-◇-) growth rates on the non-rotating wafer at a flow ratio of R=85; total flow rate F=17.2 $\frac{1}{min}$, $T_g = 750°$ und P_0=200 mbar.

Figure 3: Measured (–) and calculated (- - -) layer thickness of ceiling deposits for cooling gas compositions x_{H_2}=20%, x_{H_2}=40%, x_{H_2}=60%, and x_{H_2}=80%; F=21.5 $\frac{1}{min}$, R=53 $\frac{1}{min}$, F_{AsH_3}=50 $\frac{1}{min}$ T_g= 750°, P_0=200 mbar.

3. Numerical algorithms

The models described above are implemented into two different solution procedures:

- a finite volume method using block-structured non-orthogonal collocated grids for two-dimensional flows as described in detail in [7].

- a procedure using an unstructured grid finite volume algorithm with adaptive local grid refinement [8].

The first approach seems to be very effective in terms of accuracy and convergence rate, especially because fast solution procedures and a multigrid method based on a Full Approximation Scheme for the non-linear coupled systems of equations can be employed for a speed up of the convergence rate. The disadvantage of the block-structured approach appears with resolving the complex reactor geometry and with grid generation. In the case of unstructured grids, it is easier to introduce the complex reactor configuration, for example, provided from CAD, and automatic grid generation is possible, which makes this approach advantageous for the optimization of the reactor geometry and design. Another advantage of the unstructured grid approach is the possibility for an easy incorporation of an adaptive local grid refinement. Additional grid volumes are introduced automatically at the parts of the computational domain, where accuracy of the calculations does not satisfy a preselected criteria. However, the unstructured grid approach requires significantly more computing time and memory on a comparably fine grid than the blockstructured one . In Figure 4 the grids generated for the reactor are shown using both numerical approaches.

An optimal way to combine the advantages of both approaches is to use hybrid unstructured/structured grids. In this case it would be possible to preserve the high accuracy and convergence rate of structured algorithms and the geometrical flexibility of the unstructured approach. This will be a topic for forthcoming research.

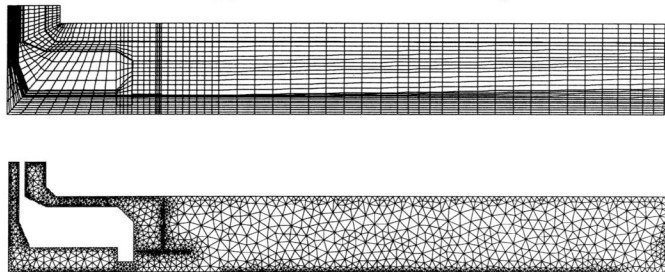

Figure 4: Comparison of structured (on top) and unstructured grids (on bottom)

References

[1] P. M. Frijlink. *J. Crystal Growth*, 93:207–215, 1988.

[2] P. M. Frijlink, J. L. Nicolas, H. P. M. M. Ambrosius, R. W. M. Linders, C. Waucquez, and J. M. Marchal. *J. Crystal Growth*, 115:203–210, 1991.

[3] C. Waucquez and J. M. Marchal. Technical report, Polyflow s. a., Louvain–la–Neuve, Belgium, 1991.

[4] T. Bergunde, F. Durst, L. Kadinskii, Yu. N. Makarov, M. Schäfer, and M. Weyers. *J. Crystal Growth*, 145:630–635, 1994.

[5] T. Bergunde, D. Gutsche, L. Kadinskii, Yu. N. Makarov, and M. Weyers. *J. Crystal Growth*, 146:564–569, 1995.

[6] L. Kadinskii, Y. N. Makarov, M. Schäfer, V. Yuferev, and M. Vasil'ev. *J. Crystal Growth*, 146:209–213, 1995.

[7] F. Durst, L. Kadinskii, and M. Schäfer. *J. Crystal Growth*, 146:202–208, 1995.

[8] A. I. Zhmakin. *V. Int. Congr. Comp. Appl. Math.*, 1994.

An Approach for Explaining Drift Phenomena in GTO Devices Using Numerical Device Simulation

S. Eicher[a], F. Bauer[b], and W. Fichtner[a]

[a]Integrated Systems Laboratory, ETH–Zürich
Gloriastrasse 35, CH-8092 Zürich, SWITZERLAND
[b]ABB Semiconductors Ltd.
Fabrikstrasse 3, CH-5600 Lenzburg, SWITZERLAND

Abstract

A possible cause for IGT drift in GTO thyristors has been identified using numerical 2D device simulation. An increase of the surface recombination velocity under the oxide between the gate and cathode contacts leads to a small degradation of the upper npn transistor gain, which in turn rises the IGT. This work focuses on the requirements on the geometrical discretization and on the procedure to extract the DC current gains of the individual transistors that form the GTO thyristor.

1. Introduction

State-of-the-art commercial GTO (gate turn-off) thyristors are subject to a drift of

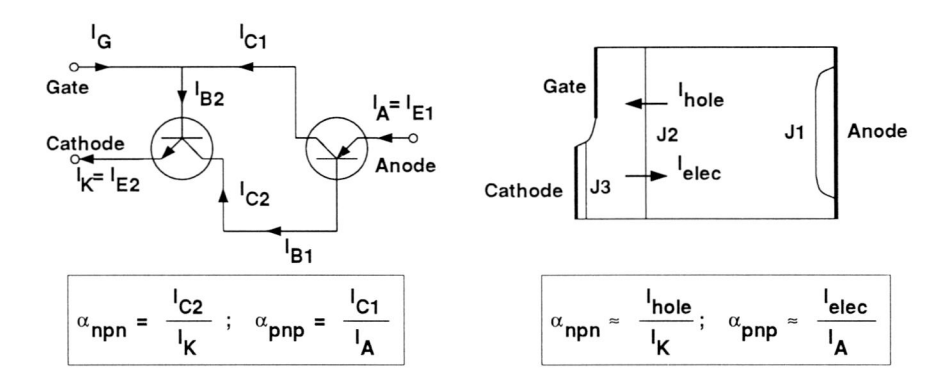

Figure 1: Definition of DC gains

the gate trigger current IGT to higher values, which may even reduce their lifetime as a high power switch. During the operation of a GTO thyristor, the oxide layer between gate and cathode is periodically stressed by high electric fields that occur in its vicinity. There results a degradation of the oxide [1] that reduces the DC gain α_{npn} of the upper transistor in Fig. 1 [2]. The measurement of the DC gains of a GTO device is limited by the following restrictions:

- Only the α_{npn} can be determined
- The measured device must have anode shorts
- The hole injection from the anode side must be close to zero

Using numerical device simulation, these restrictions can be overcome. The electron and hole currents through the center junction J2 can be determined and from that, the individual transistor current gains can be calculated. The gains must be known to determine the trigger point, since at the trigger point the equation holds:

$$\alpha_{npn} + \alpha_{pnp} + I_A \frac{\partial \alpha_{pnp}}{\partial I_A} = 1.$$

2. Requirements on the geometrical discretization

At very small forward biases of the junctions J1 and J3, the forward current density and therefore the gains α_{pnp} and α_{npn} are dominated by recombination processes in the very narrow space charge region. Hence, the grid lines of the geometrical discretization have to be very dense in these areas. Figure 2 shows a comparison of the calculated recombination rates at low current densities of devices with three different grids. The value on the x-axis defines the distance from the junction J3. Grid lines exist at the position of the symbols. The figure clearly shows that the grid lines around the

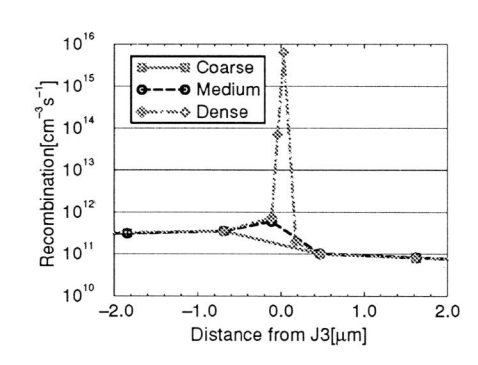

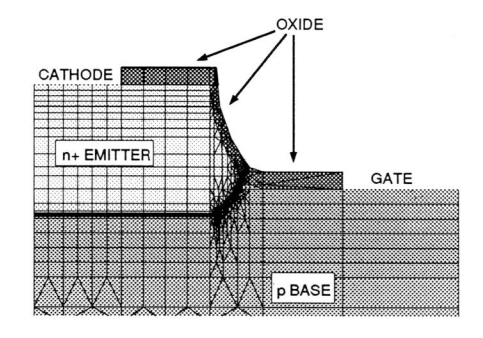

Figure 2: Recombination for different grids

Figure 3: Proper discretization of junction J2

junctions must not be further apart than about $100 nm$. Otherwise the recombination is not calculated properly and the derivation of the current gains is not correct. The grid shown in Fig. 3 leads to an accurate calculation of the recombination processes across the junction J3. This grid was generated using MDRAW [3].

3. Derivation of the transistor current gains

In general, it is not possible to measure the transistor current gains in a GTO thyristor since they depend by definition on internal currents. Devices with shorts on the anode side, however, can be regarded as npn transistors as long as the injection from the anode side is negligible. This measurement is restricted to the α_{npn} and to small forward biases. There exist no such restrictions when using numerical device simulation, because it is possible to determine the values of internal currents.

The following procedure was carried out for the determination of the individual current gains:

1. Using the device simulator DESSIS [4], a positive anode bias was applied to the GTO thyristor before the gate current was ramped up to the trigger point. During the ramping of the gate current, contour plot files were saved at regular distances.

2. The electron and hole current densities through the center junction J2 were extracted from the contour plot files and integrated over the width of the device, thus obtaining I_{elec} and I_{hole} (see Fig. 1).

3. The DC current gains were approximated as

$$\alpha_{npn} \approx \frac{I_{elec}}{I_{cathode}}, \quad \alpha_{pnp} \approx \frac{I_{hole}}{I_{anode}}.$$

Making use of a script language, all these steps were carried out automatically. This method is applicable to GTO devices with anode shorts as well as GTO thyristors with homogeneous anode.

Figure 4 shows typical gains as a function of the conduction current densities for a device with a shorted anode (left side) and a homogeneous anode without shorts (right side). While the α_{npn} look alike in both cases, the α_{pnp} are quite different. Due

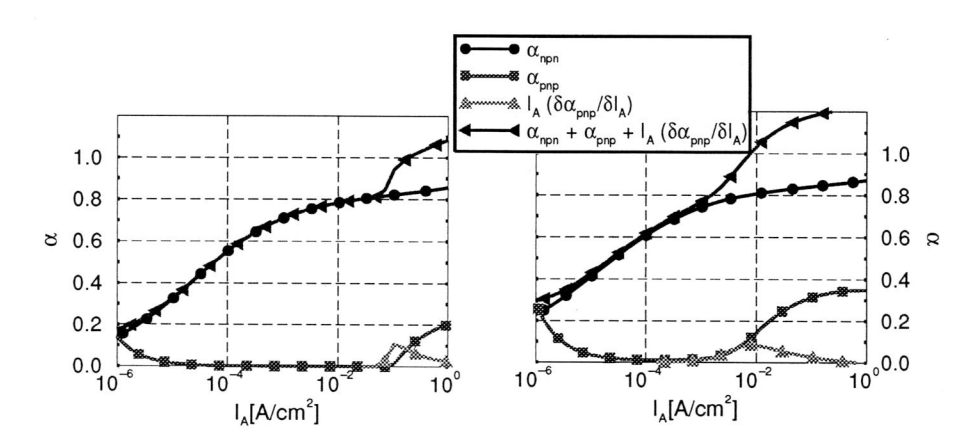

Figure 4: DC gains of shorted (left) and un-shorted (right) GTO during triggering

to its homogeneity, the anode of the device without shorts starts to inject at much smaller current densities and also the trigger point is reached at a smaller current density.

4. Drift simulation results and discussion

The degradation of the oxide layer between gate and cathode, which results from the periodical stress by high electric fields, was modeled in the simulation by varying the recombination velocity at the interface between silicon and oxide (see Fig. 3). The influence on the gain of the npn section of the shorted GTO was then calculated. Furthermore, the impact of the gain variation on the gate trigger current (IGT) and the holding current (IH) was studied. The results are summarized in Tab. 1. The α_{npn} was sampled at an anode current density of $100 mA/cm^2$, making sure that the device was not yet triggered.

Table 1: Influence of variations of the surface recombination velocity on DC gain, IGT and IH

Recombination Velocity		$0 cm/s$	$2 \cdot 10^3 cm/s$	$5 \cdot 10^3 cm/s$	$1 \cdot 10^4 cm/s$
GTO with anode shorts	α_{npn}	0.822	0.818	0.814	0.807
	$IGT[mA]$	1030	1060	1100	1164
	$IH[A]$	175.1	183.1	187.1	203.1
GTO without anode shorts	α_{npn}	0.740	0.736	0.730	0.722
	$IGT[mA]$	25.8	26.9	28.5	31.2
	$IH[A]$	4.03	4.35	4.41	4.98

Table 1 shows that the degradation of the oxide (increase of the surface recombination velocity from $0 cm/s$ to $1 \cdot 10^4 cm/s$) leads only to minor changes in the α_{npn} of less than 3%. The device without anode shorts shows more pronounced changes because of the lower absolute current level. The monotonous decrease of the gain with higher surface recombination velocities can be attributed to the higher recombination current, which adds up to the gate current and therefore lowers the gain.

The impact on the device parameters is much larger than that on the gain. The gate trigger current IGT rises by 13% and by 21% for the shorted and the non-shorted device, respectively. Due to its much lower absolute current level, the relative drift of the non-shorted GTO thyristor is substantially higher than that of the shorted device. Nevertheless, such a GTO is less sensitive in an application, because of its very small absolute value of the IGT drift. The simulated values coincide reasonably well with the increase of the IGT resulting from drift, wich is typically measured during the lifetime of a GTO device. Numerical device simulation thus contributes significantly to the analysis and understanding of reliability issues in modern high power GTO thyristors.

References

[1] H. S. Momose et al., IEEE Trans. on Electron Devices, vol. 41, pp 978–987, 1994
[2] P. Taylor, *Thyristor Design and Realization*, John Wiley & Sons, Chichester. 1987
[3] ISE Integrated Systems Engineering AG, *MDRAW - User's Guide*, Switzerland, 1994
[4] ISE Integrated Systems Engineering AG, *DESSIS - Reference Manual*, Switzerland, 1994

Parallel 3D Finite Element Power Semiconductor Device Simulator Based on Topologically Rectangular Grid

A.R. Brown, A. Asenov, S. Roy and J.R. Barker

Department of Electronics and Electrical Engineering
Glasgow University
Glasgow, G12 8QQ, Scotland, UK

Abstract

Here we report on the development of a new parallel, scalable and portable 3D finite element power semiconductor device simulator. The emphasis in the design of this simulator is placed on the FE grid generation, on the optimised parallel generation and assembly of the discretization matrices, and on the development of a suitable, scalable linear solvers. For discretization use topologically rectangular FE grid based on non-rectangular bricks.

1. Introduction

The cellular structure of most power devices requires a 3D solution of the basic semiconductor equations. The octagonal or hexagonal shape of typical power MOSFET, IGBT or MCT cells [1] and their complex doping distributions require a finite element (FE) discretization. In many cases more than one cell should be included in the simulations in order to obtain an adequate description of the device behaviour. The size and the computational complexity of the problem make it a distinguished candidate for massively parallel implementation. Only recently has a more systematic approach been applied to the design of parallel device simulation codes [2]. To achieve better results the design of the parallel simulation software should reflect the architecture of the parallel platforms.

Here we report on the development of a parallel, scalable and portable 3D finite element power semiconductor device simulator. It is based on a spatial decomposition of the simulation domain over an array of processors [3]. This approach minimises the interprocessor communications by reducing the ratio between the bulk and the surface of the partition subdomains. The emphasis is placed on the generation of topologically rectangular FE grids amenable to the domain decomposition approach, on the optimised parallel generation and assembly of the discretization matrices, and on the development of suitable, scalable linear solvers.

2. Spatial Device Decomposition

Our parallel power semiconductor device simulator is designed for Multiple Instructions Multiple Data (MIMD) parallel computers with distributed memory. It is based on the spatial device decomposition approach.

To enhance the portability of the simulator, the code is split into two distinct parts: a hardware dependent communications harness and the simulation engine. The communications harness provides all global and local communications between the processors necessary for the proper operation of the simulation engine. The simulation engine is designed to operate on an arbitrary 1D, 2D or 3D array of processors including a single processor. This means that the solvers are virtually independent of the processors topology if the necessary global and local communications are provided.

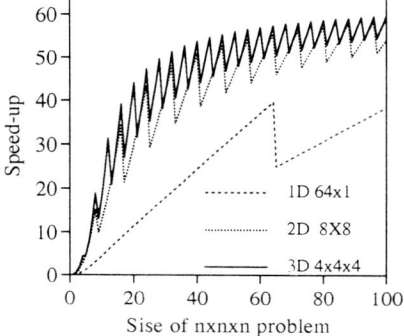

Figure 1: Theoretical speed-up of a hypothetical 3D linear solver based on a 1D, 2D and 3D array of processors

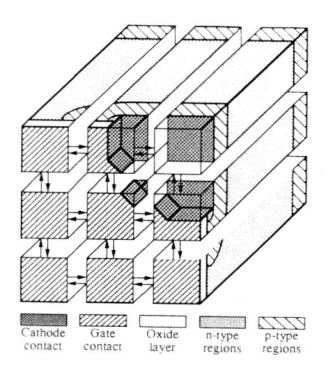

Figure 2: Partition of the 3D device simulation domain on a 2D array of mesh connected processors

The simulator can work with both a rectangular finite difference grid and a topologically rectangular finite element grid. This simplifies significantly the partitioning of the solution domain on the array of processors and the design of the communications harness. It is clear that the best processor configuration for spatial device decomposition of the topologically rectangular 3D grid is a 3D array of mesh connected processors. This is illustrate in Fig 1 where the theoretical speed-up of a hypothetical linear solver [4] is plotted as a function of the size of a $nxnxn$ rectangular grid. The grid is partitioned on 64 processors organised in three different configurations: a 64 processor pipeline, an 8x8 2D array and a 4x4x4 3D array of processors. However we are restricted to a 2D array of processors on our Parsytec parallel computers. The spatial device decomposition of a 3D device on a 2D array of processors is illustrated in Fig 2. The solution domain is automatically decomposed into NxM subdomains in one grid plane (i,j). All corresponding grid point in the third grid direction k lie on the same processor. To achieve better speed-up each subdomain in the (i,j) plane should be as square as possible.

3. Solution Domain and Grid Generation

The solution domain and the generated grid depend on the structure of the simulated device and the doping concentrations inside. The grid generation in the solution domain proceeds on a single processor. After the grid generation the doping profile is assigned to the grid points, the material type is assigned to each finite element and the boundary conditions are identified. The generated grid with the doping, material and boundary conditions information is than distributed over the processor array. The distributed semiconductor solver is universal and independent of the particular device structure.

The grid generation approach is illustrated on an example a typical octagonal IGBT. Because of the symmetry the simulation domain involves only one quarter of the cell.

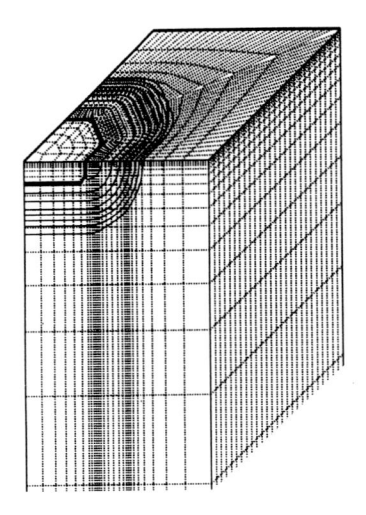

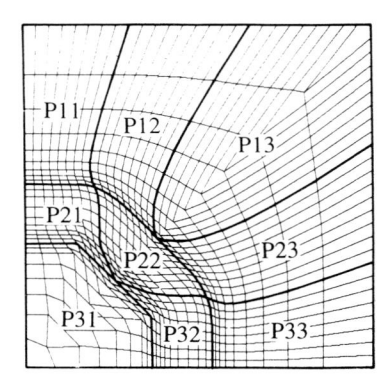

Figure 3: Solution domain and discretisation of octagonal IGBT. The boron profile in the cathode region is indicate

Figure 4: Partition of the (i,j) plane of the domain from Figure 3 on an array of 3x3 processors

The finite element grid is a topologically rectangular grid. It keeps the number of grid points along the grid lines constant in each one of the index directions i and j. The grid is based on distorted bricks. The grid generation process is determined by specified contours in the solution domain. In this particular example the guiding contours are the shape of the gate electrode and the metallurgical p-n junctions in the device. An example of the partitioning of the (i,j) plane of the solution domain on an array of 3x3 processors is given in Figure 4.

4. Parallel Discretization and Solution

We have adopted the decoupled Gummed procedure for the solution of steady-state problems and a modification of the decoupled Mock procedure based on time dependent version of the Poisson equation. The both schemes are simple and amenable to parallelization.

The Galerkin finite element approach has been adopted to solve the Poisson equation on a finite element grid. The integration over the distorted brick finite elements during the discretization was carried out by a linear isoparametric mapping of each element into an unit cube. For the parallel matrix generation and assembly we use a node based partition of the grid and node based assembly approach in which the solution subdomain on each processor is scanned not element by element, but node by node. This leads to almost 100% efficiency when the number of nodes in the i and j directions are divisible by the corresponding numbers of processors [4]. To solve the nonlinear system arising from the discretization of the Poisson equation we have adopted a Block Newton SOR scheme. The parallel performance of the method is illustrated in Figure 5.

The discretization of the current continuity equation on the distorted brick finite element grid is more complicated. We are examining three possible approaches. The simplest one is to divide each distorted brick into six tetrahedral elements and to carry out a standard 3D control volume Gummel type of discretization. The second approach is to use a 3D analogue of the 2D Gummel like discretization developed for quadrilateral finite elements. Finally shape functions exponentially fitted to the potential distribution could be used. A parallel implementation of the BiCGSTAB(2) method has been adopted for the current continuity equation.

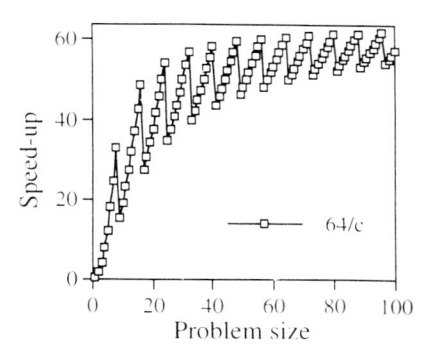

Figure 5: Speed-up of the Block Newton SOR method for a cubic $n \times n \times n$ problem on an 8x8 array of transputers

Figure 6: Electric field distribution in aoctagonal cell IGBT at 600V anode voltage

Finally an example of the electrical field distribution in a cellular IGBT at 600V anode voltage is illustrates in Fgure 6.

5. Conclusions

In this work we have presented our systematic approach to the design of a parallel finite element 3D power semiconductor device simulator. Our attempt was to build a portable and scalable parallel code which runs with high efficiency on variety of parallel platforms. To achieve this goal special measures were undertaken at each stage of the software development.

Acknowledgements

This work is funded by the EPSRC under grant GR/H23085 as part of a LINK/PEDDS project. We would like to thank Peter Waind and David Crees of GEC Plessey Semiconductors for their information and helpful discussions.

References

1. Brown, A R, Asenov A, Barker, J R, Jones, S, Waind, P, "Numerical simulation of IGBTs at elevated temperatures", *Proc. Int. Workshop on Computational Electronics*, Ed C. M. Snowden, University of Leeds Press, pp. 50-55, 1993.
2. T.F. Pana, E.L Zapata and D.J. Evans, "Finite element simulation of semiconductor devices on multiprocessor computers" *Parallel Computing*, vol. 20, pp. 1130-1159, 1993.
3. Asenov, A, Barker, J R, Brown, A R, Lee, G L: "Scalable parallel 3D finite element nonlinear Poisson solver", *Massively Parallel Processing Applications and Development*, Eds. L. Dekker, W. Smit and J.C. Zuidervaart, Elsevier, pp 665-672, 1994.
4. 7. Asenov A, Reid, D, Barker, J R: "Speed-up of scalable iterative linear solvers implemented on an array of transputers", *Parallel Computing*, Vol 20, pp 375-387, 1994.

Investigation of Silicon Carbide Diode Structures via Numerical Simulations Including Anisotropic Effects

E.Velmre[a], A.Udal[a], F.Masszi[b] and E.Nordlander [b,c]

[a] Tallinn Technical University, Inst. of Electronics,
Akadeemia tee 1, EE-0026 Tallinn, ESTONIA, evelmre@le.ttu.ee, audal@le.ttu.ee
[b] Scanner Lab, Electronics Dept., Inst. of Technology, Uppsala University,
P.O.Box 534, S-75121 Uppsala, SWEDEN, ferenc@sim.teknikum.uu.se
[c] Univ. College of Gävle-Sandviken, Dept. of Technology,
P.O.Box 6052, S-80006 Gävle, SWEDEN, enr@hgs.se

Abstract

The Poisson's and continuity equations based SiC-DYNAMIT-1DT/2DT simulators with anisotropic material analysis capabilities were developed. A comparison of experimental and simulated forward I/V curves for three 6H-SiC P^+NN^+ diodes is presented and the related model parameter adjustment problems are discussed. The influence of the strong electron mobility anisotropy on carrier distributions is investigated and the existence of a "mobility anisotropy induced anomalous charge accumulation effect" is demonstrated.

1. Introduction

The progress in SiC device technology is enhancing the development of the simulators in order to link material data to the electrical behaviour of devices. Recently, a 4.5kV 6H-SiC diode has been reported [1] with acceptable electrical performance. However, in this work, like in others, e.g. [2,3] the measured results, particularly the high forward current voltage drops, disagree with simulated results. This emphasises the need of a correct physical description, i.e. first of all the development of temperature and doping-dependent lifetime and mobility models. Another problem for simulator development is the material anisotropy, especially related to 6H-SiC (power applications), and less to 4H-SiC (high-frequency applications). In the following, a remarkable influence of the strong electron mobility anisotropy on the forward-biased pn-junction behaviour is demonstrated.

In connection with the present work, the 1D and 2D nonisothermal simulators SiC-DYNAMIT-1DT/2DT, developed at Tallinn Technical University, were also tested against the recent version of MEDICI [4] with anisotropic material simulation capabilities (but without any physical default data). If the same input data and mesh were carefully specified for forward-biased tasks like the ones described below, then a fair coincidence of results was obtained.

2. Comparison of experimental and simulated forward I/V characteristics

In the present 1D-simulations the $n_i(T)$ [5], $\mu_{n||},\mu_p(N_D+N_A,T)$ [6], $\varepsilon_{||}$=10.03 [7] (|| denotes "parallel to c-axis") and the conventional single-level SRH recombination model were used. The doping-dependent bandgap narrowing, Auger recombination, electron-hole scattering, mobility field-dependence and incomplete impurity ionization models were omitted, mainly due to the lack of reliable data. The impact ionization [5] was not applied either in the present forward-biased simulations. Fig.1 shows results for a realistic 6H-SiC diode structure [3].

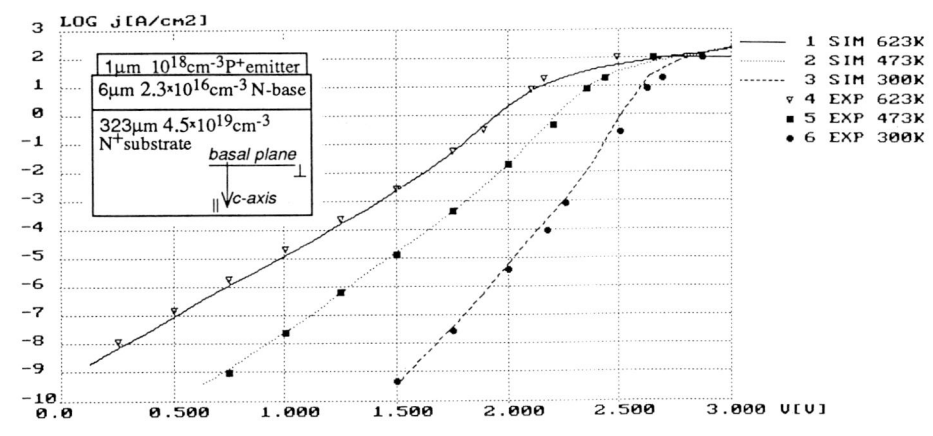

Fig.1 Simulated and experimental forward I/V characteristics at 3 temperatures

To achieve a better fit in Fig.1, the mobility temperature-dependence exponent value of -1.8 was added to the Caughey-Thomas like doping-dependence formulae from a recent work of Cree Corp. researchers [6] (authors of [6] suggest values -2.07...-1.8). Greater exponent absolute values would too strongly decrease mobilities at higher temperatures, and it would cause the I/V curves crossing point to move down to current densities not matching present experiments. For SRH capture times the values $\tau_{po}(300K)=\tau_{no}(300K)=0.2$ns (emitters) and 2ns (base) were used together with the temperature-dependence exponent of 1.5 (applied only for the base region). However, it should be pointed out, that to achieve better fit, considerably lower lifetime values have been used here in comparison with the 100ns range lifetimes observed in [1].

Fig.2 gives comparison results for two analogous diode structures, for which the matching of simulation and experiment is quite poor. These 4.5kV and 2.2kV structures were specified according to refs. [1] and [2], respectively. In the simulations $\tau_{po}=\tau_{no}=1$ns was taken for all regions, except N-bases where $\tau_{po}=430$ns [1] (4.5kV device) and $\tau_{po}=100$ns (2.2kV device) were set.

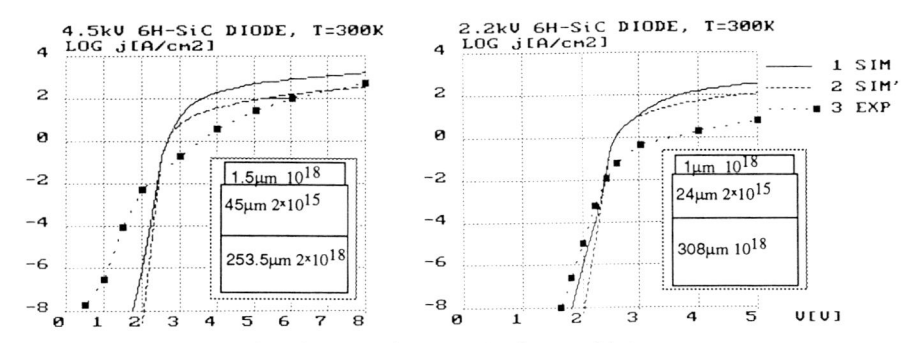

Fig.2 Some additional comparison results for two high-voltage diode structures

Main discrepancies between simulation and experiment can be observed at higher current densities, where simulated voltage drops remain remarkably lower than experimental ones. This may be due to additional contact or probe resistances in experiments, but, for instance, can be explained also as a result of declining lifetime injection-dependence caused by non-midgap recombination centers. The dashed lines in Fig.2 correspond to simulations, where the SRH recombination model parameters n_1, p_1 were specially chosen to achieve a 10-fold lifetime decrease at high injection. Mismatch in the low-current region of the 4.5kV device may be caused by device self-heating during the experiments.

3. Two-dimensional simulations with anisotropic effects

To investigate the anisotropy influence, a special 16μm×16μm 2D-structure was constructed (see Fig.4). The doping profile of this abrupt-junction diode has been taken by the diode from Fig.1 (but including only 1μm of the thick N⁺substrate). Furthermore, it has a total geometrical symmetry with respect to the x=y line. The right and bottom contacts were grounded and the voltage was applied to the P⁺emitter contact in the upper left part of structure. The dimensions were selected so that 1A current corresponds to an average current density ≈1A/cm² for all electrodes. Thus any anisotropy influence could clearly be observed as a difference between the right and bottom contact currents or as a symmetry distortion in physical quantity distributions.

While not paying attention to breakdown operation modes (strong impact ionization anisotropy) and nonisothermal heat transfer problems (heat conductivity anisotropy ≈0.7 for 6H-SiC), it can be stated for 6H-SiC, that beside the low anisotropy of the dielectric permittivity $\varepsilon\perp/\varepsilon\|$=9.66/10.03 [7], the μ_n anisotropy $\mu_{n\perp}/\mu_{n\|}$=4.8 is already a remarkable number and it stays quite constant over a wide temperature and doping range [6]. Reverse-biased simulations show that the weak 4% ε anisotropy causes only ≈2% difference in the field penetration depth and maximum field, following a square root law.

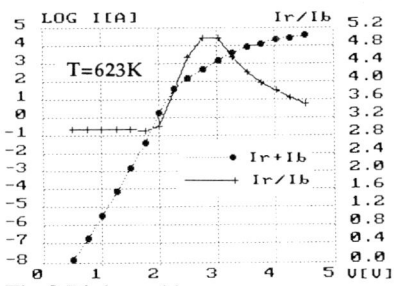

Figs.3,4 present the main results of the forward-biased simulations to investigate the mobility anisotropy influence. The ratio of horizontal (i.e. direction with greater mobility) to vertical current moves from a constant level at low currents to the peak at medium currents. At high currents, the transition to a more uniform current distribution may be observed again (see also the flowlines in Fig.4). As numerical experiments show, these changes are defined both by mobility anisotropy and structure geometry.

Fig.3 Right and bottom contact current sum and ratio versus forward bias

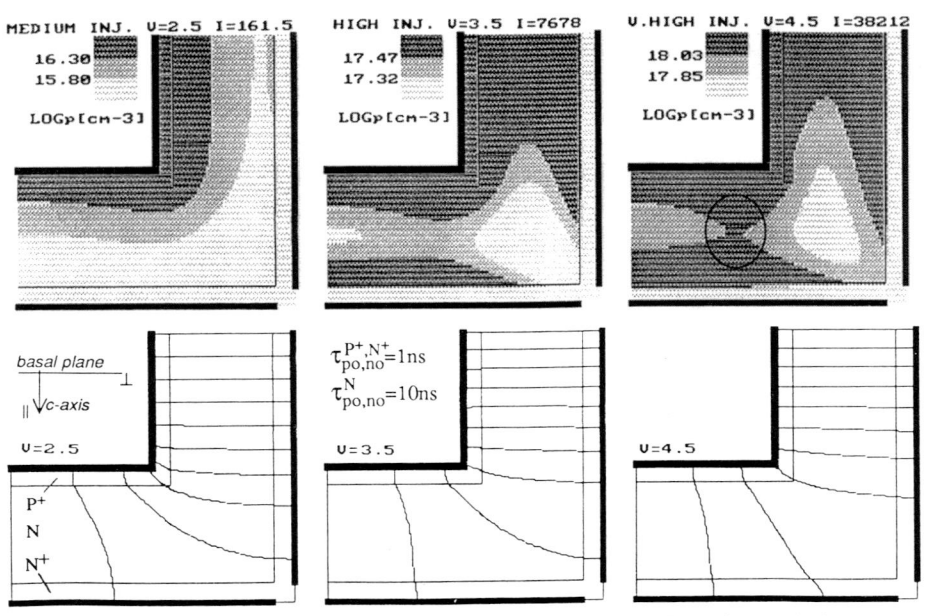

Fig.4 Hole density and total current flowline maps at three forward bias

4. The anomalous charge accumulation effect

Fig.4 shows that the "normal" p(x,y) distributions (where higher carrier density corresponds to higher mobility, like in the quasi-1D-regions near to the top and left boundaries of the structure) obtain an anomalous character by the transition to high injection mode near the pn-junction corner. An explanation for this "mobility anisotropy induced anomalous charge accumulation" is given on the basis of pn-junction and base voltage drops balance analysis in Fig.5. This theory assumes high injection in the N-base, spatially constant hole densities $p_{||}, p_\perp$ and as a quite rough estimation, the equality of vertical and horizontal current densities. The last assumption is confirmed by flowline maps in Fig.4 and by the fact that the considered area dimensions are of the same range as the charge carrier diffusion lengths. Compared to the predicted $p_{||}/p_\perp = 2.8$, Fig.4 gives a value ≈ 1.5. This can be explained by the inequality of vertical and horizontal current densities (greater horizontal current is equivalent to smaller A in formulae of Fig.5).

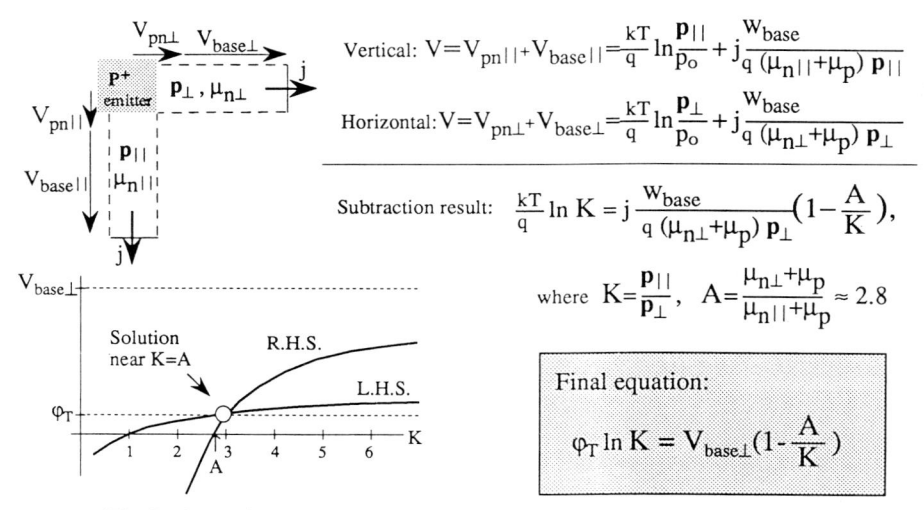

Vertical: $V = V_{pn||} + V_{base||} = \frac{kT}{q} \ln \frac{p_{||}}{p_o} + j \frac{W_{base}}{q(\mu_{n||}+\mu_p) p_{||}}$

Horizontal: $V = V_{pn\perp} + V_{base\perp} = \frac{kT}{q} \ln \frac{p_\perp}{p_o} + j \frac{W_{base}}{q(\mu_{n\perp}+\mu_p) p_\perp}$

Subtraction result: $\frac{kT}{q} \ln K = j \frac{W_{base}}{q(\mu_{n\perp}+\mu_p) p_\perp}(1 - \frac{A}{K})$,

where $K = \frac{p_{||}}{p_\perp}$, $A = \frac{\mu_{n\perp}+\mu_p}{\mu_{n||}+\mu_p} \approx 2.8$

Final equation:
$$\varphi_T \ln K = V_{base\perp}(1 - \frac{A}{K})$$

Fig.5 Theoretical explanation of the anomalous charge accumulation effect

5. Acknowledgement

This work has been supported by the Estonian Science Foundation (Project No.568), The Royal Swedish Academy of Sciences (Contract No.1489), the Swedish Institute and the Swedish National Board for Industrial and Technical Development (NUTEK Project P2305-1).

References

[1] O.Kordina, "Growth and Characterization of Silicon Carbide Power Device Material", *Ph.D.Thesis, Linköping Studies in Science and Techn. Dissertation No.352*, Linköping University, Sweden, 1994.
[2] P.G.Neudeck, D.J.Larkin, J.A.Powell and L.Matus, "2000V 6H-SiC p-n junction diodes grown by chemical vapour deposition", *Appl.Phys.Lett.* 64(11), 14 March 1994, pp.1386-1388.
[3] J.A.Edmond, H.-S. Kong and C.H. Carter Jr., "Blue LEDs, UV photodiodes and high-temperature rectifiers in 6H-SiC", *Physica*, v.B185, 1993, pp.453-460.
[4] TMA MEDICI, Two-Dimensional Device Simulation Program, Version 2.0 (including the Anisotropic Material Advanced Application Module), Sept. 1994, TMA Inc., Palo Alto, USA.
[5] E.Velmre and A.Udal, "Numerical Simulation of a Silicon Carbide Diode", *Proc. of the 4th Biennial Conf. (Baltic Electronics Conf. 1994)*, Tallinn(Estonia), Oct.9-14, 1994, pp.559-566.
[6] Schaffer W.J., Negley G.H., Irvine K.G.and Palmour J.W., "Conductivity anisotropy in epitaxial 6H and 4H SiC", *MRS Spring Meeting*, San Francisco, Apr.4-8, 1994, 6 p.
[7] L.Patrick and W.J. Choyke, "Static Dielectric Constant of SiC", *Phys.Rev.*, v.B2, 1970, pp.2255-2256.

A new physical compact model of CLBTs for circuit simulation including two-dimensional calculations

D. Freund[a] and A. Kostka[b]

[a]Braun AG, T-EAE,
Frankfurter Strasse 145, D-61476 Kronberg, GERMANY
[b]Solid-State Electronics Laboratory, Technical University of Darmstadt,
Schlossgartenstrasse 8, D-64289 Darmstadt, GERMANY

Abstract

A new, physics-based compact model for CMOS-compatible, lateral bipolar transistors is presented. For the calculation of DC collector and base currents including the influence of gate-voltage, Early- and Late-effect and the extended base resistance, a specially developed, fully 2-dimensional analysis is used resulting in closed-form analytical equations that only need device geometry and technological data as parameters. In this way, parameter extraction is drastically simplified. Together with state-of-the-art AC components, the complete model has been implemented in ELDO; its predictive abilities favourably compare with measurements on devices fabricated in different technologies, where standard models tend to fail.

1. Introduction

CLBTs (CMOS-compatible lateral bipolar transistors, see Figure 1) have gained an important position in analog circuit design during the last years. They can be found to be used in numerous applications all over the field of MOS-based analog low-power circuitry. In contrast to their favoured use, an appropriate physical compact model has not been presented for state-of-the-art simulation tools. Model approaches by Ankele [1] and Arreguit [2] are neither as physics-based as necessary to reflect the complete device behaviour nor do they offer the possibility of an inclusion in other than in-house simulators. In this paper, we present a new compact model of CLBTs which is able to describe all the important, mostly 2-dimensional, features of device operation in an analytical, physics-based form and can be incorporated in the commercial simulator ELDO.

2. Outline of the model conception

To stay related to the internal device physics and geometry, the equations for the *collector currents* in flatband operation have to solve the underlying 2D-boundary problem. For this reason, in [1] the authors introduced a special application of conformal mapping techniques to the base region of the transistors which give rise to the definition of geometrical factors, depending only on process and geometry parameters in the following generalized form for npn-CLBTs (see Fig.1):

$$I_{CLFB} = en_i^2 \frac{W_E}{N_B} K_L(W_E, N_B, g, h, \alpha, x_j) \, e^{\frac{V_{BE}}{V_T}} \tag{1}$$

$$I_{CV} = en_i^2 \frac{W_E}{N_B} K_V(W_E, N_B, g, h, \alpha, x_j) \, e^{\frac{V_{BE}}{V_T}} \tag{2}$$

The two-dimensional aspects of the device operation are described by K_L and K_V, where the transistor geometry and the bias voltages V_{BE}, V_{BCL} and V_{BCV} influence the space-charge extensions and hence g and h (see Figure 1). To include the effect of a *variable gate-voltage* on the lateral collector current in (1), an analytical solution of Poisson's equation applied to the MOS-structure has been derived. For this, it can be shown that a one-dimensional approach is sufficient for all realistic device configurations and can finally be combined with (1) in the following way (Debye-length $L_D = \sqrt{\frac{\varepsilon_{Si} U_T}{eN_B}}$, (3) and (4) still hold for npn-CLBTs):

$$\psi_S = -U_T \ln\left(\frac{\varepsilon_{ox} L_D}{\sqrt{2}\varepsilon_{Si} d_{ox}} \left(\frac{V_{GB} - V_{FB}}{U_T}\right)^2 + 1\right) \tag{3}$$

$$I_{CL} = I_{CLFB} + 4\frac{en_i^2 W_E D_N L_D}{\sqrt{2} \, gN_B} \left(e^{\frac{\psi_S}{U_T}} - 1\right) e^{\frac{V_{BE}}{U_T}} \tag{4}$$

A 2D equation for the *base current* component due to diffusion in the emitter again results from conformal mapping calculations. The recombination current in the base volume is calculated by solving the corresponding differential equations analytically under use of specially derived mathematical decomposition techniques. Altogether, this allows the introduction of new coefficients in the equation for the base current which include all two-dimensional aspects of diffusion (K_{EB}) and recombination (K_{BB}) in the device.

$$I_B = en_i^2 (\frac{W_E}{N_E} K_{EB}(W_E, \alpha, x_j) + \frac{1}{\tau_{rec} N_B} K_{BB}(W_E, h, g)) \, e^{\frac{V_{BE}}{V_T}} \tag{5}$$

Emitter doping N_E and base minority lifetime τ_{rec} are the controlling parameters.

Early- and Late-Effect have a comparatively high influence on the collector currents due to the low well (base) doping. This causes a strong bias-dependence of both, leading to a variable output conductance (Early-Effect) and a variable transconductance (Late-Effect). In our model, these features are taken into account by employing appropriate equations for the extensions of the space-charge regions at emitter and both collectors into the base, which both affect the geometrical factors K_L and K_V and hence the collector currents.

Again due to the extended configuration and the low doping of the base in CLBTs, the *base resistance* has to be modelled significantly different to the usual approaches developed for vertical devices. The model uses new equations for the resistance including spreading effects in a typical rectangular or concentric CLBT-layout as well.

These model equations describe the transistor action by only using process- and layout-related parameters without the inclusion of numerical fitting factors at all. Thus, the conception of our model ensures that the two-dimensionality of the device operation is reflected as accurate as analytical approaches allow, leading to an outstanding performance related to existing other approaches which are not extensive enough to obtain the same thorough and powerful predictive abilities.

3. Model implementation and performance in ELDO

The complete model is implemented by adding existing model components to the features from section 2. Figure 2 shows the equivalent circuit for the CLBT. The currents I_{CE}, I_{EC}, I_{CVN}, I_{CVI} represent the lateral and vertical collector currents in active and inverse operation, respectively. I_{BEN}, I_{BCI}, I_{BBN}, I_{BBI} denote the base current components due to diffusion and recombination. R_{BD}, R_{BM}, R_{BI} add up to the base resistance. The equations for these model elements employ the analytical results of the calculations outlined in Section 2. The capacitances and the G/R-components of the base current are modelled comparable to SPICE; the series resistances at emitter and both collectors are chosen to be lumped elements without relation to the underlying process environment.

The model has been implemented in the simulator ELDO using the CFAS-interface[4]. According to Figure 2, all in all 60 parameters are needed. Compared to Ankele's [1] and Arreguit's [2] approaches, in which the CLBT is described by a parallel combination of two Gummel/Poon- models for the BJT-components and an additional MOS model for the gate effect, our model needs less parameters with reduced extraction effort due to the two-dimensional conception, so that fitting is only necessary to adjust the high-current behaviour of the device via an appropriate determination of series resistances and high-injection parameters.

To show the technology-dependent predictive performance of the model, measurements on CLBTs fabricated in different processes have been performed to obtain the DC-characteristics (Fig. 3-5). Evidently, a close correspondence of measured data and model predictions has been obtained. Due to the basic conception of the model even the fitting results for the high-current parameters remain physically reasonable.

The simplified extraction procedure demonstrates the success of the physics-related outline of the new model, which ensures that the parameters establish a tight correlation between electrical device behaviour and process- and layout-parameters, hence being predictively able to keep track with changes in both, layout and technology.

4. Conclusion

In this paper, we presented a new process- and layout-related compact model for CLBTs under use of two-dimensional calculations which led to analytical, closed-form equations for the DC-behaviour of the devices. Parameter extraction for our model is very easy. Due to the strong links to device physics, their majority can be taken straight from process- and layout data without fitting, making the model very useful for analysis tasks as well as for an application in layout-generating synthesis tools. Good accordance between model predictions and measurements could be shown for different technologies. The implementation in ELDO requires less parameters than other approaches, but describes the device operation more thoroughly. In the AC-domain, up to the moment, our model does not differ from others. By deriving more physics-based equations for the capacitances, the number of parameters can be finally restricted even further, resulting in a model with a minimum of needed parameters and very easy extraction rules.

References

[1] B.Ankele, F.Schrank, *Proc. EuroAsic 91*, p. 192-197, 1991
[2] X.Arreguit, "Compatible Lateral Bipolar Transistors in CMOS Technology", Dissertation EPFL Lausanne, 1989

[3] D.Freund, A.Klös, A.Kostka, in: *Proc. SISDEP 93*, pp. 425-428, Springer-Verlag, Wien (1993)

[4] ELDO-CFAS User's Manual, Rel.: 1.0, Anacad GmBH, 1994

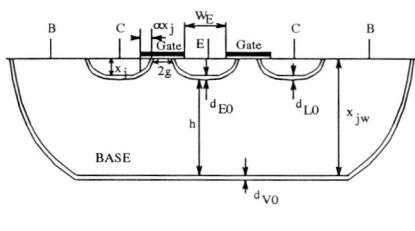

Fig. 1: Cross-section of a CLBT. d_{L0}, d_{V0}, d_{E0} denote the space-charge extensions into the base region at collectors and emitter, α the lateral diffusion coefficient.

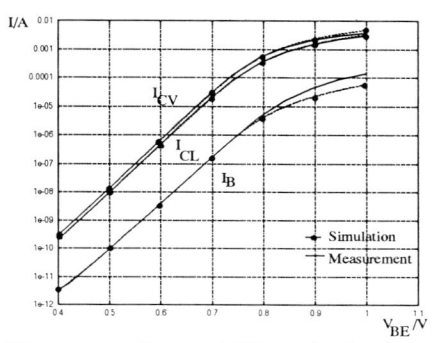

Fig. 3: Gummel-Plot of the base and collector currents in a npn-CLBT ($V_{CLB} = V_{CVB} = 3V, V_{GB} = -3V$)

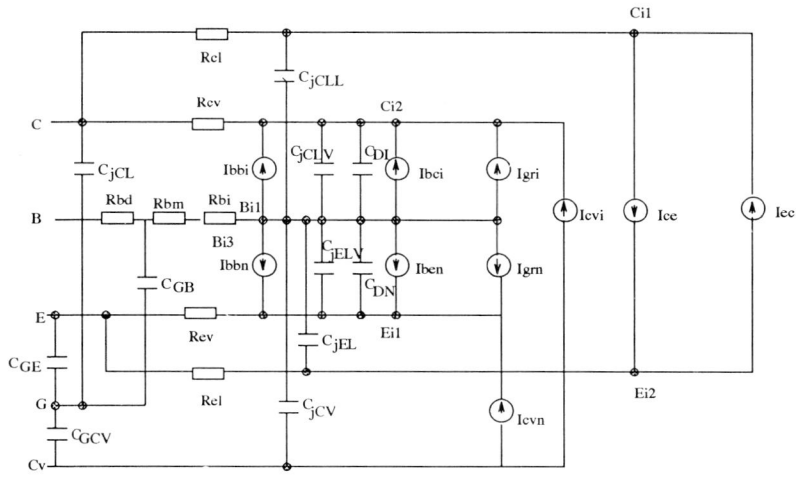

Fig. 2: Equivalent circuit for a CLBT

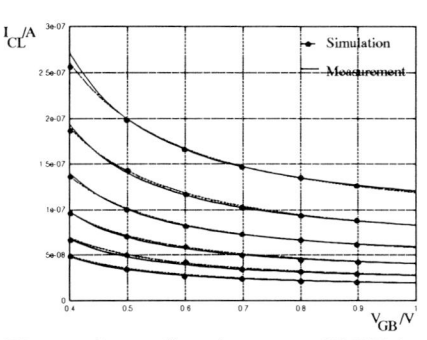

Fig. 4: Gate effect in a pnp-CLBT ($0.45V < V_{EB} < 0.5V, V_{CLB} = -3V, V_{CVB} = -3V$)

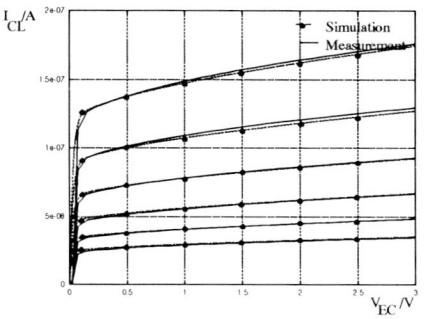

Fig. 5: I_C/V_{EC}-curves of a pnp-CLBT ($0.5V < V_{EB} < 0.55V, V_{GB} = 3V, V_{CVB} = -3V$)

Combining 2D and 3D Device Simulation with Circuit Simulation for Optimising High-Efficiency Silicon Solar Cells

Gernot Heiser[a], Pietro P. Altermatt[b], James Litsios[c]

[a]School of Computer Science and Engineering,
The University of New South Wales, Sydney 2052, AUSTRALIA
[b]Centre for Photovoltaic Devices and Systems, UNSW, Sydney 2052, AUSTRALIA
[c]Integrated Systems Lab, ETH-Zurich, 8092 Zurich, SWITZERLAND

Abstract

This paper reports on simulation techniques developed for the modelling and optimisation of complete, $2\times2\,\text{cm}^2$, high-efficiency silicon solar cells. We use three-dimensional (3d) device simulation to extract a J-V curve of an interior section of a cell. 2D simulations of the cell perimeter are then used to correct the J-V curves for the loss of carriers across the cell boundary. The resulting characteristics are input to a circuit simulation which connects the various cell sections into a model of a full cell. The J-V curve which results from that simulation can be directly compared to measured data. We find excellent agreement between simulation and measurement.

1. Introduction

The *passivated emitter, rear locally diffused* (PERL) silicon solar cell (Fig. 1), developed at the University of New South Wales (UNSW), features the highest independently confirmed efficiency, 24.0 %, of any silicon solar cell under unconcentrated terrestrial illumination conditions [1]. The goal at UNSW is to improve PERL cell efficiency to 25 %.

In the last two years 2d device simulation has been used extensively at UNSW to characterise PERL cells and has lead to a detailed understanding of the effects limiting internal cell efficiency [2]. A first 3d study [3] performed a year ago showed that 3d effects, resulting from the point contact pattern at the rear of the cell (Fig. 1), do not have a significant effect on efficiency if a single *irreducible* section of the cell (see below) is considered in isolation. However, a detailed examination of the losses due to a cell's serial resistance, R_s, showed that losses resulting from the resistivity of the front contact grid (Fig. 2) were larger than anticipated [4]. These losses are mostly a result of the so-called *non-generation loss*: Owing to the voltage drop along the contact grid, different parts of the cell operate under slightly different bias. Therefore, not all parts of the cell can operate under maximum power point (MPP) conditions at a fixed external bias, resulting in a degradation of conversion efficiency. This effect had previously been underestimated. The study, which was based on a combination of

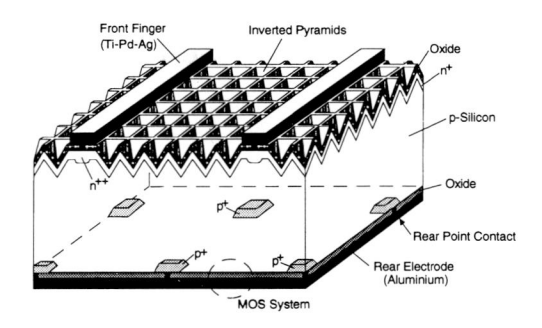

Figure 1: The UNSW PERL Si solar cell (partial view).

measurements and simulation, has led to a change in the front contact grid geometry, which contributed significantly to the latest efficiency improvement from 23.5 % to 24.0 %.

From this experience it is obvious that in order to obtain a fully optimised PERL design it is necessary to model the cell in its entirety. So far our 3d simulations were restricted to the *irreducible* section given as '*waver thickness*' × '*half the front contact grid spacing*' × '*half the rear point contact spacing*', 370μm × 400μm × 133μm for a typical PERL cell. This is sufficient under the assumption of negligible metal resistance and negligible perimeter effects.

While the study by Altermatt [4] showed that the first of these assumptions is not valid, the second one is not valid either. We know from simplified analytical models that perimeter effects degrade the MPP voltage by about 4mV [2], a small but noticeable effect. However, no detailed studies of perimeter effects in solar cells have been reported to date.

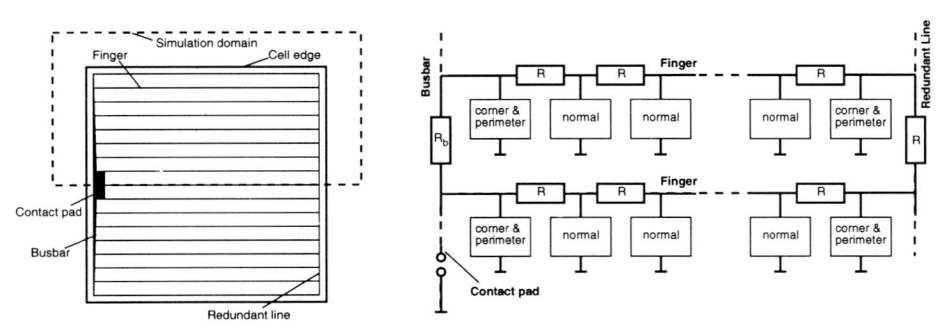

Figure 2: Schematic view of the front contact pattern of a PERL cell (left) and equivalent circuit (right).

2. Full Cell Simulations

As a full 3d simulation of a PERL cell is beyond the limits of present hardware and simulation technology, we chose an approach which combines device with circuit simulation. A J-V curve is extracted from a standard 3d simulation of an irreducible

Figure 3: Comparison of 2d simulations of unit cell (left) and perimeter region (below), the widths of the simulation domains are 400μm and 3067μm respectively. The region covered by the unit cell simulation corresponds to the leftmost part of the perimeter simulation. The plots show electron current density (light colours imply strong current) and electron current flow at MPP conditions.

section of the cell. The whole cell can be thought as consisting of many of these *unit cells*, connected by the metal contacts. Under the assumption of negligible current flow across the boundaries of the unit cells, the effect of the metal can be lumped into ohmic resistors connecting the unit cells (Fig. 2). We can then model the operation of the whole cell using a circuit simulation.

The intersections between the contact fingers and the collecting *busbar*, or the *redundant line* at the opposite end (cf. Fig. 2), require special treatment, as these areas cannot be reasonably modelled by one of our unit cells. We therefore performed a separate 3d simulation to yield a J-V curve of such *corner cells*.

The perimeter regions present a different problem: To avoid efficiency losses resulting from the damage created by sawing, the cells remain embedded in the wafer for measurement. An accurate model needs to account properly for the loss of carriers across the edge of the cell. Due to the huge minority carrier diffusion lengths (several mm in PERL cells) the active perimeter region is quite large, too large for a 3d simulation. However, since the perimeter region is essentially two dimensional, 2d simulations are adequate for the determination of the effect of the perimeter region. To this end we perform a 2d simulation of the perimeter region, as well as a 2d simulation of the unit cell. We take the difference of the two J-V curves to represent the perimeter effect. This is then added to the J-V curve of a unit cell to yield a 3d result which is *corrected* for perimeter effects; such a corrected unit cell can then be used to represent a section of the edge region of the PERL cell. The current densities across the perimeter are several orders of magnitude smaller than the current densities within the illuminated area, hence errors introduced by this correction scheme are small compared to the correction itself.

A typical PERL cell has a finger spacing of 800μm and a rear contact spacing of 250μm. With a total size of $2\times2\,\text{cm}^2$, the cell can be thought to consist of approx. 1000 unit cells, which, together with a similar number of connecting resistors, is quite a large number of devices for a circuit simulation. However, it is not necessary to simulate that many devices. The internal unit cells along a contact finger are all identical and several of them can be lumped into a single device (properly scaled). We found that 10 internal unit cells per finger were sufficient to obtain a converged result. With 13 fingers (for symmetry reasons only half the cell needs to be simulated, cf. Fig. 2) this leads to a circuit comprising 156 unit solar cells and a similar number of resistors.

3. Implementation and Results

In order to simplify the feeding of device simulation results into the subsequent circuit simulation, we extended the mixed-mode device and circuit simulator DESSIS_ISE [5] by a new circuit device type, vi, which is a voltage-controlled current source. This device is defined by an I-V curve in tabular form, which we obtain as an output from our device simulations, also performed using DESSIS_ISE. We therefore require the following for the simulation of a full PERL cell: one 2d unit cell simulation (5k grid points, 20 min CPU time), one 2d perimeter simulation (25k, 3h), one 3d unit cell simulation (65k, 12h), and one 3d corner cell simulation (110k, 24h), plus the circuit simulation (312 circuit elements, 1h). CPU times quoted are for a full J-V curve on a 60MHz Sun SS-20. To determine R_s we require a second J-V curve for a slightly different light intensity [4].

Fig. 4 shows the simulated voltage profile of a PERL cell as compared to measurement. The model parameters for the simulations where based on experimentally determined values wherever possible. The voltage profiles are clearly in excellent agreement. An extensive parameter study with the aim of optimising the contact grid design is presently under way.

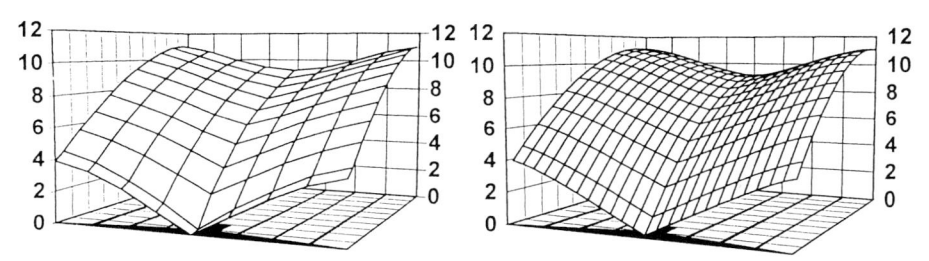

Figure 4: Measured (left) and simulated (right) voltage profile [mV] of the front contact grid of a PERL cell.

Acknowledgement

This work is supported by a grant from the Australian Research Council.

References

[1] J. Zhao, A. Wang, P. P. Altermatt, S. R. Wenham, and M. A. Green. 24% Efficient silicon solar cells. In *1st World Conf. Photovoltaic Energy Conversion*, Waikoloa, Hawaii, USA, December 1994. IEEE. To be published.

[2] A. G. Aberle, P. P. Altermatt, G. Heiser, S. J. Robinson, A. Wang, J. Zhao, U. Krumbein, and M. A. Green. Limiting loss mechanisms in 23-percent efficient silicon solar cells. *J. Appl. Physics*, 1995. To appear.

[3] G. Heiser and A. G. Aberle. Numerical modelling of non-ideal current-voltage characteristics of high-efficiency silicon solar cells. In *5th Int. W. Num. Modeling Proc. and Dev. for Integr. Circ.*, pages 177–80, Honolulu, USA, June 1994. IEEE.

[4] P. P. Altermatt. Two-dimensional numerical modelling of high-efficiency silicon solar cells. Diploma thesis (physics), University of Constance, Germany, July 1994. Thesis project carried out at UNSW, Sydney, Australia.

[5] ISE Integrated Systems Engineering AG, Zurich, Switzerland. *DESSIS 1.3.6: Manual*, 1994.

A New Quasi-two Dimensional HEMT Model

C. G. Morton, C. M. Snowden and M. J. Howes

Department of Electronic and Electrical Engineering,
University of Leeds, Leeds, UK.

Abstract

A Quasi-two Dimensional HEMT model is presented which for the first time describes accurately the I-V characteristics close to device pinch-off and at high drain current. The model uses a new analytical model for describing the injection of charge into the buffer material. This analytical description works in conjunction with an asymptotic boundary condition in the charge control solution to drastically increase the computational efficiency of the model. Electron temperature is also included in the charge control model to calculate the degeneracy factor of the electron gas under the gate where velocity overshoot occurs.

1 Introduction

The needs of the microwave engineer have changed dramatically over the last few years. Present-day MMICs and MMMICs generally make use of GaAs-based MESFETs or HEMTs which need to be characterized thoroughly at DC and RF in order to carry out a 'single point' circuit design. The availability of foundry information on process tolerances also provides the possibility of design centering for yield optimization [1]. It is in this area where physical models offer great potential, being able to build up a picture for the active device performance variations across the wafer from theoretical rather than experimental results [2]. Furthermore, the predictive nature of physical models makes it possible to optimise the transistor design as well as the circuit design for a specific application.

2 Model Description

The model described here uses a 'Quasi-two Dimensional' approach for modelling HEMTs in which it is assumed that the driving force for electron transport (electric field) is essentially in the x-direction only (source to drain) [3]. The carrier conservation, momentum and energy balance equations are used to describe electron dynamics in the conducting channel. This calculation is coupled to a solution for the electron sheet density in the y-direction (surface to substrate) using a charge-control look-up table. A highly developed charge-control model [4] is used to generate the look-up table which solves the coupled Poisson-Schrödinger equations in the y-direction but also includes the variation of x-directed field, dE_x/dx, to maintain consistency between the decoupled simulations[5]. The charge-control model makes use of an asymptotic boundary condition,

$$E_y(y) = \mp \left(\frac{2q}{\epsilon_0 \epsilon_r}\right)^{\frac{1}{2}} \left(N_c F_{\frac{3}{2}}(V) - \bar{N}V + c\right)^{\frac{1}{2}} \quad (1)$$

where E_y is the electric field in the y-direction and V is electrostatic potential. This boundary condition is used in conjunction with the Shooting Method and enables the solution domain end to be located at the lower heterointerface (see Figure 1). The boundary condition is derived by integrating Poisson's equation and describes an exact asymptotic solution to the flat-band condition in the buffer when the constant of integration, c, is evaluated for $E_y = 0, V = V_{fb}$. The term, $\bar{N}$, accounts for both net doping density and dE_x/dx in the buffer layer.

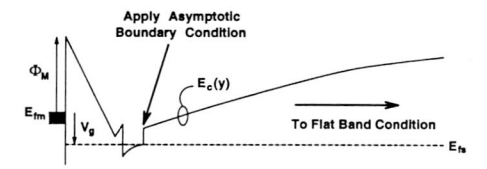

Figure 1: The asymptotic boundary condition.

The inclusion of dE_x/dx in the charge-control calculation describes elegantly the injection of electron charge into the buffer [6]. An analytical model is used to describe the buffer injection [6] which greatly improves the computational efficiency of the solutions and gives excellent pinch-off at high drain voltage : this problem has plagued previous Quasi-two Dimensional HEMT simulations.

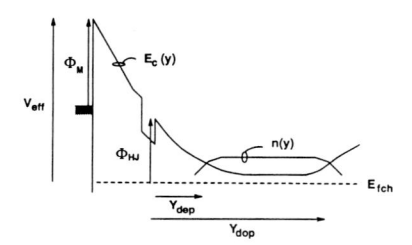

Figure 2: The analytical substrate injection model.

Figure 2 shows the implementation of the analytical substrate model. The potential, Φ_{HJ} at the lower heterojunction is first evaluated either analytically (when the channel is depleted of carriers) or numerically from the charge-control simulation. The depletion region penetration can then be calculated from the lower heterojunction and used to obtain the sheet electron charge in the buffer using the abrupt junction approximation.

Electron temperature is included in the charge control model via the Fermi-Dirac Function to calculate the degeneracy factor of the electron gas under the gate using,

$$\gamma = \frac{1}{n_s} \int_{channel} \frac{F_{\frac{3}{2}}(\eta)}{F_{\frac{1}{2}}(\eta)} n(y) dy \qquad (2)$$

where n_s is the net sheet electron density and F corresponds to the family of well known Fermi-Dirac integrals. In this calculation it is assumed that electrons reside predominantely in the Γ-valley under the gate electrode. The degeneracy factor is then used in the energy equation [7] via the sink term,

$$S_w = \frac{w - \frac{3}{2} k T_0 \gamma}{\tau_w(w)} \qquad (3)$$

where T_0 is the lattice temperature and $\tau_w(w)$ is the energy relaxation time.

3 Simulation Results

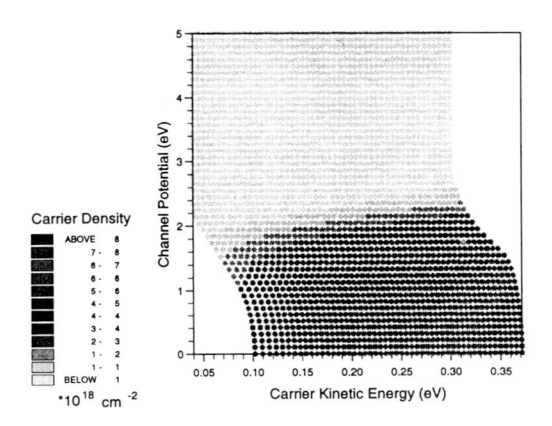

Figure 3: Shift in carrier kinetic energy due to degeneracy.

Figure 3 shows the shift in kinetic energy of the electron gas as the sheet electron density increases beyond the non-degnerate boundary. The maximum kinetic energy shift corresponds to around 0.06eV above the classical thermal energy of an ideal gas.

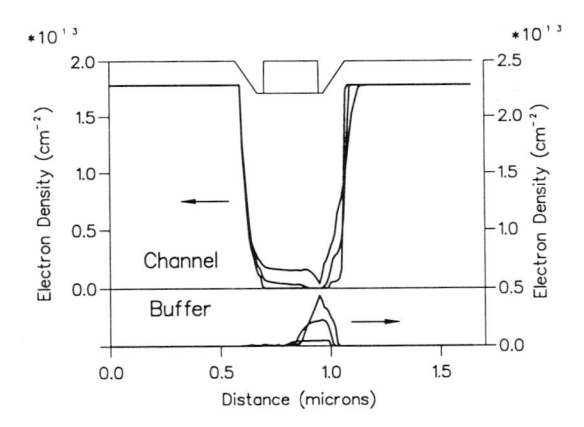

Figure 4: Carrier density profile variation as V_{GS} changes from 0V to device pinch-off.

Figure 4 shows the sheet electron density profile in the channel and in the buffer which are calculated from the charge control model and analytical substrate penetration models respectively. The diagram shows the evolution of the profiles as the gate bias is swept from zero volts to device pinch-off. Clearly the penetration of the conduction path into the buffer plays an important role in describing both the DC *and* RF operation of the HEMT. This point is reflected in Figure 5 which shows excellent pinch-off of the DC I-V characteristics and also the ability of the model to predict device breakdown at high currents and voltages.

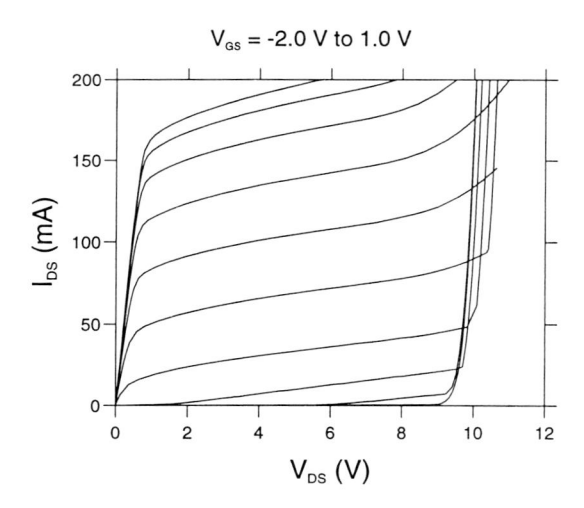

Figure 5: DC I-V characteristics for 0.25×150 μm pseudomorphic HEMT.

4 Conclusions

The new implementation of a Quasi-two Dimensonal HEMT model presented in this paper has for the first time made it possible to predict device pinch-off at high drain bias. The execution time of the simulations remains extremely fast, making the model well suited to consideration of large signal device/circuit design.

References

[1] M.D. Meehan and J. Purviance. *Yield and Reliability in Microwave Circuit and Sytem Design*. Artech House, 1993.

[2] C. M. Snowden and R. R. Pantoja. GaAs MESFET physical models for process-oriented design. *IEEE Trans. Elec. Dev.*, 40(7):1401–1409, July 1992.

[3] C. M. Snowden and R. R. Pantoja. Quasi-two-dimensional MESFET simulations for CAD. *IEEE Trans. Elec. Dev.*, 36(9):1564–1574, September 1989.

[4] C.G. Morton and J. Wood. MODFET versus MESFET: The capacitance argument. *IEEE Trans. Elec. Dev.*, 41(8):1477–1480, August 1994.

[5] R. Drury and C.M. Snowden. Fast and accurate HFET modelling for microwave cad applications. *Proceedings of the Third International Workshop on Computational Electronics*, May 1994.

[6] C.G. Morton and C.M. Snowden. Quasi-two dimensional modelling of HEMTs. *Proceedings of the 8th GaAs Simulation Workshop*, October 1994.

[7] T. Shawki, G. Salmer, and O El-Sayed. MODFET 2-D hydrodynamic energy modeling: Optimization of subquarter-micron-gate structures. *IEEE Trans. Elec. Dev.*, ED-37(1):21–30, January 1990.

Simulations of the forward behaviour of hybrid Schottky-/pn-diodes

U. Witkowski, D. Schroeder

TU Hamburg-Harburg, Techn. Electronics,
Eissendorfer Str. 38, D-21071 Hamburg, GERMANY

Abstract

Device simulations of a hybrid diode – an integrated combination of a Schottky diode and a pn-diode – are presented. It is shown that under low forward bias the device acts like a Schottky diode, while under strong forward bias the behaviour turns into that of a pn-diode. Geometric variations allow to make a trade-off between the small Schottky diode forward voltage drop and the low on-resistance of a pn-diode.

1. Introduction

We performed two-dimensional simulations of hybrid Schottky-/pn-diodes [1]. These diodes consist of weakly-doped n-silicon with a highly-doped guard ring of p-Si at the surface (see Fig. 1 for a display of the simulated part of the structure). A metallic contact covers the surface including the pn-junction. The weakly doped material forms a Schottky contact with the metal, and the highly-doped guard ring forms an ohmic contact. Between these regions, the contact characteristic must change smoothly from ohmic to rectifying. The whole structure acts as an integrated – yet interacting – combination of a Schottky diode and a pn-diode.

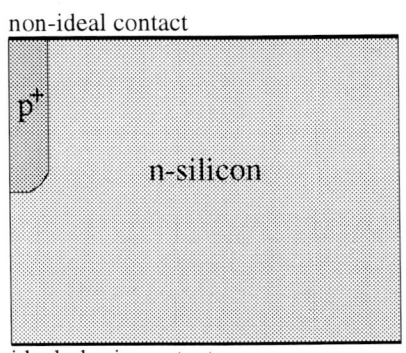

Figure 1: Simulated structure

Because of the varying contact type, this structure cannot be simulated with the conventional models of ohmic and Schottky contacts. Therefore, we used our model of non-ideal metal-semiconductor contacts [2]. The model considers both tunneling and thermionic emission current across the contact and allows the simulation of contacts on very low to very highly doped material with a single model.

We implemented the model into the device simulator PARDESIM [3] as boundary conditions for the Poisson equation and the electron and hole continuity equations. The majority carrier boundary condition models the current density across the contact, consisting of thermionic emission and tunneling currents [2]. The minority carrier boundary condition represents the thermionic emission of minority carriers [2]. The Poisson equation boundary condition is expressed as an apparent lowering of the barrier height, which is due to tunneling [2]. As soon as tunneling occurs through the whole depletion layer, the model switches automatically to the condition of charge neutrality at the contact.

While we reported in [4] on the potential and charge distribution inside the hybrid diode in equilibrium, we investigate in the present paper the electronic behaviour of the device in the forward bias regime.

2. Device simulations

The simulated structure (Fig. 1) is $10\mu m$ wide and $8\mu m$ high, the width of the p^+-region is $1\mu m$. The n-doping is $2 \cdot 10^{15} cm^{-3}$ and the p^+-doping $10^{20} cm^{-3}$. On the back-side of the device, an ideal ohmic contact has been assumed.

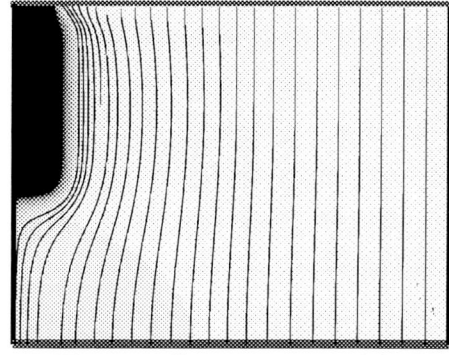

Figure 2: Current pattern at low forward bias

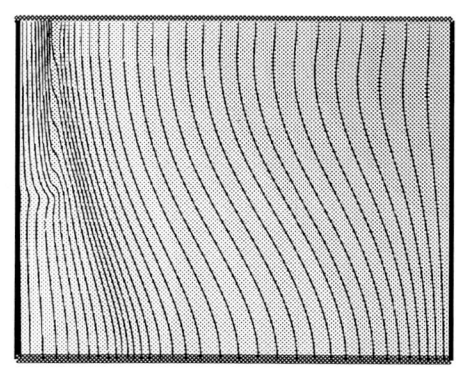

Figure 3: Current pattern at high forward bias

The simulated current distribution under a low forward bias of 0.1 V is shown in Fig. 2 [5]. The results show that in this bias regime the current flows exclusively in the Schottky diode part of the structure. The reason is that the Schottky contact is already in a conducting state, while the forward threshold of the pn-junction has not yet been reached. This situation changes drastically if we increase the bias to 1.5 V. The corresponding flow pattern is depicted in Fig. 3 [5]. Now the pn-junction is in a conducting state, too, and the current flows mainly through the p^+-part of the structure.

Figure 4 shows the simulated I/V characteristics for the forward regime. For small voltages, the Schottky diode part alone is in conduction. At 0.8 V, the pn-junction goes into the conducting state, too, and the current strongly increases. For compari-

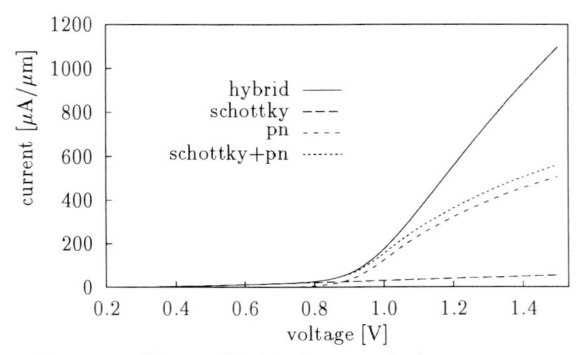

Figure 4: Forward I/V characteristics

son, we included the corresponding curves of a single pn-diode and a single Schottky diode with respective cross-sections, as well as the current of a parallel circuit of these two diodes. We note that the current of the hybrid diode is higher than the current of the equivalent combination of the single pn- and Schottky diodes. This effect is caused by an interaction of the integrated diodes; it has also been observed experimentally [6].

As the simulations show, the effect can be attributed to a reduction of the bulk resistivity by minority carriers injected from the p$^+$-region. Figure 5 displays the hole distribution in the whole structure. Despite the smallness of the p$^+$-region, the minority concentration in the whole n-region has increased from the equilibrium value of $7 \cdot 10^4 \text{cm}^{-3}$ to about $2 \cdot 10^{17} \text{cm}^{-3}$, which is even more than the equilibrium *majority* carrier concentration. In order to keep charge neutrality, a strong increase of electrons too is the consequence. This high concentration of free carriers increases the conductivity of the n-region and hence the current in the hybrid device. Since in the equivalent parallel circuit of the single devices an injection of the pn-junction into the Schottky diode cannot occur, the total current in this case is lower.

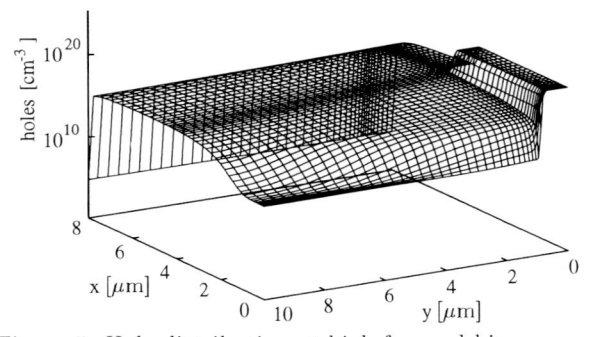

Figure 5: Hole distribution at high forward bias

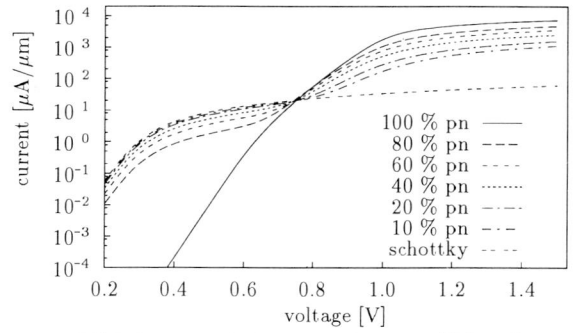

Figure 6: I/V characteristics at various pn/Schottky ratios

3. Investigation of geometric variations

Finally, we investigated the effect of various area ratios between the Schottky- and the pn-fraction. For this purpose, we varied the size of the p$^+$-region in Fig. 1 from zero to the full width of the structure, thus obtaining a Schottky resp. a pn-diode in the limits, and a number of hybrid diodes inbetween. The resulting current-voltage relationships are shown in Fig. 6.

We observe that the Schottky diode has a low forward voltage drop, but a rather large on-resistance. The pn-diode, in turn, has a high forward voltage drop and a small on-resistance. The hybrid diodes combine these features, thus having a low voltage drop *and* a a small on-resistance for higher voltages. Hence, the area ratio allows to make a trade-off between forward voltage drop and on-resistance. This behaviour also agrees well with the experimental results [6].

In conclusion we note that with the presented simulation model simulation-based optimizations of the device for a specific application become possible.

References

[1] R.A. Zettler, A.M. Cowley. P-n junction-Schottky barrier hybrid diode. IEEE Trans. Electron Devices, vol. ED-16 (1969), p. 58

[2] D. Schroeder. *Modelling of interface carrier transport for device simulation.* Springer, Wien 1994.

[3] O. Kalz, D. Schroeder. PARDESIM – A parallel device simulator on a transputer based MIMD-machine. In S. Selberherr, H. Stippel, E. Strasser, editors, *Proc. 5th Int. Conf. on Simulation of Semiconductor Devices and Processes (SISDEP'93), Sept. 7-9, 1993, Vienna*, Springer, Wien 1993, pp. 245-248.

[4] D. Schroeder, T. Ostermann, O. Kalz. Non-ideal contacts – Schottky diode soft-breakdown and hybrid diode with contact over pn-junction. In *Int. Workshop on Numerical Modeling of Processes and Devices and and for Integrated Circuits (NUPAD V), June 5-6, 1994, Honolulu*, pp. 75-78.

[5] Picture created with *Picasso* (developed by ETH Zurich and ISE AG, Zurich).

[6] J. Olsson, H. Norde, U. Magnusson. Investigation of the current-voltage behavior of a combined Schottky-p-n diode. Solid-State Electron., vol. 35 (1992), pp. 1229-1231.

HFET Breakdown Study by 2D and Quasi 2D Simulations: Topology Influence

Y. Butel, J. Hédoire, J. C. De Jaeger, M. Lefebvre, G. Salmer

IEMN-DHS, U.M.R C.N.R.S 9929
Cité scientifique - Avenue Poincaré - B.P. 69
59652 Villeneuve D' Ascq CEDEX - FRANCE

Abstract

The study of breakdown phenomena is very important for power devices. Indeed, it constitutes a great limitation for the performance. This paper proposes to study this phenomenon by two means: a two dimensional energy model and a quasi two dimensional model. The aim of this work is the optimization of the shape of the gate-recess in order to improve the breakdown voltage, taking into account the microwave performance.

1. Introduction

The breakdown phenomenon is one of the most limitative effect for power devices. The proposed study concerns the influence of specific parameters such as the gate length or the gate-recess configuration, on the breakdown voltage. The corresponding microwave device performance is also investigated for different structures such as conventionnal, or pseudomorphic or δ-doped layer AlGaAs/GaInAs/GaAs HFETs, for power applications.

2. Modeling description

The proposed study is based on two different physical simulation tools:

2.1. The two dimensional hydrodynamic energy model

This model takes into account a large part of the physical phenomenon which occur in HFETs. It is based on a set of equations deduced from Boltzmann's transport equation: continuity, energy and momentum equations combined with Poisson's equation [1]. These equations are solved numerically using a finite difference method with non-uniform meshes and variable time steps. The main advantage of this model is the accuracy of the results, but it needs very large computing time due to a large number of mesh points and the use of small time steps. Recent improvements have been brought into the simulation i.e. the breakdown phenomenon and the possibility to study real gate recess topologies.

2.2 The quasi-two dimensional model

In this model simplifying assumptions are introduced, but it accounts for the non-stationary electron dynamic effects that are of particular importance for submicron gate devices [2]. The model is based on successive resolutions along the transversal and longitudinal axes. In a first time, the charge control law of the device is computed (transversal axis) and in a second time, the average values of the physical parameters are calculated using the current equation, momentum and energy conservation equations and Poisson's equation (longitudinal axis). The main recent modifications concern the minority carrier consideration and the introduction of a generation term in the equations for hole and electron currents. In order to simplify our calculation, the hole effect is neglected in the charge control law.

Note that the two simulations, associated for the HFET breakdown study, present each their own advantages. The two dimensional model main characteristic is the physical accuracy but it needs large computing time. The validation of the quasi two dimensional model results is made by the physical simulation. Then many different structures and topologies are investigated by the quasi 2D model which needs smaller computing time.

3. Main results

The gate recess offset constitutes the main parameter which makes it possible to improve the breakdown conditions. Its influence is studied on the physical behaviour of the transistor i.e. on the main parameters of the small signal equivalent circuit and on the breakdown voltage. The physical quantities (charge concentration, total energy and potential distributions) are represented in figure 1 for a 50 nm gate recess offset device corresponding to a δ-doped layer HFET structure made by Thomson TCS [3]. A high carrier concentration in the well, a large electric field and energy domain can be remarked at the edge of the recess on the drain side. Figure 2 shows the carrier concentration for a similar device. It can be noticed the carrier injection in the buffer, the charge accumulation in the well and at the exit of the gate, and also the depleted zone under the gate recess. The average energy in the channel is represented figure 3 for a 0.3µm gate length device. The evolutions show that the distance X for energies over 0.7 eV (supposed to be the minimum value for which ionisation phenomenon appears) decreases when the gate recess offset increases. So the breakdown voltage will be improved with wider gate recess offsets as shown in figure 4. The influence of the gate length is shown in figure 5. A decrease of the breakdown voltage is noted for 0.15 µm gate length devices due to a larger average energy in the channel. A more complex topology can also be investigated by the two dimensionnal model, for instance the double recess structure. Figure 6 represents the corresponding total energy distribution. For this device, the energy domain is able to spread along the second recess and as a consequence, a better breakdown voltage will be obtained.

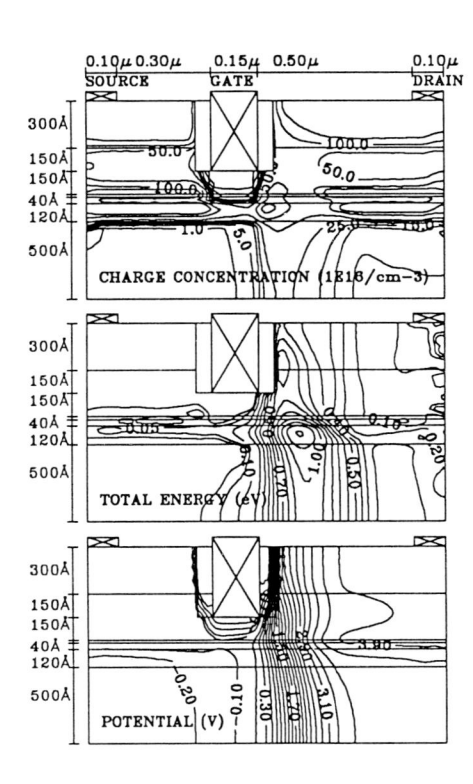

Fig.1 : Charge, energy and potential distributions
δ-doped layer HFET
(R = 50 nm, Vds = 4 V, Vgs = 0 V)

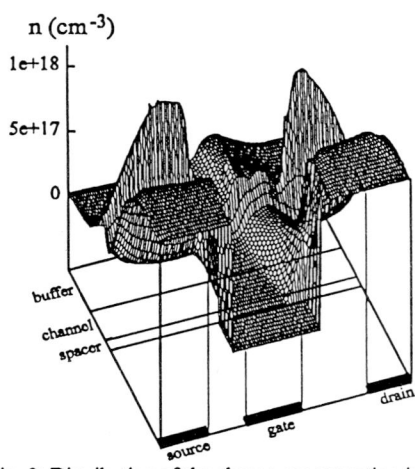

Fig. 2: Distribution of the charge concentration in a gate recess PM-HFET
(R = 100 nm, Vds = 4 V, Vgs = 0 V, Lg = 0.3 µm)

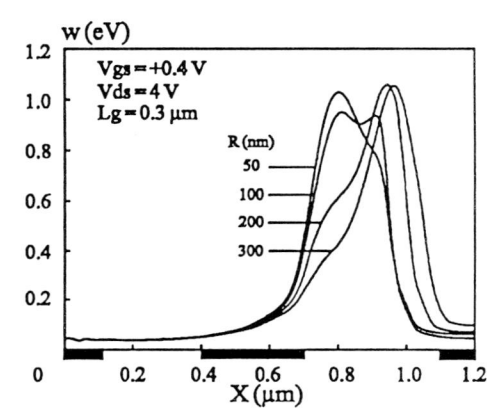

Fig. 3: Evolution of the average total energy in the channel for different gate-recess offsets

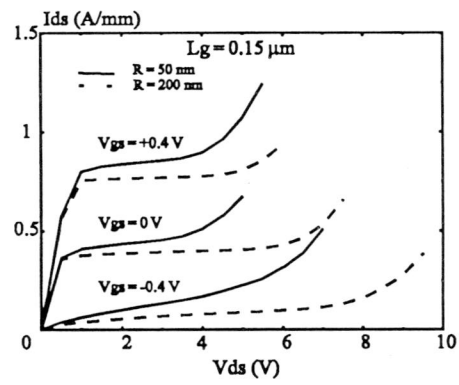

Fig. 4: Current voltage characteristics for two gate recess offsets

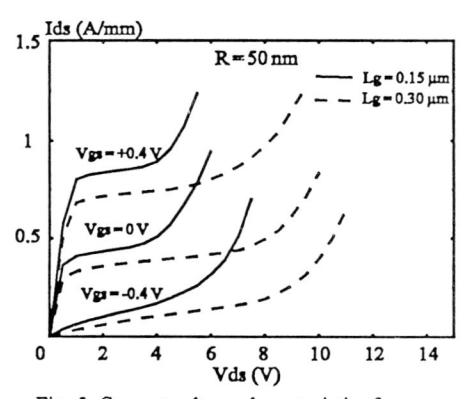

Fig. 5: Current voltage characteristics for two different gate lengths

In order to optimize power devices, the study of microwave performance is also developed by considering structures with larger gate recess offsets. The evolution of the main parameters of the equivalent circuit and the device cut-off frequency are studied for different structures. Figure 7 shows for example the intrinsic current gain cut-off frequency evolutions for two different gate recess offsets. A decrease of f_{ci} is observed for wider gate recess offsets mainly due to a decrease in transconductance.

Fig. 6: Energy distribution in a double-recess PM-HFET (R = 100 nm + 300 nm, Vds = 4 V, Vgs = 0 V, Lg = 0.3 µm)

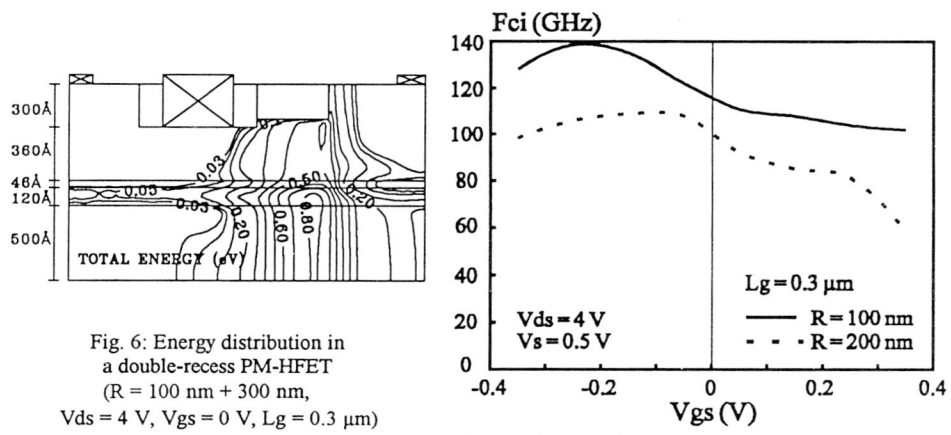

Fig. 7: Influence of the gate recess offset on the intrinsic cut-off frequency

4. Conclusion

This study, obtained by two different simulation tools makes it possible to describe the conditions to respect for power devices optimization. It shows that a compromise between an improvement of the breakdown voltage and the microwave capabilities has to be found. In particular a gate recess offset close to 0.1 µm seems to be the optimum value.

Work supported by the European Commission under the Esprit Project 6016 CLASSIC

References

[1] T. Shawki, G. Salmer and O. El Sayed, IEEE Trans. on E.D. Vol. 37, p20-21, n°1, 1990.

[2] H. Happy, G. Dambrine, J. Alamkan, F. Danneville, F. Kaptche-Tagne and A. Cappy.
Int. J. of Micro. and mm-Wave Comp.-Aided Eng. Vol. 3, n° 1, 1993.

[3] 4th Esprit-Classic report, March 1994.

Investigation of GTO Turn-on in an Inverter Circuit at Low Temperatures using 2-D Electrothermal Simulation*

Y. C. Gerstenmaier and E. Baudelot[a]

Corporate Research and Development, Siemens AG
Otto-Hahn-Ring 6, D-81730 Munich, GERMANY
[a]Drives and Standard Products, Siemens AG
G.-Scharowski-Strasse 2, D-91058 Erlangen, GERMANY

Abstract

GTO (gate-turn-off-thyristor) turn-on failure in inverter circuits for traction drives is investigated by mixed-mode 2D electrothermal device and circuit simulation. Whereas GTO turn-off failure has already been analysed extensively in the past, in this paper a novel turn-on failure mechanism at low temperatures is analysed. The turn-on failure is due to a decreased carrier lifetime at low temperature with resulting increased latching current. As a consequence at low load current and low triggering current large parts of the device area do not latch while other parts with somewhat higher carrier lifetime or deviating doping concentration have to sustain a very high current density at high voltages. It is essential for this destruction mechanism to take into consideration the distributed resistance of the gate metallisation and its contact resistance.

1. Introduction

Traction drives at low temperatures (230 - 280 K, i.e. -43°$C - 7°C$) sometimes suffer from sudden GTO-thyristor failure during switching operation. Some evidence points to the occurence of GTO failure during turn-on. This is surprising, because usually in the test circuit turn-off failure occurs but turn-on failure is not observed as long as the admissible dI/dt for the load current is not surpassed.

In this paper a novel GTO turn-on failure mechanism at low temperatures is revealed by 2-D numerical electrothermal simulation. The device simulator [1] solves the complete semiconductor equations (Poisson and continuity equations) together with the heat flow equation for the dynamic development of the lattice temperature within a general external network. Additionally thermal resistors and capacitors can be included in order to allow for the cooling of the device by the packaging.

Fig.1 and fig.2 show the inverter circuit, which is fed by a dc-voltage source. By means of the GTO-switches 1-4 an ac load current is formed with approximate sine waveform (Fig.3). For forming the first positive ac-current half-period, the switches 3 and 4 are not necessary and are therefore omitted in the simulation. Even with this reduction the problem is of severe complexity; so the calculation of only few GTO switching cycles takes considerable CPU-time. Each GTO has attached to it a RCD-protection (snubber) circuit, a freewheeling diode and gate-drive circuit.

---

*This work was supported by **eupec**, a company of AEG and Siemens

2. GTO failure

GTO turn-*off* failure has already been analysed extensively in the literature [2, 3, 4, 5, 6, 7]. During turn-off essentially two destruction mechanisms have to be dealt with for high power GTOs: current-filamentation during the spike voltage period and dynamic avalanche generation during the tail-phase. On a wafer scale many individual GTO-Segments (> 2000) in parallel contribute to current transport. Due to inhomogenities between different cells in the lateral dopant or carrier lifetime distribution over the device area, current redistribution during turn-off takes place [6] (independently of dynamic avalanche) and may destroy those segments, which carry the most heavy load.

Because of the excessive number of grid-points it is not possible to simulate the whole GTO wafers in the circuit. Therefore a simulation technique [3] is applied, where a homogeneous 4.5kV/3kA GTO (half cell scaled to full wafer size of $36cm^2$ with respect to current) has in parallel a one segment GTO ($0.03cm^2$ area) representing a local perturbation on the wafer by a higher n-emitter doping of 15% and slightly higher carrier lifetime. For simulation the GTO no.2 of fig.1 was substituted by the coupled system of large area homogeneous GTO and small area perturbed segment GTO representing one inhomogeneous GTO (Fig.4).

3. Results and Discussion

Fig.5 shows a simulation result for normal operating temperature of 400 K, where the GTO no.1 is replaced by a closed switch, while the inhomogeneous GTO no.2 is turned on and off every millisecond. The gate triggering unit supplies 20 A for $20\mu s$, 5 A for $5\mu s$ and 2 A for the rest of the on-state period ($300\mu s$). At t=0 the GTO is triggered on and the circuit load current may be adjusted to any value (in our case -10A). I_L rises at a steepness determined by the the ratio of dc-voltage source (2.8 kV) and load inductance (2mH) i.e. with $1.4A/\mu s$.

At the beginning the GTO has a homogeneous temperature of $400°K$. The maximum local temperature after 2ms amounts to $405°K$ and is due to turn-off. The device behaviour is essentially isothermal and no dramatic or dangerous temperature increase occurs.

In fig.6 the starting temperature for the same system is lowered to 280 K. In the simulation a carrier lifetime reduction by a factor 2 compared to 400 K is assumed following the work of [8, 9]. This results in an increased latching current for the GTO under consideration. Therefore, as can be seen from fig.6, the GTO is not fully latched when the gate trigger current reduces to 2A. As a consequence the anode voltage rises to 2kV. On the other hand the perturbed segment of the GTO has a lower latching and triggering current per area than the homogeneous part and has to sustain a high current density at 2kV anode voltage. This leads to a sharp temperature increase in the perturbed segment above the silicon melting point of 1700 K and the device will be destroyed after 0.2 ms, already.

It is essential for this destruction mechanism, to take into consideration the seperate gate-resistors for the homogeneous wafer and the perturbed cell, as indicated in fig.4, which is due to the gate metallisation and contact resistance. The magnitude of the resistors scales inversely proportional to the respective GTO area. Without distributed resistors the gate current of the perturbed cell becomes negative (gate current is extracted from the cell) and adds up to the gate current of the homogeneous GTO part, so that its gate current exceeds 2A considerably. This compensates for the inhomogenity of the GTO to a large extent, so that in the absence of distributed gate resistors the device will not fail in that period. Once the load current in the circuit has exceeded the latching current of the GTO the GTO will turn on safely. Therefore the failure described occurs only for small load currents.

The simulation results reveal the essential meaning of the latching current for GTO turn-on failure, which has to be smaller than the circuit load current during the gate triggering period.

References

[1] Technology Modeling Associates. Inc. Palo Alto, California, USA MEDICI user's manual. March 1992.

[2] Y.C. Gerstenmaier, ISPS'92 Proceedings, 1992, Czech Technical University in Prague, p.49.

[3] Y.C. Gerstenmaier, SISDEP'93 Proceedings, Springer Verlag, p.53-56, 1993.

[4] I. Omura and A. Nakagawa, Proceedings of 1992 International Symposium on Power Semiconductor Devices & ICs, Tokyo, pp.112-117.

[5] H. Ohashi and A. Nakagawa, IEEE IEDM-81 Tech. Dig., p.414, 1981.

[6] K. Lilja and H. Grüning, IEEE PESC Dig., p.398, 1990.

[7] G.K. Wachutka, IEEE Trans. on Electron Devices, Vol.38, p.1516, 1991.

[8] M.S.Tyagi, R.van Overstraeten, Solid State Electronics Vol.26,pp.577-597, 1983.

[9] Y.C. Gerstenmaier, Proceedings of 6th ISPSD'94, pp.271-274, 1994.

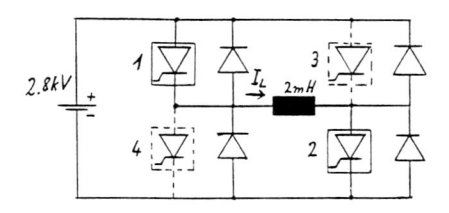

Fig.1: Voltage fed inverter circuit. Each box comprises a GTO-thyristor with snubber and gate circuit as shown in fig.2

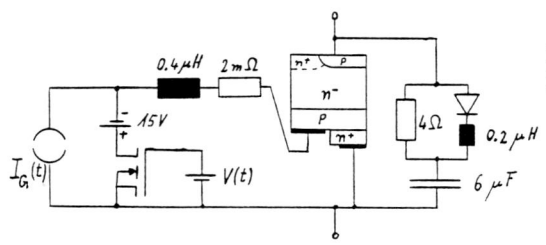

Fig.2: Switch (box) from fig.1, built up from GTO-thyristor with snubber- and gate-circuit. Turn-off is controled by V(t) and turn-on by $I_G(t)$.

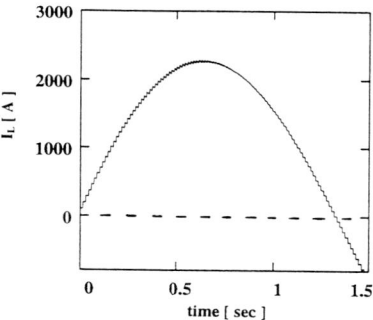

Fig.3: Locomotive-inverter load current at beginning of movement

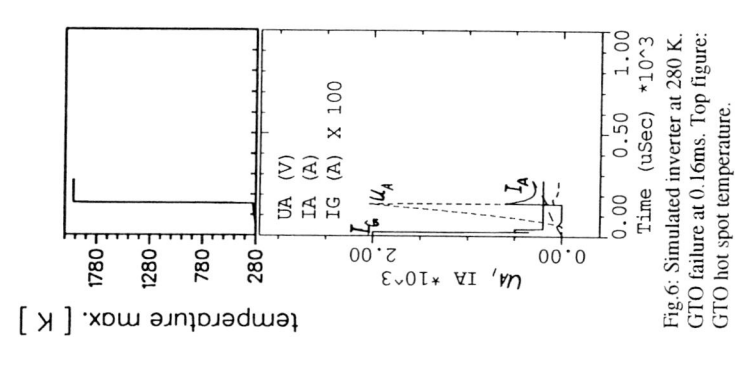

Fig.4: Inhomogeneous GTO, represented for simulation by a homogeneous GTO-wafer (35cm² area) and a small perturbed segment with deviating doping and carrier lifetime of 0.03cm² area.

Fig.5: Simulated inverter at 400 K. Dashed line: GTO anode voltage U_A; solid lines: GTO gate and anode current I_G, I_A; long dashed line: load current I_L.

Fig.6: Simulated inverter at 280 K. GTO failure at 0.16ms. Top figure: GTO hot spot temperature.

Large Scale Thermal Mixed Mode Device and Circuit Simulation

J. Litsios[a,b]   B. Schmithüsen[a,b]   W. Fichtner[a,b]

[a]Integrated Systems Laboratory, ETH–Zürich, Switzerland
[b]Integrated Systems Engineering, ETH–Zürich, Switzerland
E-mail: litsios@ise.ch, bernhard@ise.ch, fw@ise.ch

Abstract

The thermal mixed-mode capabilities of the device and circuit simulator DESSIS are presented. The mode allows both electrical and thermal netlists to interconnect physical and circuit devices. As an example, the full electrothermal simulation of IGBT power module is presented.

A thermal mixed-mode extension to the DESSIS device and circuit simulator has been developed to simulate thermoelectric systems of yet unprecedented complexity. The originality of this work lies in its generality. Whereas existing mixed-mode simulators make a distinction between the use of temperature in physical semiconductor devices versus SPICE-like circuit devices, DESSIS does not, and allows full thermal coupling between any type of device. Links between the electrical and thermal networks can be done either through physical devices (with either thermodynamic or hydrodynamic transport models) or through circuit devices.

Fig. 1 shows how the existing versions of DESSIS (formerly SIMUL[1]) was extended. The new addition to the simulator are the support for thermal netlists and mixed thermoelectric circuit models. These improvements build upon the programs existing features, such as, the capability of mixing 1D, 2D and 3D physical and circuit devices (using circuit models from BONSIM [2]), advanced physical models and an extensive set of solvers.

	Electrical	Thermal	Thermo-electric
Device	Iso-thermal devices	Die, heat sync, packaging	Self-heating high-power devices
Circuit	Electrical netlist	Thermal netlist	Coupling of T-E netlists

Figure 1: New thermal mixed mode features added to DESSIS (shaded)

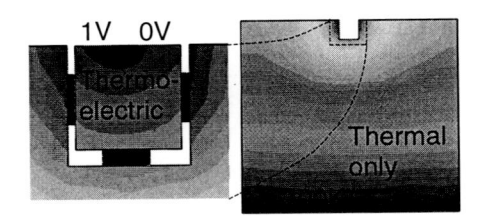

Figure 2: Thermoelectric simulation of a resistive device

An important use of the thermal mixed mode is to reduce the work necessary to simulate the large thermal environment of power devices. A typical elementary power device is many times smaller than its package size. While electrically it makes sense to limit the range of the device simulation and use circuit models for most of the devices, this simplification is much more difficult for a realistic thermal simulation because of the much larger range in the thermal interactions. The thermal mixed mode solves this problem in two ways. First, a large device that could formerly only be computed with a total thermoelectric simulation can be subdivided into a thermoelectric part and a thermal only part. A simple example of this is shown in fig. 2 where a resistive device is simulated with the full set of thermoelectric equations over a small device which is thermally connected through three thermal links to a much larger device where only the thermal equation is computed.

Another way the thermal mixed mode can be used is to connect the physical devices and the circuit devices through a thermal network. Fig. 3 shows the two ways this type of connection can be done in DESSIS, by directly coupling to temperature variables in the circuit models or by using an interface device to access the temperature parameter of the models. This latter method is valid in most cases because the temperature usually varies quite slowly.

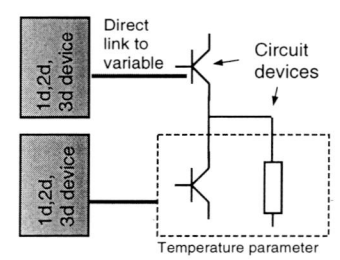

Figure 3: Thermal connection between physical devices and circuit device can be done in two ways

The definition of the system to be simulated is straightforward with only a small extension over the previous DESSIS input syntax: the thermal nodes of the netlist must be declared as such. This allows both electrical nodes and thermal nodes to share physical device contacts. Whether the connection is electrical or thermal is determined by the type of the node.

Fig. 4 shows how the thermal mixed mode can be used to simulate a large power module with an IGBT and SPEED diode. Here the module structure is broken down into four parts: the IGBT and SPEED diode device simulated as 2d physical devices (1415 and 568 vertices) using the thermodynamic model, the gate control circuitry is simulated with circuit models and the package is simulated with a 2d device (150 vertices) with only thermal models. Only a very rough approximation of the geometry of the package was considered with as size of 6x4x4cm.

The thermal connectivity of this example is done as follows: The IGBT is connected thermally to the package directly through its collector and resistively through its emitter. The SPEED diode is connected in a similar manner. Each circuit element can be thermally connected to specific point of the package but for the sake of simplicity all circuit element were attached to a single thermal contact in the package. During the solve process, a decoupled iteration is done between the full coupling of the temperature related equations and the Poisson, continuity and circuit equations.

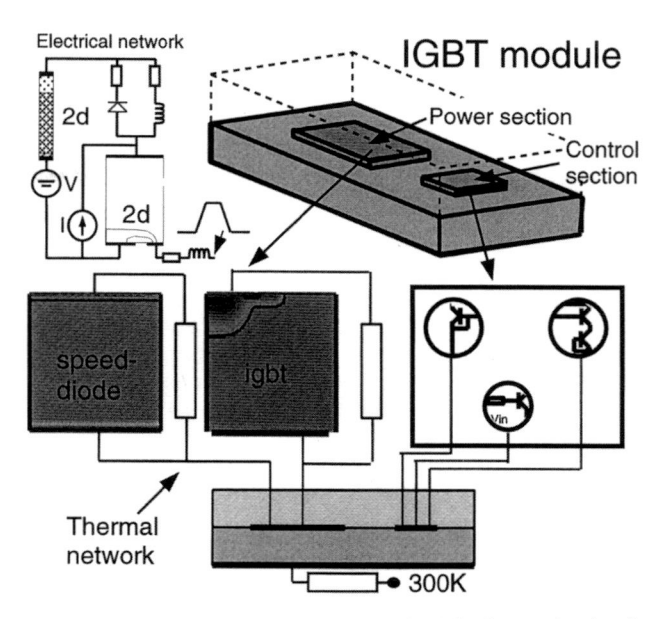

Figure 4: IGBT module structure simulated with thermal mixed mode

The following preliminary results are given for the study of the turn-on turn-off behaviour of a 800V IGBT. A first transient simulation was run starting with a general temperature of 300K. Given the size of the package a large number of cycles would be necessary to reach thermal steady state, thus only one cycle was simulated from which a mean power loss was extracted (about 200W). This value was used to re-initialize the package by computing its temperature distribution subject to a fixed heat supply. A new transient simulation was recomputed with this new initial solution resulting in the following results: Fig. 5 shows the temperature distribution through the package. Of interest is the temperature at the substrate of the IGBT ($\approx$400K) and the gradient of the temperature over the control circuit section. Note that these values are realistic but will be improved with a more exact description of the package. Both Fig. 6 and 7 show the behaviour of the IGBT for a turn-on, turn-off cycle. Fig. 6 shows the short circuit thermal failure that results when the the IGBT is fixed to 800V and the pulse is long (80μs). Fig. 7 shows the behaviour of the IGBT when connected to the SPEED diode and subject to a fixed 75A load. Here a longer pulse will fail to turn off due to the heating of the device.

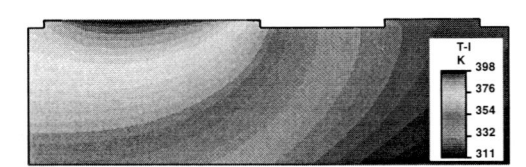

Figure 5: Near steady state distribution of temperature in package

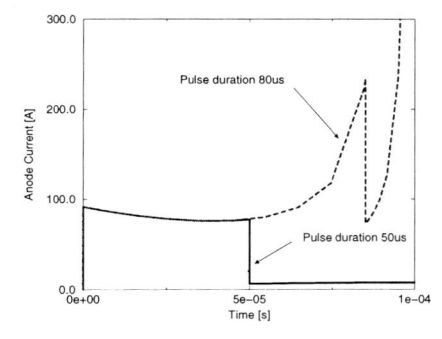

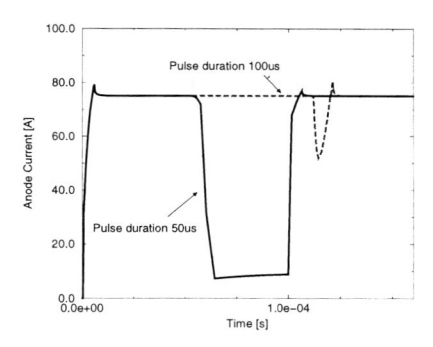

Figure 6: Short-circuit thermal runaway of the IGBT

Figure 7: Turn-off failure of the IGBT

In a general, a full electrothermal simulation of such a large system is not necessary because of the the fundamentally different time spans between the thermal and electrical equations. Solving the thermal equation for the package during the transient computation is not necessary because its temperature varies very much more slowly in the package than in the power devices. On the other hand, resolving the package adds no noticible amount of CPU time. In practice, to efficiently obtain the asymptotic solution it is best to iterate between solving the package in steady-state and the devices in transient.

Work still needs to be done to ease the creation of the thermal devices and networks. A layout to circuit tool approach [3] is necessary to assure a precise thermal modeling. On the numerical side work can be done to take advantage to the fundamentally different time spans between the thermal and electrical equations.

This work has been financially supported by the ESPRIT-6075 (DESSIS) Project.

References

[1] J. Litsios, S. Müller, and W. Fichtner, "Mixed-mode multi-dimensional device and circuit simulation," in *SISDEP-5*, (Vienna), pp. 129–132, Sept. 1993.

[2] Robert Bosch GmbH, Reutlingen, Germany, *BONSIM, Bosch Netzwerk - Simuation, Version 2.1*, 1992.

[3] B. H. Krabbenborg, H. de Graaf, A. J. Mouthan, H. Boesen, A. Bosma, and C. Tekin, "3d thermal/electrical simulation of breakdown in a bjt using a circuit simulator and layout-to-circuit extraction tool," in *SISDEP-5*, (Vienna), pp. 57–60, Sept. 1993.

Scaling of Conventional MOSFET's to the 0.1-μm Regime

M.J. van Dort, J.W. Slotboom, and P.H. Woerlee

Philips Research Laboratories,
Prof. Holstlaan 4,
5656 AA Eindhoven, The Netherlands

Abstract

As fundamental limits of MOSFET's are being explored, new device structures have been proposed in order to maintain good short-channel behaviour in the deep submicron regime. These advanced transistors usually require complex channel and source/drain engineering, and will probably not be excepted by industry if the conventional way of scaling is still feasible. The conventional MOSFET is the benchmark for semiconductor industries. This paper addresses some of the issues which are important when conventional MOSFET's are scaled down to the deep submicron regime.

1. Introduction

In the past decades CMOS processes have been successfully scaled down to submicron dimensions. Intensive research efforts have been put into the issue of how to best scale the devices from one generation to the next one. Various scaling laws have been proposed (see e.g. [1, 2, 3]). Each of these scaling scenario's successfully miniaturize the CMOS processes, but they have different philosophies regarding issues like for instance the internal electric fields, the current density and the power consumption. In particular, an important boundary condition for the scaling factor for voltage was set by the immunity against hot-carrier degradation. Straightforward down scaling turned out to be impossible, and different drain structures have been proposed and implemented in order to meet the 10-year lifetime criterion against hot-carrier degradation. The best-known of the changes in the drain configuration are the introduction of the LDD and the LATID implantation, but also more advanced structures have been investigated. All the structures that modify the gate/drain overlap have in common that they improve the lifetime of the transistor at the expense of the DC or AC performance of the devices.

The aim of this paper is to discuss two important boundary conditions for device scaling in the deep-submicron regime, the hot-carrier degradation and the threshold voltage. Non-local carrier heating is very important in the sub 0.1-micron regime, making it possible to abandon the LDD structure for the low power supply voltages (V_{dd}) needed for these devices and return to the conventional S/D structure. This issue will be addressed in Sec. 2. Of course, the threshold voltage V_t has to be scaled in accordance with the decrease of V_{dd}. The low V_t's required for proper circuit operation make it difficult to turn off the transistor completely. The reason is that it is not possible to scale the subthreshold swing accordingly. This puts a severe demand on λ_V, the scaling factor for voltages. The alternative is to relax λ_V. This will however have a severe consequence for the performance of the future CMOS generations, as it will increase the power dissipation and the current density to unacceptable high values [2]. Optimization in the deep-submicron regime is therefore likely to be dominated by clever power management, a proper choice for the scaling factor for the (threshold) voltage and the inevitable poor off-state performance of these MOSFET's. A thorough understanding of the mechanisms determining the V_t in the deep submicron regime is thus extremely important. This issue will be addressed in Sec. 3. Short-channel related issues are finally discussed in Sec. 4.

2. Hot-carrier immunity

For design rules smaller than 2.0 μm hot-carrier degradation is a severe problem. The lateral electric fields at the drain are high and some of the electrons gain energies high enough to surpass the Si-SiO$_2$ barrier. Subsequent trapping of these hot carriers in the gate oxide degrades the MOSFET's to an extent where they cannot be used any more.

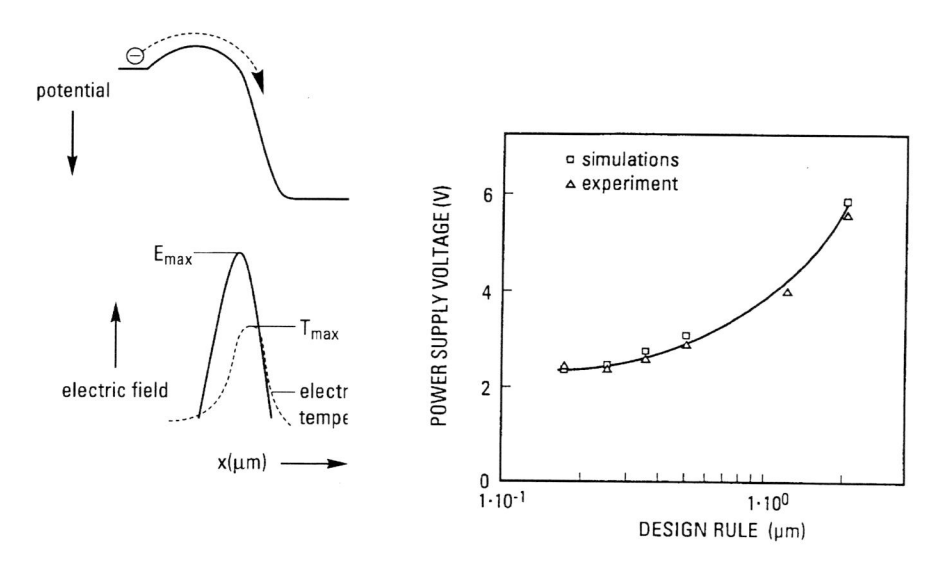

Figure 1: *Non-local carrier heating. The electron temperature is lagging behind the electric field due to the finite energy relaxation length. The energy of the electrons can be lowered by reducing the <u>width</u> of the electric field as well as by a reduction of $E_{\max}$.*

Figure 2: *Measured and simulated maximum power supply voltage $V_{DD,MAX}$ as a function of the design rule of the process. The leveling-off of $V_{DD,MAX}$ is caused by non-local carrier heating [7].*

The energy distribution of the electron population depends on the magnitude of the electric field and on the shape of the electric field peak. If the width of the field peak is large ($\geq$ 100 nm), the energy distribution of the electrons stays approximately in equilibrium with the local electric field. In this situation the degradation of the MOSFET can be avoided by a reduction of the maximum field at the drain (figure 1). This is accomplished with the introduction of the LDD or LATID implantations. These drain structures are used for process generation with design rules down to 0.35 μm. For generation with minimum feature lengths of 0.25 μm or less, the width of the electric field peak is so narrow that we begin to notice the effect of the finite energy relaxation length. In this situation, the electrons never reach the energies 'belonging to' the maximum electric field (see figure 1). We are then in the regime of non-local carrier heating: a reduction of the width of the electric field causes the average energy of the electrons to drop. We can benefit from this physical effect by making the S/D profile as steep as possible [4].

Another effect which helps to reduce the energy of the electrons is the presence of the Si-SiO$_2$ interface. In a properly scaled device the shallow source/drain junctions are used and the current path of the electrons is very close to the interface. The electrons notice the presence of this interface as an extra scattering mechanism. This in itself makes it more difficult for the electrons to gain enough energy to cause damage [5].

The energy distribution, or the electron temperature, can be modeled by solving the hydro-dynamical equation in addition to the normal drift-diffusion equations. A fully self-consistent solution is CPU intensive and often not necessary. An efficient post-processing method to calculate the substrate currents incorporating both the surface impact ionization as well as the non-local carrier heating has been presented in [6].

Figure 2 depicts the maximum power supply voltage $V_{DD,MAX}$ as a function of the design rules for devices scaled according to the quasi-constant-voltage scaling laws [7]. Significant scattering of $V_{DD,MAX}$ has been reported in the literature for 0.1-μm devices, but all values for $V_{DD,MAX}$ are well above 1.5 V. The maximum allowed power supply voltage is therefore likely to be above the value required according to the a realistic scaling scenario.

3. Threshold voltage

In the previous section it was demonstrated that hot-electron degradation is not an important issue for devices in the deep-submicron regime with low V_{dd}. The scaling of the threshold voltage V_t on the other hand is less straightforward. The problem we encounter is that the subthreshold slope can not be scaled. Too low a V_t will imply that the transistor can not be completely turned off: a significant leakage current will flow with 0 Volts applied to the gate. This is for most applications not acceptable and this will probably lead to a different scaling factor for the threshold voltage than the one used for the power supply voltage. Low values of V_t are acceptable when the subthreshold slope is as steep as possible and the spread in V_t is kept to an absolute minimum value. In this section the threshold voltage is discussed in more detail.

CMOS processes scaled down to the 0.1-μm regime need high substrate doping levels in order to maintain good short-channel behaviour. The use of high concentrations of channel doping will introduce new effects on the threshold voltage. Firstly, the quantization of the inversion layer becomes noticeable when the electrons are exposed to high normal electric fields. Secondly, the statistical distribution of the dopant atoms in the channel wil affect the mean value as well as the spread of the threshold voltage.

3.1. Quantization effects

An accurate modeling of the threshold voltage is essential for the comparison of the different scaling scenarios. A systematic deviation has been found between the observed V_t and the V_t simulated with a conventional device simulator (figure 3). The simulated V_t is always lower and this difference increases with increasing substrate doping level.

This deviation has been attributed to the quantum-mechanical splitting of the energy level in the conduction band. At the onset of strong inversion an analytical solution of the Schrödinger equation is available [8] and it is quite easy to calculate the threshold voltage with inclusion of the quantization effects. A simple model that accounts for this QM effect has been presented in [9] and compared with the selfconsistent solution [10]. The result is

$$V_t^{QM} \approx V_t^{CLAS} + \Delta\Psi_S \left(1 + \frac{1}{2C_{ox}}\sqrt{\frac{\varepsilon_{Si}qN_A}{\phi_B}}\right), \tag{1}$$

with

$$q\Delta\Psi_S \approx \frac{13}{9}\beta\left(\frac{\epsilon_S}{4qu_T}\right)^{1/3}E_y(0)^{2/3}. \tag{2}$$

In this formula, $E_y(0)$ is the perpendicular electric field at the Si-SiO$_2$ interface, u_T the thermal voltage and $\beta = 4.3 \times 10^{-8}$ eVcm. The perpendicular electric field at the onset of strong inversion is given by $E_y(0) \approx qN_AW_m/\epsilon_S$.

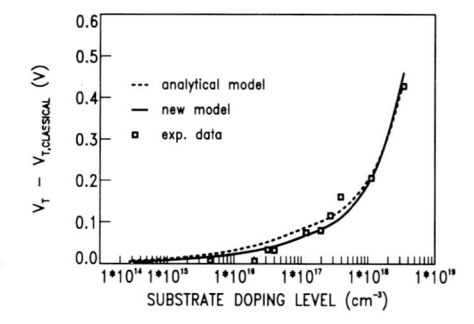

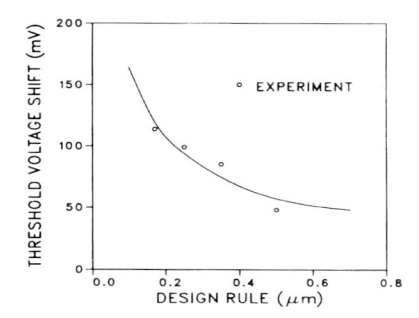

Figure 3: Deviation between the classically-simulated and the measured long-channel V_t as a function of the doping concentration ($t_{ox} = 14$ nm) [9].

Figure 4: Deviation between the classically-simulated and the measured long-channel V_t as a function of the design rule. The measurements have been done on devices which were scaled according to the QCV scaling rules. Solid line is the theoretical curve for this scaling scenario [9].

This model has been implemented in the device simulator MINIMOS, and is used throughout this paper to calculate the threshold voltage. Figure 3 shows the V_t- shift as a function of the doping concentration at fixed oxide thickness t_{ox}. In a properly scaled device, the oxide thickness decreases as a function of the design rule. Figure 4 shows the measured and simulated V_t shift for devices scaled according to the quasi-constant-voltage scaling rules. From this figure we observe that the quantization effect can not be neglected in the deep-submicron regime.

3.2. Statistical variations of the channel dopants

Another problem we encounter when conventional MOSFET's are scaled to deep submicron dimensions is of a more statistical nature. If the active area ($W \times L$) of a MOSFET is small, the depletion layer charge consists of relatively few dopant atoms. The number of atoms building up this depletion layer will fluctuate. In addition, these charges will not be distributed uniformly. These two effects have an effect on the threshold voltage. It has been shown that the average value of the threshold voltage, $< V_t >$, drops due to the microscopic distribution of dopant atoms [11]. More importantly, it has a significant effect on the spread σ_{V_t} of the V_t. Significant fluctuations of the threshold voltage have recently been reported [12].

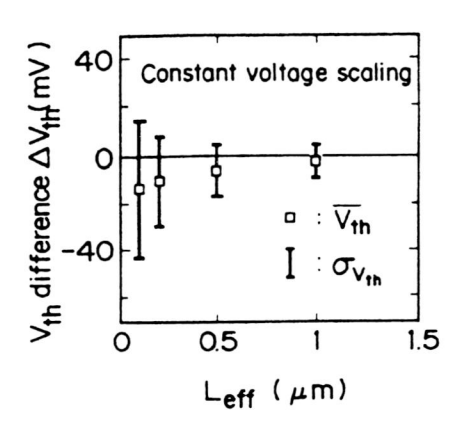

Figure 5: V_t distribution due to the N_A distribution. Constant voltage scaling laws are used. Figure taken from ref. [11].

Figure 5 shows the V_t distribution as a function of the design rule for devices scaled according to the constant-voltage scaling laws. Miniaturization of the spread is mandatory for circuit operation when the devices are operating at low values of V_{dd} and V_t. This effect could well be a limitation for ultimate conventional MOSFET's and might have a severe impact on the way the devices are scaled.

4. Short-channel devices

The limits of the conventional MOSFET can further be explored by examining the control of the short-channel behaviour, or the V_t roll off. A key parameter in the various scaling scenario's is the junction depth D_j. It determines the drain-induced barrier lowering of the short-channel MOSFET's. A minimum junction depth has been realized in experiments by Noda et al.[13] using two subgates to induce inversion layers acting as drain extensions (see inset of figure 6). These two inversion layers mimick infinitesimal shallow junctions (inversion layers are typically 30 Å thick). The larger junction depths used in figure 6 are oridinairy S/D constructions. Accurate simulations of the D_j dependence is important. Figure 7 shows the original data as well as the MINIMOS simulations that we have performed on these data. The original data can be reproduced using device simulations and the model for V_t described in Sec. 3.

Now that we have verified the simulation tools, we can estimate the limits of the conventional MOS scaling. The long-channel threshold voltage $V_{t,L}$ has been varied and the minimum effective channel length for which we still get good short-channel behaviour has been determined. As a criterion for good short-channel behaviour we have assumed that the V_t of the MOSFET with the minimum gate length is at least $0.75 \times V_{t,L}$, when the voltage applied to the drain is to $V_D = 1.5$ V. This means that we test the punch-through behaviour.

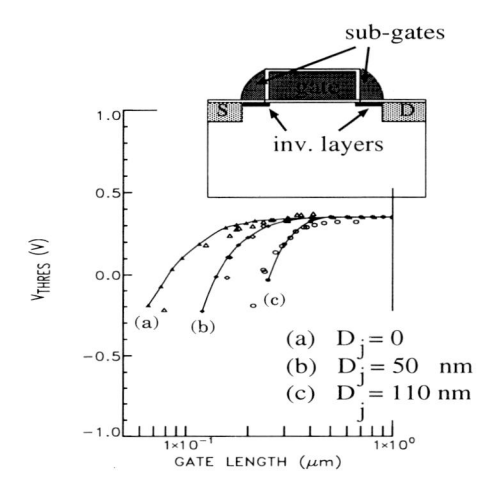

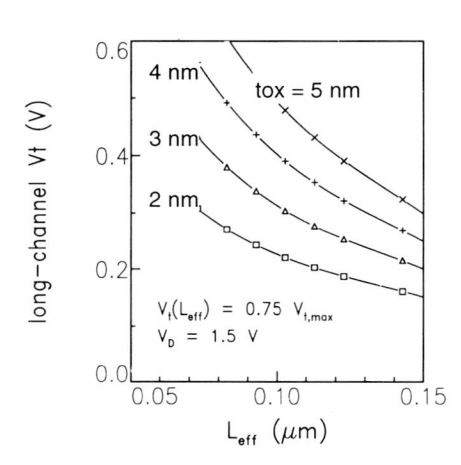

Figure 6: Threshold voltage V_t versus the gate length for different junction depths D_j. Open symbols are the data from [13], solid symbols are the results of our MINIMOS simulations.

Figure 7: Long-channel V_t versus the minimum effective channel with good short-channel behaviour.

We have implicitly assumed that the power supply voltage of deep-submicron generation will be 1.5 V. This is a reasonable assumption and has been used in many experimental studies in the 0.1 μm regime. The results of the simulations are displayed in figure 7.

For instance, if we aim a MOS generation with $L_{eff,min} = 0.10$ μm and $V_{t,L} = 0.35$ V, we see from figure 7 that the oxide thickness should be 3.5 nm or less. This value for the oxide thickness is expected to be the lower limit for conventional SiO_2. Below 3.5 nm direct tunneling through the oxide becomes too important.

4.1. Reverse short channel effect (RSCE)

The MOSFET is an intrinsic 2D device, and an accurate simulation of the 2D doping profile is essential. One of the most important phenomena for short-channel devices is RSCE (figure 8), which is caused by anomalous diffusion effects near the edge of the gate. RSCE can be caused by 2D oxidation-enhanced diffusion due to the gate reoxidation [14], or 2D transient-enhanced diffusion due to the implantation of the LDD or the source and drain [15].

Although these effects can in principle be simulated with a 2D process simulator, an accurate prediction of RSCE is still difficult. This is mainly caused by the complex nature of the poin-defect dynamics, especially for TED, and the inability to directly measure the 2D doping profile. The diffusion coefficient of boron can be approximated by $D_B \approx D_B^* \times C_I/C_I^*$ where $*$ denotes the equilibrium value and C_I the 2D

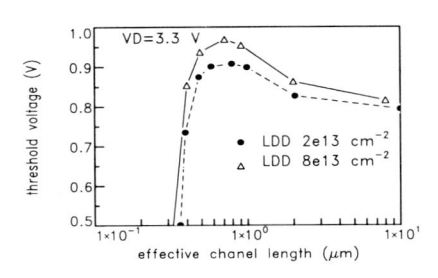

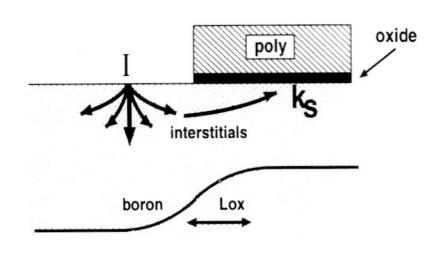

Figure 8: *Reverse short channel effect. This figure illustrates the RSCE due to 2D TED. The LDD dose was varied in this experiment.*

Figure 9: *Simulation of RSCE requires the simulation of C_I back to thermal equilibrium. Excess interstitials are injected (rate I) during oxidation or implantation and absorbed at the interface with rate k_S.*

interstitial profile. Simulation of the RSCE thus requires the interstitial distribution.

Key parameters for the simulations of the evolution of the interstitial profile are models for the injection I and k_S/D_I, the ratio of the recombination rate at the interface and the diffusion coefficient of interstitials (figure 9). Injection of point defects is in this case caused by oxidation or by implantation.

Special test structures have been designed to investigate the 2D boron profile after 2D TED, showing an important role in the 2D point-defect dynamics for the extended defects formed during armorphizing implantations [16]. The modeling of the initial conditions for TED and an efficient inclusion of the extended defects in the process simulations is still a challenging problem.

5. Conclusions

Some of the issues concerning the scaling of MOSFET's have been discussed. From the point of view of device operation, there seems to be no fundamental limit. Quasi-conventional devices an effective channel length as small as 0.05 μm have been fabricated [17]. In the deep-submicron regime there will be less emphasis on the hot-carrier performance. The scaling of the V_t and V_{dd} is more important. For MOSFET's in the 0.1-μm regime with low V_t's, it is absolutely mandatory to minimize the spread in the transistor parameters, especially for the V_t. Techniques to identify the process parameters responsible for the fluctuations in the V_t are available. This issue will be discussed elsewhere [18]. It is expected that quantization effects and the statistical distribution of the impurities will become more important. It is still an open question of how to implement physical models to account for these effects efficiently in the device simulators. For MOSFET's in the 0.1 -μm regime, it is further expected that anomalous diffusion effects will dominate the process simulations. Although progress has been made in this field in the last couple of years, it is still an important research topic. Conventionally scaled MOSFET are expected to be operational for process generations designed for effective channel lengths below 0.1 μm.

References

[1] P. Chatterjee, W. Hunter, T. Holloway, and Y. Lin, *IEEE Electron Dev. Lett.*, vol 10, p. 220 (1980).

[2] G. Baccarani, M. Wordeman, R. Dennard, "Generalized Scaling Theory and Its Application to a 1/4 Micrometer MOSFET Design", *IEEE Trans. Electron Devices*, vol 31, p. 452 (1984).

[3] H. Hu, J. Jacobs, L. Su, and A. Antoniadis, "A Study of Deep-Submicron MOSFET Scaling Based on Experiment and Simulation", *IEEE Trans. Electron Devices*, vol. 42, p. 669 (1995).

[4] J. Slotboom, et al., "Non-Local Impact Ionization in Silicon Devices", *Tech. Digest IEDM*, p. 127 (1991).

[5] J. Slotboom, G. Streutker, G. Davids, and P. Hartog, "Surface Impact Ionization in Silicon Devices", *Tech. Digest IEDM*, p. 494 (1987).

[6] M. van Dort, J. Slotboom, G. Streutker, and P. Woerlee, "Lifetime Calculations of MOSFET's using Depth-Dependent Non-Local Impact Ionization", *Microelectronics Journal* vol. 26, p. 301 (1995).

[7] P. Woerlee et al., "The Impact on Hot-Carrier Degradation and Supply Voltage of Deep-Submicron NMOS Transistors", *Tech. Digest IEDM*, p. 537 (1991).

[8] F. Stern, "Quantum properties of surface space-charge layers", *CRC Crit. Rev. Solid State Sci.*, p. 499, 1974.

[9] M. van Dort et al., "Quantum-Mechanical Threshold Voltage Shifts of MOSFET's Caused by High Levels of Channel Doping", *Tech. Digest IEDM*, p. 495 (1991).

[10] M. van Dort et al., 'A Simple Model for Quantisation Effects in Heavily-Doped Silicon MOSFET's at Inversion Conditions.' *Solid-State Electronics*, Vol.37, p. 411 (1994).

[11] K. Nishinohara, N. Shigyo, and T. Wada, "Effects of Microscopic Fluctuations in Dopant Distributions on MOSFET Threshold Voltage", *IEEE Trans. Electron Devices* vol 39, p. 634 (1992).

[12] T. Mizuno, J Okamura, and A. Toriumi ,"Experimental Study of Threshold Voltage Fluctuation Due to Statistical Variation of Channel Dopant Number in MOSFET's", *IEEE Trans. Electron Devices* vol 41, p. 2216 (1994).

[13] H. Noda, F. Murai, and S. Kimura, "Threshold Voltage Controlled 0.1-μm MOSFET Utilizing Inversion Layer as Extreme Shallow Source/Drain", *Tech. Digest IEDM*, p. 123 (1993).

[14] M. Orlowski, C. Mazuré and F. Lau, "Submicron Short Channel Effects due to Gate Reoxidation Induced Lateral Interstitial Diffusion", *Techn. Digest IEDM*, p. 632 (1987).

[15] C. Rafferty et al., "Explanation of Reverse Short Channel Effect by Defect Gradients", *Tech.Dig. IEDM*, p. 311, 1993.

[16] M van Dort et al., "Two-Dimensional Transient-Enhanced Diffusion and Its Impact on Bipolar Transistors", *Tech.Dig. IEDM*, p 865, 1994.

[17] A. Hori et al., "A 0.05-μm with Ultra Shallow S/D Junctions Fabricated by 5 keV Ion Implantation an Rapid Thermal Annealing", *Tech.Dig. IEDM*, p. 485 (1994).

[18] M. van Dort and D. Klaassen, "Sensitivity Analysis of an Industrial CMOS Process using RSM Techniques", *Proc. SISPAD*, 1995.

Monte Carlo Simulation of Carrier Capture at Deep Centers for Silicon and Gallium Arsenide Devices

A. Palma, J.A. Jiménez-Tejada, A. Godoy and J.E. Carceller

Departamento de Electrónica y Tecnologia de Computadores. Facultad de Ciencias.
Universidad de Granada. 18071 Granada, SPAIN

Abstract

This work provides a direct way to include multiphonon capture at deep levels in the simulation of electron transport by the Monte Carlo method. This has been possible by adding the capture probability as one more scattering probability together with lattice mechanisms. To check this probability, numerical capture cross sections for deep centers in Si and GaAs obtained by our method have been fitted with experimental measurements.

1. Introduction

The multiphonon-emission process, responsible for carrier capture in bulk deep centers, has been included in the framework of a standard Monte Carlo simulation [1] of the electron transport in Si and GaAs semiconductors doped with deep neutral impurities. Multiphonon-emission probability has been added to the simulation lattice-scattering mechanisms. This work comprises a first step in the implementation of this process in device simulators.

At the limit of low temperatures and weak coupling, the probability of emission of p phonons with energy $\hbar\omega$, W^{mph}, has been calculated on the basis of the adiabatic formalism developed by Ridley [2, 3]:

$$W^{mph} = \frac{16\pi^2 \omega S^p e^{-S}(\nu_T a^*)^3}{V(2\pi p)^{1/2}(p/e)^p} p[0.4 + 0.18(p-1)](n+1)^p e^{-2nS}, \quad (1)$$

where S is the Huang-Rhys factor (S<<p), V the crystal volume, and a^* the effective Bohr radius in the semiconductor. The delta function model has been used to model the center bound state, $\phi_T(r) \propto r^{-1} e^{-r/\nu_T a^*}$. Dependence with the deep energy level, E_T, is through ν_T as follows:

$$\nu_T^2 = \frac{E_H^*}{E_T}, \quad (2)$$

where E_H^* is the effective Rydberg energy (32 meV in Si and 5.3 meV in GaAs), and dependence with the temperature T is included in the occupation factor of phonons,

$n = (e^{\hbar\omega/k_B T} - 1)^{-1}$, where k_B is the Boltzmman constant. In this eq., we considered the emitted phonons to be the non-polar LO phonons ($\hbar\omega$=63 meV) in Si and ($\hbar\omega$=30 meV) in GaAs. Therefore, the Huang-Rhys factor is the only free parameter in Expression 1. In order to verify the capture probability, in the next section we have numerically calculated the thermal dependence of the capture cross sections for several deep centers in Si and in GaAs.

2. Numerical procedure

The task of incorporating this probability in the Monte Carlo simulation was solved by restricting the carrier motion to one impurity space. Therefore, in a sample with an empty impurity concentration of N_T, the real probabilities included were:

$$< W^{mph} > = W^{mph} V N_T, \qquad (3)$$

which means that we are simulating the motion of one electron in the average volume corresponding to one trap. The third body [3] and the screening effects were neglected.

In the Monte Carlo simulation, all phonon mechanisms were included and the non-parabolicity effects were accounted for both in Si and in GaAs. The numerical procedure was as follows: One electron is introduced with the thermal energy corresponding to the lattice temperature and is allowed to move without including for the multiphonon process in order to avoid dependencies with the initial carrier state. After a certain number of scatterings (15000), the multiphonon mechanism is activated. At precisely the moment when this mechanism is stochastically chosen, the electron is considered captured, and the mean velocity and the time spent since the multiphonon probability inclusion, the so-called capture time, are recorded. This one-electron procedure is repeated for a very large number of carriers, and the average values of the mean velocities, $<v>$, and the capture times, $<\tau_c>$, are used to calculate the average thermal capture cross section for electrons, σ_n:

$$\sigma_n = \frac{1}{N_T <v><\tau_c>}. \qquad (4)$$

3. Results

In order to check whether the average capture time is the proper value to use in Eq. 4, the distribution of electrons, N_e, with the capture-time interval is plotted is Fig. 1. This figure shows one example of the exponentiality of this numerical magnitude as the Schokley-Read-Hall statistic predicts. We have fitted our results with experimental data [4, 5] of capture cross sections vs. temperature for the acceptor level of Au in silicon (Fig.2) and the A defect in gallium arsenide (Fig.3). The Huang-Rhys factors obtained in these fittings agree with theoretical and experimental measurements [2, 6].

In summary, the inclusion of the capture mechanism with the rest of the lattice mechanisms allows a better understanding of this phenomena, showing as it does the relationship between the multiphonon process and the scattering mechanisms. The agreement between our results and experimental data has been achieved with physical values of the different parameters of Eq. 1, resulting in an easy way of implementing this capture mechanism in carrier transport simulators.

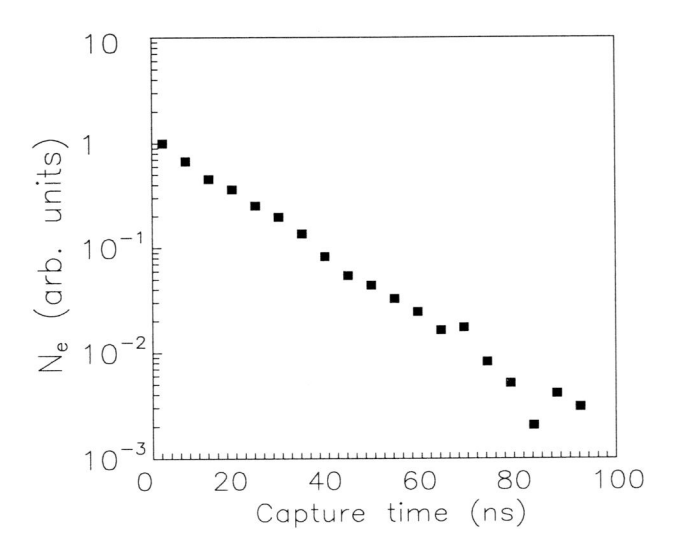

Figure 1: Histogram of N_e vs. capture time. Simulation for Pt acceptor level in Si with $N_T = 5 \cdot 10^{14}$ cm^{-3} at T=80 K

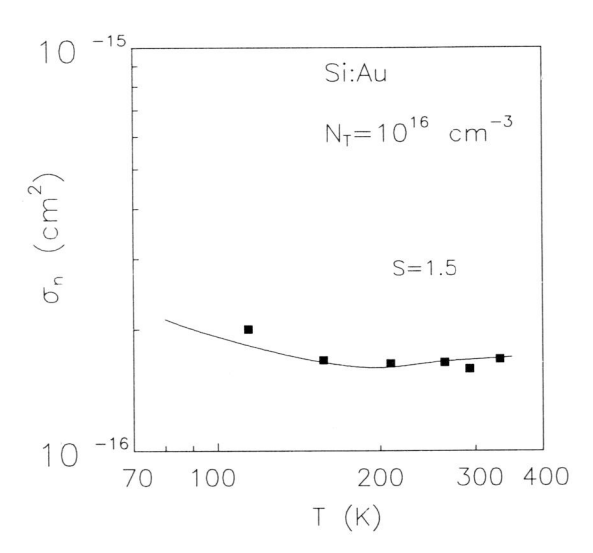

Figure 2: Comparison between experimental[4] (symbols) and numerical (line) σ_n of electrons for Au acceptor level in Si

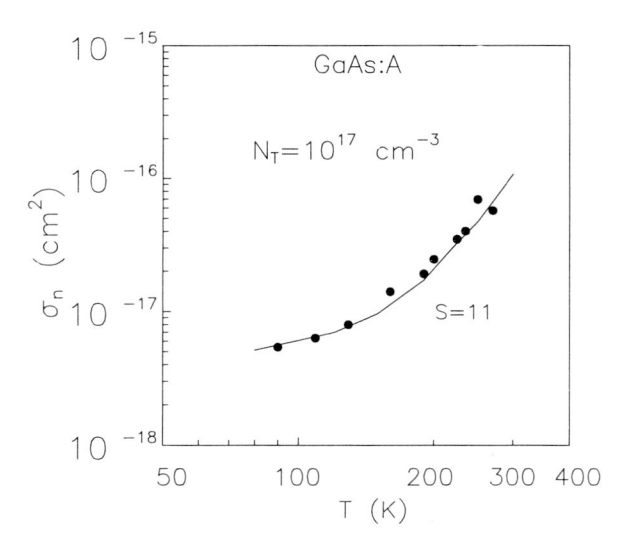

Figure 3: Experimental[5] (symbols) and numerical (line) σ_n for A center in GaAs

References

[1] C. Jacoboni and L. Reggiani, "The Monte Carlo method for the solution of charge transport in semiconductors with applications to covalent materials," *Rev. Mod. Phys.*, vol. 55, no. 3, pp. 645-705, 1983.

[2] B.K. Ridley, *Quantum Processes in Semiconductors*, Clarendon Press, Oxford, 1993

[3] B.K. Ridley, "On the multiphonon capture rate in semiconductors," *Solid-State Electron.*, vol 21, pp. 1319-1323, 1978

[4] D.V. Lang, H.G. Grimmeiss, E. Meijer and M. Jaros, "Complex nature of gold-related deep levels in silicon," *Phys. Rev. B*, vol. 22, no 7, pp. 3917-3934, 1980

[5] C.H. Henry and D.V. Lang, "Nonradiative capture and recombination by multiphonon emission in GaAs and GaP," *Phys. Rev. B*, vol. 15, no. 2, pp. 989-1016, 1977

[6] S.T. Pantelides, *Deep Centers in Semiconductors*, Gordon and Breach Science Publishers, New York, 1992

A New Statistical Enhancement Technique in Parallelized Monte Carlo Device Simulation

K. Shigeta, K. Tanaka, T. Iizuka, H. Kato[a] and H. Matsumoto

Microelectronics Res. Labs., NEC Corp., [a]ULSI Device Dev. Labs., NEC Corp.
1120, Shimokuzawa, Sagamihara, Kanagawa 229, Japan

Abstract

A new statistical enhancement technique (split-and-remove technique) in Monte Carlo device simulation, which is suitable for parallel processing, has been developed. By using this technique, an accurate energy distribution function near the drain edge can be obtained within a reasonable CPU time.

1. Introduction

On the advent of parallel computing, Monte Carlo(MC) device simulation is about to recover its dream of being a practical and accurate tool. Even in such a rosy view, a statistical enhancement technique is mandatory for sampling rare events, such as hot carriers, often jammed with stochastic noise due to a limited number of samplings. Split-and-gather (SG)[1](Fig. 1(a)) and multiple refresh (MR)[2] are such techniques that manipulate MC particles belonging to a region #i, defined in phase space $(\vec{r}, \vec{k})$, through adjusting particle's statistical weight w. This manipulation allows more MC particles with "light" weight to be loaded in a region of interest than in others, which leads to detailed sampling in the region.

The SG technique has two drawbacks in its gathering process; (i) momentum and energy cannot be conserved simultaneously, which causes a distortion of carrier distribution in phase space, (ii) searching a partner is CPU timewise expensive. The MR technique avoids such difficulties. However, it would be unsuitable for parallel computing. The MR technique refers to the distribution of particle weights in each region to readjust the number of MC particles to a desired population N_i in the region. This requires fetching particle data handled by other processors, which should be avoided because interprocessor communication is usually a bottleneck in parallel processing. In this paper, the authors propose a "split-and-remove" (SR) technique (Fig. 1(b)), with particular attention to parallel computing.

2. Split-and-Remove Technique

The SR technique comprises two processes; splitting and removing MC particles. In the splitting process, a much "heavier" particle than the desired weight W_i given by $(\sum_{\text{in } \#i} w)/N_i$ is split into $int(w/W_i)$ particles. In the removing process, a much "lighter" particle than W_i, is randomly removed with probability $r(0 < r < 1)$, since it comes into a region of little interest and hence it should be labeled as surplus particle to be removed for saving the CPU time. The weight of the unremoved surplus particle is scaled up by a factor of $1/(1-r)$ (Fig. 2). Therefore, the obtained carrier distribution in phase space is preserved in a statistical manner.

In the removing process, the total charge is not exactly conserved because of random process while its expected value is conserved. Although the total charge conservation could be realized by adding a further adjustment process that refers to all the particles in each region, the SR technique omits this additional process on purpose in parallel processing to reduce the interprocessor communication. In this way, complete parallelization with respect to particles is achieved.

The previous MR technique randomly chooses as many particles as there should be within each region among all existent particles according to their weights, which allows any particle to be removed or be split by chance. In contrast, the random removal in the SR acts only on the "light" particles labeled as surplus. Therefore, the SR is more robust against stochastic noise than the MR.

The number of MC particles can be kept close to N_i in each region without *a priori* knowledge of the distribution, since W_i is automatically updated along with the transit in the carrier distribution. The characteristics of the three SR, SG and MR techniques are summarized in Table 1.

3. Simulation Results and Discussion

Fig. 3 shows influences of the gathering process in the SG and the removing process in the SR on energy distribution. In the gathering process, two MC particles within the same energy region are joined satisfying momentum conservation. The removing process is not accompanied by any distortions in the energy distribution that is observed for the gathering process. In addition, device simulation results reveal that our method is 5 times faster than the SG technique because finding partners to be gathered is unnecessary.

The SR approach is implemented in our 2-carrier self-consistent MC device simulator. Fig. 4 shows MC particle distribution in MOSFET obtained by the simulation using parameters shown in Table 2. This parameter set is intended for allocating MC particles mainly to the drain edge region (#3). In fact, MC particles placed in #3 is about 40 times as many as the case without the technique. Statistical enhancement in energy space is also designed by dividing region #3 into 81 subregions as indicated under the x-axis in Fig. 5 and by allocating the same number of particles in each subregion.

As shown in Fig. 5, the accurate energy distribution function is obtained by using the SR technique (b), while it is jammed with stochastic noise without the SR technique (a), although almost the same number of MC particles are used. Moreover, the higher energy tail over 1.7eV with the SR is more accurate than the reference (dashed line) which consists of almost the same number of particles sampled during a much longer sampling period (4.0ps = 4000 time steps). This shows that the SR technique not only in real space but also in energy space is important for statistical enhancement of higher energy tail near the drain.

In our numerical experiments, the SR process in every time step takes up only 10% of the total CPU time. On a parallel machine Cenju-3[3](16PEs) which has a VR4400SC (75MHz) processor and 64MB local memory in each PE, the MC simulation for 1.0 ps (1,000 time steps) takes about 50 minutes.

4. Conclusion

The split-and-remove (SR) technique was proposed and implemented in our parallelized Monte Carlo device simulator. Its efficiency in statistical enhancement was demonstrated in the energy distribution function in a MOSFET. This technique drives MC device simulator toward a daily-use tool using a parallel machine.

References

[1] F. Venturi, R. K. Smith, E. C. Sangiorgi, M. R. Pinto and B. Ricco, "A General Purpuse Device Simulatior Coupling Poisson and Monte Carlo Transport with Applications to Deep Submicron MOSFET's," *IEEE Trans.Computer-Aided Design*, Vol. 8, pp. 360–369, 1989.

[2] R. Thoma, J. Peifer, W. L. Engl, W. Quade, R. Brunetti and C. Jacoboni, "An Improved Impact-ionization Model for High-energy Electron Transport in Si with Monte Carlo Simulation," *J. Appl. Phys.*, Vol. 69, pp. 2300–2311, 1991.

[3] K. Muramatsu, S. Doi, T. Washio and T. Nakata, "Cenju-3 Parallel Computer and its Application to CFD," in *Proc. 1994 Int. Symp. on Parallel Architectures, Algorithms and Networks (ISPAN)*, Kanazawa, pp.318–325, 1994.

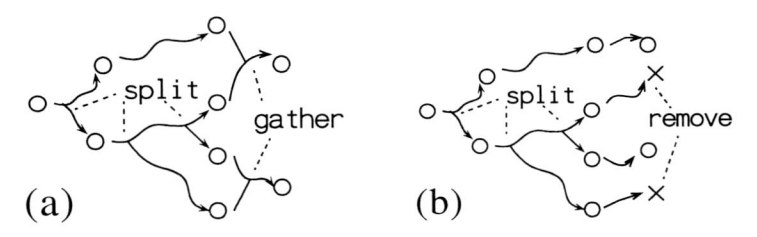

Figure 1: (a) Split-and-gather technique. The particles which are much "heavier" than the desired weight W_i are split, while much "lighter" particles are gathered into fewer particles. (b) Split-and-remove technique. Instead of gathering, "lighter" particles are randomly removed as illustrated in Fig. 2.

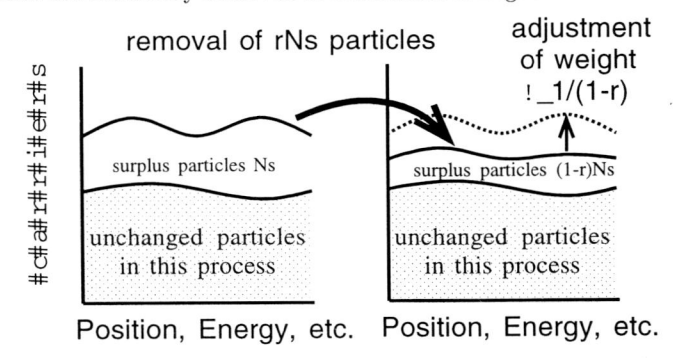

Figure 2: The removing process. This process is applied only to the surplus MC particles. They are randomly removed with probability r, and then the weights of the remaining surplus particles are scaled by a factor of $1/(1-r)$. Since rN_S particles are removed without changing the distribution of the surplus particles, the distribution of the total particles is obviously unchanged.

	SR	SG	MR
accuracy	○	×	○
calculation time	○	×	○
parallel processing	○	×	×
total charge conservation	△*	○	○

Table 1: Comparison of the three techniques. * The expected value of total charge is conserved.

Figure 3: Influence of particle reduction on energy distribution. Only energy space is taken into account.

number of particles	
electron:	~ 220,000 (adjusted weight)
hole:	~ 110,000 (constant weight)
number of regions	
real space:	4 (denoted as #1, #2, etc.)
energy space:	6 (in #1), 11(in #2), 81(in #3), 11(in #4)
distribution of particles	
real space: $N'_{\#1} : N'_{\#2} : N'_{\#3} : N'_{\#4}$ =1:1:10:1 ($N'_{\#i}$ =number of particles in #i)	
energy space: $N'_{\#i}$ divided by number of energy space division for #i	
removing probability	$r = 0.05$
time step width	$\Delta t = 1 fsec$

Table 2: Simulation parameters. The SR is applied only to electrons.

Figure 4: MC particles' distribution.

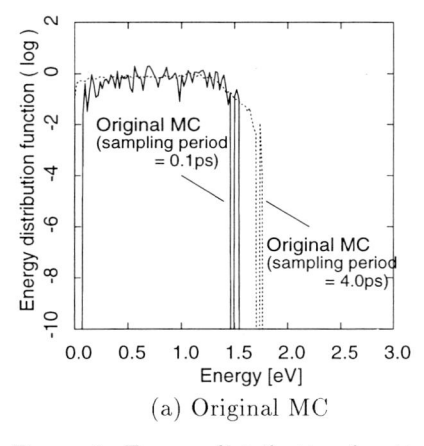

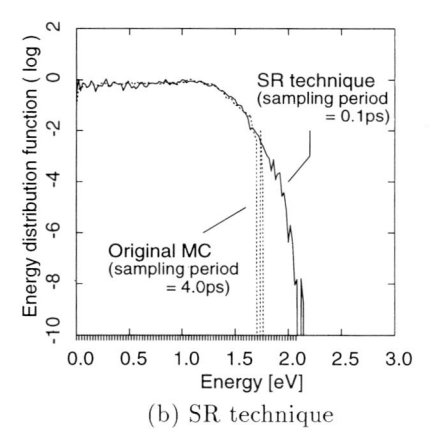

(a) Original MC

(b) SR technique

Figure 5: Energy distribution function (a) without and (b) with the SR technique, sampled for a period of 0.1ps. The dashed line is reference data without the SR, sampled for 4.0ps. Before sampling, both required about 1.0ps to reach steady state.

Stability Issues in Self-Consistent Monte Carlo-Poisson Simulations

Andrea Ghetti[a], Xiaolin Wang[*], Franco Venturi[b], and Francisco A. Leon[*]

[a]Dipartimento di Elettronica Informatica e Sistemistica
Viale Risorgimento 2, 40136 Bologna, Italy
[*]Technology CAD, Intel Corporation
2200 Mission College Boulevard, 95092 Santa Clara, California
[b]DII, University of Parma,
Via delle Scienze, 40300 Parma, Italy

Abstract

This paper investigates the time stability of self-consistent Monte Carlo-Non Linear Poisson simulations (MC-NLP). A simplified analytical stability theory has been developed and verified by means of extensive simulations. The properties of the MC-NLP scheme are compared to those of Monte Carlo-Linear Poisson (MC-LP) scheme. The influence of statistics collection and charge assignment algorithms is also analyzed.

1. Introduction

Self-consistent Monte Carlo-Poisson simulation is typically based on the linear form of Poisson equation (LP). As recently demonstrated [1], this approach can lead to instability, hence to unphysical results, unless the time step between successive Poisson solutions (Δt) is appropriately chosen. Stability forces to choose very small Δt, resulting in long CPU times.

In principle, the non linear formulation of the Poisson equation (NLP), in which charge concentration is expressed as an exponential function of potential and pseudopotentials, helps to alleviate these problems due to the damped sensitivity of potential to charge fluctuations, typical of Monte Carlo simulation [2]. In the following, the time stability of coupled Monte Carlo-Non Linear Poisson simulation will be studied in detail by means of an analytical theory and extensive simulations.

2. Analytical theory of MC-NLP time stability

Following the guide example of [1], we first derived a linearized analytical theory of MC-NLP stability for a uniformly doped semiconductor at zero applied field. All relevant quantities (concentration n, field E, velocity u) are the sum of a steady state value and a perturbation ($A = A_{DC} + \tilde{A}(x,t)$). The perturbation is expressed as $\tilde{A}(x,t) = \tilde{A}e^{ikx}e^{-i\omega t}$ where $k = \frac{2\pi}{\lambda}$, λ being the perturbation wavelength. In addition we have $n_{DC} = N_D$, $E_{DC} = 0$, $u_{DC} = 0$. The system is described by the first two moments of the Boltzmann equation and by the non linear form of the Poisson equation. Neglecting space discretization and assuming a constant effective mass m^*, these equations can be easily linearized and discretized in the time domain. In particular, the discretized non linear Poisson equation reads:

$$ikE_n = -\frac{q}{\epsilon_s}\left[n - \frac{n_{DC}}{ikV_T}(E_n - E_{n-1})\right], \qquad (1)$$

where q is the electron charge, ϵ_s the semiconductor dielectric constant, V_T the thermal voltage, and $E_n = E(t_n)$ is the perturbation field. The system of equations can be solved exactly between t_n and $t_n + \Delta t$ under a constant perturbation field E_n, thus obtaining a third order characteristic equation in the quantity $z = e^{-i\omega\Delta t}$

$$z^3 + z^2\left[\left(\frac{\alpha-(1-\delta)}{\eta^2}-\beta\right)\gamma-(1+\delta)\right] + z\left[\delta - \frac{\alpha\delta-(1-\delta)}{\eta^2}\gamma + \beta\gamma(1+\delta)\right] - \beta\gamma\delta = 0, \quad (2)$$

where $\nu_c = q/\mu_0 m^*$ is the scattering rate (related to low field mobility μ_0), $\omega_p = \sqrt{q^2 n_{DC}/\epsilon_s m^*}$ is the plasma frequency, $\eta = \nu_c/\omega_p$, $\alpha = \nu_c\Delta t$, $\delta = e^{-\alpha}$, L_D is the Debye length, $\beta = (\lambda/L_D)^2$, $\gamma = 1/(1+\beta)$. The main differences between Eq.(2) and the corresponding one for MC-LP [1] are: 1) the equation is cubical in z (instead of being quadratic) since the NLP equation (1) depends on the field at the previous iteration; 2) the perturbation wavelength λ (i.e. β) never cancels out. Fig.1 compares the stability domain ($|z| < 1$) of MC-NLP to that of MC-LP for a few values of β. Since β can be assumed to have an uniform spectrum, the instability domain of MC-NLP appears to be larger than that of MC-LP.

3. Simulations

The above analysis is based on a linear approximation neglecting space discretization and the non linear dependence of scattering rates on energy. To verify its accuracy, we employed a simplified 1D MC code featuring one parabolic band, optical and acoustic phonon scattering and periodic boundary conditions. Transport parameters were tuned to reproduce drift velocity [3] and mean kinetic energy [4] in the low field regime. For self-consistent simulations we employed a uniform grid with spacing equal to the Debye length. First, particles are placed uniformly in space with Maxwellian energy distribution; then, their motion is simulated according to the total scattering rate, until the Monte Carlo iteration ends. Finally, the charge is assigned to each grid node, and the Poisson equation is solved.

Fig.2 reports simulation results of a uniformly doped bar for different values of ν_c/ω_p and $\omega_p\Delta t$. It shows that while the analytical theory is reasonably accurate in predicting the stability of the simulation conditions (filled symbols), unpredicted stable solutions (open symbol) can be obtained in the proximity of the limit $\beta \to 0$, representing the boundary of the region stable for any β. Fig.3 shows the time evolution of mean kinetic energy (W) for parameters corresponding to MC-NLP instability. Notice that identifying stable simulations is not always as simple as Fig.3 may suggest. As an example, Fig.4 shows W for a few simulations featuring large $\omega_p\Delta t$. The one performed with MC-NLP is apparently stable, while the MC-LP ones are clearly unstable. However, the electrostatic field energy increases continuously for both schemes (Fig.5). Thus, both simulations are actually unstable [1].

The stability domains of Fig.1 imply severe restrictions on the choice of Δt for both MC-LP and MC-NLP, with significant increase of the CPU time spent for solving Poisson equation. To explore ways of relaxing these tight constraints, we investigated different methods for collecting statistics: Before Scattering (BS) and Ensemble Monte Carlo (EMC) [5] to collect data; Nearest Grid Point (NGP) and Cloud in Cell (CIC) [6] to assign charge to the grid. EMC advances particles synchronously and collects statistics at the end of each iteration, while BS moves particles asynchronously and collects statistic just before each scattering. NGP assigns the whole charge of a particle to the nearest mesh point, while CIC spreads the charge over the cell and assigns it to its vertices according to their distance from the particle. Fig.6 shows results obtained using MC-LP, short time step and two extreme configurations: BS+NGP and EMC+CIC. As can be seen, EMC+CIC provides stable solutions for $\omega_p\Delta t$

twice as large as that of BS+NGP. Eq.(2) cannot predict this result because it neglects space discretization.

To further investigate the properties of MC-NLP, we simulated a 1-D $n^+ - n - n^+$ diode ($N_D = 10^{18} - 10^{17} - 10^{18}$) with different methods. We compared the average device current obtained with the configuration featuring the largest stability domain (EMC+CIC+LP+ short Δt) to that one computed using (BS+NGP+NLP+long Δt), as in [2]. Fig.7 and Fig.8 show the corresponding results as a function of the simulation time for the same total CPU time. Although parameters are such that the MC-NLP simulation should be unstable, the NLP algorithm damps oscillations within limits only slightly larger than those of LP, while it still provides the same average current ($\simeq 1.8 mA/\mu m^2$). On the other hand, MC-NLP does not reproduce the details of the initial velocity overshoot due to the long Δt, but collects enough statistics for comparable standard deviation in less CPU time and memory occupation than MC-LP, thus providing a significant performance advantage.

4. Conclusion

We have analyzed the performance trade-offs of different self-consistent Monte Carlo-Poisson solution schemes, showing that: a) EMC+CIC+LP provides the largest stability domain and reproduces time dependent physical effects at the expenses of very large CPU and memory requirements; b) BS+NGP+NLP has a smaller convergence domain, but the uncertainty on terminal currents and other physical quantities is often acceptable even in unstable conditions. Hence, with respect to LP solution schemes much larger time steps or less particles can be chosen, with a significant reduction of CPU and memory requirements.

References

[1] P.W. Rambo et al., *IEEE Trans. on CAD*, vol.12, Nov.1993, p.1734.
[2] F. Venturi et al., *IEEE Trans. on CAD*, vol.8, Apr.1989, p.360.
[3] C. Canali et al., *Phys. Rev. B*, 12:2265-2284, 1975.
[4] M.V. Fischetti et al., *IEEE Trans. on Electron Devices*, vol. 38, Mar. 1991, p.634.
[5] C. Jacoboni and P. Lugli, *The Monte Carlo Method for Semiconductor Device Simulation*, Springer-Verlag, 1989.
[6] R.W. Hockney and J.W. Eastwood, *Computer Simulation Using Particles*, New York: Adam Hilger, 1988.

Fig.1 Stability regions as a function of collisionality ν_c/ω_p and normalized time step $\omega_p \Delta t$ for linear Poisson [1] and non linear Poisson (NLP) with different normalized perturbation wavelength $\beta = (\lambda/L_D)^2$.

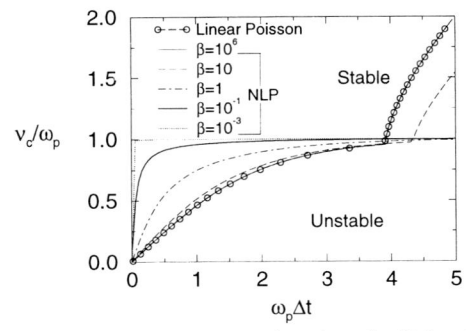

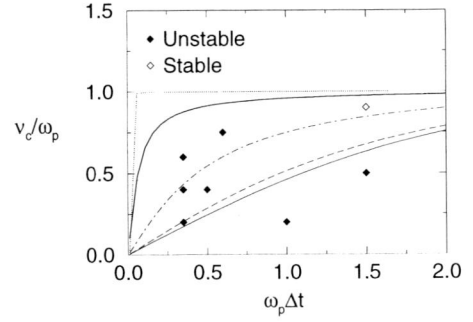

Fig.2 Numerical stability of MC-NLP as a function of collisionality ν_c/ω_p and normalized time step $\omega_p \Delta t$. Lines represent the analytic thresholds from Fig.1. Markers represent simulation results: open for stable, filled for unstable.

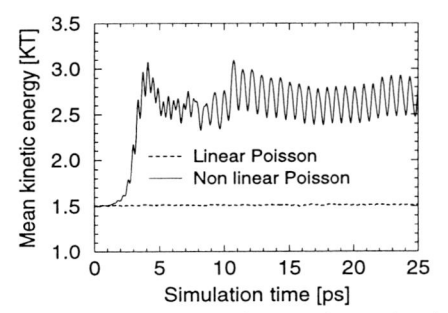

Fig.3 Mean kinetic energy in KT units as a function of the simulation time with $\nu_c/\omega_p = 0.2$ and $\omega_p \Delta t = 0.35$. As expected from Fig.1, the MC-NLP solution is unstable.

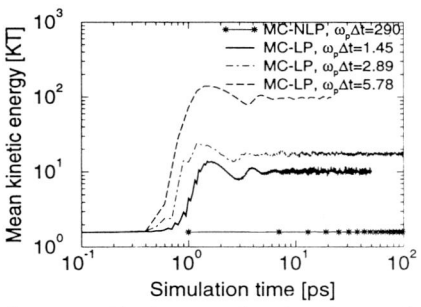

Fig.4 Mean kinetic energy in KT units as a function of the simulation time for a few simulations of a uniformly doped bar featuring large $\omega_p \Delta t$.

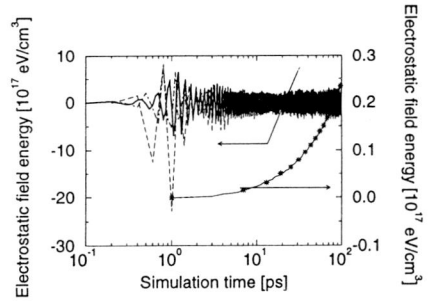

Fig.5 Electrostatic field energy as a function of the simulation time for the same simulations and using the same symbols of Fig.4.

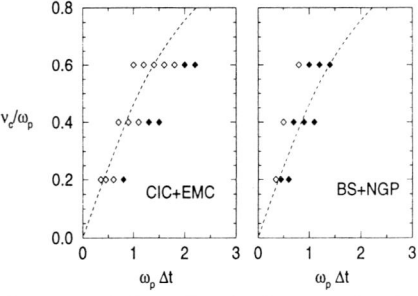

Fig.6 Numerical stability of MC-LP as a function of collisionality ν_c/ω_p and normalized time step $\omega_p \Delta t$ for two differents methods of computing statistics. Dashed line is the analytic stability threshold. Markers represent results of simulations: open for stable, filled for unstable.

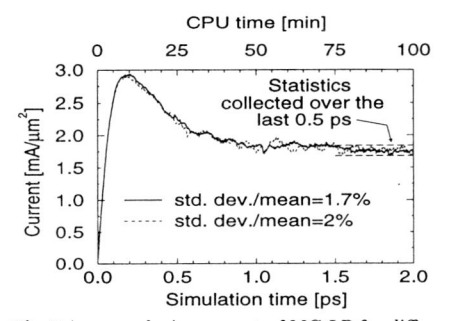

Fig.7 Average device current of MC-LP for different number of simulated particles. Solid line: 2×10^5 part.; dashed: 1×10^5 part. The ratio (standard deviation/mean value) is computed over the data of the last $0.5ps$ of simulation. $\Delta t = 0.002ps$. The CPU scale refers to the 2×10^5 particles case.

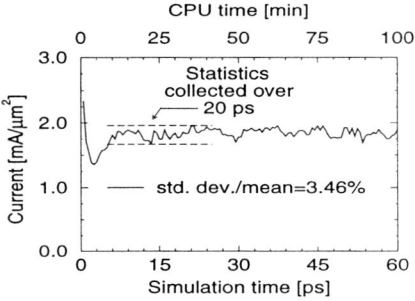

Fig.8 Average device current of MC-NLP. The ratio (standard deviation/mean value) is computed over the data of $20ps$. $\Delta t = 0.5ps$. The number of simulated particles is 5×10^4.

The Path Integral Monte Carlo Method for Quantum Transport on a Parallel Computer

C. Schulz-Mirbach

Arbeitsbereich Hochfrequenztechnik
Technische Universität Hamburg-Harburg
Wallgraben 55, D-21071 Hamburg, Germany

Abstract

Based on the Feynman path integral formulation for the time evolution amplitude, we compute the quantum mechanical transition probability for a charge carrier in a semiconductor crystal. Our implementation is performed on a parallel computer (Parsytec GC 64). We discuss the ability of the method to achieve a spatially resolved probability amplitude which is necessary for the analysis of quantum electronic devices. Macroscopic observables are evaluated using this probability function. It complements the conventional distribution function which results from the solution of the semi-classical Boltzmann transport equation.

1. Introduction

It has been suggested that the Feynman path integral formulation of quantum mechanics is an appropriate method for describing quantum transport in ultrasmall electronic devices [1], [2]. One of the advantages of the method is that the coupling of charge carriers to phonons and other scattering processes, non-constant electric fields which are not necessarily sinusoidal, as well as complicated band and quantum well structures can easily be included, which are encountered in hetero-structure devices such as Resonant Tunneling Diodes. Thanks to these promising facts and despite of the drawbacks of the method like high CPU time requirements and the difficulty to estimate the error of the approximations, we aim to derive a simulation technique which is intended to overcome the limitations of the Boltzmann transport equation. Due to the wave-like character of the charge carriers, there is a limit in the characteristic length of devices above which carriers can be described as localized particles. Hetero-structure devices are an example for this situation.

2. Path integrals on a discretized time axis

The probability amplitude of a physical system to evolve from time t_a to t_b is given by the matrix elements of the time-evolution operator $U(t_b, t_a) = exp(i\mathcal{H}(t_b - t_a)/\hbar)$, where $\mathcal{H}$ is the Hamiltonian of the system [3]. In order to study the momentum distribution function for homogeneous material it is appropriate to expand $U(t_b, t_a)$ in momentum eigenfunctions, where the following expression evolves [1]

$$(\vec{p}_b, t_b | \vec{p}_a, t_a) = <\vec{p}_b | U(t_b, t_a) | \vec{p}_a>$$
$$\approx \prod_{n=1}^{N} \left[\int_{-\infty}^{\infty} \frac{d\vec{p}_n}{2\pi\hbar} \right] \prod_{n=1}^{N+1} \left[\int_{-\infty}^{\infty} d\vec{r}_n \right]$$
$$\times \exp \left\{ \frac{i}{\hbar} \sum_{n=1}^{N+1} \left[-\vec{r}_n(\vec{p}_n - \vec{p}_{n-1}) - \frac{(t_b - t_a)}{N+1} \mathcal{H}(\vec{p}_n, \vec{r}_n, t_n) \right] \right\} \quad (1)$$

$$t_0 = t_a, \quad t_{N+1} = t_b$$

This expression means that the exponential has to be evaluated for all momentum-space paths $\vec{p}_a \rightsquigarrow \vec{p}_b$ with the resulting terms being summed up. The spatial integrals refer to time points $t_1 \ldots t_{N+1}$, so all possible spatial end points are summed up. No spatial point $\vec{r}_0$ occurs. [2]

The momentum time evolution amplitude $(\vec{p}_b, t_b | \vec{p}_a, t_a)$ equals the conditional probability amplitude for a particle to come to a momentum state $|\vec{p}_b>$ at time t_b after starting with $|\vec{p}_a>$ at time t_a. Assuming knowledge of the wavefunction $\Psi(\vec{p}, t_a)$, the probability of measuring a momentum $\vec{p}$ at time t_b is given by the distribution function $f(\vec{p}, t_b)$

$$f(\vec{p}, t_b) = |\Psi(\vec{p}_b, t_b)|^2 = \left| \int \frac{d\vec{p}_a}{(2\pi\hbar)^3} (\vec{p}_b, t_b | \vec{p}_a, t_a) \Psi(\vec{p}_a, t_a) \right|^2 \quad (2)$$

Defining cells at time t_a around the momentum values $\{\vec{p}_a^1, \vec{p}_a^2, \vec{p}_a^3, \ldots\}$ and at time t_b, $\{\vec{p}_b^1, \vec{p}_b^2, \vec{p}_b^3, \ldots\}$, the following calculation scheme is set up

$$f(\vec{p}_b^1, t_b) = |(\vec{p}_b^1, t_b | \vec{p}_a^1, t_a) + (\vec{p}_b^1, t_b | \vec{p}_a^2, t_a) + \ldots |^2$$
$$f(\vec{p}_b^2, t_b) = |(\vec{p}_b^2, t_b | \vec{p}_a^1, t_a) + (\vec{p}_b^2, t_b | \vec{p}_a^2, t_a) + \ldots |^2$$
$$\vdots \quad (3)$$

In our calculations we assume

$$\Psi(\vec{p}_a, t_a) = <\vec{p}_a | \Psi(t_a)> = \sum_{i}^{M} (2\pi\hbar)^3 \delta(\vec{p}_a - \vec{p}_i^{MB})/\sqrt{M}, \quad (4)$$

where the $\vec{p}_i^{MB}$ are drawn from a Maxwell-Boltzmann-distribution. The consequence of the δ-function is, that one has to consider only momentum values $\vec{p}_i^{MB}$ as path beginning values. It is important to mention that by the same way any momentum distribution function resulting from semiclassical methods (e.g. Monte Carlo solution of the Boltzmann transport equation) could be used. That is a possible way of combining semiclassical and quantum regime transport.

The Hamiltonian includes an imaginary potential term for the scattering of the charge carriers and the energy term due to the electric field as well as a term describing the conduction band structure. The imaginary potential term leads to a factor $\exp(-\sum_{n=1}^{N+1}/2\tau(\vec{p}_n))$ in the integral, stochastically damping out momentum configurations being associated with high scattering rates $1/\tau(\vec{p})$. Thus the momentum

---
[1] The equality holds only for $N \to \infty$. In the numerical treatment N has to be finite.
[2] The Dirac bra-ket notation is used. The subindex of the eigenfunctions $|\vec{p}_n>$ denotes them belonging to the corresponding time point t_n.

integrals can be performed using the Metropolis algorithm (cf. [4]). The space integrals are reduced to integrals over boxes accompanying the classical space path. The box size is chosen to be of range of the characteristic length, which is considered to be a measure for the size of the quantum mechanical fluctuations. This method corresponds to the windowing technique described in [2]. From the point of view of quantum device simulations, the interesting regions are those where no classical paths exist. The estimate for the size of the integration boxes is of crucial importance.

The momentum axis lying in the direction of the electric field is subdivided into cells. For each path ending in one of the cells, the corresponging exponential is summed up by its real and imaginary part separately, yielding the quantum mechanical phase.

3. Spatially resolved distribution functions for quantum devices

For the understanding of the quantum interference phenomena in heterostructures one is interested in a spatially resolved momentum distribution function. By expanding the time evolution operator in space eigenfunctions yields the more familiar path integral

$$(\vec{r}_b, t_b | \vec{r}_a, t_a) \approx \prod_{n=1}^{N} \left[\int_{-\infty}^{\infty} d\vec{r}_n \right] \prod_{n=1}^{N+1} \left[\int_{-\infty}^{\infty} \frac{d\vec{p}_n}{2\pi\hbar} \right]$$
$$\times \exp\left\{ \frac{i}{\hbar} \sum_{n=1}^{N+1} \left[\vec{p}_n (\vec{r}_n - \vec{r}_{n-1}) - \frac{(t_b - t_a)}{N+1} \mathcal{H}(\vec{r}_n, \vec{p}_n, t_n) \right] \right\}. \qquad (5)$$

In this expression the initial position $\vec{r}_0 = \vec{r}_b$ is assumed to be given. By defining space cells (in the direction of the electric field) $x^{cell\,1}, x^{cell\,2}, \ldots$ for the end point of time the probability of reaching a cell and the velocity v equals

$$Prob(cell\ i) = \left| (x^{cell\,i}, t_b | x_a, t_a) \right|^2, \quad v = \frac{\left| (x^{cell\,i}, t_b | x_a, t_a) \right|^2 \frac{(x^{cell\,i} - x_a)}{t_b - t_a}}{\sum_i \left| (x^{cell\,i}, t_b | x_a, t_a) \right|^2}. \qquad (6)$$

For each spatial cell we define momentum cells and sum up the probabilities for the momentum occurring in it. This is how to get a spatially resolved momentum distribution function, e.g. to analyse the region behind the barriers in a Double-Barrier-RTD. [3]

4. Importance Sampling on the Parallel Computer

Our implementations are performed on a parallel computer containing 64 nodes with two Power PC plus processors each (80 MFlops, 16 MB memory per node). A unit of 16 nodes is connected in a fixed topology with the other units. We use the Parix runtime-system with the programming language C. We define virtual link connections from one node to all the others, thus obtaining a farming model [5]. The farmer organizes the distribution of the starting data and the collection of the results.

The program calculates the momentum distribution function for a fixed time interval. Every processor generates one chain of paths, thus the communication between the

---

[3] The momentum distribution for a spatial region contains less phase information than the one derived above, since the contributing paths are taken from a small spatial region. The interference with paths ending in other regions is not accounted for in this sum.

nodes is kept to a minimum. In order to save memory, the momentum distribution was calculated only in the direction of the electric field. Two other grids were set up to register the velocities and the energies occurring at the end time. For each cell, the real and imaginary parts of the exponential were summed up. The probabilities for the various processors are calculated locally and then sent to the master. The average of the probabilities is calculated for every cell by the master. The decrease of the standard deviation can be controlled by the master processor at synchronisation points. For the numerical solution of the spatial distribution functions, every processor receives a space cell. The end points of all the generated paths on one processor are kept at this value. The occurring momentum end points are registered, as well as the energies and the velocities. Apart from the scattering potential, the approximation error depends on the discretization of the time interval and on the sizes of the cells.

5. Results and Conclusions

We calculate the energy, velocity, and wave vector distribution as a function of time. We use a continuous parameterization for the non-parabolic three valley conduction band structure over the Brillouin zone. The imaginary scattering potential is derived from first-order scattering rates. In Figure 1, one can see the probability of an electron to be encountered in wave-vector states for different time intervals elapsed since the field was turned on and it started to propagate. The results obtained from standard Monte Carlo technique are reproduced in their tendency, although not in their accuracy. The probability to be found in an upper L-valley increases with time, while the Gamma-valley probability decreases. Due to the size of $\hbar$ the method should be applied for an energy and time range that fulfill $\epsilon(\vec{p})/\hbar \cdot (t_a - t_b) \approx 1$ otherwise leading to large oscillations that consume a huge amount of cpu-time.

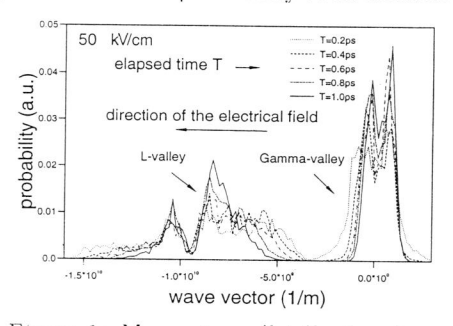

Figure 1: Momentum distribution function for bulk GaAs and constant electric field.

Acknowledgements: The author gratefully acknowledges the Deutsche Forschungsgemeinschaft for financial support.

References

[1] Massimo V. Fischetti and D. J. DiMaria, "Quantum Monte Carlo Simulation of High-Field Electron Transport: An Application to Silicon Dioxide," *Phys. Rev. Lett.*, vol. 55, pp. 2475 - 2478, 1985.

[2] L. F. Register, M. A. Stroscio and M. A. Littlejohn, "Efficient path-integral Monte Carlo technique for ultrasmall device applications," *Superlatt. M.*, vol. 6, pp. 233 - 243, 1989.

[3] Hagen Kleinert, "Pfadintegrals in Quantum Mechanics, Statistics and Polymer Physics", BI-Wiss.-Verl., 1993

[4] Malvin H. Kalos and Paula A. Whitlock, "Monte Carlo Methods", Wiley, 1986

[5] Dimitri P. Bertsekas and John N. Tsitsiklis, "Parallel and Distributed Computation", Prentice-Hall, 1989

A Monte Carlo Transport Model Based on Spherical Harmonics Expansion of the Valence Bands

H. Kosina, M. Harrer, P. Vogl[a], S. Selberherr

Institute for Microelectronics, TU Vienna
Guhausstrasse 27-29, A-1040 Vienna, Austria
[a]Walter Schottky Institut, TU Munich
Am Coulombwall, D-85748 Garching, Germany

Abstract

To represent the valence bands of cubic semiconductors a coordinate transformation is proposed such that the hole energy becomes an independent variable. This choice considerably simplifies the evaluation of the integrated scattering probability and the choice of the state after scattering in a Monte Carlo procedure. In the new coordinate system, a numerically given band structure is expanded into a series of spherical harmonics. This expansion technique is capable of resolving details of the band structure at the Brillouin zone boundary and hence can span an energy range of several electron-volts. Results of a Monte Carlo simulation employing the new band representation are shown.

1. Introduction

Efforts on numerical modeling of hot carrier transport published to date deal mainly with hot electrons. One reason might be that for electrons some important transport properties are readily revealed by assuming simple effective-mass band models. For holes, however, an effective mass approximation is poor even very close to the Γ-point. Non-parabolicity is very pronounced and cannot be described by simple analytic expressions. The warped-band model [3], which is essentially parabolic, cannot be implemented in the Monte Carlo technique without additional simplifications [2].

The representation of the valence bands we present is specifically tailored to the needs of Monte Carlo transport calculations. These needs include efficient calculation of the scattering integrals and a straightforward algorithm for the choice of the state after scattering.

2. Representation of the Bandstructure

To obtain the total scattering rate the transition probability given by Fermi's Golden rule has to be integrated in the three-dimensional k-space. Because of the energy-conserving δ-function in the transition probability a coordinate transformation is desirable such that energy becomes one of the integration variables. Assume that the band structure is given in polar coordinates: $\epsilon = \mathcal{E}(k,\Omega)$. We now introduce a coordinate transformation $(k,\Omega) \to (\epsilon,\Omega)$ by inverting the function $\mathcal{E}(k,\Omega)$ with respect to k. The result of such an inversion is a function $\mathcal{K}$ describing equi-energy surfaces in k-space: $k = \mathcal{K}(\epsilon,\Omega)$. Inversion of a function is possible only in an interval where the function is monotonous. By inspection of the full band structure one finds that both the heavy hole and split-off bands can entirely be represented by such functions $\mathcal{K}$. Above a hole energy of $E_X(3.04eV)$ inversion of the light hole band is no longer unique.

In this work, we represent the function $\mathcal{K}$ as a series of spherical harmonics.

$$\mathcal{K}_b(\epsilon,\Omega)^3 = \frac{3}{4\pi} \sum_{l=0}^{\infty} \sum_{m=0}^{l} a_{b,lm}(\epsilon) P_l^m(\cos\theta) \cos m\phi, \qquad b = \text{H, L, SO} \qquad (1)$$

Derivation of the scattering rates is considerably eased by taking the third power of $\mathcal{K}$ as the function to be expanded. For symmetry reasons non-vanishing coefficients only exist for even values of l and for m being a multiple of 4. With (1) a set of functions $a_{b,lm}(\epsilon)$ contains the whole band structure information.

The density of states of a band represented by (1) is solely determined by the zero order coefficient.

$$g_b(\epsilon) = \frac{1}{4\pi^3} \frac{d}{d\epsilon} a_{b,00}(\epsilon), \qquad b = \text{H, L, SO} \qquad (2)$$

3. Scattering Rates

Wihthin this framework, we derived the scattering rates for acoustic deformation potential (ADP) scattering in the elastic approximation, optic deformation potential (ODP) scattering and ionized impurity scattering (ION) in the Brooks and Herring formalism.

$$\lambda_{ij}^{ADP}(\epsilon) = \frac{D_A^2 k_B T_L}{8\pi^2 \hbar \rho v_s^2} \frac{d}{d\epsilon} a_{j,00}(\epsilon) \qquad (3)$$

$$\lambda_{ij}^{ODP}(\epsilon) = \frac{3 D_o^2}{2^4 \pi^2 \rho \omega_{op}} \left(\frac{N_{op}}{N_{op}+1} \right) \frac{d}{d\epsilon} a_{j,00}(\epsilon \pm \hbar\omega_0) \qquad (4)$$

$$\lambda_{ij}^{ION}(\epsilon,\Omega) = \frac{Z^2 N_I e^4}{4\pi^2 \hbar (\epsilon_0 \epsilon_r)^2} \frac{1}{(2k_i \langle k_j \rangle)^2} \sum_{l=0}^{\infty} h_l^{ij}(\epsilon) \sum_{m=0}^{l} \frac{d}{d\epsilon} a_{j,lm}(\epsilon) P_l^m(\cos\theta) \cos m\phi \qquad (5)$$

All these mechanisms induce both intraband and interband transitions. Other than for electrons, overlap integrals cannot be neglected for holes. The used approximations are of the form $\mathcal{G}_{ii} = \frac{1}{4}(1 + 3\cos^2\beta)$ and $\mathcal{G}_{ij} = \frac{3}{4}\sin^2\beta$. The Coulomb scattering rate, which additionally depends on the solid angle of the wave vector, is expressed as a series of spherical harmonics. In Eq. (5), $\langle k_j \rangle$ denotes an average value over the solid angle, which is defined as $\langle k_j \rangle = (\frac{3}{4\pi} a_{j,00}(\epsilon))^{1/3}$. The coefficients $h_l^{ij}(\epsilon)$ being a result of integration can be expressed in terms of Legendre functions of the second kind.

The distribution functions of the solid angle after scattering are given as spherical harmonics series. In a Monte Carlo procedure. the after scattering state can be chosen according to these distributions by a simple rejection technique.

4. Results and Discussion

In this work, we use the series expansion (1) to represent the heavy and light hole bands up to $\epsilon_{hole} = 3.04 eV$, which is the band-energy at the X-points. The numerical band structure has been computed by a nonlocal empirical pseudopotential method.

The functions $a_{b,lm}(\epsilon)$ are represented numerically by means of a finite element method. To ensure continuous derivatives shape functions of third order have been chosen. The unknowns associated with the nodes of the energy grid have been determined by a variational approach. From numerical band data the functions $a_{b,lm}(\epsilon)$ can well be computed for non-vanishing hole energies, but not for an energy of zero. To obtain the $a_{b,lm}(0)$ we expand the expression for the warped band approximation. In this way, our band model combines the warped band approximation in the vicinity of the Γ-point where not enough numerical data points are available, and the numerical band structure for higher hole energies.

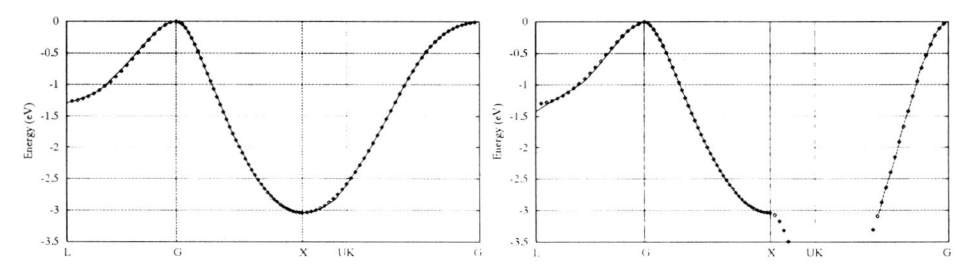

Figure 1: Comparison of numerical band structure (symbols) and the spherical harmonics expansion (lines) for the heavy hole (left) and the light hole (right) bands.

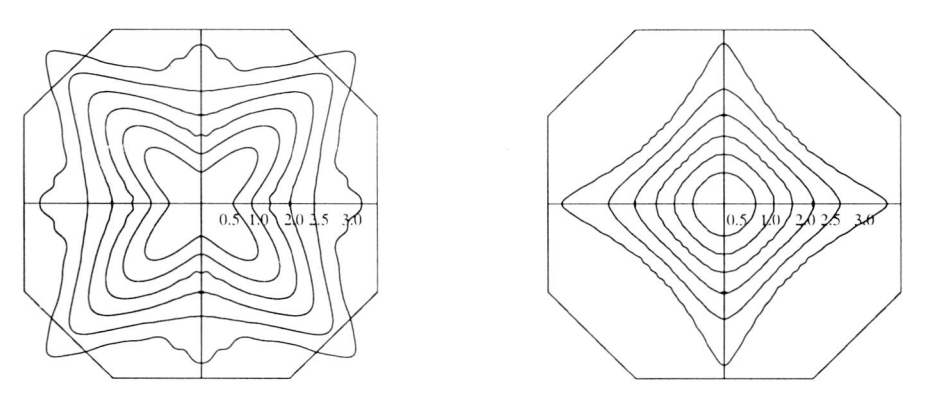

Figure 2: Cross section through the heavy hole (left) and light hole (right) bands from $0.5eV$ to $3.0eV$ in $0.5eV$ steps. The surrounding octagon indicates the boundary of the Brillouin zone

Fig. 1 shows the band diagrams for the bands under consideration. Symbols refer to the data points of the numerical band structure, solid lines to the series expansion. In Fig. 2 equi-energy lines are plotted in k-space. It turned out that at low energies less harmonics are required than at high energies. Therefore, we make the number of harmonics a function of energy. For instance, for the light hole band $l_{max} = 20$ at $0.5eV$ and $l_{max} = 60$ at $3.0eV$. The weak ripples at $3.0eV$ indicate that some higher order harmonics are still missing. In general, the higher the number of harmoncis, the better the details of the band structure can be resolved at the boundary of the Brillouin zone. On the other hand, for hole energies below E_L ($1.27eV$), where the band structure does not yet touch the zone boundary, a lower value of l_{max} is sufficient (typically $l_{max} \leq 28$).

As can be seen in Fig. 2 the series representation provides states outside the first Brillouin zone which do not exist in reality. These artificial states yield an increased density of states and hence increased scattering rates. In the Monte Carlo procedure, scattering events to such artificial states outside the Brillouin zone are rejected and self-scattering is performed instead.

In Fig. 3 the simulated drift velocity is compared to measured data [1]. Fig. 4 depicts the average hole energy as function of the electric field. In this simulation, the split-off band has been neglected.

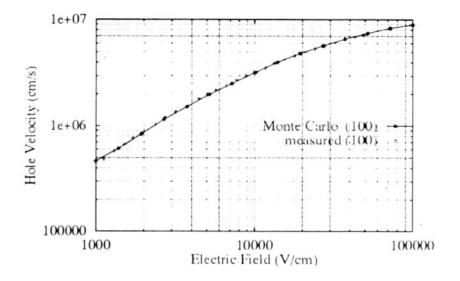

Figure 3: Comparison of simulated and measured [1] hole drift velocities as function of an electric field in (100) direction at 300K.

Figure 4: Average hole energy as function of an electric field in (100) direction at 300K.

5. Conclusion

A new method to represent numerical valence band data for Monte Carlo transport calculations has been developed. A function basically describing equi-energy surfaces in k-space is expanded into a series of spherical harmonics. Depending on the energy range accounted for and the number of harmonics invoked the model can be considered either as an improved analytical band model or as a full-band model. In this work we demonstrated the full-band capabilities for hole energies up to $E_X(3.04eV)$.

References

[1] C. Canali, G. Ottaviani, and A.A. Quaranta, Drift Velocity of Electrons and Holes and Associated Anisotropic Effects in Silicon, *J.Phys.Chem.Solids*, 32(8):1707–1720, 1971.

[2] C. Jacoboni and L. Reggiani, The Monte Carlo Method for the Solution of Charge Transport in Semiconductors with Applications to Covalent Materials, *Rev.Mod.Phys.*, 55(3):645–705, 1983.

[3] K. Seeger, *Semiconductor Physics*, Springer, 1989.

Full-Band Monte Carlo Transport Calculation in an Integrated Simulation Platform

U. Krumbein[a], P.D. Yoder[a], A. Benvenuti[a,b], A. Schenk[a], W. Fichtner[a]

[a]Integrated Systems Laboratory, ETH–Zürich,
Gloriastrasse 35, CH–8092 Zürich, Switzerland
[b]SGS-Thomson Microelectronics
Via Olivetti 2, Agrate Brianza (MI), I–20041 Italy

Abstract

We present a hierarchical CAD environment for realistic silicon device simulation, combining the utility of process, drift-diffusion/hydrodynamic, and Monte Carlo simulation in a unified platform. Monte Carlo simulation results are presented for the cases of an NIN diode and a 40nm LDD-MOSFET, using information given by a hydrodynamic pre-processing step. In addition we compare drift-diffusion, hydrodynamic and Monte Carlo results for an 0.5μm MOSFET whose geometry and doping profiles were generated by a 2-dimensional process simulation.

1. Introduction

The Monte Carlo method of charge transport simulation offers the possibility to extract information about all quantities derivable from the semiclassical distribution function, whose accuracy is limited explicitly by statistical convergence and implicitly by the quality of the physical models. To date, much effort has been devoted to improving models for band structure and scattering mechanisms, such as electron-phonon scattering [1, 2, 3, 4], impact ionization [5, 6, 7] and other carrier-carrier scattering [8, 9]. However, the practical usefulness of Monte Carlo device simulation has not entirely lived up to its promise, as evidenced by the observed propensity to simulate simplified device structures.

2. Degas – A Combined Hydrodynamic and Monte Carlo Simulator

A unique device simulation environment has been developed which unites the capabilities of process, drift-diffusion/hydro, and Monte Carlo simulation into a single platform. One may use DIOS$_{\text{ISE}}$ [10] to begin with a process simulation. Drift-diffusion or hydrodynamic simulations can be performed with the mixed-mode multi-dimensional device simulator DESSIS$_{\text{ISE}}$ [11, 12] as a preprocessing step. The full-band Monte Carlo simulator VEGAS was embedded into DESSIS$_{\text{ISE}}$ by a window technique, which is called DEGAS$_{\text{ISE}}$ [13]. The domain of the Monte Carlo simulation may be chosen either as the entire device, or as a rectangular sub-domain. When the Monte Carlo simulation is invoked, it uses the precise device structure which has been generated by the process simulation. Former implementations of the window technique used the *drift-diffusion* information as initial and boundary conditions [14, 15]. In this work, carrier densities, velocities and temperatures are extracted from *hydrodynamic* simulation as calculated by DESSIS$_{\text{ISE}}$, and passed to VEGAS for use as initial and boundary conditions. Monte Carlo simulation may be performed self-consistently or using a frozen field provided by DESSIS$_{\text{ISE}}$, either in one or two dimensions.

3. Examples

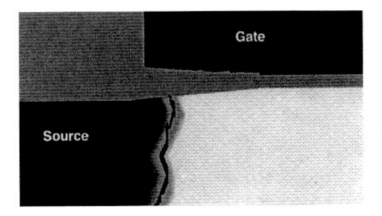

Figure 1: 0.5μm nMOSFET with arbitrarily shaped Si-SiO$_2$ interface from process simulation.

We present three examples: a 0.5μm MOSFET, a 40nm MOSFET and a 0.5μm NIN structure.

The 0.5μm nMOSFET was fabricated and measured by Fujitsu. The process was simulated with DIOS$_{\text{ISE}}$, and resulted in a non-planar Si-SiO$_2$ interface (Figure 1). Figure 2 shows that the drift-diffusion ansatz completely fails in this example.

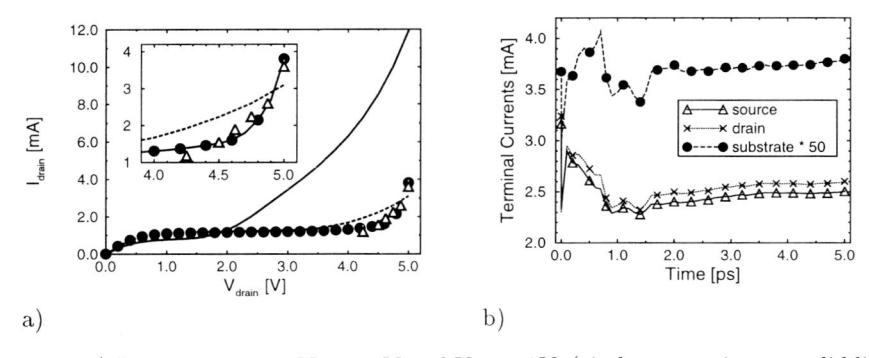

Figure 2: a) Drain current at V$_{\text{subs}}$=0V and V$_{\text{gate}}$=2V. (circles: experiment, solid line: drift-diffusion, dashed line: hydrodynamic, triangles: Monte Carlo) b) Convergence of terminal current during the Monte Carlo simulation at V$_{\text{drain}}$=4.875V.

The hydrodynamic simulation leads to good agreement with the measurements until about 4V drain voltage. Only the Monte Carlo method predicted the breakdown. The terminal currents are evaluated by a powerful domain integration technique. The terminal current convergence as a function of simulation time is shown in Figure 2.b. After less than two picoseconds convergence is obtained also for the substrate current.

The second example is a 40nm LDD-MOSFET. In Figure 3 the hydrodynamic electron temperature and the Monte Carlo electron average energy are compared. The rectangle denotes the boundary of the Monte Carlo simulation domain. While the hydrodynamic solution shows the highest temperatures at the highly doped drain edge, the largest Monte Carlo energies are at the bottom of the LDD implant and the region of hot carriers is much more extended into the drain.

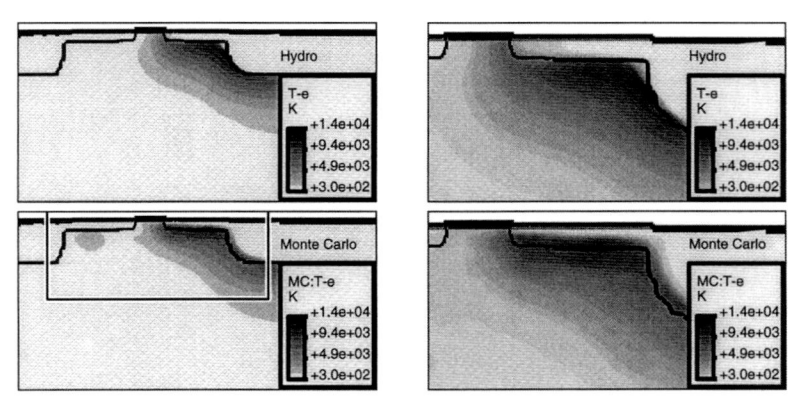

Figure 3: Comparison of electron temperatures computed by the hydrodynamic and the Monte Carlo method. (V_{gate}=2V, V_{drain}=4V, V_{sub}=0V) The plots on the right are zooms into the drain region.

The third example consists of an NIN structure with doping concentrations of 5×10^{17} and $2 \times 10^{15} cm^{-3}$, where the intrinsic region has a length of 0.5 μm. In Figure 4 the need for Monte Carlo simulations is demonstrated by the impact ionization rate of the hydrodynamic in comparison with the Monte Carlo result. The hydrodynamic rate, a function of the carrier temperatures, cannot satisfactorily account for the non-locality needed in this example. Even when hydrodynamic temperatures and the average energies from Monte Carlo agree quite well, more detailed information about the non-local hot electron distribution is needed than the hydrodynamic formulation can model.

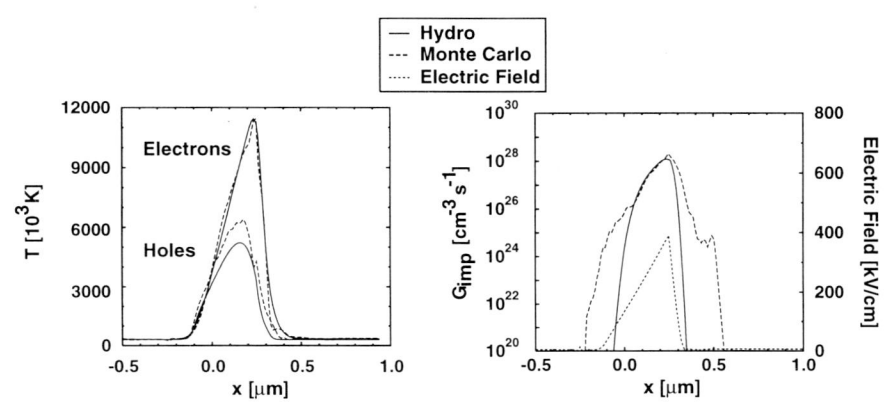

Figure 4: Carrier temperatures/average energies, avalanche generation, and electric field at 10V applied bias in the NIN example.

The coupling which has been presented between the hydrodynamic ansatz and the Monte Carlo method within the same software environment enables the user not only to simulate deep submicron devices very accurately, but even to verify and adjust parameters of the hydrodynamic model. The mixed-mode and multi-device capabilities of DESSIS$_{-ISE}$ are not limited.

Acknowledgment

The authors are grateful to Dr. N. Sasaki from Fujitsu, Atsugi (Japan) for providing experimental data. This work has been financially supported by the Swiss Research Program LESIT and the ESPRIT-6075 (DESSIS) Project.

References

[1] M.V. Fischetti, J. Higman, "Theory and Calculation of the Deformation Potential Electron-Phonon Scattering Rates in Semiconductors" in "Monte Carlo Device Simulation: Full Band and Beyond", Kluwer Academic Publishers, Boston, editor Karl Hess, 1991.

[2] P.D. Yoder, V.D. Natoli, R.M. Martin, "Ab-Initio Analysis of the Electron-Phonon Interaction in Si", J. Appl. Phys., vol. 73, pp. 4378-4383, 1993.

[3] P.D. Yoder, "First Principles Monte Carlo simulation of transport in Si", Semicond. Sci. Technol., vol. 9, pp. 852-854, 1994.

[4] T. Kunikiyo, M. Takenaka, Y. Kamakura, M. Yamaji, H. Mizuno, M. Morifuji, K. Taniguchi and C. Hamaguchi, "A Monte Carlo simulation of anisotropic electron transport in silicon including full band structure and anisotropic impact-ionization model", J. Appl. Phys., vol. 75, pp. 297-312, 1994.

[5] J. Bude, K. Hess and G.J. Iafrate, "Impact ionization in semiconductors: Effects of high electric fields and high scattering rates", Phys. Rev. B., vol. 45, pp. 10958-10964, 1992.

[6] E. Cartier, M.V. Fischetti, E.A. Eklund and F.R. McFreeley, "Impact Ionization in Silicon", Appl. Phys. Lett., vol. 62, pp. 3339-3340, 1993.

[7] Y. Kamakura, H. Mizuno, M. Yamaji, M. Morifuji, K. Taniguchi, C. Hamaguchi, T. Kunikiyo, M. Takenaka, "Impact ionization model for full band Monte Carlo simulation", J. Appl. Phys., vol. 75, pp. 3500-3506, 1994.

[8] M. Fischetti, S. Laux, "Monte Carlo analysis of electron transport in small semiconductor devices including band-structure and space-charge effects", Phys. Rev. B, vol. 38, pp. 9721-9745, 1988.

[9] A. Abramo, R. Brunetti, C. Jacoboni, F. Venturi, E. Sangiorgi, "A multiband Monte Carlo approach to Coulomb interaction for device analysis", J. Appl. Phys., vol. 76, pp. 5786-5794, 1994.

[10] N. Strecker, T. Feudel, and W. Fichtner, DIOS : Manual, Technical report, ETH Zurich, Integrated Systems Laboratory, ETH Zentrum, 1992.

[11] S. Müller, K. Kells A. Benvenuti, J. Litsios, U. Krumbein, A. Schenk, and W. Fichtner, DESSIS 1.3.6: Manual, Technical report, ISE Integrated Systems Engineering AG, 1994.

[12] J. Litsios, S. Müller, and W. Fichtner, "Mixed-mode multi-dimensional device and circuit simulation", In SISDEP-5, pp. 129–132, Vienna, Sept. 1993.

[13] P. D. Yoder, U. Krumbein, DEGAS$_{-ISE}$ Manual, Technical report, ISE Integrated Systems Engineering AG, 1994. Zentrum, 1995.

[14] D. Cheng, C. Hwang, and R. Dutton, "Pisces-MC: A multiwindow, multimethod 2-d device simulator", IEEE Trans., CAD, vol. 7, pp. 1017–1026, 1988.

[15] Hans Kosina and Siegfried Selberher, "A hybrid device simulator that combines monte carlo and drift-diffusion analysis", IEEE Trans., CAD, vol. 13, pp. 201–210, 1994.

On Particle–Mesh Coupling in Monte Carlo Semiconductor Device Simulation

S.E. Laux
IBM Research Division; T.J. Watson Research Center
P.O Box 218; Yorktown Heights, NY 10598 USA

Abstract

Improved NGP and CIC particle-mesh schemes are suggested, and a NEC scheme proposed, to help reduce self forces in Monte Carlo semiconductor device simulation. An attempt to design a scheme with reduced self forces for unstructured triangular meshes is unsuccessful.

1. Introduction

A proper coupling between charged particles and Coulombic forces is required to maintain temporal stability and spatial accuracy in self-consistent Monte Carlo device modeling. Particle-mesh (PM) coupling can be broken into four steps[1]:

1. assign particle charge to the mesh;
2. solve the Poisson equation on the mesh;
3. calculate the mesh-defined forces; and
4. interpolate to find forces on the particles.

The usual charge assignment and force interpolation schemes employed are nearest-grid-point (NGP) or cloud-in-cell (CIC). Both schemes guarantee zero self force, *i.e.*, the force a charge exerts upon itself due to numerical artifacts. As stated in [1]:

> At best, the presence of the self force presents a nonphysical restriction on the time step and, at worst, it is disastrous.

The classical NGP and CIC schemes depend on two severe assumptions: (a) constant permittivity, and (b) tensor product mesh with uniform spacing in x- and y-axis directions. While it is straightforward to exercise NGP and CIC schemes for non-uniformly spaced tensor-product meshes and/or spatially dependent permittivity, it is also incorrect: self forces will not be zero, or even necessarily small.

After discussing the importance of self forces, this paper describes improvements to the NGP/CIC charge assignment/force interpolation schemes which do not require assumption (a) above, and a new nearest-element-center (NEC) scheme which, in addition, relaxes assumption (b). This leads naturally to a discussion of designing a scheme for unstructured triangular meshes, and how finding such a scheme remains an open question. The discussion here assumes two space dimensions throughout, although the extension to three dimensions is obvious.

2. Self-Forces and Their Importance

Discussing self forces is only relevant if the Coulombic forces are obtained self-consistently in time with particle position. Any smoothing of the electric field in time (*e.g.*, [2]) or using a time-invariant ("frozen") field (*e.g.*, [3]) renders a discussion of self forces irrelevant, as the force does not track instantaneous particle motion.

In its simplest embodiment, self-forces arise due to the failure of a non-uniformly spaced mesh to resolve symmetrically the potential singularity of a single charge (in 2D, a line charge). Differencing this potential and interpolating back to the charge position yields an erroneous, non-zero force exerted by the charge upon "itself". But note, self forces may locally be small compared to the total force. Self forces are expected to be most serious in regions of a device where internal fields are "low".

3. Improved NGP/CIC and the NEC Scheme

An improved NGP/CIC scheme can be obtained by altering step 3. These schemes (denoted NGP* and CIC*) pertain to the case when permittivity depends on position, yet the mesh spacing still obeys assumption (b). First, some notation is required: consider a uniformly-spaced, tensor product mesh with meshlines x_i, $i = 1, \ldots, \mathcal{N}_x$ and y_j, $j = 1, \ldots, \mathcal{N}_y$. Permittivities are considered constant within each mesh element (for simplicity only) and are denoted ϵ_{ij}.[1] Define centered finite-differences of the potential ψ in the x- and y-axis directions at element midpoints as follows:

$$\Delta^x_{k+\frac{1}{2},\ell} \doteq \frac{\psi_{k+1,\ell} - \psi_{k\ell}}{x_{k+1} - x_k}, \qquad \Delta^y_{k,\ell+\frac{1}{2}} \doteq \frac{\psi_{k,\ell+1} - \psi_{k\ell}}{y_{\ell+1} - y_\ell}.$$

For the NGP* and CIC* schemes, the new step 3 becomes:

3*. calculate the mesh-defined electric field at the four element vertices (k, ℓ), $\vec{E}_{k\ell} = [E^x_{k\ell}; E^y_{k\ell}]$, as $E^x_{k\ell} = (\epsilon_{k-1,j}\Delta^x_{k-\frac{1}{2},\ell} + \epsilon_{kj}\Delta^x_{k+\frac{1}{2},\ell})/(2\epsilon_{ij})$, $E^y_{k\ell} = (\epsilon_{i,\ell-1}\Delta^y_{k,\ell-\frac{1}{2}} + \epsilon_{i\ell}\Delta^y_{k,\ell+\frac{1}{2}})/(2\epsilon_{ij})$, for $k = i, i+1$ and $\ell = j, j+1$.

The standard NGP/CIC schemes use $E^x_{k\ell} = (\Delta^x_{k-\frac{1}{2},\ell} + \Delta^x_{k+\frac{1}{2},\ell})/2$ and $E^y_{k\ell} = (\Delta^y_{k,\ell-\frac{1}{2}} + \Delta^y_{k,\ell+\frac{1}{2}})/2$, which are incorrect if permittivity is spatially dependent.

Before describing the NEC scheme, a short digression is necessary: unfortunately, the NEC scheme only yields "approximately zero" self forces. To understand this, consider the canonical self force Gedanken experiment: a single line change Q is placed upon a two-dimensional mesh of infinite extent ($-\infty < x, y < \infty$). With no other charges present, no boundaries where induced charges may reside, and a constant permittivity everywhere, this charge should not experience any acceleration (the force on Q, *i.e.*, the self force, should be zero). This requires the charge assignment/force interpolation schemes be the same *and* forces be equal and opposite between two charges[1]. This latter condition can be restated: if Q is located on a mesh node p, then the force at any other mesh node o depends only on the directed distance separating p and o. This is obvious, and can be rigorously proven for a tensor-product mesh with uniform spacings in the x- and y-axis direction[1]. Physically the force behaves exactly this way; for this category of meshes the discrete case reproduces this behavior. To my knowledge this is not rigorously true for unstructured meshes nor for non-uniformly spaced tensor-product meshes; however, *assume this is* approximately *true for all two-dimensional meshes in what follows*[2]. The extent to which it is not true in the

---
[1]Elements, and elemental quantities like permittivity, are indexed by the minimum i and j indices of the element.
[2]Computations indicate this "approximation" is remarkably true, to within a 1% error; can this be shown rigorously?

discrete case is responsible for nonzero instead of truly zero self forces. I shall call this the "well-behaved forces" assumption (WBF).

The nearest element center (NEC) charge assignment/force interpolation scheme will now be described. This scheme attempts to reduce self forces in the presence of non-uniformly spaced tensor-product meshes and/or spatially-dependent permittivity. In addition, the NEC scheme can be utilized in one axis direction (where local mesh spacing is non-uniform) and the CIC* scheme can be utilized in the other (where local mesh spacing is uniform). Such hybrid schemes, dubbed NEC-x–CIC*-y or NEC-y–CIC*-x, offer smoother assignment/interpolation on the mesh compared to pure NEC.

Consider a line charge ρ residing at (x,y) in a rectangular mesh element (i,j) with permittivity ϵ_{ij}. The new steps of the (pure-)NEC PM scheme are:

1′. assign the line charge *equally* to the four mesh points of the element (i,j);

3′. calculate the fields $\Delta^x_{i+\frac{1}{2},\ell}$, $\ell = j, j+1$, and $\Delta^y_{k,j+\frac{1}{2}}$, $k = i, i+1$;

4′. interpolate the field $\vec{E}(x,y) = [E^x; E^y]$, according to $E^x = (\Delta^x_{i+\frac{1}{2},j} + \Delta^x_{i+\frac{1}{2},j+1})/2$, $E^y = (\Delta^y_{i,j+\frac{1}{2}} + \Delta^y_{i+1,j+\frac{1}{2}})/2$.

The NEC designation derives from the appearance, in step 1′, of moving the charge to the center of its element and applying a CIC assignment scheme. The NEC scheme involves *only* one mesh element and its four nodal values of potential. This locality makes the method well-suited to non-uniform mesh spacings and permittivity.

The NEC-x–CIC*-y scheme involves only (i,j) elemental quantities for x-directed fields, but not for y-directed fields. The new steps of the NEC-x–CIC*-y scheme are:

1″. conceptually move the charge from its location (x,y) to $(0.5(x_i + x_{i+1}), y)$ and apply the CIC charge weighting scheme (denote these CIC weights as w_{ij}, $w_{i+1,j}$, $w_{i,j+1}$ and $w_{i+1,j+1}$; note $w_{ij} = w_{i+1,j}$ and $w_{i,j+1} = w_{i+1,j+1}$);

3″. calculate fields $\Delta^x_{i+\frac{1}{2},j}$ and $\Delta^x_{i+\frac{1}{2},j+1}$ as in pure-NEC; calculate fields $E^y_{k\ell}$, $k = i, i+1$ and $\ell = j, j+1$ as in CIC*;

4″. interpolate the field $\vec{E}(x,y) = [E^x; E^y]$ according to $E^x = (w_{ij} + w_{i+1,j})\Delta^x_{i+\frac{1}{2},j} + (w_{i,j+1} + w_{i+1,j+1})\Delta^x_{i+\frac{1}{2},j+1}$, $E^y = \sum_{k=i,i+1;\ell=j,j+1} w_{k\ell}E^y_{k\ell}$, where the w_{ij} are the CIC weights from step 1″.

The other mixed scheme, NEC-y–CIC*-x, is defined analogously.

The mixed schemes provide for smoother assignment and interpolation in the "CIC" direction. In a device simulation context, the decision whether to apply NEC or CIC* in the x-axis direction (for example) in the vicinity of element (i,j) is: *if* the local x-axis mesh spacing is constant (*i.e.*, $x_i - x_{i-1} = x_{i+1} - x_i = x_{i+2} - x_{i+1}$), *and* the local permittivity is constant (*i.e.*, $\epsilon_{i-1,j} = \epsilon_{ij} = \epsilon_{i+1,j}$), *and* none of the nodes (k,ℓ), $k = i-1, \ldots, i+2$ and $\ell = j, j+1$ are contact nodes, *then* CIC* should be used; *otherwise*, NEC should be employed. Applying analogous rules in the y-axis direction leads to four outcomes for element (i,j): pure NEC or CIC*, or the two hybrid schemes.

4. What About Unstructured Triangular Meshes?

Consider a triangular element $\mathcal{T}$ with nodes i, j, k in an unstructured triangular mesh of infinite spatial extent. A line charge ρ resides at (x,y) within $\mathcal{T}$. Assume the potential is a linear function of position within $\mathcal{T}$, *i.e.*, $\psi(x,y) = \psi_i N_i(x,y) + \psi_j N_j(x,y) + \psi_k N_k(x,y)$, where the $N_\ell(x,y)$ are the linear shape functions associated with $\mathcal{T}$.[3] Will

---

[3]Linear shape functions on triangular elements are the simplest shape functions which are local to the element, making them a natural choice[4].

any charge assignment scheme yield a PM coupling with zero self force (in the WBF sense) for a simple PM scheme? Consider the four steps:

$1^\triangle$. Charge assignment is left unspecified; assign charge to nodes as $\varrho \doteq [\rho_i, \rho_j, \rho_k]^T = \rho[c_i, c_j, c_k]^T \doteq \rho\mathbf{c}$, where $\sum_{\ell=i,j,k} c_\ell = 1$ by charge conservation.

$2^\triangle$. Solve the Poisson equation; the charges at the nodes will result in a potential $\varphi = \mathbf{G}\varrho$, where $\varphi = [\psi_i, \psi_j, \psi_k]^T$ and $\mathbf{G} = (g_{ij})$ is a 3×3 matrix extracted from the inverse of the Laplacian on the infinite mesh. The WBF assumption dictates the form of $\mathbf{G}$: diagonal entries are all equal and the matrix is symmetric. Therefore,

$$\mathbf{G} = \begin{pmatrix} g_{ii} & g_{ij} & g_{ik} \\ g_{ji} & g_{jj} & g_{jk} \\ g_{ki} & g_{kj} & g_{kk} \end{pmatrix} \stackrel{\text{from}}{\underset{\text{WBF}}{=}} \begin{pmatrix} g_{ii} & g_{ij} & g_{ik} \\ g_{ij} & g_{ii} & g_{jk} \\ g_{ik} & g_{jk} & g_{ii} \end{pmatrix} = g_{ii} \begin{pmatrix} 1 & \alpha & \beta \\ \alpha & 1 & \gamma \\ \beta & \gamma & 1 \end{pmatrix} \doteq g_{ii}\tilde{\mathbf{G}},$$

where $0 < \alpha, \beta, \gamma < 1$ and α, β and γ are not equal in general.

$3^\triangle$–$4^\triangle$. The electric field in $\mathcal{T}$ is the constant vector $\vec{E} = -\nabla\psi = -\sum \psi_\ell \nabla N_\ell$. Since the field in $\mathcal{T}$ is everywhere the same, field interpolation is automatically accomplished.

A *zero* field on the original line charge $\rho(x,y)$ is sought; therefore, the field in step $4^\triangle$ should be zero. Obtaining $\vec{E} = \vec{0}$ can only occur if $\psi_i = \psi_j = \psi_k$ (straightforward to show, but not proven here). A charge weighting scheme $\mathbf{c}$ that gives equal nodal potentials obeys:

$$\mathbf{G}\varrho = g_{ii}\rho\tilde{\mathbf{G}}\mathbf{c} = \varphi \propto [1\,1\,1]^T \rightarrow \tilde{\mathbf{G}}\mathbf{c} \propto [1\,1\,1]^T.$$

This has a complicated solution ($\mathbf{c} \propto \tilde{\mathbf{G}}^{-1}[1\,1\,1]^T$) and a simple solution ($c_i = c_j = c_k = \frac{1}{3}$, if $\alpha = \beta = \gamma$). To be useful, the charge weights $\mathbf{c}$ should be local functions of the element $\mathcal{T}$; since α, β and γ are global functions of the entire mesh, $\mathbf{c}$ should be independent of α, β and γ. This requires the complicated solution be discarded. The simple solution obtains $\mathbf{c}$ independent of α, β and γ as required, but only when $\alpha = \beta = \gamma$. This implies the mesh be composed of identically-sized equilateral triangles, which is not a very useful "unstructured" triangular mesh! Can a useful PM coupling for unstructured triangular meshes be designed which yields zero self forces (in the WBF sense)? To my knowledge this important question remains unanswered.

5. Conclusions

The NEC*, CIC* and NEC PM schemes are proposed to reduce self forces. Hybrid combinations of these methods are described. An attempt to design a PM scheme for unstructured triangular meshes is successful only for meshes of equilateral triangles.

References

[1] R.W. Hockney and J.W. Eastwood, *Computer Simulation Using Particles*, Adam Hilger, New York, 1988.
[2] K. Throngnumchai, K. Asada and T. Sugano, "Modeling of 0.1-μm MOSFET on SOI Structure Using Monte Carlo Simulation Technique" *IEEE Trans. Electron Devices*, vol. 33, no. 7, pp. 1005–1011, 1986.
[3] J.M. Higman, K. Hess, C.G. Hwang, and R.W. Dutton, "Coupled Monte Carlo–Drift Diffusion Analysis of Hot-Electron Effects in MOSFET's", *IEEE Trans. Electron Devices*, vol. 36, no. 5, pp. 930-937, 1989.
[4] O.C. Zienkiewicz, *The Finite Element Method*, McGraw-Hill, New York, 1977.

T²CAD: Total Design for Sub-um Process and Device Optimization with Technology-CAD

Hiroo Masuda

Hitachi, Ltd.
2326 Imai, Ome-shi, Tokyo, Japan

Abstract

"T²CAD (Total-Technology-CAD)" is proposed as an application technology of process & device simulators, which provides useful and important data for VLSI process design and optimization. It includes (1) reliability & accuracy of TCAD simulations and (2) process and device optimization method with TCAD, as well as (3) application methodology to VLSI chip design. A sub-um CMOS technology for Mega-bit DRAMs has been optimized based on T²CAD.

1. Introduction

Trends of Dynamic Random Access Memory (DRAM) and MOST feature size in VLSI's are well-known. As for DRAM integration, the bit density has doubled every three-years. To achieve this high integration in VLSI's, process and device feature sizes have been scaled-down to 0.3 um in 1994. On the way toward this sub-half micron process & device technology, many physical phenomena for TCAD application have been highlighted as shown in Fig. 1. All of these critical phenomena required intensive TCAD simulations and careful experimental verifications. However, the use of TCAD in process and device design and optimization for sub-um technology looks less successful because of difficulties in achieving simulation accuracy and experimental verification.

In this paper, a TCAD strategy towards a practical application tool will be discussed for updated VLSI process, device and circuit development [1]. To achieve better prediction with TCAD, a process database for low-temperature impurity diffusion has been constructed [2], incorporating compact modeling of the TED (Transient Enhanced Diffusion) effect. Drain current database has also been developed to calibrate current driving capability of sub-um CMOS devices [3]. Process and device optimization was conducted by the full use of advanced Response Surface Method using the experimental global calibration approach [4,5]. Based on the optimized process and device, circuit model parameters were generated to support a worst case circuit design before fabrication using a test structure.

These basic TCAD technologies were utilized successfully to design a sub-um CMOS process, device and circuit in Mega-bit DRAM development. Experiments show good agreement with predicted device characteristics, and the fabricated DRAM demonstrates good performance and yield.

2. Basic Technology

2.1 Process and Device Database

Process and device databases used in the TCAD calibrations are important to obtain reliable prediction data from TCAD. In the process database, we focused on TED phenomena, which is increasingly important in shallow junction formation with lowered annealing temperature. In the device database, we summarized drain current in sub-um CMOS. It is noted that an anomalous degradation of the CMOS current driving capabilities have been clarified even for the Vdd=3.3V condition. The physical mechanism of the degradation is found to be caused by carrier velocity saturation effects even at the source end of channel.

(A) Process database
Over 300 samples of different implanted dose/energy and annealing temperature/time were fabricated for Phosphorus, Boron, BF2 and Arsenic as dopants. The final depth profiles were measured by SIMS analysis using the CAMECA-ims4f. To parameterize the TED effect, we extracted a compact "TED" parameter of effective diffusivity (D*/Do) based on the experimental doping profile. The TED coefficients for phosphorus is shown in Fig. 2, as functions of implant doses and annealing temperature. It shows that the TED parameter is constant with an error <1.5% when the dose is higher than $1 \times 10^{14}/cm^2$, and no transient enhanced diffusion is observed when the implant dose is under $1 \times 10^{13}/cm^2$. It is noted that the TED parameter is closely independent to implant energy and annealing time.
Experimental samples used in constructing the TED database varied as follows:
- Phos.: Dose (1013-2x1015 cm-2), Energy (30K-3M eV), anneal Temp. (850-1000 C),
- Boron : Dose (5x1012-1016 cm-2), Energy (20K-3M eV), anneal Temp. (900-1000 C),
- BF2 : Dose (2x1013-2x1015 cm-2), Energy (5K-100K eV), anneal Temp. (900-1000 C),
- As : Dose (1.5x1015-3x1015 cm-2), Energy (40K-100K eV), anneal Temp. (850-950 C).
In summary on the TED database, phosphorus implant and diffusion showed the highest TED effect especially at lowered temperature annealing. The TED effect has to be taken into the account for boron implant and diffusion for accurate simulation. Arsenic diffusion is much more complex due to clustering phenomena, which cannot be parameterize with the simple TED model. We are studying further the RTA process for sub-quarter um CMOS process.

(B) Drain current database
One of the vital issues for high-speed VLSI's is current driving ability of component devices. Historically, a simple scaling-down approach has been believed to enhance the MOS device performance as well as to achieve high density VLSI. However, we found an anomalous degradation of sub-um MOS device performance based on a study of intrinsic drain current which eliminates geometrical-effects and two-dimensional field-effects. To characterize this effect, a drain-current database has been developed, and analyzed based on a compact model with carrier velocity saturation. In Fig. 3, the database characterizing the degraded drain current is shown for a NMOS. It clearly exhibits performance loss in shorter channel length when evaluating the normalized (W/L=1) drain current "Idso" at a constant effective gate bias (Ve=Vg-Vt). Note that the degradation occurs even for the device with L=2um at Vdd=3.3V, and lower Vdd cause a less performance loss in shorter channel devices. This data can be used in calibrating device simulation results in terms of the drain characteristics (I-V curves) prediction for sub-um CMOS, which is an essential requirement to provide a reasonable parameter-set for I-V curves with prediction error <3%.
To understand the physical mechanism of the degraded current driving shown in Fig. 3, we newly formulated the effect based on the fact that weak velocity-saturation of carrier

dominates the carrier flow at the source end of channel. The model leads to good agreement with experiments in sub-um NMOS, showing the Ids~$L^{-0.54}$ relation.

2.2 Optimization Technology

A new methodology in simulation-based CMOS process designs has been proposed, using a hierarchical RSM (Response Surface Method) and efficient experimental calibrations. The new design method has been verified in half-um CMOS process/device development, which results in reliable prediction of the threshold voltage (Vth) and drain current (Ids) within 0.01V and 0.84% error, respectively.

The basic idea behind RSM is shown in Fig. 4. In NMOS Vth design, first design-parameters have to be specified, such as gate oxide thickness (Tox) and gate length (Lg) as variables. To obtain systematic data of Vth (Tox, Lg) at specific bias conditions, one can get a Vth data-matrix from nine-point TCAD simulations, just like experimental device measurements of test structures with various fabricated wafers.

The response (Vth) to variables (Tox, Lg) can be approximated with a RSF (response Surface Function) in quadratic form, which can be used in Vth design and optimization.

We fully use the RSM to optimize and design sub-um CMOS process and device. However, in practice, two major drawbacks are found that need to be overcome to give reliable predictions. Since physical models used in process/device simulators are known to be insufficient, even if we use the process database, direct use of RSM based on TCAD simulation may cause significant errors in predicted results of Vth and Ids. To overcome this drawback, we have proposed an efficient calibration method, after the formation of the RSF with minimal number of experimental data. Another drawback is that the design table has to be fixed before conducting a series of RSM design. In practice, one often finds another variable which turns out to be of interest after the completion of a set of simulations. To improve this situation, we have developed a new design strategy, which provides an effective hierarchical design in the above case [6].

Note that in conducting RSM design, (1) pre-conditioning of the variable and response is extremely important to achieve a reliable RSF, which require intensive variable & response transformations to get a linearized relationship between the two, (2) systematic global calibration based on reliable experiments is unavoidable especially for Vth design and analysis, and (3) the range of the variable has to be considered carefully, since the RSF is less reliable if the variable exceeds the range defined in the design table.

Fig. 5 shows an example of the design table and TCAD-responses to Vth and Ids(max). A composite design matrix is used in this case and all the variables and Ids(max) are transformed into a "log(x)" form. After 27-case numerical process and device simulations, a RSF for the Vth is derived, and compared with original simulation results as shown in Fig. 6. As demonstrated in the figure, the RSF coincides with original data within the average error of 0.026V. However, if we compare with experimental data of fabricated devices, significant discrepancy between the RSF and measurement occurred as demonstrated in Fig. 7. It is noted that this discrepancy is not a special case but observed in most cases in sub-um CMOS Vth analysis. It may be caused by insufficient two-dimensional diffusion modeling and unexpected process-condition error. To overcome this problems, a systematic global calibration on the RSF has to be conducted, which results in a good fit between a new RSF and experiments as shown in Fig. 8. The resulting calibrated RSFs for the Vth and Ids is shown in Fig. 9. The error of the new Vth-RSF is less than 0.02V.

Once we get this RSF, efficient process and device design can be done in sensitivity analysis, process optimization and parametric yield prediction, including scaled down device design. An example of process sensitivity analysis on Vth is shown in Fig. 10. Assuming process variation of +/- 10% for Tox, Lg, etc., δTox will cause the largest Vth variation as shown in the figure. However, if one thinks of real control specifications for the process, δVth will be most sensitive to δLg (more than 40% of total Vth variation). Fig. 11 shows Vth analysis based on the RSF. It shows a reasonable Vth-Lg surface for the variables of Tox, Ncd and Vbb. By using Monte Carlo calculation on the RSF, we can get quite easily predicted distributions of Vth and Ids as shown in Fig. 12. As demonstrated in the figure, the Vth distribution shows a half-Gaussian distribution which reflects a sharp Vth drop in Vth-Lg characteristics. Scaled devices can be also optimized based on the RSF, as shown in Fig. 13. After a 0.65 um NMOS is optimized in terms of Vth=0.35V and Vth-Lg curve, one can optimize a scaled 0.55 um NMOS with the same Vth specification as shown in the figure. Expected Ids increase with the scaled device is 10% in this case.

As demonstrated above, various process and device optimizations have been achieved based on the experimentally calibrated RSF, which gives us many good design choices as well as better quality in process and device optimization in sub-micron CMOS technology.

3. Generation of CKT model parameters

To achieve concurrent design of the process and circuit, precise prediction of circuit model parameters is one of the essential issues in VLSI memory development. Circuit model parameters, such as MOST model parameters, junction and interconnect capacitance, have to be accurately obtained without test chip fabrications. We have developed a practical T^{2}CAD focused on the new global experimental calibrations, which generates process recipe and device performance as well as complete model parameters in circuit simulation for sub-um VLSI memories. A couple of tens of parallel TCAD simulation were conducted based on the RSM statistical design table with optimum variable transformation. The resulting RSF for the objective design parameters (Vth, Ids, BV, etc.) were generated. RSF-design on process and device characterization were performed extensively using sensitivity analysis techniques, parametric yield estimation and global optimization, as mentioned previously. Finally, optimized process and devices were analyzed in detail to generate I-V (C-V) data for a parameter extraction system for circuit simulators.

Fig. 14 shows an example of generated MOS model parameters for a 0.5 um CMOS DRAM. Model fitting error against original (simulated and calibrated) I-V data is 0.85%. Experimental verification of worst case simulation on the predicted Vth-Lg and Ids-Lg characteristics shows reasonable accuracy as shown in Fig. 15. As shown in the figure, fabricated devices shows good agreement with prediction based on the T^{2}CAD within the errors of process-fluctuation limits for the Vth and Ids. The T^{2}CAD used 200hrs CPU on the HP/9000, and took one month in design.

References

[1] H. Masuda et al.; 'TCAD strategy for predictive VLSI memory development', Technical Digest of IEDM'94, pp. 153-156, Dec. 1994.
[2] H. Sato, K. Tsuneno and H. Masuda; 'Evaluation of two-dimensional transient enhanced diffusion of Phosphorous during shallow junction formation,' IEICE Trans. Electron., Vol. E-77-C, No. 2, pp. 106-111, Feb. 1994.
[3] K. Tsuneno, H. Sato and H. Masuda; 'Modeling and simulation of anomalous degradation of submicron NMOS's current-driving due to velocity-saturation effect,' IEICE Trans. Electron., Vol. E-77-C, No. 2, pp. 161-165, Feb. 1994.
[4] G. E. P. Box and N. R. Draper; 'Empirical model building and response surface, John Wiley and Sons, 1987.
[5] H. Masuda, F. Otsuka, Y. Aoki, and S. Satoh, S. Shimada ; 'Response surface method for submicron MOSFETs characterization with variable transformation technology,' IEICE Trans., E-74, Vol. 6, pp.1621-1633, June 1991.
[6] H. Sato et al.; 'A new hierarchical RSM for TCAD-based device design to predict CMOS development', Technical Digest of ICMTS'95, pp. 299-302, March 1995.

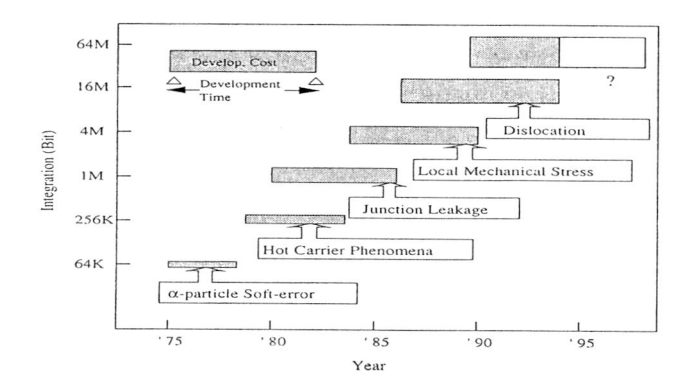

Fig. 1 DRAM trends and TCAD applications.

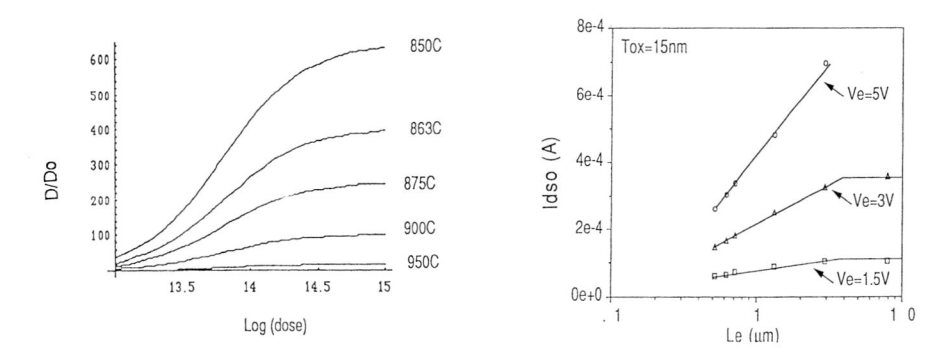

Fig. 2 TED database for Phosphorus.

Fig. 3 Drain current degradation in NMOS.

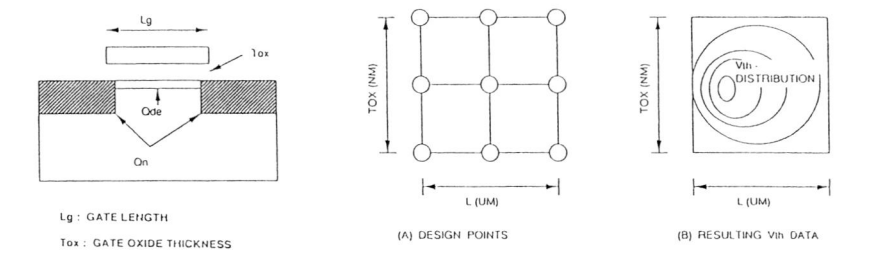

Fig. 4 Basic idea of RSM.

	X1	X2	X3	X4	Lg(um)	Tox(nm)	Qde (/cm2)	Qn· (/cm2)	Ids(5V) (mA)	Vth(V)
1	1.000	1.000	1.000	1.000	1.450	19.570	2.42e+12	1.21e+13	2.487	1.525
2	1.000	1.000	1.000	-1.000	1.450	19.570	2.42e+12	8.30e+12	2.405	1.535
3	1.000	1.000	-1.000	1.000	1.450	19.570	1.65e+12	1.21e+13	3.072	1.202
4	1.000	1.000	-1.000	-1.000	1.450	19.570	1.65e+12	8.30e+12	3.003	1.208
5	1.000	-1.000	1.000	1.000	1.450	13.370	2.42e+12	1.21e+13	4.286	0.930
6	1.000	-1.000	1.000	-1.000	1.450	13.370	2.42e+12	8.30e+12	4.098	0.825
7	1.000	-1.000	-1.000	1.000	1.450	13.370	1.65e+12	1.21e+13	4.968	0.704
8	1.000	-1.000	-1.000	-1.000	1.450	13.370	1.65e+12	8.30e+12	4.790	0.707
9	-1.000	1.000	1.000	1.000	0.990	19.570	2.42e+12	1.21e+13	3.167	1.436
10	-1.000	1.000	1.000	-1.000	0.990	19.570	2.42e+12	8.30e+12	2.589	1.450
11	-1.000	1.000	-1.000	1.000	0.990	19.570	1.65e+12	1.21e+13	3.788	1.090
12	-1.000	1.000	-1.000	-1.000	0.990	19.570	1.65e+12	8.30e+12	3.536	1.108
13	-1.000	-1.000	1.000	1.000	0.990	13.370	2.42e+12	1.21e+13	5.070	0.857
14	-1.000	-1.000	1.000	-1.000	0.990	13.370	2.42e+12	8.30e+12	4.371	0.870
15	-1.000	-1.000	-1.000	1.000	0.990	13.370	1.65e+12	1.21e+13	5.820	0.620
16	-1.000	-1.000	-1.000	-1.000	0.990	13.370	1.65e+12	8.30e+12	5.208	0.633
17	2.000	0.000	0.000	0.000	1.750	16.200	2.00e+12	1.00e+13	3.203	1.099
18	-2.000	0.000	0.000	0.000	0.820	16.200	2.00e+12	1.00e+13	4.668	0.879
19	0.000	2.000	0.000	0.000	1.200	23.600	2.00e+12	1.00e+13	2.265	1.680
20	0.000	-2.000	0.000	0.000	1.200	11.070	2.00e+12	1.00e+13	5.756	0.480
21	0.000	0.000	2.000	0.000	1.200	16.200	2.92e+12	1.00e+13	3.200	1.335
22	0.000	0.000	-2.000	0.000	1.200	16.200	1.37e+12	1.00e+13	4.460	0.795
23	0.000	0.000	0.000	2.000	1.200	16.200	2.00e+12	1.46e+13	4.089	1.025
24	0.000	0.000	0.000	-2.000	1.200	16.200	2.00e+12	6.80e+12	3.633	1.036
25	0.000	0.000	0.000	0.000	1.200	16.200	2.00e+12	1.00e+13	3.870	1.032

Fig. 5 Composite-Design table example.

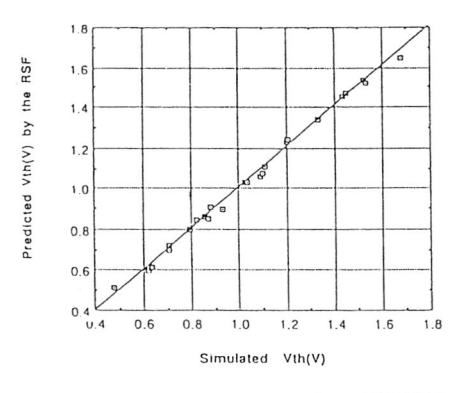

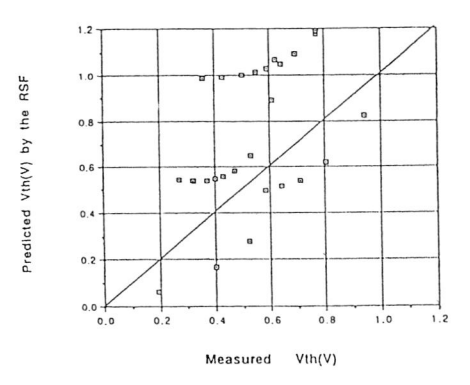

Fig. 6 RSF obtained by a series of TCAD.

Fig. 7 Experimental error of original RSF.

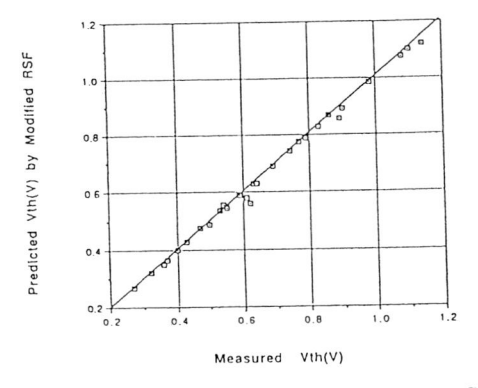

Fig. 8 Experimental verification of calibrated RSF.

$$\text{Vth(V)} = 0.59 + 0.955x_1 + 0.1057x_2 + 0.1348x_3 + 0.0003x_4 - 0.0080x_1^2 + 0.00167x_2^2$$
$$+ 0.050x_3^2 + 0.000062x_4^2 + 0.0105x_1x_2 - 0.0237x_1x_3 + 0.018x_1x_4 + 0.0121x_2x_3$$
$$- 0.004x_2x_4 + 0.003x_3x_4 + \varepsilon$$

$$\ln(\text{Ids.max(mA)}) = 0.4832 - 0.0473x_1 - 0.0636x_2 - 0.0399x_3 + 0.0163x_4 - 0.0002x_1^2$$
$$- 0.0029x_2^2 - 0.026x_3^2 - 0.0004x_4^2 - 0.0048x_1x_2 + 0.0024x_1x_3 - 0.0146x_1x_4$$
$$- 0.0051x_2x_3 - 0.0002x_2x_4 + 0.051x_3x_4 + \varepsilon$$

Fig. 9 Obtained RSFs for Vth and Ids.

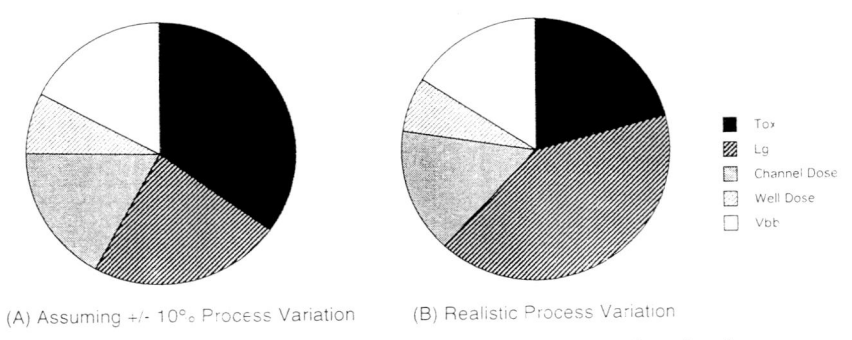

(A) Assuming +/- 10% Process Variation   (B) Realistic Process Variation

Fig. 10 Process sensitivity analysis of 0.5um NMOS. Region size denotes degree of sensitivity.

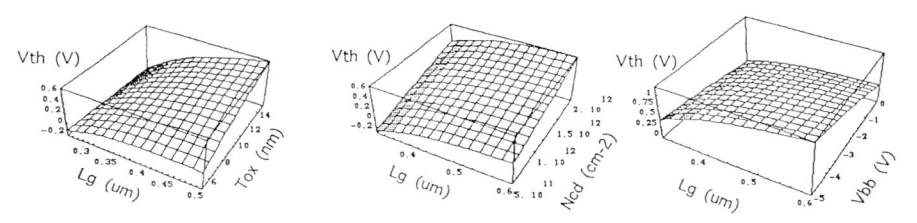

Fig. 11 Predicted Vth-Lg surface based on new RSF.

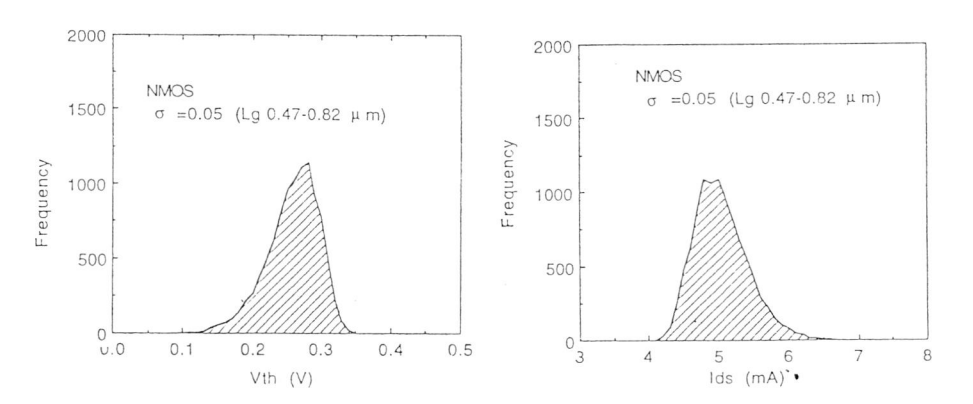

Fig. 12 Monte Carlo calculation on Vth and Ids distributions.

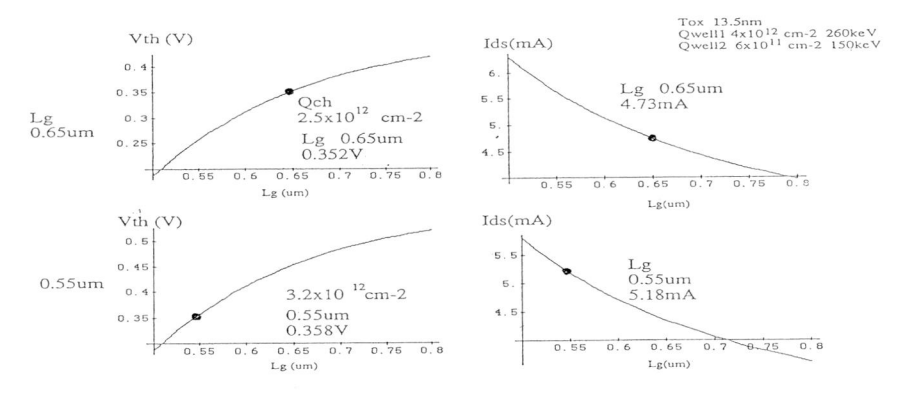

Fig. 13 Example of Vth and Ids design for a scaled device.

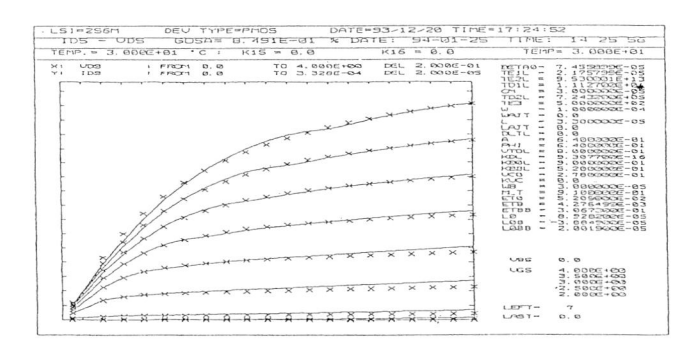

Fig. 14 Generated MOS model parameters based on drain current simulated by the T²CAD.

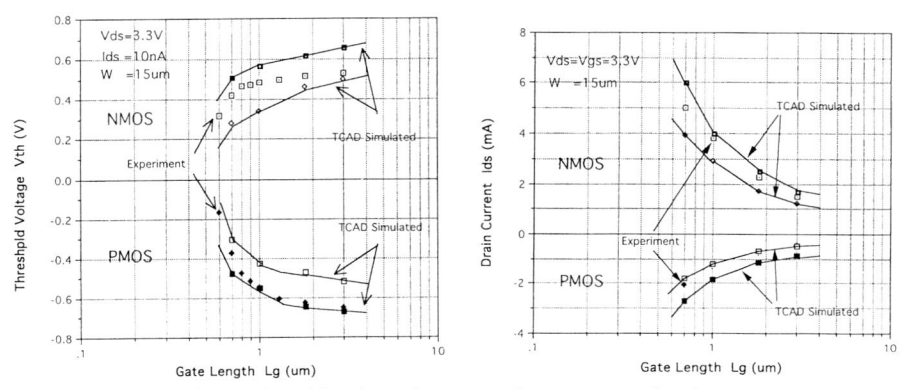

Fig. 15 Experimental verification of generated worst-case circuit parameters. The two TCAD simulations indicate the upper and lower boundary of the worst case parameters

Modelling Impact-Ionization in the Framework of the Spherical-Harmonics Expansion of the Boltzmann Transport Equation with Full-Band Structure Effects

M. C. Vecchi, M. Rudan

Dipartimento di Elettronica, Università di Bologna,
viale Risorgimento 2, 40136 Bologna, ITALY

Abstract

Band-structure effects have been incorporated in the framework of the Spherical-Harmonics Expansion (SHE) of the Boltzmann Transport Equation (BTE) for electrons in silicon [1], using the density of states (DOS) and the group velocity (GV) obtained from the full-band system [2]. In this paper an impact-ionization model is presented along with the numerical results. The model is consistent with the full-band system mentioned above and is able to fit the impact-ionization coefficient, the impact-ionization quantum yield, and the data from soft x-ray photoemission spettroscopy available in recent literature (e.g., [3]).

1. Physical model

The SHE of the BTE has been tested successfully in a wide range of problems in the field of electron transport simulation [4, 5]. The main advantage of this method is the large dynamic range of its deterministic solution and the ability of predicting the electron distribution, in both the spatially homogeneous and non-homogeneous cases, without the heavy computational burden typical of stochastic methods. Full-band structure effects are incorporated through the DOS and GV independently calculated from the full-band system [2] by means of a suitable averaging procedure. The framework of the SHE method in steady state provides the differential equation [4]

$$\begin{aligned}
& - q^2 F^2 \frac{\partial}{\partial E}\left[\tau\, g(E)\, u_g^2(E)\, \frac{\partial f_0}{\partial E}\right] = \\
& + 3 c_{op}\, g(E) \left\{ g^+(E) \left[N_{op}^+ f_0^+(E) - N_{op} f_0(E)\right] - g^-(E)\left[N_{op}^+ f_0(E) - N_{op} f_0^-(E)\right]\right\} \\
& - 3 c_{ii}\, g^2(E)\, f_0(E) + 3 g(E) \int A(E', E)\, f_0(E')\, g(E')\, \mathrm{d}E'. \quad (1)
\end{aligned}$$

The symbols have the following meaning: $g(E)$ is the DOS, $u_g(E)$ the modulus of the GV, τ the total scattering rate, F the electric field, c_{op} a constant proportional to the optical-phonon coupling constant, N_{op} the optical-phonon occupation number, $N_{op}^+ = N_{op} + 1$, $g^\pm(E) = g(E \pm \hbar\omega_{op})$, where $\hbar\omega_{op}$ is the optical-phonon energy, and similarly for $f_0^\pm(E)$. Impact ionization is also considered: $c_{ii}g(E)$ is the total impact-ionization scattering rate and $A(E', E)$ is a suitable kernel [4]. The non-linear optimization code PROFILE [6] has been used to obtain the best set of scattering parameters by fitting suitable average quantities (mean velocity, energy, impact-ionization coefficient) provided by the Monte Carlo code DAMOCLES [2] and experimental measurements in spatially-homogeneous conditions. The fitting procedure based on the full-band structure, but still using the impact-ionization model of [4], provides the impact-ionization scattering rate shown in Fig. 1. It is seen that the adoption of a full-band structure brings the result of SHE closer to that of Monte Carlo analysis, also shown in the figure. Although the agreement between SHE and Monte Carlo data of Fig. 1 is fair, it can considerably be improved by a sounder description of the impact-ionization mechanism, as shown below.

2. Impact-Ionization model

A three-threshold model is worked out. In order to avoid the simulation of the electrons in the valence band, the latter is assumed flat and full of electrons. The scattering matrix, derived in the Born approximation [7], is:

$$S^{ii}(\mathbf{k}, \mathbf{k}', \mathbf{k}'') = \sum_{j=1}^{3} S_j^{ii}(\mathbf{k}, \mathbf{k}', \mathbf{k}'') =$$

$$= \sum_{j=1}^{3} b_j^{ii} [a_j^2 + (\mathbf{k}' - \mathbf{k})^2]^{-2} \delta(E - E' - E'' - E_{Gj}), \quad (2)$$

where E_{Gj} is the ionization threshold, a_j the inverse screening length, and b_j^{ii} a normalizing constant. The values of the parameters have been determined in order to reproduce the scattering rate presented in [3]. Such scattering rate, in turn, is consistent with the experimental data of [8] and [9]. The scattering rate of [3] is shown in Fig. 2 along with the scattering rate obtained by SHE using 2, while the values of the adopted parameters are reported in Table I.

Table I: impact-ionization model parameters			
Description	Symbol	Value	Unit
Energy threshold	E_{G1}	1.2	eV
Energy threshold	E_{G2}	1.8	eV
Energy threshold	E_{G3}	3.45	eV
Inverse screening length	a_1	5.709×10^7	cm^{-1}
Inverse screening length	a_2	7.032×10^7	cm^{-1}
Inverse screening length	a_3	1.08×10^8	cm^{-1}
Normalization constant	b_1^{ii}	2.96×10^{-5}	cm$^2 \times$eV/sec
Normalization constant	b_2^{ii}	1.04×10^{-3}	cm$^2 \times$eV/sec
Normalization constant	b_3^{ii}	2.92×10^{-1}	cm$^2 \times$eV/sec

It is worth adding that this calculation dealt only with impact ionization, namely, the other parameters mentioned in the previous section have been left unchanged. The impact-ionization coefficient is shown in Fig. 3 and compared with experimental data [10, 11] in a large interval of electric fields: the good agreement in the low-field region is related to the presence of a soft threshold in (2). Fig. 4 shows the effect of the impact-ionization model on the electron energy-distribution function at 200 kV/cm: the high-energy tail computed with the new model (2) is a few orders of magnitude lower than the one obtained with the old model, due to the higher scattering rate provided by the new model at high energies (compare Figs. 1 and 2). These results emphasize the importance of a correct description of the band structure and impact ionization especially in the analysis of carrier transport at high energies. On the other hand, they also show the ability of the SHE scheme to efficiently incorporate the features of the transport mechanisms to a rather general extent, and reproduce the results of state-of-the-art stochastic methods.

Acknowledgments

This work was part of ADEQUAT (JESSI BT11) and was funded as ESPRIT Project 8002.

References

[1] M. C. Vecchi, D. Ventura, A. Gnudi, G. Baccarani, "Incorporating full band-structure effects in the spherical harmonics expansion of the Boltzmann transport equation," *Proc. of the* NUPAD V *Conf.*, Honolulu, HI, 1994.

[2] M. V. Fischetti, S. E. Laux, *Phys. Rev. B*, "Monte-Carlo analysis of electron transport in small semiconductor devices including band-structure and space-charge effects," vol. 38, pp. 9721–9745, 1988.

[3] E. Cartier, M. V. Fischetti, E. A. Eklund, F. R. McFeely, "Impact Ionization in Silicon," *Appl. Phys. Lett.*, vol. 62, no. 25, 1993.

[4] A. Gnudi, D. Ventura, G. Baccarani, "Modeling Impact Ionization in a BJT by Means of Spherical Harmonics Expansion of the Boltzmann Transport Equation," IEEE *Trans. on* CAD *of* ICAS, vol. 12, no. 11, pp. 1706–1713, 1993.

[5] N. Goldsman, L. Henrickson, J. Frey, "A Physics–Based Analytical/Numerical Solution to the Boltzmann Transport Equation for Use in Device Simulation," *Solid-St. Electr.*, vol. 34, no. 4, p. 389, 1991.

[6] G. J. L. Ouwerling, "Non-destructive Measurament of 2D doping profiles by inverse modelling," *Proc. of the 6th Int.* NASECODE *Conf.*, p. 534, Dublin, 1989.

[7] E. Schöll, W. Quade, "Effect of Impact Ionization on Hot-Carrier Energy and Momentum Relaxation in Semiconductors," *Jour. Physics C*, vol. 20, p. L 861, 1987.

[8] D. J. DiMaria, T. N. Theis, J. R. Kirtley, F. L. Pesavento, D. W. Dong, S. D. Brorson, "Electron heating in silicon dioxide and off-stechiometric silicon dioxide films," *Jour. of Appl. Phys.*, vol. 57, no. 4, p. 1214, 1985.

[9] E. A. Eklund, P. D. Kirchner, D. K. Shuh, F. R. McFeely, E. Cartier, "Direct determination of impact-ionization rates near threshold in semiconductors using soft-x-ray photoemission," *Phys. Rev. Lett.*, vol. 68, no. 6, p. 831, 1992.

[10] R. Van Overstraeten, H. De Man, "Impact ionization of hot electrons in silicon," *Solid-St. Electr.*, vol. 13, pp. 583–608, 1970.

[11] I. Takayanagi, K. Matsumoto, J. Nakamura, "Measurements of electron impact ionization coefficient in bulk silicon under a low electric field," *J. of Appl. Physics*, vol. 72, no. 5, pp. 1989–1992, 1992.

Fig. 1

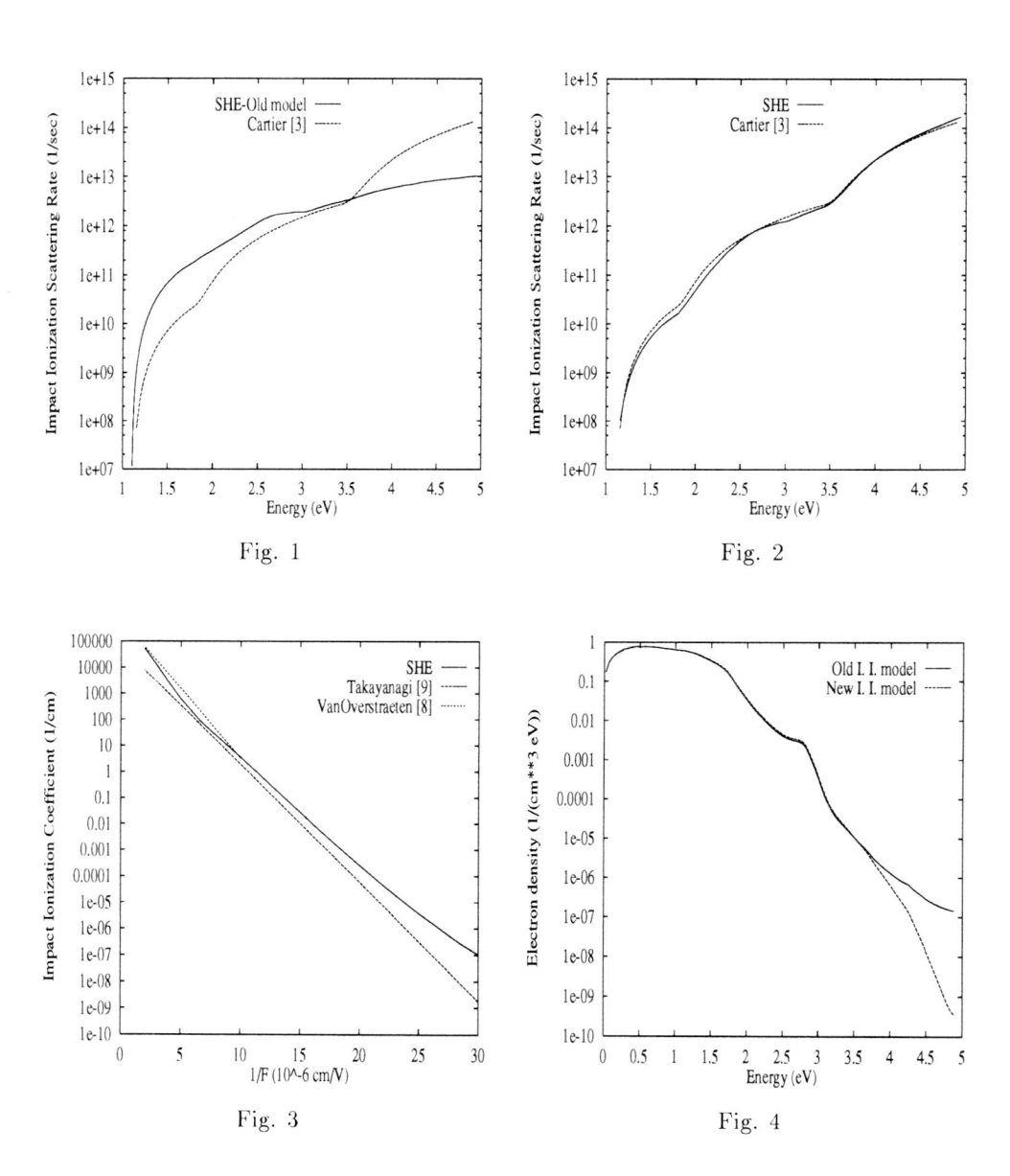

Fig. 2

Fig. 3

Fig. 4

Impact Ionization Model Using Second- and Fourth-Order Moments of Distribution Function

K. Sonoda[a,b], M. Yamaji[a], K. Taniguchi[a], and C. Hamaguchi[a]

[a]Department of Electronic Engineering, Osaka University,
Yamada-Oka 2-1, Suita City Osaka, 565 JAPAN
[b]ULSI Laboratory, Mitsubishi Electric Corporation,
Mizuhara 4-1, Itami City Hyogo, 664 JAPAN

Abstract

This paper describes an impact ionization model suitable for calculation of an impact ionization rate in inhomogeneous electric field. The model is formulated using second- and fourth-order moments of an electron energy distribution function. A set of model equations for carrier transport in semiconductor devices is also presented to perform practical device simulation with the impact ionization model. The calculation result with the new models agrees to Monte Carlo simulation result.

1. Introduction

Scaling-down the dimensions of silicon devices without proportional decrease of power supply voltage induces high electric field, which causes hot carrier effects. Device degradation caused by hot carriers has been main concern from the reliability point of view. Because secondary-generated carriers created by impact ionization (I.I.) have great influence on the degradation of gate oxide, accurate modeling of I.I. is necessary.

An ionization coefficient, which is the number of I.I. event per unit length, has been conventionally expressed as a function of electric field[1]. It has also been formulated using average carrier energy[2] to take the effect of non-uniform electric field into account. In the past few years, it has been reported that the average energy is still insufficient to describe nonlocal nature of I.I. in non-uniform electric field[3][4].

We propose an I.I. model which is formulated using second- and fourth-order moments of an electron energy distribution function. A set of model equations to calculate the fourth-order moment is also presented to perform device simulation with the I.I. model.

2. Impact Ionization Model

To investigate the I.I. phenomena in inhomogeneous field, we use the Monte Carlo (MC) simulation program with analytical multi-valley band structure, in which phonon scattering rates [5] and the impact ionization rate [6] are implemented as a function of electron energy.

Calculated average energy, $\langle \varepsilon \rangle$, and impact ionization coefficient, α, in the inhomogeneous electric field (Fig. 1(a)) are shown in Figs. 1(b) and (d), respectively. The symbol, $\langle A \rangle$, means $\int A f d\mathbf{k} / \int f d\mathbf{k}$ hereafter. The ionization coefficient, α, is obtained from the relation, $G_{ii} = n|\langle \mathbf{u} \rangle|\alpha$, where G_{ii} is the electron-hole pair generation rate caused by I.I., n is the electron density, and $\mathbf{u}$ is the group velocity of an electron. Note that n, and $\langle \mathbf{u} \rangle$ are zeroth-, and first-order moments, respectively. These figures show that in spatially varying electric field, the impact ionization coefficient is no longer determined by the average energy alone, but strongly depends on the high energy tail of distribution function (Fig. 2).

In order to express impact ionization coefficient precisely, we use a fourth-order moment of the distribution function, $\langle \varepsilon^2 \rangle$, in addition to second-order moment, $\langle \varepsilon \rangle$. The fourth-order moment is parameterized in a normalized form, $\xi = ((3/5)\langle \varepsilon^2 \rangle)^{1/2}/\langle \varepsilon \rangle$. The factor $(3/5)^{1/2}$ is introduced so as to $\xi = 1$ when the distribution function is Maxwellian. Fig. 1(c) shows the parameter, ξ, calculated using MC simulation. The figure shows that ξ starts to increase where the field decreases, which coincides with the fact that the high energy tail of the distribution function remains in spite of the rapid decrease of electric field and average energy.

Fig. 3 shows calculated ionization coefficient for several maximum field ($E_{\max} = 200, 300, 400, 500 \text{kV/cm}$) in Fig. 1(a) as a function of the inverse of the average energy with several ξ's as a parameter. The figure shows that ionization coefficient, α, is expressed as $\alpha_0 \exp(-\varepsilon_c/\langle \varepsilon \rangle)$ for a given ξ, where α_0 is a constant and ε_c depends on ξ. The coefficient, ε_c, is plotted as a function of ξ in Fig. 4 to show the relation, $\varepsilon_c \propto \exp(-\gamma \xi)$, where γ is constant. From Figs. 3 and 4, the impact ionization coefficient, α, is modeled as

$$\alpha = \alpha_0 \exp\left(-\frac{\varepsilon_{c0} \exp(-\gamma \xi)}{\langle \varepsilon \rangle}\right), \qquad (1)$$

where $\alpha_0 = 1.4 \times 10^7 \text{cm}^{-1}$, $\varepsilon_{c0} = 82.3 \text{eV}$, and $\gamma = 2.56$. In the homogeneous field case, $\xi = 0.88$ and $\varepsilon_c = 8.7 \text{eV}$ from the MC simulation.

3. Moment Conservation Equations

Both second-order moment, $\langle \varepsilon \rangle$, and fourth-order one, $\langle \varepsilon^2 \rangle$, are required to use the new I.I. model (1) in device simulation, The fourth-order moment is numerically calculated from the conservation equations of the moment incorporated in a hydrodynamic model[7]. The equations are derived from the Boltzmann transport equation (BTE) to be

$$\nabla \cdot (n \langle \mathbf{u} \varepsilon^2 \rangle) = -2q \mathbf{E} \cdot \mathbf{S} - n \frac{\langle \varepsilon^2 \rangle - \langle \varepsilon^2 \rangle_0}{\tau_{\langle \varepsilon^2 \rangle}} - U_{\langle \varepsilon^2 \rangle} \qquad (2)$$

$$n \langle \mathbf{u} \varepsilon^2 \rangle = \frac{\tau_{\langle \mathbf{u} \varepsilon^2 \rangle}}{\tau_{\langle \mathbf{u} \rangle}} \frac{7}{3} \left(\langle \varepsilon^2 \rangle \frac{\mathbf{J}}{-q} - \frac{k T_n}{q} n \mu \nabla \langle \varepsilon^2 \rangle \right), \qquad (3)$$

where $\mathbf{E}$ is the electric field, $\mathbf{J} \equiv -qn\langle \mathbf{u} \rangle$ is the electron current density, $\mathbf{S} \equiv n \langle \mathbf{u} \varepsilon \rangle$ is the electron energy flux, $\langle \varepsilon^2 \rangle_0$ is the fourth-order moment at thermal equilibrium, $U_{\langle \varepsilon^2 \rangle}$ is the net loss rate of $\langle \varepsilon^2 \rangle$ due to generation-recombination process, $\tau_{\langle A \rangle}$ is the relaxation time of $\langle A \rangle$, μ is the electron mobility, and T_n is the electron temperature defined by $3kT_n/2 \equiv \langle \varepsilon \rangle$. The parameters, $\tau_{\langle \varepsilon^2 \rangle} = 0.29 \text{ps}$, $\tau_{\langle \mathbf{u} \varepsilon^2 \rangle}/\tau_{\langle \mathbf{u} \rangle} = 0.59$ are extracted from MC simulation in homogeneous electric field.

4. Results and Discussions

In order to verify the new I.I. model, we calculate I.I. generation rate, G_{ii}, in an n^+nn^+ structure using moment equations with different I.I. models. They are compared with MC result in Figs. 5(a) and (b). Parameters used in the I.I. models are calibrated to provide the same I.I. coefficient as that obtained from MC simulation in homogeneous electric field. In the MC simulation, the BTE and Poisson equation are solved self-consistently. In the MC simulation, generated carriers by impact ionization are ignored because the number of created carriers has little effect on the total number of carriers in the calculation condition here. Generation terms in the moment equations are also ignored by the same reason.

The calculated results with a local I.I. model, $\alpha(E)$, in which the coefficient is determined by the electric field, overestimate maximum I.I. rate nearly one order of magnitude. Moreover, it underestimates the generation rate in the decreasing field region. Although the maximum I.I. rate is improved with $\alpha(\langle\varepsilon\rangle)$, in which α is expressed as a function of average energy only, this model still underestimate G_{ii} in the region where the electric field decreases. In contrast with the previous two models, the new model ($\alpha(\langle\varepsilon\rangle, \xi)$) provides the generation rate which agrees with the MC result better than that with previous two models, especially in the field decreasing region.

The electric field of the n^+nn^+ structure (Fig. 5(a)), which shows rapid increase and decrease, is similar to the previous field profile (Fig. 1(a)). The large generation rate in the field decreasing region attributes to the fact that the energetic carriers are still exist in spite of the low electric field and the low average energy as shown in Figs. 1 and 2. The parameter, ξ, incorporated with average energy, $\langle\varepsilon\rangle$, is an indicator for the high energy portion of the distribution function which contributes to impact ionization so that the new I.I. model can predict the generation rate better than previous I.I. models.

5. Conclusion

We proposed the I.I. model including second- and fourth-order moments of the distribution function, which is applicable for spatially varying electric field. The validity of the new model was verified through the comparison between the numerical calculation based on the generalized moment conservation equations and MC simulation in the n^+nn^+ structure.

References

[1] A. G. Chynoweth, *Phys. Rev.* vol. 109, p. 1537, 1958.
[2] M. Fukuma et al., *IEEE Electron Device Lett.*, vol. EDL-8, p. 214, 1987.
[3] J.-G. Ahn et al., *1993 VPAD*, p. 28, 1993.
[4] P. Scrobohaci et al., *IEICE Trans. Electron.*, vol.E77-C, p. 134, 1994.
[5] T. Kunikiyo et al., *J. Appl. Phys.*, vol. 75, p. 297, 1994.
[6] Y. Kamakura et al., *J. Appl. Phys.*, vol. 75, p. 3500, 1994.
[7] R. Thoma et al., *IEDM Tech. Dig.*, p. 139, 1989.

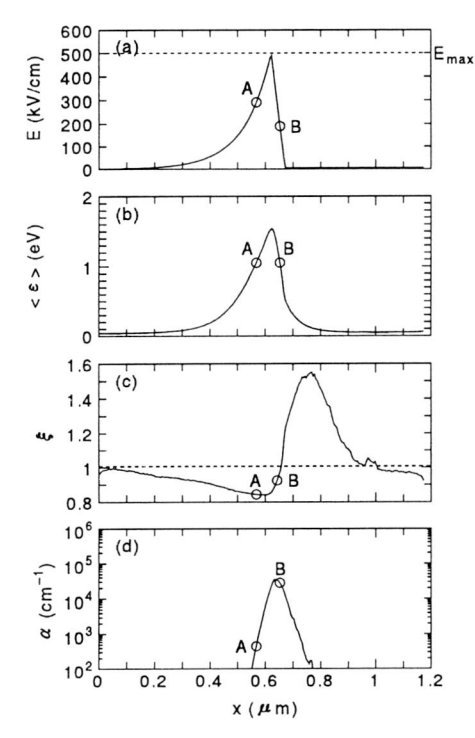

Figure 1: Results of Monte Carlo simulation at a given electric field profile. (a) Electric field, E (increases exponentially and decreases linearly), (b) Average energy, $\langle\varepsilon\rangle$, (c) $\xi \equiv \sqrt{(3/5)\langle\varepsilon^2\rangle}/\langle\varepsilon\rangle$, (d) Impact ionization coefficient, α.

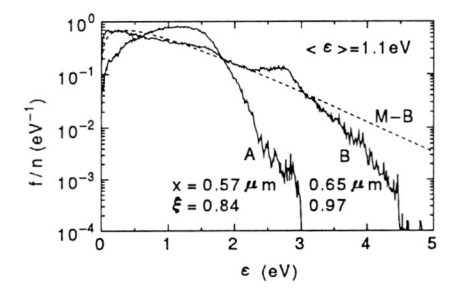

Figure 2: Electron energy distribution functions at points A and B in Fig. 1 where the average energy $\langle\varepsilon\rangle = 1.1\,\text{eV}$. The dotted curve means Maxwell-Boltzmann distribution function for the same average energy.

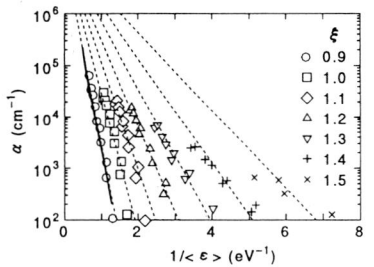

Figure 3: Impact ionization coefficient as a function of inverse average energy for different ξ values. Dashed lines are obtained from a least square fit to the data. Solid line indicates the value in homogeneous electric field.

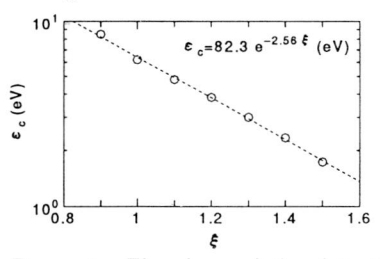

Figure 4: The slope of the data in Fig. 3, ε_c, as a function of the parameter, ξ. Dashed line indicates the least square fit to the data.

$$\varepsilon_c = 82.3\, e^{-2.56\,\xi}\ (\text{eV})$$

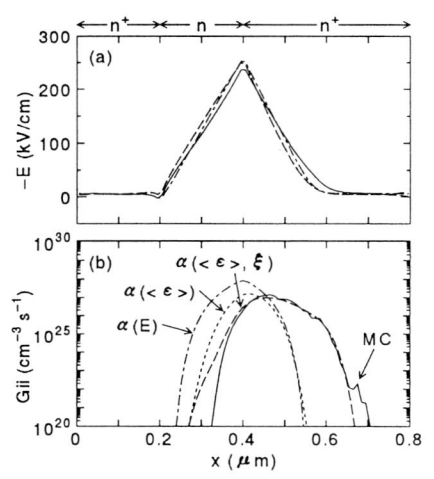

Figure 5: Calculated electric field and impact ionization rate in an n^+nn^+ structure ($n^+ = 2 \times 10^{17}\,\text{cm}^{-3}$, $n = 5 \times 10^{15}\,\text{cm}^{-3}$). Applied voltage is 5V. (a) Electric field, (b) Impact ionization generation rate.

An Accurate NMOS Mobility Model for 0.25μm MOSFETs

S. A. Mujtaba[a,b], M. R. Pinto[b], D. M. Boulin[b], C. S. Rafferty[b], R. W. Dutton[a]

[a]Center for Integrated Systems, Stanford University
Stanford, California 94305, USA

[b]AT&T Bell Laboratories
Murray Hill, New Jersey 07974, USA

Abstract

A physically-based, semi-empirical, local mobility model for 2D device simulation is presented that is accurate in both the intrinsic (channel) as well as the extrinsic (parasitic) regions of LDD MOSFETs. It is demonstrated that for deep submicron MOSFETs the gate-voltage-dependent extrinsic series resistance is poorly modeled by existing mobility models. A systematic methodology is presented for the calibration and validation of the new model with experimental data. Broad applicability of the new model is established with excellent agreement for a range of experimental data (subthreshold, linear, and saturation) including channel lengths from 20.0μm down to 0.25μm.

1. Introduction

The various components that contribute to resistance in an LDD MOSFET include inversion-layer resistance (which occurs in the channel), accumulation-layer resistance (which occurs in the overlap region between the gate and the LDD diffusion), spreading resistance (which occurs near the end of the accumulation layer owing to current crowding), bulk sheet resistance, and contact resistance. As MOSFET dimensions continue to scale, reductions in channel length and oxide thickness have led to a decrease in the inversion-layer resistance. However, hot-carrier reliability concerns have not allowed the LDD resistance to scale proportionally with channel resistance.

Traditionally, channel resistance was the dominant factor limiting current transport in a MOSFET. Naturally, most of the effort was devoted to modeling mobility in the inversion layer. This is no longer the case in deep submicron MOSFETs which tend to have non-negligible LDD resistance; simulations that use conventional mobility models whose formulation is based only on the inversion-layer over-predict the drain current and fail to capture the slope of the I_d-V_{gs} curve as well. Figure 1 illustrates this point for a 0.25μm MOSFET. Although the value of the contact resistance (which is obtained experimentally) is supplied to the simulator, incorrect calculation of mobility in the accumulation-layer leads to significant errors in the I-V curves.

To address the above concerns, we present a new mobility model that accurately models both the accumulation and inversion layer components. For purposes of calibrating the new model with experimental data, we present a systematic methodology that involves coupled process and device simulations, and emphasizes the need for extraction of parameters from standard test structures that are critical for benchmarking of mobility. We demonstrate broad applicability of the model by presenting excellent agreement between simulation and experimental data for devices from 20μm down to 0.25μm.

2. New Model

The new model builds upon earlier work that treated inversion-layer mobility in a comprehensive fashion [1]. The distinguishing features between this new model and the earlier work is the treatment of various scattering mechanisms in the 2D accumulation layer. Figure 2 illustrates how the new model is partitioned based on the "dimensionality" and the nature of the scattering mechanisms.

It was experimentally shown by Sun and Plummer [2] that both the accumulation-layer and inversion-layer electrons follow the same universal mobility curve (UMC). Since the model reported in our previous work was calibrated using the UMC, it provides a natural platform for extension to the accumulation layer. The model for inversion-layer phonon scattering has a channel-doping-dependent term, which is required in order to reproduce all the properties of the UMC. To extend this model to the accumulation layer, the doping dependence is modified to account for both donors (which occur in the accumulation layer) and acceptors (which occur in the inversion-layer). Thus, the model for 2D phonon scattering is given by:

$$\mu_{ph}^{2D} = \frac{A}{E_\perp} + \frac{B(N_A + N_D)^\gamma}{T \cdot E_\perp^{1/3}} \tag{1}$$

Fundamentally, Coulombic scattering in the inversion layer differs from that in the accumulation layer because attractive charge centers (donors) are more effective in scattering electrons than repulsive charge centers (acceptors). The ratio of attractive to repulsive Coulombic mobility has been modeled by Klaassen [3], and it appears as the function $G(P)$ in Klaassen's formulation. Inversion-layer Coulombic mobility, which was modeled in our previous work [1], is scaled with $G(P)$ to obtain the accumulation-layer Coulombic mobility, as shown below:

$$\mu_{acc}^{2D} = \left[\mu_{inv}^{2D}(N_A = N_D)\right] \cdot G(P) \text{ where } \mu_{inv}^{2D} = max\left[D_1 \frac{n^\kappa}{N_A^\nu}, \frac{D_2}{N_A^\nu}\right] \tag{2}$$

In the case of surface roughness scattering, the model for the inversion layer also applies to the accumulation layer unless the accumulation-layer interface is known to be different from that of the inversion-layer. Hence, the following expression holds for both the accumulation and the inversion layer:

$$\mu_{sr} = C/E_\perp^2 \tag{3}$$

Total phonon scattering is obtained by taking the *minimum* of 2D phonon and 3D phonon scattering, where 3D phonon scattering is given by:

$$\mu_{ph}^{3D} = \mu_{max} \cdot \left[\frac{300}{T}\right]^\theta \tag{4}$$

2D Coulombic mobility is given by the Matthiessen's sum of accumulation-layer and inversion-layer mobility given in (2). To get the total Coulombic mobility, the 2D term and the 3D term are pieced together through a transition function:

$$\mu_C = f(\alpha)\mu_C^{3D} + [1 - f(\alpha)]\mu_C^{2D} \text{ where } \alpha = \alpha(E_\perp) \tag{5}$$

and the 3D term for Coulombic mobility is taken from [3].

Total transverse field mobility is obtained via a Matthiessen's summation of phonon, surface roughness, and Coulombic scattering. The transverse field model is coupled with Hansch's [4] longitudinal-field degradation model to arrive at the complete model.

Compared with our earlier model for inversion-layer electrons [1], no new calibrating parameters are introduced in the formulation of the new model, since the extension from the earlier model has been performed on a physical basis.

3. Validation Methodology

In order to compare the new mobility model with experimental data obtained from *actual* devices that have highly non-uniform doping profiles, the validation procedure necessarily requires coupled process and device simulations. The outline of the validation procedure is shown in Fig. 3. The first step is to supply the process recipe to AT&T's process simulator, PROPHET, which then generates 2D doping and geometry information. Before any I-V curves can be generated by the device simulator using the new model, it needs to be supplied with the values for effective oxide thickness ($T_{ox,eff}$) and contact resistance (R_{co}). $T_{ox,eff}$ is obtained from low-frequency C-V rather than ellipsometric measurements since quantum and poly-depletion effects invariably make electrical thickness to appear larger than the physical thickness. R_{co} is calculated from measurements made on 4-probe Kelvin test structures.

After the specification of the device structure is complete, the validation methodology proceeds in a step-wise manner. First, the linear I-V characteristics of a long channel device (20µm) are compared with device simulations to validate the accuracy of the inversion-layer part of the model. Next, the linear I-V characteristics of the short channel devices are examined. Fig. 4 illustrates that excellent agreement is obtained in the linear region across four short channel devices. On comparison with Fig. 1, Fig. 4 confirms that the accumulation-layer mobility is modeled accurately. At this stage, the transverse-field part of the model has been conclusively established, the next step is to examine the model in saturation. As Fig. 5(a) and (b) illustrate, the model provides excellent fits in saturation as well for devices ranging from 20µm to 0.25µm, thus establishing broad applicability and accuracy of the model over a wide range of channel lengths.

4. Conclusions

We have presented a new MOSFET mobility model for 2-D device simulation that is accurate in both inversion and accumulation layers. Based on a systematic calibration methodology, we demonstrated broad applicability of the new model by showing excellent agreement with data over a wide range of channel lengths down to 0.25µm.

5. Acknowledgments

This work was supported by AT&T Bell Labs and the Semiconductor Research Corporation. The authors gratefully acknowledge the help of Stephen Moccio in I-V measurements, and thank Kathleen Krisch for providing C-V data.

6. References

[1] S. A. Mujtaba et. al., *NUPAD-IV*, Honolulu, Hawaii, June 5-6, 1994, pp. 3-6
[2] S. C. Sun and J. D. Plummer, *IEEE Trans. Elec. Dev.*, vol. 27, p. 1497, 1980
[3] D. M. B. Klaassen, *Solid State Electronics*, vol. 35, no. 7, pp. 953-959, 1992
[4] W. Hänsch and M. Miura-Mattausch, *NASECODE-IV*, Dublin, 1985

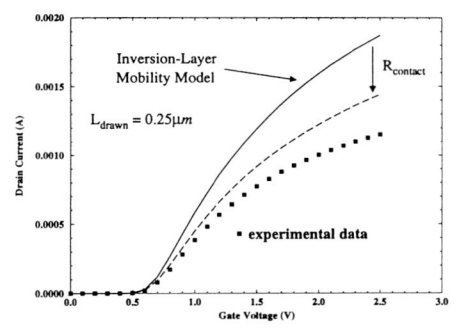

Figure 1. Mobility models formulated for the inversion-layer incorrectly calculate mobility in the accumulation layer, leading to significant errors in I-V curves in the case of deep submicron LDD MOSFETs, which have non-negligible accumulation-layer resistance.

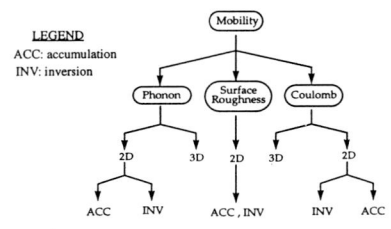

Figure 2. Mobility model hierarchy illustrating the scope of the new model. Essential features of the new model include the addition of the accumulation terms.

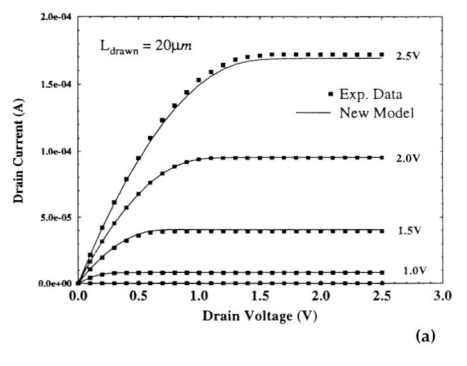

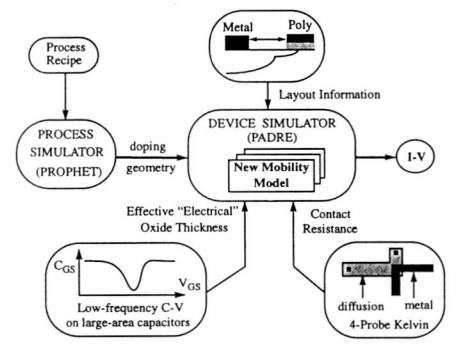

Figure 3. Validation Methodology involves coupled process and device simulations. The process recipe is fed to the process simulator to get the 2D doping profiles. Effective oxide thickness and contact resistance values are supplied to the device simulator from independent measurements. Together with the layout information, simulations are performed with the new model to generate I-V curves, which are then compared with experimental data from devices fabricated based on the process recipe.

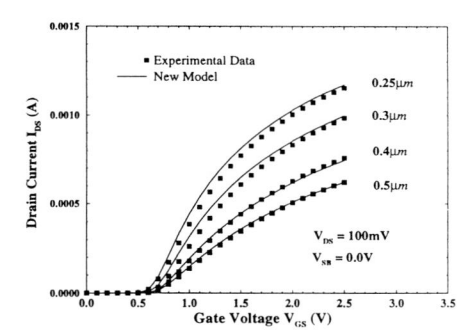

Figure 4. Comparison between the new mobility model and experimental data across four short channel devices is presented, establishing the accuracy of the new model.

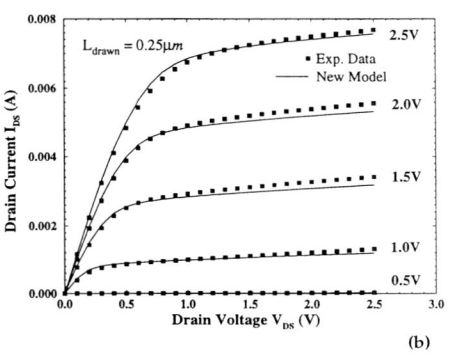

Figure 5. Broad applicability of the new model is also demonstrated in saturation, where comparison between the new model and experimental data is shown for a 20μm and a 0.25μm device. Channel lengths between 20 and 0.25μm exhibit excellent fits as well.

A 2-D modeling of Metal-Oxide-Polycrystalline Silicon-Silicon (MOPS) structures for the determination of interface state and grain boundary state distributions.

A-C. Salaün, H. Lhermite, B. Fortin, O. Bonnaud

Groupe de Microélectronique et Visualisation, URA CNRS 1648,
Université de RENNES I, 35042 RENNES Cedex, FRANCE

Abstract

The aim of this work is the study of the Metal-Oxide-Polycrystalline Silicon-Silicon (MOPS) structure. This study is carried out by means of the two-dimension numerical resolution of Poisson's equation with a model of state continuum in the bandgap and at the Si/SiO$_2$ interface. By fitting experimental C(V) data with computed results, both grain boundary state and interface state distributions are determined.

1. Introduction

The Metal-Oxide-Polycrystalline Silicon-Silicon (MOPS) structure can be an active zone of a thin film transistor fabricated onto an integrated circuit (3-D architecture). A numerical modeling is used for the study of the interface and polysilicon grain boundary trap distributions. This study is based on the resolution of the Poisson's equation to calculate the electrostatic potential variation induced by a gate voltage in a polysilicon MOS capacitor. We perform a numerical integration of this equation with a proper choice for the geometrical modeling of the polycrystalline silicon layer. Inside the grain boundary and at the Si/SiO$_2$ interface, we consider a U-shape distribution of states, with three kinds of traps: band tail states, dangling bond states and discrete states with proper parameters for each distribution. By fitting experimental data with numerical data, bulk state and interface state parameters are determined. We firstly explain the numerical method we use to solve the Poisson's equation, then we specify the simulation domain and finally we present and discuss the results of the simulation.

2. Resolution of Poisson's equation

The modeling is based on the two dimensional numerical solution of Poisson's equation:
$$\text{div } \varepsilon \text{ grad } \Phi = q\,(n - p + N_A^- - N_D^+ + N_{TA}^- - N_{TD}^+)$$
where ε is the local permittivity (silicon or oxide), Φ the electrostatic potential, n and p the electron and hole concentrations expressed in Maxwell-Boltzmann's statistics. The ionised trap density $N_{TA}^- - N_{TD}^+$ is computed with the Shockley, Read and Hall [1] model for the tail states and discrete states and with the Sah-Shockley [2] model for the dangling bonds states. N_A^- et N_D^+ are acceptor and donor ionised doping impurity densities. The method used to solve the Poisson's equation is the finite-difference method.

The resulting capacitance value can be expressed as a function of Φ from $C = \dfrac{dQm}{dVg}$, where Vg is the applied voltage and Qm the electric charge given by a two numerical integration:
$$Qm = \varepsilon_{ox} \cdot \iint_{\Sigma} [\overrightarrow{\text{grad}\,\Phi}]_{x=0} \, \vec{n} \, dS$$

3. Simulation domain

The polycrystalline silicon layer is modeled as a juxtaposition of monocrystalline grains separated by a 1 nm thick amorphous region perpendicular to the oxide-polycrystalline silicon interface (Figure 1).

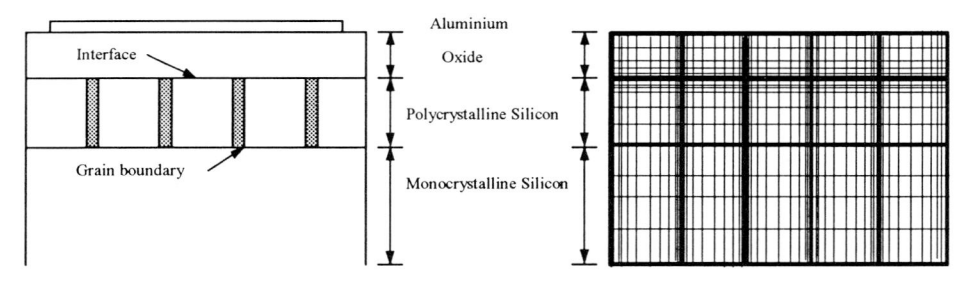

Figure 1: Schematic representation of the structure   *Figure 2: Mesh of the structure*

The n-type polycrystalline silicon (300 nm thick, 2×10^{16} cm^{-3}) is deposited and *in-situ* doped by Very Low Pressure Chemical Vapor Deposition (VLPCVD) on a highly doped monocrystalline silicon substrate (0.02 Ω.cm). The oxide layer (150 nm thick) is obtained by thermal oxidation at 1050 °C. Subsequent aluminium deposit and photolithography give the final structure. Mean grain size (parallel to interface) is 300 nm. Figure 2 shows the mesh used for the discretization of the geometrical structure.

4. Trap distribution

Inside the grain boundary and at the Si/SiO$_2$ interface, we consider a U-shape distribution of states, with three kinds of traps: band tail states, dangling bond states and discrete states (Figure 3) with proper parameters for each distribution.

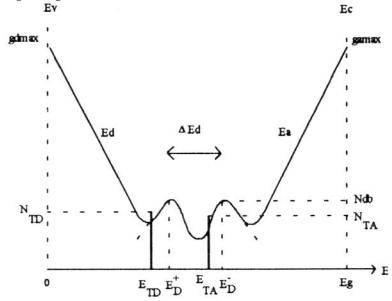

Fig 3: Schematic representation of state distributions

Band tails are caused by localisation of carriers within the wells and hills of potential fluctuations [3]. This exponential distribution may be derived from a Boltzman distribution of isolated defect centres. The trap density due to band tails is given by N_{TD} for the valence band tail and N_{TA} for the conduction band tail:

$$N_{TD} = \int_{Ev}^{Ec} f_D(E) g_D(E) \, dE \quad \text{and} \quad N_{TD} = \int_{Ev}^{Ec} f_A(E) g_A(E) \, dE$$

where $f_A(E)$ and $f_D(E)$ are the acceptor and the donor occupation functions and $g_A(E)$ and $g_D(E)$ the acceptor and donor distribution functions respectively. E_C and E_V are the conduction and valence band edge energies.

Dangling bonds are located in the middle of the bandgap and are characterised by their amphoteric character. These defects come from the polycrystalline silicon structure and correspond to no satisfied Si-Si bonds. The energy distribution is gaussian. The total density of ionised states is given by $N_{DB} = N_{D+} - N_{D-}$ where N_{D+} and N_{D-} are the donor and acceptor dangling bonds density respectively. These dangling bond state densities are given by:

$$N_{D+} = \int_{Ev}^{Ec} n_D(E) f^+(E) \, dE \quad \text{and} \quad N_{D-} = \int_{Ev}^{Ec} n_D(E) f^-(E) \, dE$$

where $f^+(E)$ et $f^-(E)$ are the distribution functions and $n_D(E)$ is the gaussian trap states distribution function.

Discrete states are traps due to uncontrolled impurities, which can give localised states in the band gap with either an energy E_{TA} for acceptor-like traps or either an energy E_{TD} for donor-like traps.

5. Numerical results

Numerical values for the physical parameters as the permittivity, the electron affinity, the densities of states in the conduction and valence band edge, the gap energies in the crystallite (1.12 eV) and in the grain boundary (1.7 eV) are taken from the literature. Fitting parameters are the doping of the polysilicon layer and the distribution of tail states and dangling bonds both at grain boundary and at Si/SiO$_2$ interface. Figure 4(a) shows the comparison of experimental and computed data for the high frequency and quasi-static C(V) plots. This fit is possible by adjusting both the interface states distribution Dit (b) and the grain boundary states distribution Dgb (c).

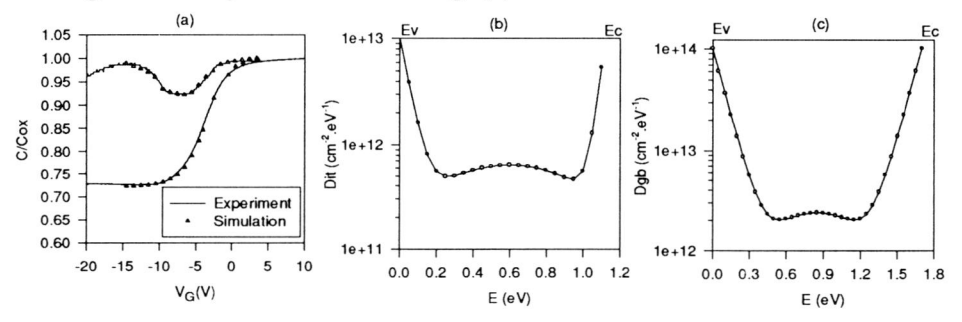

Figure 4: Example of the high frequency and quasi-static C(V) plots simulation (a), Distribution of interface state density (b), of grain boundary state density (c).

6. Conclusion

The polysilicon device performances depend on the polysilicon layer quality. Traps resulting from defects at grain boundary as well as at polysilicon/oxide interface affects the electronic behaviour of the device. By fitting experimental C(V) data with computed results, both grain boundary state and interface state distributions are determined. The differentiation of these distributions is needed if we want to qualify the efficiency of process like hydrogenation used to lowered these two trap distributions.

References

[1] W. Shockley and W. T. Read, "Statistics of the recombinations of holes and electrons", Phys. Rev., vol. 87, no. 5, pp. 835-842, 1952.

[2] C. T. Sah and W. Shockley, "Electron-hole recombination statistics in semiconductors through flaws with many charge conditions". Phys. Rev., vol. 109, no.4, 1958.

[3] J. H. Werner, M. Peisl, "Exponential band tails in polycrystalline semiconductors films", Physical Review B, Vol.31, No.10, pp. 6881-6883, May 1985.

Sensitivity Analysis of an Industrial CMOS Process using RSM Techniques

M.J. van Dort and D.B.M. Klaassen

Philips Research Laboratories,
Prof. Holstlaan 4,
5656 AA Eindhoven, The Netherlands

Abstract

This paper describes a method to identify the process parameters responsible for the spread in the transistor parameters. The method consists of modeling the response surfaces for the transistor parameters and the correlations between them in terms of process variables. Incorporation of the correlations in the analysis turned out to be essential in order to correctly identify the cause for the fluctuations in the transistor parameters.

I. Introduction

Manufacturability and yield improvement are extremely important in the IC industry. For this purpose, the transistor parameters of MOSFET's in a CMOS production line are constantly monitored. A modern CMOS process consists of a very large number of process steps. Each individual process parameter (e.g. t_{ox}) will contribute to the spread in the transistor parameters (e.g. V_t). Identification of the most critical process parameters is extremely important. In this paper, we show that the combination of statistical analysis of the measured devices and Response Surface Modeling (RSM) techniques [1] in a TCAD framework is an extremely powerful tool for modeling the statistics of the transistor parameters. For the first time, we show that correct identification of the most important process parameters is only possible if the correlations between the transistor parameters are explicitly taken into account.

II. Experimental

The DC characteristics of MOSFET's in a CMOS production line were monitored over a long period of time. The MOS-MODEL9 parameters [2] were determined with direct parameter extraction and used for statistical analysis. The database contained more than 11.000 fully characterized sets of MOSFET's. Each set consists of n and p channel devices with various channel lengths. The DC characteristics were modeled using MOS MODEL9. Only the long-channel n and p-channel devices ($W=L=10$ μm) are considered in this paper. The 8 most important transistor parameters and the 6 correlations dealing with the V_t's and the gain factors were chosen for modeling purposes. The spreads of these parameters and the correlations between them form the base of our analysis.

All process parameters are subjected to short-term variations (within one batch) and long-term variations (from batch to batch). In this paper, only the variations around the average values of the parameters of the batches were used in the analysis. We therefore focus on the short-term variations.

III. Simulations

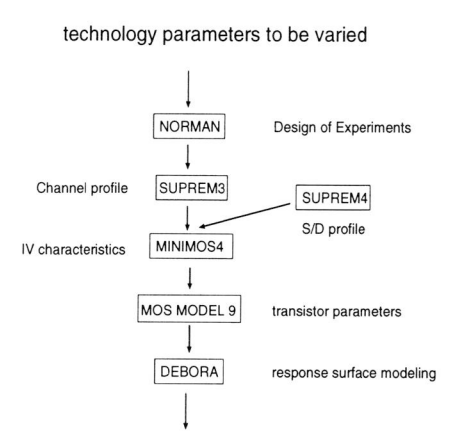

Figure 1: *Simulation chain to simulate the spreads in the transistor parameters and the correlation coefficients.*

The simulation chain is depicted in figure 1. The NORMAN/DEBORA package[3] was used for the modeling of the response surfaces. The doping profile was constructed using SUPREM3 for the channel profile and SUPREM4 for the S/D profile. Since we are interested in the modeling of long-channel devices, we have only taken the variations in the channel profile into account. The S/D profile was assumed to be constant. The 2D S/D profile was merged with the 1D channel profile to obtain a 2D doping profile for the device simulations. The IV characteristics were simulated with MINIMOS4, and then the compact model parameters were extracted using MOS MODEL9. In total 15 process parameters were varied. Included were the temperatures of all furnace anneals, the energies and doses of the implantations and the layer thicknesses. In the initial simulations, 1st-order Taylor expansions for the response surfaces were

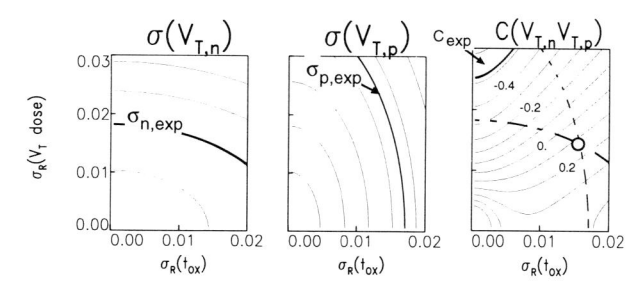

Figure 2: *-The first figure shows the simulated equi-$\sigma(V_{t,n})$ lines. The thick solid line indicates the experimental value. The second figure depicts the same information for $\sigma(V_{t,p})$. The intersection of these two equi-σ lines (for $\sigma(V_{t,n})$ and $\sigma(V_{t,p})$) gives the spread for t_{ox} and the V_t dose. -The last figure shows the equi-$C(V_{t,n}, V_{t,p})$ lines. The two dashed lines are the $\sigma(V_t)$ lines and the intersection is the simulated correlation coefficient. We find $C(V_{t,n}, V_{t,p}) \approx 0.1$, which is completely wrong because the experiments show that $C(V_{t,n}, V_{t,p}) \approx -0.53$*

used to filter out some unimportant process parameters. In subsequent sets of simulations, the response surfaces were modeled with higher accuracy for the remaining process variables using 2nd-order Taylor expansions. These 2nd-order functions were used to model the spreads in the transistor parameters and their correlations.

We were able to identify the five main process parameters responsible for the spreads in the transistor parameters. This is in accordance with principal components analysis on the database with the measured set of devices, which showed that we are dealing with a 5 dimensional parameter space. Some of these parameters are very obvious (e.g. t_{ox}), others are less trivial.

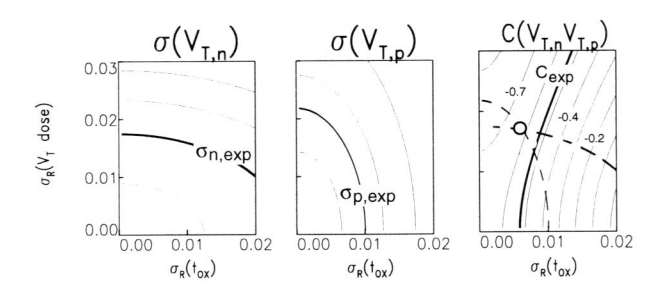

Figure 3: *Situation for the optimum fit. Note that both the experiment as well as the simulations give $C(V_{t,n}, V_{t,p}) \approx -0.53$. In this case we correctly simulate the $\sigma(V_{t,n})$, $\sigma(V_{t,p})$ and $C(V_{t,n}, V_{t,p})$. Each Equi-$\sigma(V_{t,n})$ line represents a 10 % change. For the Equi-$\sigma(V_{t,p})$ lines intervals are 20% changes (same in figure 2).*

The problem we faced was to find a unique description of the statistics in terms of the process parameters. It is not sufficient to fit the simulated spreads in the transistor parameters to the experimental ones. This is illustrated in figure 2, where we briefly discuss the dependence of $V_{t,n}$ and $V_{t,p}$ on t_{ox} and the dose of the V_t implant. This figure depicts the situation were we have fitted the spreads of all parameters, but did not pay any attention to the correlations. The correlation coefficient $C(V_{t,n}, V_{t,p})$ belonging to this solution is way off, and this strategy leads to the wrong identification of the critical process parameters. Figure 3, on the other hand, shows the situation where we have fitted the spreads as well as the correlation coefficients to the experiments. In this case we obtain good fit and this leads to different conclusions concerning the process weaknesses. The correlation coefficients are extremely important to take into account, because they are the 'fingerprints' of the process (figures 4 and 5).

It is interesting to look at the $\sigma(V_{t,n})$ and $\sigma(V_{t,p})$ in more detail, because the spread in $V_{t,p}$ is almost twice as much as the one for $V_{t,n}$ (figure 4). From our analysis, it is easy to identify the cause of this large spread. In the process under consideration, this spread is caused by the n well construction. This knowledge can in turn be used to minimize the spread in the transistor parameters.

It is also interesting to analyze the variations from batch to batch instead of the short-term variations investigated in this paper. Some of the correlations which exists on a short time scale disappear on a long time scale. The same RSM analysis can be done for the long-term variations. If a self-consistent solution can be found, the outcome will reveal the cause of these long-term fluctuations.

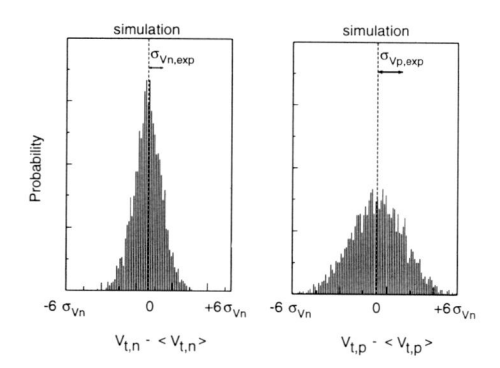

Figure 4: *Simulated distributions for $V_{t,n}$ and $V_{t,p}$. Note that the spread in $V_{t,p}$ is almost twice as large as the spread in $V_{t,n}$. These are simulation results. The experimental spreads are indicated in the figure.*

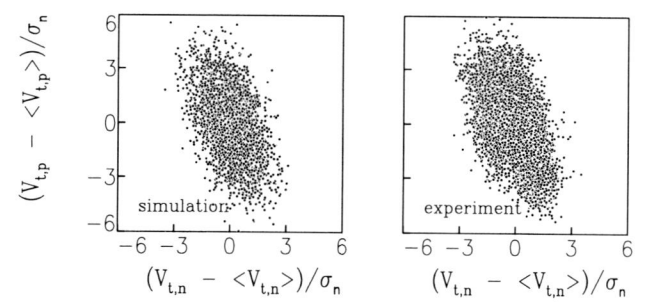

Figure 5: *Simulated and experimental spread in $V_{t,n}$ and $V_{t,p}$. Good agreement exists between the experiments and the simulations. These plots are the 'fingerprints' of the process.*

IV. Conclusion

In this paper we have shown that the RSM approach can be successfully applied to an industrial environment to simulate the spreads and correlations between the transistor parameters. Modeling of the correlations turned out to be essential in order to reveal the process weaknesses. Furthermore, modeling of the correlations between transistor parameters is also important for circuit designers to improve the design window.

The authors thank F. Postma for supply of the MOS database, R. Velghe for parameter extraction and would further like to acknowledge the support of ESPRIT 8002 project ADEQUAT for this work.

References

[1] G. Box, W. Hunter and J. Hunter, *Statistics for Experimenters*, Wiley & Sons.
[2] R. Velghe, D. Klaassen and F. Klaassen, *Tech Digest IEDM*, p. 485 (1993).
[3] R. Cartuyvels et. al. *Proc. SISDEP*, p. 29 (1993).

Process- and Devicesimulation of Very High Speed Vertical MOS Transistors

F.Lau, W.H.Krautschneider, F.Hofmann, H.Gossner, H.Schäfer

SIEMENS AG, Corp. R&D, D-81730 Munich, GERMANY

Abstract

Optical lithography does not allow the scaling of MOS transistors down to 100nm dimensions. Thus the channel length of high speed MOS devices must depend on alternative processing steps. In this work layer deposition and etching are analysed with respect to the formation of very short MOS transistors with vertical orientation. Dopant diffusion with very steep gradients are studied in epitaxial layers. Process and device engineering aspects for a vertical MOS transistor at the sidewall of an etched trench are discussed.

1. Introduction

To improve the performance of MOS transistors the structure size of the devices must be scaled down. However the decrease of the channel length below the quarter-micron dimension down to 100nm is limited by optical lithography. To realize very high speed MOS devices within a 500nm lithography environment, alternative approaches are required, in which the channel length does not depend on lithography. Layer deposition and etching allow fine adjustment on the nanometer scale. In addition, during the deposition of epitaxial layers doping profiles with very steep gradients can be performed. Etching and deposition, however, cause vertical structure modifications. Thus the MOS device must also include a vertical orientation [1]. In this work two aspects are analysed:

1. Channel length definition by **epilayer formation** (Fig.1): The layer formation with Chemical Vapor Deposition (CVD) or Molecular Beam Epitaxy (MBE) allows mimimum channel lengths in the sub 100nm region. In the process flow layer deposition with simultaneous doping formation must be followed by thermal oxidation for gate oxide growth. Depending on dopant species and on concentration regime anneals are necessary for dopant activation. We analysed whether silicon bulk diffusion models available in commercial process simulators can be applied to dopant transport in CVD and MBE layers with extrem dopant gradients [2].

2. Channel length definition by **trench etching** (Fig.2): Silicon etching is less expensive and more useful for production engineering than CVD and MBE. Process- and device simulation was used to develop and optimize a vertical MOS transistor at the sidewall of an etched trench. This was done on the basis of more conventional types of processing steps.

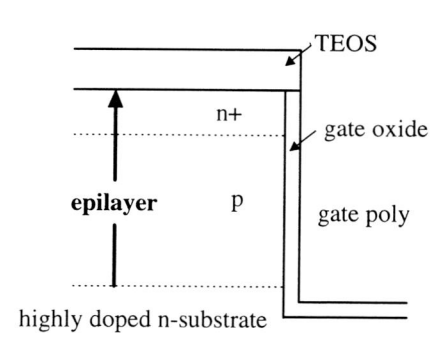

Fig.1: Vertical MOS transistor formed by epitaxial layer deposition.

Fig.2: Vertical MOS transistor formed by trench etching and n$^+$ implantations.

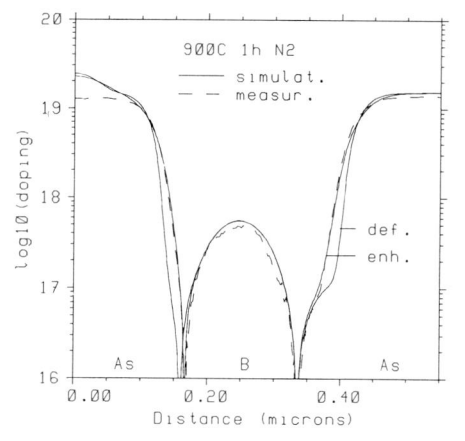

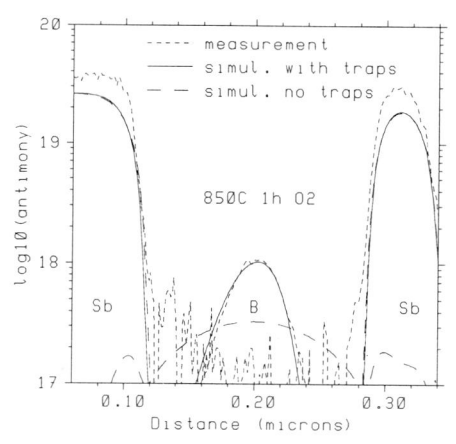

Fig.3: Doping profiles in an annealed CVD epilayer. Default diffusivities for arsenic result in an error of 2*15nm for the resulting channel length.

Fig.4: Doping profiles in an annealed MBE epilayer. Boron shows no OED.

2. Dopant diffusion in vertical MOS structures formed by CVD and MBE

The diffusion behaviour of boron and arsenic in CVD layers between 800 and 1000°C in nitrogen atmosphere is investigated. In MBE layers we studied boron and antimony between 700 and 900°C in oxidizing atmosphere. Doping profiles were measured by SIMS. As a result, the silicon bulk diffusion models can be applied to CVD layers by and large. If extrem accuracy is required for sub quarter-micron devices, the diffusivity of arsenic must be enhanced (Fig.3). In MBE layers (Fig.4) antimony diffuses as in silicon bulk. The boron profiles, however, show nearly no oxidation enhanced diffusion (OED) that would be predicted for silicon bulk. Boron diffuses mainly via interstitials which are injected during oxidation. In TEM cross sections the generation of dislocation loops is observed during thermal treatment. Dislocation loops are known to absorb free interstitials very effectively. Usually the growth and shrinkage of dislocation loops and corresponding capture and emission rates for interstitials cannot be modeled with commercial process simulators. TSUPREM4 [3], however, includes an interstitials trap model. In a first approach this model was used to regard for the capture of interstitials which are generated during oxidation. The total trap concentration was modified to find agreement between the profiles from SIMS and from simulation (equilibrium trap occupation assumed). The total trap concentration increases with temperature suggesting that the size of dislocation loops increases also with temperature. TEM cross sections show no loops before the oxidizing anneal after MBE epitaxy. In summary we conclude that thermal treatment of MBE layers which is necessary for MOS processing creates dislocation loops. These loops affect the dopant diffusion by the capture of interstitials preventing OED.

3. Process and device optimization of vertical MOS structures

In this part we study features in process and device engineering which are important for the formation of a vertical MOS transistor. The channel length is defined by trench etching (Fig.2). The simulations were performed with TSUPREM4 and MEDICI [3]. The doped regions are formed by a trench etch and by implantation. A schematic structure is shown in Fig.2. In contrast to lateral MOS transistors following aspects must be regarded: 1. The vertical profile of implanted arsenic (channeling) forming the upper n^+ region at the silicon surface affects the channel length. 2. A VT-implant is difficult to perform. The threshold voltage must be adjusted by a well. 3. The oxide thickness at the trench sidewall must mask the arsenic implantation (usually tilted) for the lower n^+ region at the trench bottom. 4. Diffusion of the lower n^+ doping around the trench corner is important for the saturation current and device symmetry between between source and drain. 5. If the corners at the trench bottom are rounded (due to isotropic components during trench etching) the saturation current decreases with increasing curvature radius.

The resulting device performance (Fig.5,6) do not differ from conventional lateral MOS transistors.

4. Conclusion

For the development of vertical MOS transistors we have analysed the dopant transport in MBE- and CVD-epilayers during thermal treatment. The results, summarized in the following table,

	boron	arsenic	antimony
CVD, N_2 ambient:	as in Si-bulk	enhanced diffusion	-
MBE, O_2 ambient:	reduced OED (disl.loops)	-	as in Si-bulk

indicate, that epilayers in general cannot be treated in the same way as silicon bulk material. Some authors use MBE layers in experimental structures to determine model parameters for pointdefects [4]. These kind of experiments need an extra analysis, which proofs the quality of the epilayer.

We have also shown that vertical MOS transistors can be fabricated in a production line without expensive processing steps. The channel length of 200nm is far below the limits given by lithography.

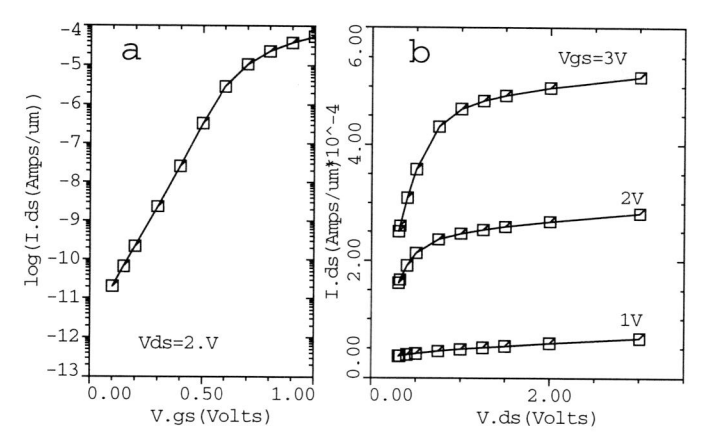

Fig.5: Gate characteristics $I_{ds}(V_{gs})$ (a) and drain characteristics $I_{ds}(V_{ds})$ (b) of a vertical MOS transistor with 200nm channel length (formed as shown in Fig.2.).

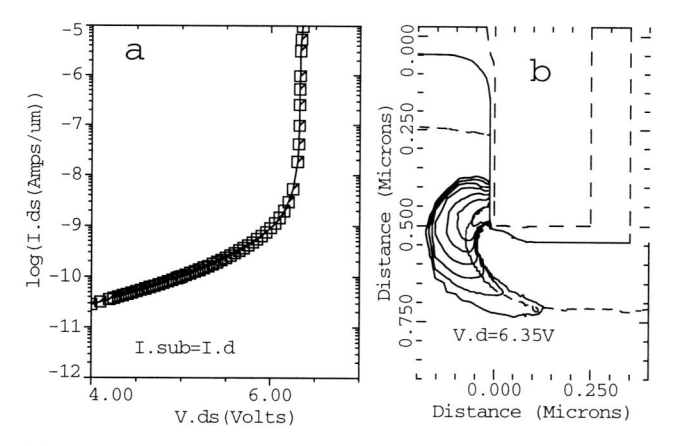

Fig.6: Breakdown characteristics $I_{ds}(V_{ds})$ (a) and contourlines of the impact ionization rate (b) for the same device as shown in Fig.2 and Fig.5.

References

[1] A.H.Perera et.al.; *IEDM Tech. Dig. 1994*, p.851
[2] W.H. Krautschneider, F.Lau, H.Gossner and H.Schaefer; to be printed in *Proceedings of the MRS 1994 Fall Meeting*, Symposium F
[3] Technology Modeling Associates, Palo Alto CA
[4] K.Ghaderi et.al.; *The Electrochem. Soc., Spring Meeting 1994*, Ext. Abstr., No.459

Two-Dimensional Transient Simulation of Charge-Coupled Devices Using MINIMOS NT

M. Rottinger, T. Simlinger, and S. Selberherr

Institute for Microelectronics, TU Vienna
Gusshausstrasse 27–29, A-1040 Vienna, Austria

Abstract

This paper will provide information on the features of the new device simulator MINIMOS NT and demonstrate its efficiency in transient simulation of complex device structures. An example will indicate the feasibility of such simulations even on a workstation.

1. Introduction

Charge-Coupled Devices (CCDs) have a broad field of application in optical imaging and analog signal processing. Since their invention in 1970 the performance of CCDs, such as noise reduction, photo-sensitivity, resolution and power consumption, has been improved continuously [1]. In contrast to this development simulation of CCDs, particular transient simulation, has not been practicable because of the high requirements on computational resources, hence only single CCD cells have been simulated [2] [3].

Using our simulator, MINIMOS NT, it is possible to simulate complex device structures with reasonable demands on computational resources. For instance, CCDs containing up to 60 gates have been simulated in two space dimensions on a workstation equipped with 128MB of main memory.

2. Description of the simulator

MINIMOS NT's software architecture is based upon up-to-date software engineering knowledge. The demands of numerous users of device simulation software were carefully evaluated over time in order to found the required features. For instance, discretization, Jacobian matrix assembly and linear system solution are linked with abstract protocolls. Thereby it is possible to account for fully different physical models in different, however, connected segments, e.g., in one segment a hydro-dynamic model can be used whereas in a neighbouring segment only a drift-diffusion model is considered. The same holds true for the various physical parameter models. Mixed numerical discretization schemes are also possible. This concept holds for two and three space dimensions. For transient integration a predictor-corrector method was evaluated to be most suitable for the semiconductor problem. Circuits containing different numerically treated semiconductor devices and the usual discrete components are also handled properly.

The Jacobian matrix is scaled fully automatically with an iterative algorithm [4]. The linear system is solved with a state-of-the-art BiCGStab algorithm where automatic switching to a Gauß-solver is performed if convergence of the iterative algorithm seems not attainable.

3. Time step estimation

The predictor used to estimate the size of the next time step is based upon a quadratic extrapolation of the potential-update norm. The ratio of the estimated step size and the previous step size is restricted to a maximum, the sectio aurea constant (1.618...), to limit the step size variations. With this restriction a quasi-uniform mesh is achieved, which gives a second order local function error. After calculation of each time step the potential update is checked. For this purpose the $L2$ norm and the infinity norm are calculated. If at least one of these norms exceeds a respective threshold, the step size is reduced by quadratic interpolation and the calculation is repeated.

In order to accurately follow the predefined contact signals it is necessary to calculate a time step at the instances used to specify the contact signals. This requires a reduction of the estimated step size to exactly match the instance and causes an increase in the total number of required time steps. Therefore the number of specified instances should be as small as possible to take full advantage of the quadratic time step estimation. To reduce the number of specified instances necessary for a sufficient representation of nonlinear contact signals a quadratic interpolation is used.

4. Transient simulation of a CCD

As an example a three-phase clock, n-channel CCD composed of fifteen gates – results of devices with more gates can hardly be visualized on paper – has been simulated in two space dimensions. Fig. 1 shows the structure of the device. The source and drain contacts were held constant at 0V, the bulk contact at -1V. The voltages applied to the gates varied between -1V and $+5$V (Fig. 3). The simulation over ten clock periods required 360 time steps. The adapted space grid consisted of approximately 10,000 points. A snapshot of the electron concentration in the channel region is shown in Fig. 3. One can nicely see the alternation of accumulation to strong inversion under the gates. As a further result of the simulation Fig. 4 shows the charge that has passed through the source, drain and bulk contact, respectively. Charge flowing into the device is counted positive.

5. Conclusion

The simulator is fully integrated into the VISTA framework. Therefore all features of the framework, e.g., coupling to the various process simulators, visualizers and animators, embeded optimizers, are readily available. The above example, which has been calculated on a DEC 3000/400, illustrates the ability of MINIMOS NT to perform transient simulations of complex device structures and demonstrates the feasibility of such simulations even on a workstation.

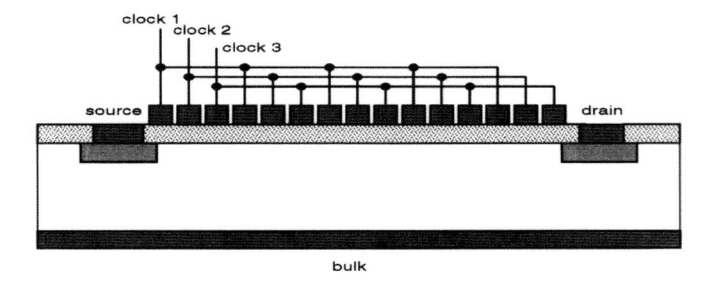

Figure 1: The structure of a 15-gate CCD.

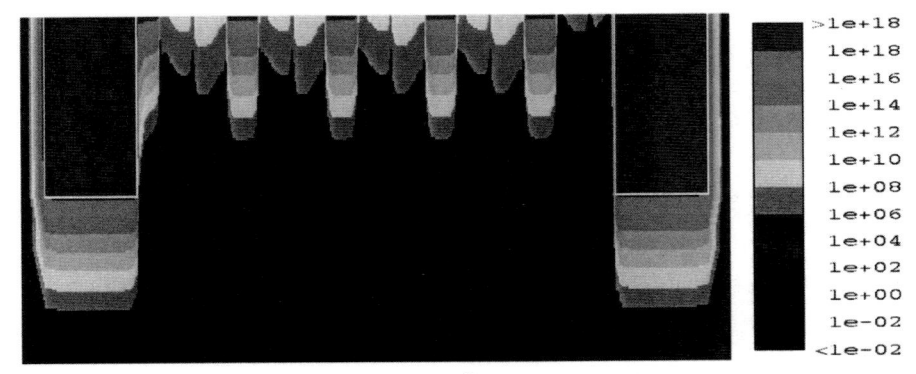

Figure 2: Electron concentration $[cm^{-3}]$ in the channel region at $t = 3.36\mu s$.

Acknowledgement

This work is supported by Digital Equipment Corporation, Hudson, USA; Motorola, Austin, USA; Philips, Eindhoven, The Netherlands.

References

[1] R. Melen and D. Buss (eds.), *Charge-Coupled Devices: Technology and Applications*, IEEE Press, 1976.

[2] N. Ula, G. A. Cooper, J. C. Davidson, S. P. Swierkowski and Ch. E. Hunt, "Optimization of Thin-Film Resistive-Gate and Capacitive-Gate GaAs Charge-Coupled Devices", *IEEE Trans. Electron Devices*, vol. 39(5), pp. 1032–1040, 1992.

[3] Ch. R. Smith and S. G. Chamberlain, "Theory and Design Methodology for an Optimum Single-Phase CCD", *IEEE Trans. Electron Devices*, vol. 39(4), pp. 864–873, 1992.

[4] C. Fischer and S. Selberherr, "Optimum Scaling of Non-Symmetric Jacobian Matrices for Threshold Pivoting Preconditioners", *Proc. Int. Workshop on Numerical Modeling of Processes and Devices for Integrated Circuits, NUPAD V*, pp. 123–126, 1994.

[5] S. Halama et al., "The Viennese Integrated System for Technology CAD Applications", *Technology CAD Systems*, pp. 197–236, 1993.

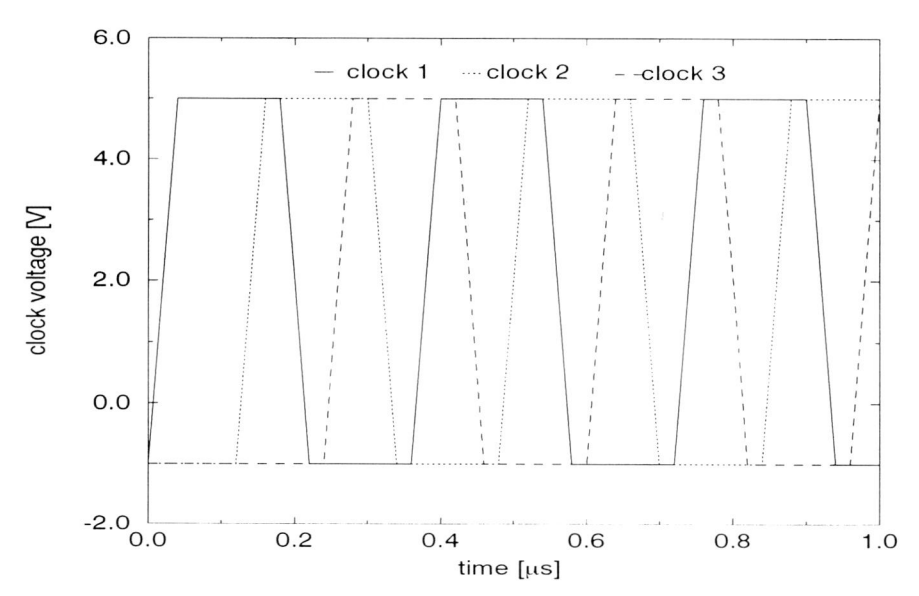

Figure 3: Clock voltages during $1\mu s$.

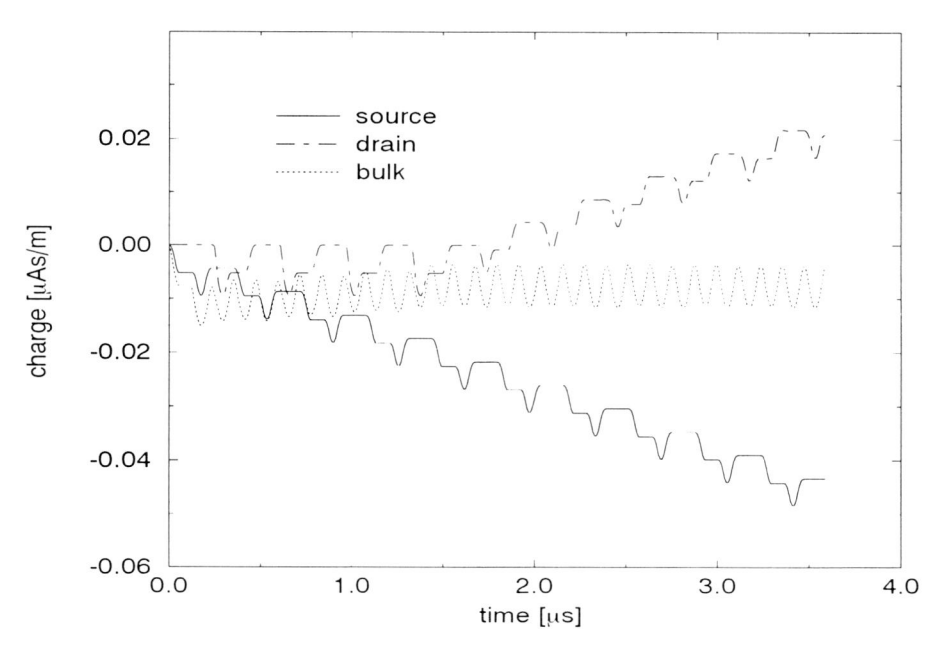

Figure 4: Charge transferred through the contacts of a 15-gate CCD during 10 clock periods.

Determination of Vacancy Diffusivity in Silicon for Process Simulation

T. Shimizu[a], Y. Zaitsu[a], S. Matsumoto[a], E. Arai[b], M. Yoshida[c], T. Abe[d]

[a]Department of Electrical Engineering, Keio University,
Hiyoshi, Yokohama 223, JAPAN
[b]NTT LSI Laboratories, Atsugi, Kanagawa 243, JAPAN
[c]Kyusyu Institute of Design, Shiobaru, Fukuoka 960
[d]Shinetsu Semiconductors, Isobe, Gunma 543, JAPAN

Abstract

Vacancy diffusivity has been determined by using the fact that Si_3N_4 films deposited on silicon can be used as an extrinsic source of vacancies. Excess vacancies generated at the backside with Si_3N_4 films migrate to the boron-implanted region at the top side and its boron diffusion is retarded. Using these data, vacancy diffusivity is determined from the simulation.

1. Introduction

Process simulation becomes more important with the shrinkage of the device dimensions in ULSI. Development of numerical method and establishment of physical model are essential for progress of process simulation. The present situation is, however, far from reality, mainly due to the lack of the reliable physical model [1] . Dopant diffusion is known to be one of the key process in ULSI technology and its accurate prediction is essential to device simulators. It has been recognized that dopant diffusion in silicon is mediated by interaction of dopant atoms with both vacancies and self-interstitials [2] . Thus many physical parameters of point defects (vacancy and silicon self-interstitial) are necessary for the simulation of dopant diffusion.

Self-interstitial diffusivity has been extensively studied because oxidation was the extrinsic source of self-interstitials. However, most parameters are still uncertain. Particulary, there are very few data about vacancy diffusivity.

Recently, we investigated the vacancy concentration in boron-diffused region in Si under both Si_3N_4 and SiO_2 /Si_3N_4 films by slow positron beam technique and found that excess vacancies were introduced in the Si substrate under Si_3N_4 films due to the thermal stress at the interface [3][4] . That is, Si_3N_4 films can be used as the extrinsic source of vacancies.

In this work, we tried to determine vacancy diffusivity based on the simple concept. The method determining vacancy diffusivity is consisting of the following three steps: (1) generation of excess vacancies at the backside with Si_3N_4 films, (2) observation of vacancy migration by boron diffusion at the top side, (3) estimation of vacancy diffusivity by the simulation.

| Si3N4 |
| SiO2 |
| Boron |

| FZ-Si |

| SiO2 |
| Si3N4 |

sample-ON

| Si3N4 |
| SiO2 |
| Boron |
x=h

| FZ-Si |
x=0

| Si3N4 |

sample-BN

| Si3N4 |
| Boron |

| FZ-Si |

| SiO2 |
| Si3N4 |

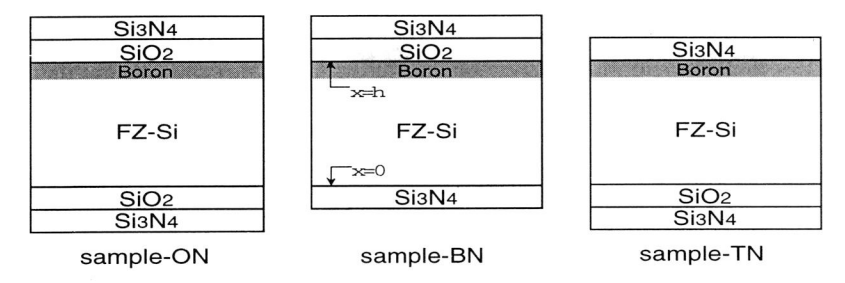

sample-TN

Figure 1: Sample structures.

2. Experimental

Based on the above concept, samples having different structures shown in Fig.1 were prepared. Starting materials were p-type FZ (100) Si single crystals with resistivity of 3-5 Ωcm and thickness of 238 μm. First, boron implanted regions were formed at the top side by ion implantation with an energy of 70 keV and a dose of 7.5×10^{13}cm^{-2} through 50 nm oxide. Samples were annealed at 900°C in N$_2$ for 30 min to restore the damage. For sample-BN, oxide at the backside was removed and for sample-TN, oxide at the top side was etched off. Then stoichiometric Si$_3$N$_4$ films were deposited by ECR plasma CVD to both sides using SiH$_4$ and N$_2$ gases. The thickness of Si$_3$N$_4$ was 50 nm. Cross section of the respective sample structures is shown in Fig.1. The sample-ON has a structure with SiO$_2$ /Si$_3$N$_4$ double layered films at both sides. Boron diffusivity obtained from sample-ON corresponds to the intrinsic diffusivity of boron [5], which is used as a reference. Using sample-BN, the migration of vacancies from the backside is monitored by the diffusion of boron at the top side. The sample-TN is necessary for the determination of excess vacancy concentration generated under Si$_3$N$_4$ films, which is used as the boundary condition as described later. Annealing was performed in N$_2$ at 1000°C for 10, 33, 70, 129 h. After etching SiO$_2$ and Si$_3$N$_4$ films, boron concentration profiles were measured by secondary ion mass spectroscopy (SIMS)(ATOMIKA 6500). The boron diffusivity of Si, $\langle D_B \rangle$, was determined by fitting measured profiles with simulated plofiles obtained by solving Fick's diffusion equation.

3. Results and Disscussion

A typical example of boron concentration profiles obtained from respective sample structures by SIMS analysis is shown in Fig.2 after annealing at 1000°C for 129 hr. Boron diffusion in sample-BN is retarded as compared with that of sample-ON. The annealing time dependence of the ratio of boron diffusivity, $\langle D_B \rangle$ to that of sample-ON(intrinsic diffusivity), D_B^* is shown in Fig.3. At 10 h, its ratio is nearly 1. It means that excess vacancies migrating from the backside recombine almost with self-interstitials in the bulk and does not reach the top side. Above 30 h, they reach at the top side and cause the undersaturarion of self-interstitials, that is, retardation of boron diffusion. In sample-TN, boron diffusion is also retarded as shown in Fig.2. The annealing time dependence of $\langle D_B \rangle / D_B^*$ is shown in Fig.3. In sample-TN, retarded diffusion of boron is observed at 10 h and its ratio decreases slightly with the increase of annealing time.

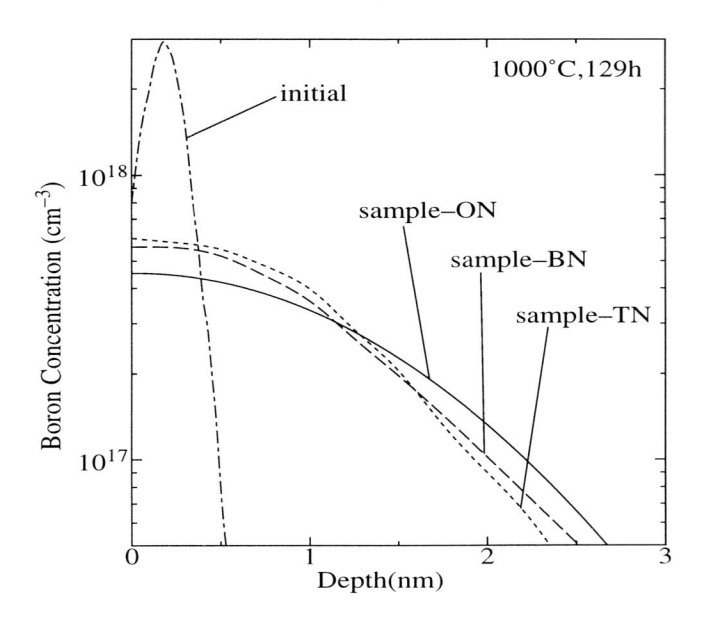

Figure 2: Boron concentration profiles of respective samples.

Simulation of D_V and k_V

1. Estimation of vacancy concentration at the top side in sample-BN by solving the diffusion equation of vacancy with the appropriate boundary conditions.

$$\frac{\partial C'(x,t)}{\partial t} = D_V \frac{\partial^2 C'(x,t)}{\partial x^2}, \quad (0 < x < h) \tag{1}$$

where $C'(x,t)$ is the normalized vacancy concentration ($= C_V(x,t)/C_V^*$), C_V and C_V^* are the vacancy concentration and its equilibrium value, respectively, D_V is the vacancy diffusivity, and h is the thickness of the substrate.

Boundary conditions;

$$\text{at the backside } (x = 0), \quad C'(0,t) = P(t) \tag{2}$$

where $P(t)$ is the excess vacancy concentration generated at the Si/Si$_3$N$_4$ interface and is given from the experimental data of sample-TN.

$$\text{At the top side } (x = h), \quad D_V \frac{\partial C'}{\partial x}\bigg|_{x=h} + k_V \left[C'(h,t) - 1\right] = 0 \tag{3}$$

where k_V is the recombination rate of vacancy at Si/SiO$_2$ interface.

2. Determination of vacancy diffusivity and recombination rate by fitting procedure.

Boron diffusivity at the top side is estimated from the following equation.

$$\left\langle \frac{D_B}{D_B^*} \right\rangle = f_I \left\langle \frac{C_I}{C_I^*} \right\rangle + (1 - f_I) \left\langle \frac{C_V}{C_V^*} \right\rangle \tag{4}$$

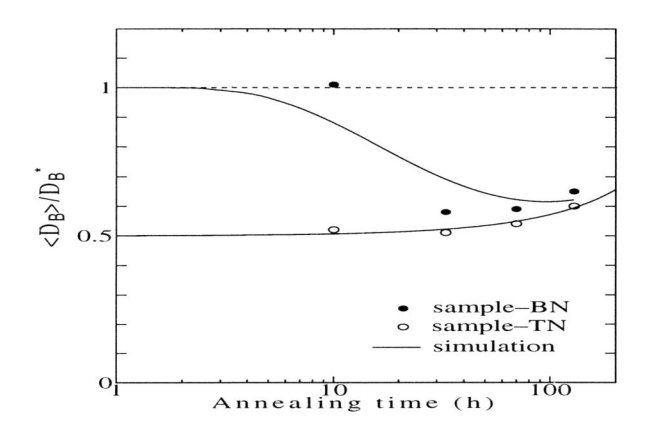

Figure 3: Annealing time dependence of $\langle D_B \rangle / D_B^*$ of sample-BN and sample-TN

where f_I is the fraction of interstitialcy mechanism. When the vacancy concentration is given from the above procedure, boron diffusivity can be caluculated as a function of time.

In the caluculation, we assume $f_I=1$ for boron diffusion. Local equilibrium condition that $C_V C_I = C_V^* C_I^*$ is also assumed. Changing both D_V and k_V, fitting the calculated with the experimental data is repeated. Two solid lines in Fig.3 are the simulated results using the values that $D_V = 4.1 \times 10^{-9} \text{cm}^2/\text{s}$ and $k_V = 8.9 \times 10^{-10} \text{cm/s}$ at 1000°C. It is concluded that physical quantities relating vacancy can be determined based on the experimental results.

4. Acknowlegement

The authors would like to thank Z.Lee and M.Sakurazawa for SIMS measurement.

References

[1] J. D. Plummer, "Process simulation in submicron silicon structures," Proc. of Symposium on PROCESS PHYSICS AND MODELING IN SEMICONDUCTOR TECHNOLOGY (The Electrochemical Society), pp. 93-102, 1993.

[2] T. Y. Tan and U. Gösele, "Point defects, diffusion processes, and swirl defect formation in silicon," Appl. Phys., vol. A37, pp. 1-17, 1985.

[3] K. Osada, Y. Zaitsu, S. Matsumoto, M. Yoshida, E. Arai and T. Abe, "Effects of stress in the deposited silicon nitride films on boron diffusion of silicon," J. Electrochem. Soc., vol. 142, no. 1, pp. 202-206, 1995.

[4] K. Osada, Y. Zaitsu, S. Matsumoto, S. Tanigawa, A. Uedono, M. Yoshida, E. Arai and T. Abe, "Direct observation of vacancy supersaturation in retarded diffusion of boron in silicon probed by monoenergetic positron beam," Extended Abstract of Solid State Devices and Materials, pp. 739-741, 1994.

[5] R. B. Fair, in Processing Technologies, D. Knang, Editor, Applied Solid State Science, Suppl., 2B, Academic Press, Inc., New York, pp. 1-103, 1981.

Precipitation phenomena and transient diffusion/activation during high concentration boron annealing

Alexander Höfler[a], Thomas Feudel[a], Arno Liegmann[a], Norbert Strecker[a], and Wolfgang Fichtner[a]
Yuji Kataoka[b], Kunihiro Suzuki[b], Nobuo Sasaki[b]

[a]Integrated Systems Laboratory
Swiss Federal Institute of Technology, ETH-Zentrum
8092 Zürich, SWITZERLAND
[b]Fujitsu Laboratories, Ltd., ULSI Research Division
Atsugi 243-01, JAPAN

Abstract

The understanding of low thermal budget transient diffusion and activation of shallow p^+ implants remains a crucial issue of technology simulation. In this paper, we apply a coupled point defect assisted diffusion/precipitation model to the redistribution phenomena observed during annealing of shallow boron implants. Comparison with experimental data shows that the precipitation model can account for unusual segregation phenomena in the highly disordered zone around the projected range of the as–implanted profile.

1. Introduction

The complexity of high concentration boron activation and transient enhanced diffusion (TED) has challenged the process modeling community for some years now [1, 2]. The trend towards low thermal budget and the inevitability of boron as a high dose p-dopant in e.g. PMOS transistor fabrication requires development and calibration of process simulation tools which mirror the physical processes involved. As far as TED phenomena are concerned, point defect assisted diffusion models have proved to be the most powerful approaches, because unlike empirical models they give a more consistent view of nonlocal diffusion effects. In the framework of sophisticated point defect models, TED is usually modeled making assumptions on the initial point defect supersaturation present in the crystal immediately after implantation. Dopant activation phenomena have been described in terms of nonequilibrium cluster models [3], or, most recently, by applying precipitation models to laser activated arsenic and phosphorus square profiles [4].

In our contribution, we will present the application of a nonequilibrium point defect/dopant precipitation model to the redistribution and activation of high–dose, inhomogeneous boron profiles during rapid thermal anneals (RTA) and furnace anneals.

The activation of boron is modeled by tracing the dynamics of a spatial dependent size distribution of boron clusters.

2. Model description

In particular, we solve a coupled drift–diffusion–reaction system, covering balance equations for interstitials, vacancies and dopants in various charge states. The fluxes **j** of dopant-interstitial pairs $(A_j I)^{(z)}$ and isolated interstitials I^z read

$$\mathbf{j}_{(A_j I)^{(z)}} = -D_{jIz} k_{jIz} \left(\frac{n}{n_i}\right)^{(-q_j - z)} \nabla \left[C_{A_j^{(q_j)}} C_{I^0} \left(\frac{n}{n_i}\right)^{(q_j)} \right] \tag{1}$$

and

$$\mathbf{j}_{I^{(z)}} = -D_{Iz} k_{Iz} \left(\frac{n}{n_i}\right)^{(-z)} \nabla C_{I^0}, \tag{2}$$

respectively, where C_X, D_X, n, n_i, k_{jIz} and k_{Iz} denote the concentration of species X, diffusivity of species X, electron density, intrinsic carrier density, equilibrium constants for interstitial and pair ionization, respectively. Exponents in parentheses denote charge states. Similar fluxes are used for vacancies and dopant–vacancy pairs.

For the precipitates, we solve a coupled reaction system at every gridpoint. Assuming charge conservation, the growth rate $R_{N \to N+1}$ of precipitates containing N boron atoms is

$$R_{N \to N+1} = k_f^N \left[C_N C_1 \left(\frac{n}{n_i}\right)^q - k_{eq}^N C_{N+1} \right], \tag{3}$$

where C_N, C_{N+1}, C_1, n and n_i denote the concentration of precipitates of size N, size $N+1$, isolated dopants (q-fold charged), electrons and intrinsic electron density, respectively, and k_f^N is the reaction rate. The reaction system of precipitates is written as

$$\frac{\partial C_N}{\partial t} = -R_{N \to N+1} + R_{N-1 \to N} \qquad 1 < N < N_{max} \tag{4}$$

$$\frac{\partial C_N}{\partial t} = R_{N-1 \to N} \qquad N = N_{max} \tag{5}$$

$$\frac{\partial C_1}{\partial t} = -\nabla \mathbf{j}_1 - R_{1 \to 2} - \sum_{N=1}^{N_{max}-1} R_{N \to N+1} \qquad N = 1, \tag{6}$$

where $\mathbf{j}_1$ denotes the flux of the particles of "size 1", which is a sum of fluxes of the type Eq. 1. The equilibrium constant k_{eq}^N is determined by the Gibbs–Thomson equation:

$$k_{eq}^N = \Theta \frac{C_L}{C_P} C_{sol} \exp\left(\frac{\Delta G_N}{k_B T}\right), \tag{7}$$

where C_L, C_P and C_{sol} are the concentrations of silicon lattice sites, boron atoms in a precipitate, and equilibrium boron solid solubility, respectively. ΔG_N is the size dependent excess free energy change due to the particle transition from the surrounding matrix to the precipitated phase. The factor Θ accounts for possible degeneracies.

The simulations are carried out with the latest version of TESIM [5], a one-dimensional multilayer process simulator.

3. Model assumptions

A modified "+1"–model is used as the initial damage assumption, which means that the interstitial distribution in the beginning of the simulation follows the as–implanted doping profile. We use an equilibrium clustering model for interstitials, to obtain the experimentally observed TED. Recently, small dissolving {113} stacking faults have been experimentally identified as the interstitial source at the intial stage of implantation/annealing steps, giving evidence to the interstitial cluster picture [6]. We started the simulations with a completely inactive profile, and a nuclei size distribution which corresponds to the local total as–implanted boron concentration. A concentration dependent nuclei size distribution above a certain nucleation threshold is given as a tentative explanation for unusual "up–hill" diffusion phenomena observed in the heavily disordered region around the projected range of implanted ions.

4. Results

Fig. 1 shows a comparison of experimental and calculated results of a 10s, 1000°C boron RTA experiment, with implantation conditions as shown in the insert. The RTA simulations included measured temperature ramps. The transient diffusion model works quite satisfactory, as can be seen from the diffusion tail. Fig. 2 illustrates the situation at 900°C during the ramp–up.

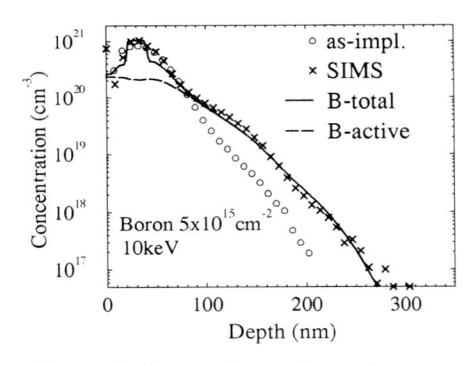

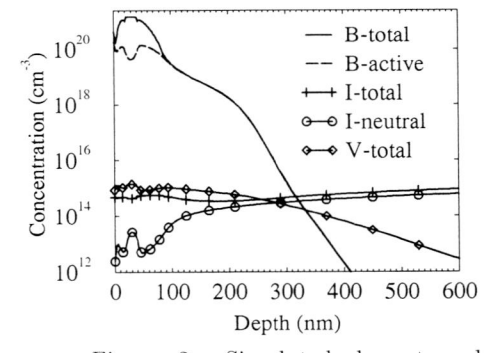

Figure 1: Comparison of experimental and simulated (lines) profiles of a 10s, 1000°C boron RTA experiment.

Figure 2: Simulated dopant and point defect concentrations during ramp–up.

The SIMS profile suggests an "up–hill" diffusion at the peak of the profile, which is reproduced by the simulation. From the simulation point of view, the reason is that large, slow–dissolving or even growing precipitates are present in this region, which acts as a boron sink. Figs. 3 and 4 show the spatial dependence of the distribution function during the ramp–up at 900°C, and after the complete thermal cycle, respectively. One can observe a dip in the active concentration (precipitates of "size 1") during ramp–up, which leads to the increase of the total concentration.

5. Conclusion

Transient enhanced diffusion and activation of shallow boron implants has been studied using a coupled pair diffusion/precipitation model. TED was described using a

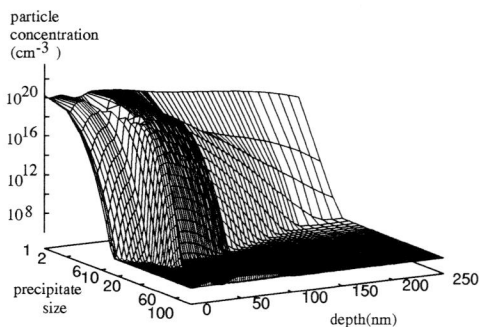

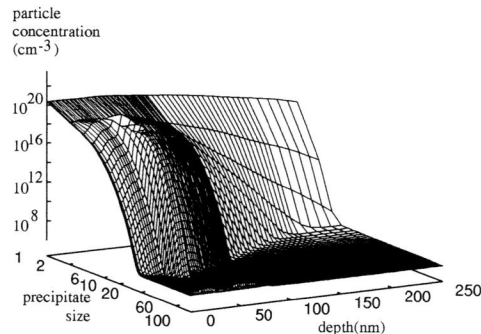

Figure 3: Spatial distribution of boron precipitates during ramp up at 900°C.

Figure 4: Spatial distribution of boron precipitates after the full RTA temperature cycle.

modified "+1"-model. Activation could be modeled successfully taking advantage of the time development of a spatially dependent size distribution function of boron precipitates.

6. Acknowledgements

This work was supported by the swiss priority program on power electronics, systems and information technology (LESIT) and Cray Research, Inc.

References

[1] S. Solmi, F. Baruffaldi, R. Canteri, "Diffusion of boron in silicon during post-implantation annealing," *J. Appl. Phys.*, vol. **69**, pp. 2135–2142, 1991.

[2] N.E.B. Cowern, K.T.F. Janssen, H.F.F. Jos, "Transient diffusion of ion-implanted B in Si: dose, time, and matrix dependence of atomic and electrical profiles," *J. Appl. Phys.*, vol. **68**, pp. 6191–6198, 1990.

[3] B. Baccus, E. Vandenbossche, "Modeling high concentration boron diffusion with dynamic clustering: Influence of the initial conditions," in *Simulation of Semiconductor Devices and Processes*, vol. **5**, pp. 133–136, 1993.

[4] S.T. Dunham, "Kinetics of dopant precipitation in silicon," in *Semiconductor Silicon/1994*, (Pennington, NJ), pp. 711–719, The Electrochemical Society, 1994.

[5] T. Feudel, *TESIM User's Guide*. ISE Integrated Systems Engineering AG, 1994.

[6] D.J. Eaglesham, P.A. Stolk, H.-J. Gossmann, and J.M. Poate, "Implantation and transient B diffusion in Si: The source of the intersititals," *Appl. Phys. Lett.*, vol. **65**, pp. 2305–2307, 1994.

Modelling of silicon interstitial surface recombination velocity at non-oxidizing interfaces

C. Tsamis and D. Tsoukalas

IMEL, Inst. of Microelectronics, NCSR "Demokritos",
15310 Aghia Paraskevi, Athens Greece

Abstract

In this work we present a model for the surface recombination velocity of silicon interstitials at non oxidizing interfaces. The model takes in to account the experimentally observed diffusion of silicon atoms through an oxide. The influence of the interfacial region in also discussed.

1. Introduction

Point defect interaction at non oxidizing interfaces is of fundamental importance for accurate simulation of thermal processes in silicon. Silicon interstitial recombination at non-oxidizing interfaces has been historically modelled as

$$F_S = \sigma_{eff}(C_I(t) - C_I^{eq}) \qquad (1)$$

where σ_{eff} is the surface recombination velocity, $C_I(t)$ and C_I^{eq} are the actual and the equilibrium concentration of interstitials at the interface. From the physical point of view this approach indicates that the interface acts as a infinite sink for interstitials. However one would rather expect that the ability of the interface to absorb interstitials is decreasing with time, probably by the filling up of the sites where the atoms can interact [1]. Experimental results obtained from backsurface oxidation-front surface stacking fault grow experiments with silicon membranes [2] indicate that σ_{eff} is a decreasing function of time, and not time independent, as it is usually assumed.

In this work we present a model for the surface recombination velocity of silicon interstitials at non oxidizing interfaces. The model accounts for the diffusion of silicon atoms through the oxide as well as the influence of the interfacial layer between the silicon and the oxide.

2. The model

In a previous work[3] we have shown that when a silicon interstitial reaches the interface with the oxide it diffuses through it. Although the exact nature of the interaction of interstitials with the oxide is not known, we face two possibilities: either silicon atoms segregate at the interface, or they react with it to form SiO molecules. Subsequently silicon atoms or SiO molecules diffuse in the oxide. It is obvious that correct modelling of the surface recombination velocity must take into account such behaviour.

We assume that silicon interstitial segregate at the silicon-oxide interface and then they diffuse in the oxide. If m is the segregation coefficient for silicon interstitials at the interface, we can express the interstitial flux at the interface as

$$F_s = \sigma_o \left(C_I^{Si} - \frac{C_I^{ox}(t)}{m} \right) \qquad (2)$$

where C_I^{Si}, C_I^{ox} are the concentrations at the interface in silicon and in the oxide respectively and σ_o is the surface reaction constant. Neglecting recombination of intestitials at the interface and assuming that interstitials diffuse in the oxide we can solve for the interstitial concentration in the oxide. For convenience we assume that interstitial concentration in silicon near the interface has a constant value and that the oxide has infinite thickness. Equating (1) and (2) and after some mathematical treatment, we find the time dependence of surface recombination velocity,

$$\sigma_{eff} = \sigma_o \exp\left(\frac{t}{t_o}\right) \text{erfc}\left(\sqrt{\frac{t}{t_o}}\right), \quad t_o = \frac{m^2 D}{\sigma_o^2} \qquad (3)$$

Fig 1 shows the time dependence of σ_{eff}. From this figure we see that surface recombination velocity remains constant for small values of t/to, and gradually decreases with time, according to a $t^{-.5}$ law. This is in agreement with the experiments with silicon membranes[2] for the silicon oxide interface for both small and large times. We note that such a time dependence can explain experimental data over a wide range of temperatures and times which predict either constant or time dependent recombination velocity. Similar time dependence is obtained if one assumes formation of SiO at the interface and diffusion in the oxide.

This above analytical model of the surface recombination velocity can be implemented in a process simulator that actually utilises eq.1, with minor code modifications. Besides that one notes that there is only one additional parameter that must be fitted properly.

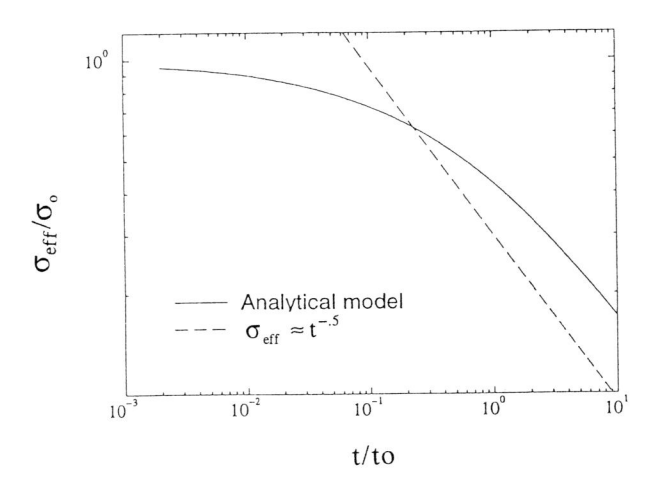

Figure 1: Time dependence of the surface recombination velocity (solid line). We note that σ_eff is constant for small values of t/to and gradually decreases with time. The dash line shows a $t^{-.5}$ dependence

However the silicon-silicon dioxide interface is much more complicated than an ideal plane that imposes boundary conditions. Experimental observations shows that there exist a buffer layer of some finite thickness, between the silicon and the oxide. The physical properties of this layer, eg width, density, stoichiometry, stress, are strongly dependent on the oxidation process as well as any other thermal treatment that takes place. The interaction of silicon atoms with this interfacial layer must have a strong influence on the surface recombination velocity.

Although the exact nature of the interaction is not known, one could assume that there exist some sites inside this layer which act as sinks or traps for silicon atoms. Assume that there exist only one type of trap. Each trap is occupied by 0 or 1 particle. Let Θ be the fraction of traps that are occupied. The diffusion equation inside the interfacial layer is modified to account for the existence of traps.

If we assume that the trapping and detrapping follows a first order kinetic with rate constants k_1 and k_2 respectively then the time evolution of traps is governed by

$$\frac{d\Theta}{dt} = k_1 C_I^{ox}(1-\Theta) - k_2 \Theta \quad (4)$$

where c_I^{ox} is the concentration of free silicon atoms in the interfacial region.

However since the problem is more complicated we can obtain only numerical solutions for the silicon / interfacial region / silicon dioxide system. Following the same procedure as above we can calculate σ_eff with the presence of traps. A typical result is shown in Fig.2 in comparison with the simple case discussed previously. The influence of the traps will be more pronounced when the flux of silicon atoms into the traps is larger compared to the diffusion in the oxide. The time dependence of surface recombination velocity in this case becomes more complicated and is dependent on the trap parameters, which are determined by the properties of the interface. However, even in this case eq. 3 could provide a fairly good approximation.

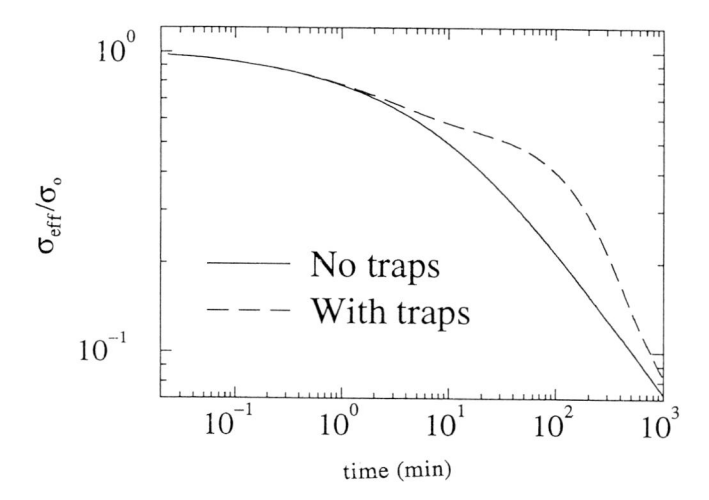

Figure 2: Surface recombination velocity as a function of time with and without traps.

The above analysis is applicable to interfaces of silicon with other materials, eg. silicon nitride. Since silicon atoms have low diffussivity in the nitride, the recombination kinetics will be be strongly dependent on the interfacial layer properties. In such cases the surface recombination velocity could fall faster with time, depending on the trap parameters. A $t^{-1.2}$ dependence has been observed for this case [2].

The authors acknowledge Dr. P. Normand of NCSR "Demokritos" for stimulating discussions.

References

[1] S. T Ahn, P.B Griffin, J. D. Scott, J. D. Plummer and W.A. Tiller, J. Appl. Phys. **62**, 4745 (1987)
[2]. W. Boyd Rogers and Hisham Z. Massoud, J. Electrochem. Soc., **138**, 3483 (1993)
[3]. D. Tsoukalas, C. Tsamis and J. Stoemenos, Appl. Phys. Lett. **63**, 3167 (1993)

Efficient Hybrid Solution of Sparse Linear Systems

A. Liegmann[a], K. Gärtner[b], W. Fichtner[a,b]

[a]Integrated Systems Engineering AG,
Im Dornacher 8, 8127 Forch, SWITZERLAND
[b]Institut für Integrierte Systeme,
ETH-Zentrum, Gloriastr. 35, 8092 Zürich, SWITZERLAND

Abstract

During a numerical simulation usually many linear systems have to be solved. Using a direct method requires to factor each coefficient matrix separately. In this paper we present a hybrid approach which combines our supernodal direct solver with an iterative solver such that the iterative solver is called as often as possible to avoid the computationally expensive factorizations.

1. Motivation

In the last few years the research efforts towards efficient solutions of sparse linear systems using direct or iterative methods have made significant progress. Unfortunately, it seems impossible to find an universal method which is optimal with respect to memory consumption and computational speed for all types of problems. Therefore, so-called *hybrid* methods have been considered which use a combination of usually stand-alone solution techniques. The idea behind such hybrid methods is that during the solution of a problem one can select the method which is known to work best on a particular phase of the solution process whereas the normal approach uses only one method for the whole solution process. Our approach uses a combination of iterative and direct techniques to solve sparse structurally symmetric linear systems as they appear in numerical semiconductor simulation.

2. The hybrid approach

Numerical semiconductor device simulation involves the solution of the discretized device equations usually by a Newton approximation algorithm where each Newton step requires the solution of a linear system describing the coupled device equations. Especially during transient simulations a large number of linear system solves are necessary in order to complete a simulation. The usual direct approach requires a factorization for each linear system to be solved. Since factoring the coefficient matrix is the most time consuming part of the solution process, it has to be avoided as often as possible. Unfortunately, direct methods provide no means to avoid the factorization. Consequently, we apply a preconditioned iterative method using a given factorization

```
call SUPER($A^{(0)}$,$x^{(0)}$,$b^{(0)}$)           else
$\overline{LU} = L^{(0)}U^{(0)}$                       call CGS($A^{(i)}$,$x^{(i)}$,$b^{(i)}$)
$\bar{x} = x^{(0)}$; $\bar{b} = b^{(0)}$               if CGS fails do
for $i = 1, 2, \ldots$ do                                call SUPER($A^{(i)}$,$x^{(i)}$,$b^{(i)}$)
   $z = \|\bar{b} - A^{(i)}\bar{x}\|/\|\bar{b}\|$          $\overline{LU} = L^{(i)}U^{(i)}$
   if $z >$ threshold do                                 $\bar{x} = x^{(i)}$; $\bar{b} = b^{(i)}$
      call SUPER($A^{(i)}$,$x^{(i)}$,$b^{(i)}$)       end if
      $\overline{LU} = L^{(i)}U^{(i)}$              end if
      $\bar{x} = x^{(i)}$; $\bar{b} = b^{(i)}$     end for
```

Algorithm 1: The supernodal hybrid solution approach.

as a preconditioner. On the other hand, we are willing to invest a factorization, if convergence of the iterative procedure is slow.

Algorithm 1 outlines the strategy of our hybrid approach which combines our sparse linear solver SUPER [1] with the preconditioned *conjugate gradient squared* (CGS) iterative method by Sonneveld [2] which is known to converge fast. As a preconditioner the most recent factorization is used. Initially, our sparse solver SUPER is called which solves the first linear system of the simulation process. Upon return from the direct solver the LU factorization of the coefficient matrix, the solution vector, and the right-hand side are saved in the data structures $\overline{LU}$, $\bar{x}$, and $\bar{b}$, respectively. There are used to estimate the norm of the matrix difference $\|A^{(i)} - \overline{LU}\|$. All further linear systems are solved by either SUPER or CGS. The iterative process is started only if condition

$$threshold \geq \|\bar{b} - A^{(i)}\bar{x}\|/\|\bar{b}\| \tag{1}$$

is satisfied. Parameter *threshold* denotes an upper limit for the relative residual norm of the current coefficient matrix and the solution and right-hand side vectors from the last exact solution. This relative residual norm is a measure how much the two vectors $\bar{b}$ and $A^{(i)}\bar{x}$ differ. If the relative residual norm is small, one can conclude that the numerical values of $A^{(i)}$ and $\overline{LU}$ do not differ too much so that CGS is expected to converge fast using $\overline{LU}$ as the preconditioner. If condition (1) does not hold, SUPER is called; otherwise CGS is invoked. On the other hand, even if condition (1) is satisfied, CGS is not guaranteed to succeed. In this case SUPER is called to compute an exact solution.

3. Preconditioned CGS

Because we want to solve sets of linear equations with continuous parameter dependent coefficient matrices, we expect the iterative procedure to be invoked only for cases with eigenvalues of the iteration matrix sufficiently close to zero. If this is true, CGS will converge fast; otherwise, one observes large oscillations in the residual norm [3]. This fact is used to interrupt the iterative procedure and to calculate a new factorization. Furthermore, we restrict the number of iterations by the parameter *maxiter* which is computed as the quotient of the measured times for a factorization and for the first CGS iteration (initially, *maxiter* is set to 1). After the first iteration *maxiter* is adjusted according to the measured times and the following formula:

$$maxiter = \left\lfloor \frac{T_{factorization}}{T_{first\_CGS\_iteration}} \times c \right\rfloor, \quad c \leq 1. \tag{2}$$

procedure CGS
$x = 0;\ r = b$
for $i = 1$ **to** $maxiter$ **do**
$\quad \rho_1 = b^T r$
$\quad$ **if** ($\rho_1 = 0$) return failure
$\quad$ **if** $i = 1$ **do**
$\quad\quad u = r;\ p = r$
$\quad$ **else**
$\quad\quad \beta = \rho_1/\rho_2;\ u = r + \beta q$
$\quad\quad p = u + \beta q + \beta^2 p$
$\quad$ **end if**
$\quad \rho_2 = \rho_1$
$\quad$ solve $\overline{LU}\hat{p} = p$
$\quad \hat{v} = A\hat{p};\ \alpha = \rho_1/(b^T\hat{v})$
$\quad q = u - \alpha \hat{v}$
$\quad$ solve $\overline{LU}\hat{v} = u + q$
$\quad u = A\hat{v};\ x = x + \alpha \hat{v}$
$\quad r = r - \alpha v$
$\quad$ **call** $check\_convergence(i)$
end for
end procedure

Algorithm 2: The preconditioned CGS algorithm.

In other words, $maxiter$ is set to reflect how many CGS iterations can be executed, not exceeding the time required for a full factorization.

After the necessary vectors and scalars have been computed, CGS requires to solve two linear systems. At this point, the preconditioner comes into play. Recall from the previous section that our preconditioner is the LU factorization of the most recent exact solve which is denoted as $\overline{LU}$. Consequently, only forward and backward substitution are required to solve the systems $\overline{LU}\hat{p} = p$ and $\overline{LU}\hat{v} = u + q$. This improves the computational efficiency of the CGS algorithm significantly.

Eventually, the solution vector x and the residual r are updated. The relative residual tolerance $z = \|r\|/\|b\|$ is then used to check convergence of the preconditioned CGS method. Depending on this value, it is decided whether CGS is considered to have failed, the solution is found, or another CGS iteration has to be performed.

At this point of the CGS algorithm, the method is considered to have failed in the following cases:

1. $z > maxnorm$
   If the relative residual norm exceeds the predefined limit given by the parameter $maxnorm$, the CGS process is expected to converge too slowly.

2. $i = maxiter/2 \wedge z > \sqrt{mintol}$
   Parameter $mintol$ specifies the accuracy z has to achieve so that the linear system is considered as solved. If we have not reached an accuracy of $\sqrt{mintol}$ after half of the iterations allowed, convergence is considered to be too slow.

3. $i \geq maxiter \wedge z > mintol$
   If CGS has not been able to solve the system with accuracy $mintol$ after $maxiter$ iterations, the iterative process is stopped.

These failure criteria are used to optimize the behavior of CGS in a way that performance loss is minimized if the CGS process has to be stopped and replaced by a direct solve.

4. Performance of the hybrid solver

Figure 1 displays the effect of our hybrid approach on the overall execution time for a 3D transient simulation of an IGBT. The simulation was performed on an

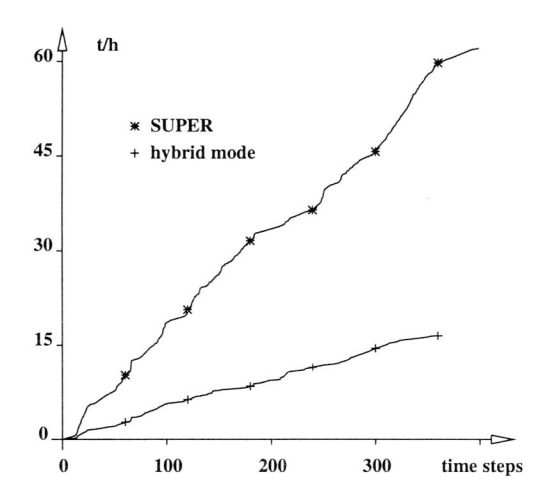

Figure 1: The effect of using a hybrid approach on the overall execution time for a 3D transient simulation of an IGBT (grid size: 7790 points) on an IBM RS/6000-590. The upper curve denoted with ∗, depicts the accumulated time when SUPER is used exclusively. The lower curve marked with + shows the accumulated simulation time using the hybrid approach.

IBM RS/6000-590 workstation. The grid size of this device was 7790 points (i.e. one coupled linear system has $n = 3 \times 7790 = 23370$ unknowns). The upper curve, marked with ∗, shows the accumulated execution time of the simulation when only the direct sparse supernodal solver SUPER was used. Here, the simulation required more than 60 hours wall clock time to complete. The lower curve, marked with +, displays the significantly smaller execution time of the same simulation using the hybrid approach. In this case, the simulation was finished after 16 hours wall clock time. This means, using the hybrid approach we were able to speed up the simulation by almost a factor of four! The reason for this speedup is that out of the 4819 linear systems solved during the simulation only 446 factorizations were required. The remaining systems could be solved by CGS.

References

[1] A. Liegmann and W. Fichtner, "The application of sparse supernodal factorization algorithms for structurally symmetric linear systems in semiconductor device simulation", In S. Selberherr, H. Stippel, and E. Strasser, editors, *Simulation of Semiconductor Devices and Processes*, vol. 5, pp. 77–80, Springer-Verlag, 1993.

[2] P. Sonneveld, "CGS, a fast Lanczos-type solver for nonsymmetric linear systems", *SIAM Journal on Scientific and Statistical Computing*, vol. 10, pp. 36–52, 1989.

[3] H.A. van der Vorst, "Lecture notes on iterative methods", Technical Report No. 838, Department of Mathematics, University, Utrecht, 1993.

Mesh Generation for 3D Process Simulation and the Moving Boundary Problem

S. Bozek, B. Baccus, V. Senez and Z.Z. Wang

IEMN Département ISEN,
41, Boulevard Vauban, 59046 Lille Cedex, FRANCE

Abstract

This paper presents the concepts of a mesh generation technique for 3D process simulation involving structure deformation. One of the main problems is the displacement of boundaries leading to a (complete) remeshing of the structure, large cpu times and complexity of the algorithms. Our approach, based on Delaunay criterion, tetrahedral elements and triangular faces, allows local remeshings of the structure.

1. Introduction

Physical phenomena applied to a structure can be divided in two sets : the ones which do not modify the shape of the structure and those which induce moving boundaries. Dopant diffusion belongs to the first set while oxidation or silicidation belong to the second one. In numerical simulation, moving boundaries induce severe constraints on the mesh generation [1]. The very large number of nodes and elements needed for realistic 3D process simulations prevents the use of a strategy that resorts to complete remeshing at each time step, due to the cpu time. As a result, it is desirable a) to define the parts of the structure that really need to be meshed and b) to investigate algorithms based on local mesh updates. Let us focus our attention on the oxidation phase (fig. 1).

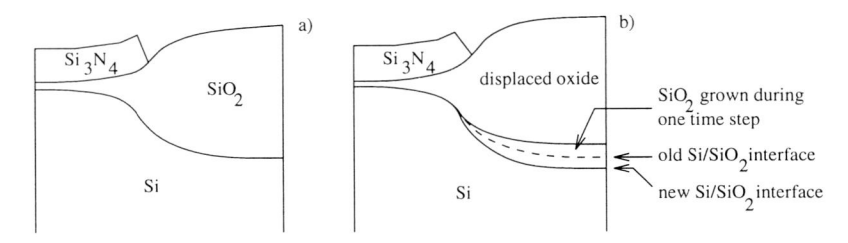

Figure 1: Schematic representation of the mechanical problem for local oxidation (LOCOS) with a) the initial structure, b) the structure after oxidation.

After one time step, the new oxide displaces the initial oxide layer upwards and the Si/SiO$_2$ interface downwards. If we assume that the displaced oxide is always well triangulated, the only part of the oxide to mesh is the new narrow band which appeared (fig. 1b). As may be seen on fig. 2, there is only a local zone of the silicon layer which is affected by Si/SiO$_2$ interface displacement. This thin band, in comparison with the Si layer, also needs to be remeshed.

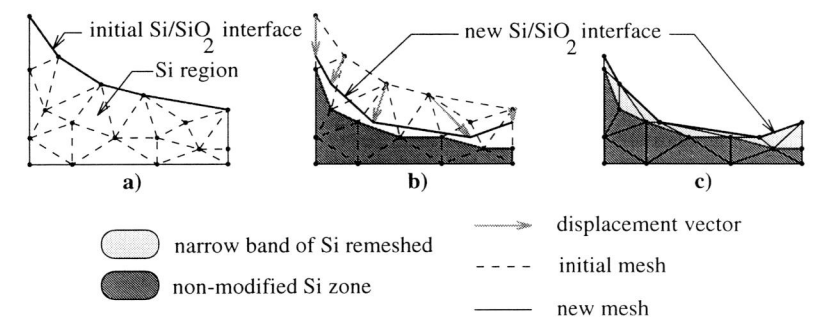

Figure 2: Displacement of the Si/SiO$_2$ interface and local remeshing of the Si layer.

As a result, this approach requires especially the meshing of narrow bands at each time step, with the following basic requirements : a) complete respect of the faces describing the hull of the region and b) no insertion or deletion of boundary nodes. While efficient methods have been reported for 3D device simulation (e.g. in [2]), they usually rely on octree, tensor-product or intersection-based algorithms, which are generaly not compatible with these requirements. In this paper, we present the basic concepts designed for this purpose and based on tetrahedral mesh generation.

2. Basic principles

In this second part, we briefly present the concepts that lead to the generation of a mesh from a set of given points [3, 4]. In our case, those points belongs to the hull of a local zone to mesh. First, the main stages are presented. Then, the local mesh modification produced by a node insertion is detailed.
Let P be the set of boundary points of R^3 and F the set of triangular faces describing the hull of the region. The algorithm can be divided in three steps. At first, a set T of tetrahedra is calculated, defining the convex hull of P. Next, the external elements are removed from T via F, in order to restore the original hull. At last, internal nodes are inserted into the mesh to fit physical needs and geometrical quality. Let's focus our attention on the first stage. Let T_j be the Delaunay-triangulated polytope (convex polyhedron) built with P_j, the j first nodes of P, and F_j the external faces of T_j. T_{j+1} is derivated from T_j by a local remeshing using a) p_{j+1}, b) the set of tetrahedra of T_j with their circumscribed sphere including p_{j+1}, and c) the elements of F_j which define a plane that strictly separates p_{j+1} and T_j (that means that p_{j+1} is on one side of the plane and all the P_i points, $i \leq j$, are on the other side or in this plane). More precisely, three cases can be considered (fig. 3). In the first case, the inserted node p is in the meshed polytope T_j (fig. 3a). Local remeshing is performed by deleting the tetrahedra with their circumscribed sphere including p, and by creating new elements with the faces of the hull defined by those elements and p. In the second case, p is out of any circumscribed sphere. Then, new tetrahedra are created with p and the separating external faces of the polytope T_j. In the third case, p is out of T_j but inside some circumscribed spheres. The mesh is updated with the

non-separating faces of the hull defined by the tetrahedra with their circumscribed sphere that includes p, and the separating external faces of T_j.

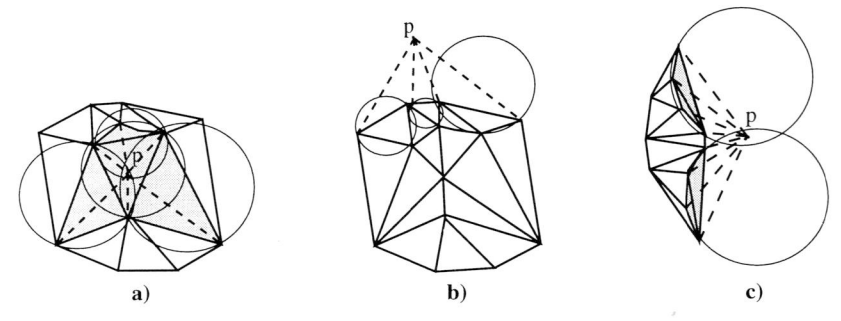

Figure 3: Different derivations from T_j to T_{j+1} using the inserted node p, after [3].

The internal node insertion is only a sub-case of the convex hull calculation, i.e. the case where p_{j+1} is inside T_j.

3. Implementation

The mesh generator has been implemented in C++ language. Although the C++ code is generaly less efficient at run-time than fortran code, the main advantages of an oriented-object approach consist in security of code and quickness to develop or modify part of program. Special attention has been devoted to the implementation of the algorithms, in particular the different cases of fig. 3. Indeed, it is very sensitive to rounding errors which can produce erroneous results.

4. Application

Hereafter, we give two first results issued from the above method. The figure 4 shows the mesh generated for a narrow band as can be obtained from the strategy reported in fig. 1-2. Despite the slow variation of the slopes, no points are inserted on the boundary, as would have generally been the case with strategies used for 3D device simulation. As can be seen, the external tetrahedra have been removed and no internal nodes have been introduced. The region contains 152 nodes, 309 tetrahedra and 300 faces.

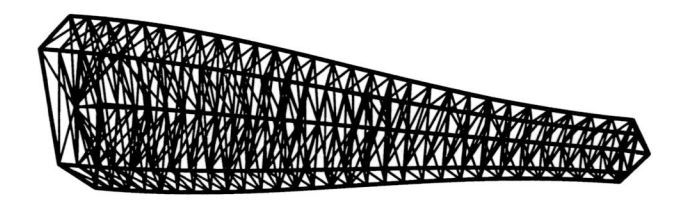

Figure 4: 3D mesh of a band as arising from the oxidation step.

The figure 5 is the result of the refinement of an initial cubic region with node insertion in accordance with the doping profile variation. The decrease in element quality is minored by the use of the local remeshing technique reported in fig. 3a. This example contains 33,793 tetrahedra, 5912 nodes and 1422 faces.

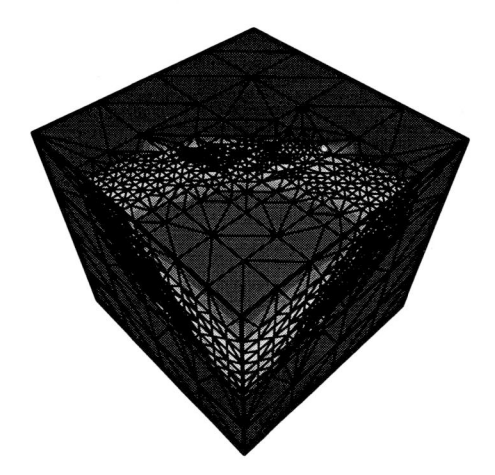

Figure 5: Arsenic contours and 3D meshing during a diffusion step, including a refinement procedure based on node insertion.

5. Conclusion

A strategy aimed in limiting the zones to be remeshed during the oxidation steps has been presented. A concept of mesh generation has been summarized and first results have been given.

6. Acknowledgements

This work is part of PROMPT (JESSI project BT8B) and was funded as ESPRIT project 8150. The authors would like to thanks ISE-AG for the use of PICASSO. Helpful discussions with P.L. George are also gratefully acknowledged.

References

[1] M.E. Law, "Grid generation for three-dimensional non-rectangular semiconductor devices", In *Proc. of SISDEP 5 Conf.*, pp. 1-8, 1993.

[2] N. Hitschfeld, *Challenges to Achieving Accurate Three-Dimensional Process Simulation* , PhD thesis, ETH Zurich, 1993. Published by Hartung-Gorre Verlag, Konstanz, Germany.

[3] P.L. George, F. Hermeline, "Delaunay's mesh of a convex polyhedron in dimension d. Application to arbitrary polyhedra", *Int. J. Num. Meth. Eng.*, vol. 33, pp. 975-995, 1992.

[4] P.L. George, F. Hecht, and E. Saltel, "Automatic 3D mesh generation with prescribed meshed boundaries", *IEEE Trans. Magn.*, vol. 26, no. 2, pp 771-774, 1990.

Three-Dimensional Grid Adaptation Using a Mixed-Element Decomposition Method

E. Leitner and S. Selberherr

Institute for Microelectronics, TU Vienna
Gusshausstrasse 27-29, A-1040 Vienna, Austria

Abstract

A new method for adaptive refinement of unstructured grids has been developed. This method ensures preservation of the element quality and the structural anisotropies of the initial grid. The flexibility of unstructured grids in combination with the implemented error estimation provides a powerful basis for three-dimensional simulation of time dependent processes.

1. Introduction

Efficient and accurate simulation of transient three-dimensional redistribution processes requires adaptive gridding methods. The computational grid is responsible for both the accuracy of the solution as well as for the simulation efficiency. In order to meet with these requirements throughout a transient simulation, the grid has to be adapted as the distribution of the solution changes.

One approach to solve this problem is to start with a coarse initial grid and to adapt the local grid density by means of recursive element refinement. On one hand the low required density of the initial grid allows a fast generation and, on the other hand, the recursive refinement can also be carried out efficiently. Thus, the overall computational effort for the grid handling is kept low.

2. The mixed-element decomposition method

Our adaptation algorithm starts from a coarse unstructured grid, which resolves the computational domain and may consist of tetrahedrons and octahedrons. In contrast to octree based methods (e.g. [1]) the alignment of the grid elements is not restricted to a rectangular bounding box. Thus, also oblique interfaces and boundaries can be resolved optimally. The elements of this initial grid are refined recursively until the desired accuracy is reached. As the diffusion advances further adaptation steps may be required. Then already refined elements are either refined again or replaced by their parent element to achieve the required grid density.

For a recursive refinement algorithm it is important to preserve the essential grid properties, i.e., the grid quality and the structural anisotropies. Therefore we developed the mixed element decomposition method:

We divide a tetrahedron into four tetrahedra of the same shape and one octahedron. The four tetrahedra are located at the parent's corners and the octahedron is placed in the center (Fig. 1). An octahedron is divided into six octahedra of the same shape and eight tetrahedra. The six octahedra are located at the parent's corners and the remaining tetrahedral parts have a common node in the center (Fig. 2). In order to discretize an octahedron, we split it into eight tetrahedron, each of which has one face of the octahedron as ground plane and the octahedral center as opposite node.

To evaluate the effectivity of the method, it is of interest, how much the grid quality is decreased by the refinement. In order to compare the element quality of the elements generated by the refinement with the element quality of the parent element, we use (1) as a measure for the element quality, where V is the volume and h_{max} is the maximum size of the element (see [2]).

$$Q_e = \frac{V}{h_{max}^3} \qquad (1)$$

The first refinement step introduces elements with a new aspect ratio. The elements generated by all following refinement steps have either the shape of the tetrahedra or the shape of the octahedra which exist after the first refinement step (see Fig. 3). Thus, the element quality is affected only by the first refinement step.

Taking into account the discretization of the octahedron permits a reasonable comparison of the octahedron and the tetrahedron: we compare the element quality of the tetrahedral parent with the element quality of the tetrahedra used for discretization of the octahedron. It can be shown, that the degradation of the element quality after (1) is limited to a factor of 1/2 for the tetrahedron and 1/4 for the octahedron.

As the refinement is always done locally, unrefined elements may be adjacent to refined ones. These neighbouring elements are called incompatible elements, and we define the order of incompatibility as the difference of the refinement levels of two adjacent elements. In our algorithm the order of incompatibility is restricted to one. A two dimensional example of such an incompatible situation is shown in Fig. 4. In order to estimate the grid quality at a compatible node between incompatible elements, we use $Q_n = \min(V_i)/\max(V_i)$, where V_i are the volumes of all elements incident to this node (see [2]). It can be shown, that the degradation of this nodal grid quality is limited to a factor of 1/4 for the tetrahedron and 1/8 for the octahedron.

3. Error Estimation

For the practical use of the mixed element decomposition method, an error estimation was implemented, which is based on a gradient smoothing of a finite element solution (see [3]). It allows to compute the gradient error as well as the local dose error. We use a linear combination of both as grid density criterion, where the weights can be chosen independently. All elements which are not reaching the desired accuracy are refined. On the other hand, elements with a very small discretization error are replaced by their parent elements (coarsening).

A proper discretization of the incompatible elements within one parent has to account for the C_0-continuity condition, which is a common requirement for the standard finite-element method. The function values for the incompatible node are determined by the interpolation equation which is the shape function of the parent element. For consistency reasons the matrices for the elements which are incident to the incompatible node are preassembled locally, and the equations for the incompatible nodes are replaced using the interpolation function. This results in a reduced matrix which we assemble to the global system matrix.

4. Example

We applied our algorithm to the silicon block of a conventional LOCOS-structure (Fig. 5). Firstly, we performed a Boron channel-implant, which we computed by a Monte-Carlo ion implantation simulation module[4] with an energy of 20keV and a dose of 1e14cm^{-2}. Then we adapted the initial grid according to the Boron profile, where the discretization error limit was set to 3% for the dose error and to 10% for the gradient error. Fig. 6 shows the resulting grid which consists of 9146 elements and 4285 nodes.

5. Conclusion

The mixed-element decomposition method combines the high flexibility of fully unstructured grids and the fast adaptation capability through recursive element refinement. From the shape preserving property it follows, that our algorithm preserves the boundaries and interfaces, and the structural anisotropy of the grid. Additionally, the quality degradation caused by the algorithm is limited to a constant factor. Thus, our grid adaptation method provides a powerful basis for three-dimensional process simulation.

Acknowledgement

This work is supported by Digital Equipment Corp., Hudson, USA; and Philips B.V., Eindhoven, The Netherlands.

References

[1] N. Hitschfeld and W. Fichtner, "3D Grid Generation for Semiconductor Devices Using a Fully Flexible Refinement Approach", In S. Selberherr, H. Stippel, and E. Strasser, editors, *Simulation of Semiconductor Devices and Processes*, pages 413–416. Springer-Verlag Wien New York, September 1993.

[2] R.E. Bank, *PLTMG: A Software Package for Solving Elliptic Partial Differential*, SIAM, Philadelphia, 1990.

[3] O.C. Zienkiewicz, *The Finite Element Method*, McGraw-Hill, 1989.

[4] W. Bohmayr and S. Selberherr, "Trajectory Split Method for Monte Carlo Simulation of Ion Implantation Demonstrated by Three-Dimensional Poly-Buffered LOCOS Field Oxide Corners", In *Int.Symposium on VLSI Technology, Systems, and Applications*, Taipei, 1995.

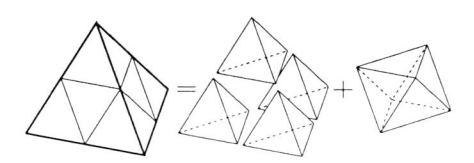

Figure 1: Tessellation for a tetrahedron

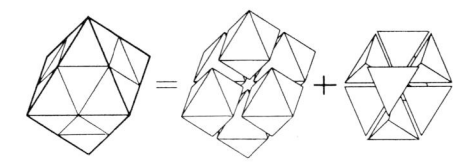

Figure 2: Tessellation for an octahedron

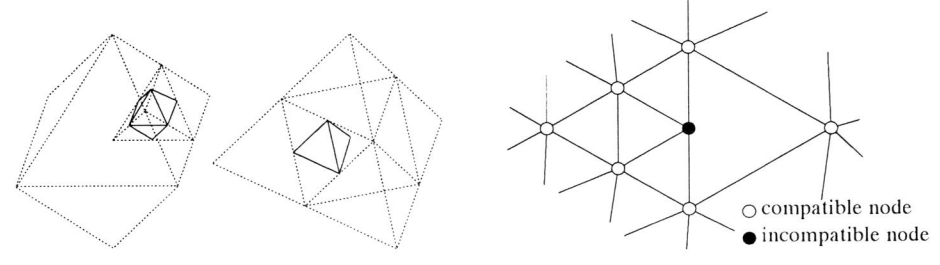

Figure 3: Shape preservation for recursive refinement

Figure 4: Incompatible elements

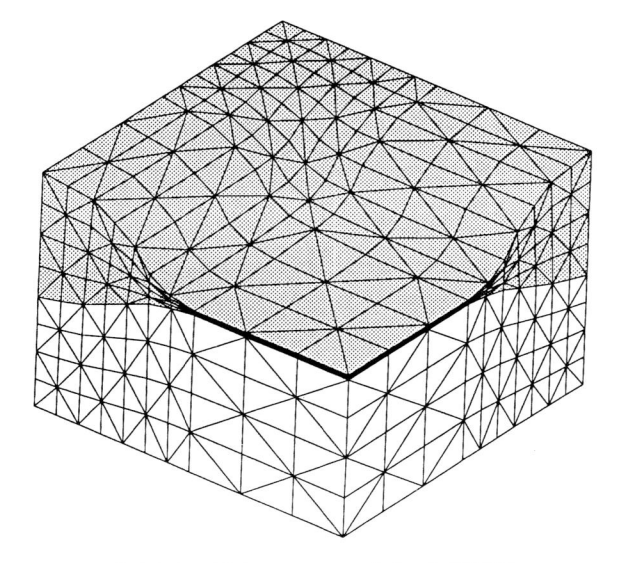

Figure 5: Initial grid for corner of the LOCOS structure

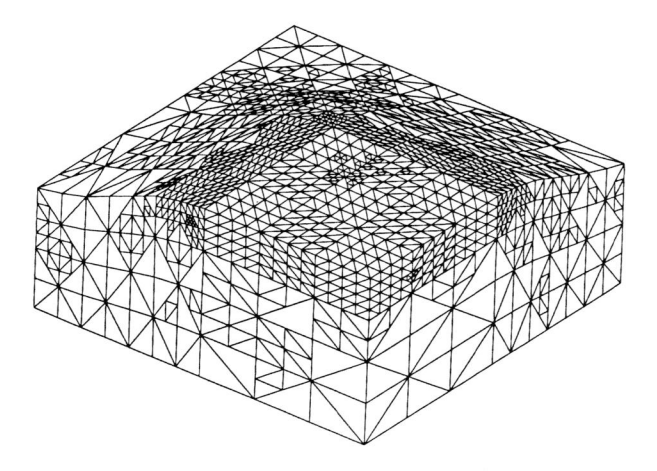

Figure 6: Grid of the silicon block adapted to the implanted Boron profile

Unified Grid Generation and Adaptation for Device Simulation

G. Garretón   L. Villablanca   N. Strecker   W. Fichtner

Integrated Systems Laboratory, ETH–Zürich, Switzerland
Phone: +41 1 632 49 50, FAX: +41 1 252 09 94
E-mail: gilda@iis.ee.ethz.ch

Abstract

This paper describes the design and development of a dimension-independent grid generator suitable for device simulation. The purpose of this work is to describe a modular, flexible and dimension-independent approach for the generation of grids with complex boundary restrictions.

1. Introduction

The increasing complexity of modern semiconductor devices and the need for powerful simulation have led to stringent requirements for grid generation. Without doubt, the grid becomes one of the most critical issues in the device simulation environment. A proper and suitable mesh is the key for success in any device simulation.

In the past, several techniques have been applied based on structured and unstructured meshes. For structured meshes, qtrees in 2-D and octrees in 3-D are the typical techniques. Delaunay-type algorithms, bisection-type approaches and advancing-front methods have been utilized for unstructured meshes.

Some of the techniques used for generating grid suitable for device simulation include octrees, modified 2-4-8 trees [1] and mixed elements [2] grid generators. However, many complex non-planar non-convex geometries can not be treated with these concepts.

Complex geometries could be treated if a set of basic properties is considered in the generation of the grid. Among others, these properties are an intersection-based algorithm, n-irregular elements and the construction of the first coarse grid for complex devices (initial grid) [3].

In the past few years, we have developed a series of grid generators for one, two and three dimensions (GRID1D [4], MESHBUILD [5] and OMEGA [6]). Our experience with these grid generators together with our successful research on those basic properties have motivated the unification of mesh-building algorithms in a modular and dimension-independent procedure, presented on this paper.

The basic steps and algorithms described here have been designed and grouped in a set of exchangeable modules. A modular implementation allows us to incorporate new basic elements and properties without losing generality. These modules can be arranged depending on the grid requirements for the specific application. This contribution describes: 1. A set of important properties and concepts to build up a powerful and flexible grid generator. 2. The effort spent on the design and development of MESH-LIB, a modular and dimension-independent grid generator. 3. Suitable meshes for the device simulator DESSIS [7] and the thermo-mechanical simulator SOLIDIS [8] obtained using MESH-LIB. 4. Preliminary results for 3-D modules.

2. General Concepts for an unified grid generator

From a large series of different 1-D, 2-D and 3-D examples, we are able to define some key elements which must be present in order to have success with any arbitrary geometry. These elements allow us to define the general algorithms to be implemented as the kernel of the tool (see Fig. 1).

These elements or properties have been identified as: 1. Construction of an initial grid. 2. Intersection-based and iterative algorithms. 3. The choice of the proper set of basic elements to use. 4. Adequate representation and adaptation to user inputs. 5. Templates management for boundary conditions and point propagations. 6. The definition of final grid properties depending on the next application to be used (Delaunay meshes for process simulators or Box-Method conditions in device simulators).

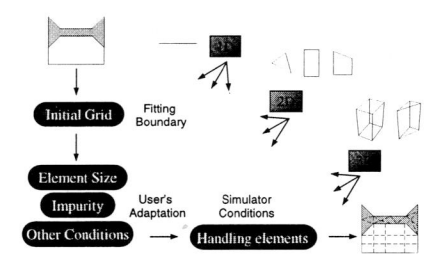

Figure 1: General overview of an unified and modular grid generator MESH-LIB.

3. Construction of the initial grid

The initial grid is the first coarse mesh which exactly fits the boundaries of the geometry. The quality of final grids depends strongly on this initial grid. Tensor-product grids or octree approaches are not sufficient to handle complex geometries.

We propose a set of heuristics to fit the given boundary description using a set of segments in 1-D, polygons in 2-D and polyhedra in 3-D.

4. Grid Adaptation

The conditions for grid adaptation are given by desirable points densities according to the supplied density functions. At this point, suitable oriented meshes (meshes with locally anisotropic point densities) are required.

Grid adaptation can be performed conforming to data obtained either from process or device simulation. Results in 1-D and 2-D confirm that the adequate accuracy and quality can be achieved using this technique (see Fig. 3.a).

5. Intersection-based Algorithms

In order to avoid the shortcomings of bisection-based approaches, intersection-based algorithms allow to refine elements at any arbitrary point. This property adds flexibility to grid generation since the best point can be chosen at each refinement step. The approach is useful for readaptation because it permits better mesh orientation. A good intersection-based algorithm reduces the overall point propagation considerably.

In contrast to previous grid generators [2], MESH-LIB handles n-connected elements which are the base for intersection-based algorithms [3]. These elements allow to have more than one neighboring element per edge in 2-D or per face in 3-D.

6. The Proper Set of Basic Elements and Templates

The main concept is to manage different types of elements according to the need of a particular physic problem. To illustrate this with an example in two dimension, the actual version of MESH-LIB has the flexibility to generate triangles and arbitrary quadrilaterals. For an application in semiconductor device simulation, this generality has been reduced to triangles and rectangles. For the multi-dimensional thermo-mechanical simulator SOLIDIS [8], MESH-LIB produces quadrilaterals but not triangles.

After the initial grid generation and grid adaptation, it is necessary to create the final elements from the n-connected elements. This last step is typically called *handling green points* (non-vertex points) or *handling n-connected elements*. The appropriate use of templates in this step allows to limit point propagation and to avoid redundant grid points in inactive regions of the device.

7. Comparison Between Two Approaches for the Initial Grid in 3-D

The Figure 2 shows the difference of having a tensor-product as initial grid as in OMEGA and having a intersection-based ones as in MESH-LIB. A simple Manhattan geometry displays the redundant refinements using tensor-product approach.

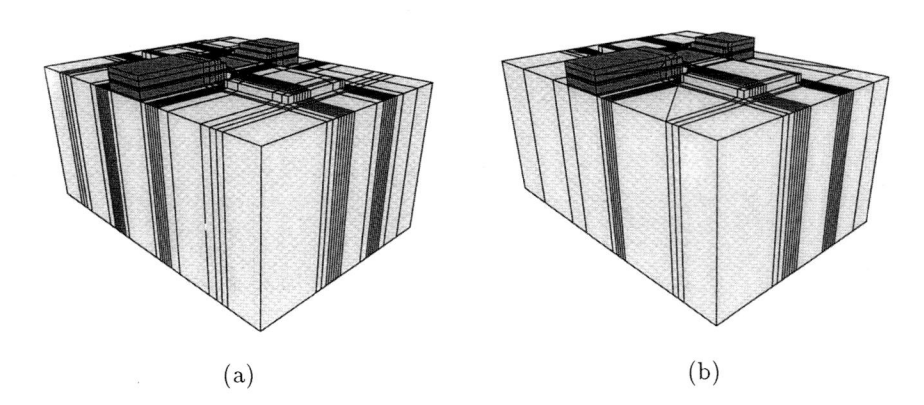

(a) (b)

Figure 2: (a) Tensor-product approach: 1254 elements. (b) Intersection-based approach: 805 elements

And finally the Figure 3.b shows part of a complex 3-D EEPROM. This geometry can not be handled using the tensor-product available in OMEGA. The new approaches in MESH-LIB allow us to fit more complex geometries as the ones shown in 3.b.

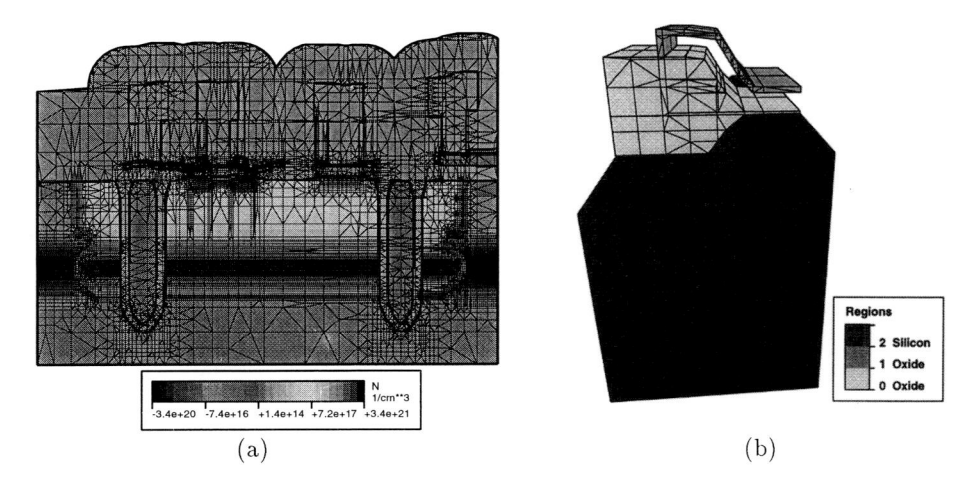

Figure 3: (a) 2-D adaptation according to data from process simulation obtained by MESH-LIB. (b) Fitting a complex 3-D geometry with MESH-LIB: 948 elements.

Acknowledgments

This work has been financially supported by the ESPRIT-6075 (DESSIS) Project and by the ESPRIT-8150 (PROMPT) Project.

References

[1] P. Conti, M. Tomizawa, and A. Yoshi, "Generation of oriented Three-Dimensional Delaunay Grids Suitable for the Control Volume Integration Method," *Int. J. Numer. Methods Eng.*, vol. 37, pp. 3211–3227, 1994.

[2] N. Hitschfeld, *Grid Generation for Three-Dimensional Non-Rectangular Semiconductor Devices*. PhD thesis, ETH Zurich, Switzerland, 1993. publ. by Hartung-Gorre Verlag, Konstanz, Germany.

[3] G. Garretón, L. Villablanca, N. Strecker, and W. Fichtner, "A new approach for 2-d mesh generation for complex device structures," in *NUPAD V - Technical Digest*, (Honolulu, USA), June 1994.

[4] ISE Integrated Systems Engineering AG, *GRID1D - User's Guide*, 1994.

[5] S. Müller, K. Kells, and W. Fichtner, "Automatic Rectangle-based Adaptive Mesh Generation without Obtuse Angles," *IEEE Trans. Computer Aided Design*, vol. 11, pp. 855–863, July 1992.

[6] ISE Integrated Systems Engineering AG, *OMEGA - User's Guide*, 1994.

[7] ISE Integrated Systems Engineering AG, *DESSIS - Manual*, 1994.

[8] J. Funk, J. Korvink, M. Bächtold, G. Garretón, and H. Baltes, "An efficient simulation toolbox for coupled field analysis," in *XV CILAMCE Proc.*, (Belo Horizonte, Brasil), pp. 181–189, 30. 11 - 2. 12 1994.

Platinum Diffusion at Low Temperatures

M. Jacob[a], P. Pichler[a], H. Ryssel[a,b], and R. Falster[c]

[a]Fraunhofer-Institut für Integrierte Schaltungen,
Schottkystrasse 10, 91058 Erlangen, GERMANY
[b]Lehrstuhl für elektronische Bauelemente, Universität Erlangen-Nürnberg,
Cauerstrasse 6, 91058 Erlangen, GERMANY
[c]MEMC Electronic Materials SpA,
Viale Gherzi 31, 28100 Novara, ITALY

Abstract

In a series of experiments, the diffusion of platinum in silicon was investigated at low temperatures in the range from 700 °C to 800 °C. Depth profiles measured in as-grown float zone (FZ) silicon were found to differ strongly from those in as-grown Czochralski (CZ) silicon. These differences as well as different profiles measured in various FZ wafers after identical processing can be attributed to different initial concentrations of intrinsic point defects. In general, platinum depth profiles were found to agree qualitatively with the predictions of standard diffusion theories, but not quantitatively. Therefore, parameters for platinum and point defect diffusion reported in the literature were modified to describe consistently the diffusion in both kinds of material.

1. Introduction

Platinum forms deep electronic levels in silicon which act as recombination centers for carriers. Because of this feature platinum is used to adjust lifetime in fast switching diodes or thyristors. Another application of platinum in silicon technology is the formation of ohmic and Schottky contacts with platinum or platinum silicide, but the unintended diffusion of the transition metal may affect device performance. For both cases, quantitative knowledge of the diffusion of platinum in silicon is required. On the one hand, if platinum is used to adjust lifetime it is necessary to determine the most adequate diffusion conditions like diffusion time and temperature, and on the other hand, for the use of platinum as contact material one has to prevent unintended contamination of the silicon substrate.

It is generally assumed that platinum diffuses in silicon predominantly as interstitial atoms. In equilibrium, however, platinum atoms occupy mainly substitutional sites where they are assumed to be immobile. The change of an impurity from an interstitial position to a substitutional site can take place by two different reactions: Either, an interstitial platinum atom recombines with a lattice vacancy (Frank-Turnbull mechanism [1]), or, it generates a self-interstitial when occupying a substitutional site (kick-out mechanism [2]). Based on these considerations, a system of partial differential equations for the substitutional platinum atoms, the interstitial platinum atoms, the silicon self-interstitials, and the vacancies can be derived [3] describing the redistribution of the platinum atoms and the intrinsic point defects.

Prabhakar et al. [4] investigated the diffusion of platinum into silicon from PtSi/Si interfaces at temperatures between 300 °C and 800 °C. Mantovani et al. [5] performed similar experiments with epitaxially grown silicon substrates at temperatures between 700 °C and 850 °C. In both cases, the observed concentrations of substitutional platinum after diffusions at 700 °C and 750 °C for half an hour in depths up to a few microns were around 10^{12} cm^{-3}. The depth profiles in the region close to the surface were interpreted using the kick-out mechanism. On the contrary, Zimmermann et al. [3] found much higher concentrations on the order of 10^{14} cm^{-3} after diffusions at 700 °C and 770 °C. They concluded that for temperatures below approximately 850 °C, the Frank-Turnbull or dissociative mechanism governs platinum diffusion. Our goal was to investigate the diffusion of platinum at low temperatures and for short times in different kinds of silicon substrates, namely in FZ and CZ silicon.

2. Experimental

The experiments were performed on as-grown, dislocation-free, (100)-oriented p-type silicon substrates. Both, FZ wafers with diameters of 100 mm, thicknesses of 535 μm, and resistivities of 4-6 Ωcm, and CZ wafers with diameters of 150 mm, thicknesses of 675 μm, and resistivities of 60 Ωcm were used. One- and double-sided diffusion was investigated. The back-sides of the wafers for investigation of one-sided diffusion were covered by low-temperature oxide. After a short dip in fluoric acid and water rinsing, platinum was deposited at the oxide-free surfaces with a thickness of the order of one atomic layer to prevent point-defect generation by silicidation. Drive-in diffusions were performed in a horizontal furnace in nitrogen ambient. Furnace and sample holder were preheated to guarantee that the wafers reach the nominal process temperature as fast as possible. Already four minutes after loading, the sample temperature was only 4 °C below the final value. Depth profiles of substitutional platinum were obtained from DLTS measurements on bevelled specimens.

3. Results

In general, in as-grown, commercially available FZ wafers, platinum concentrations around 10^{14} cm^{-3} were measured after short time and low temperature diffusion.

In one of these experiments, platinum was diffused from the front side into three different FZ wafers at 730 °C for 20 min (Fig. 1). Remarkable is that the profiles do not correspond although the processing conditions were identical. This indicates a non-negligible influence of the initial concentrations of the intrinsic point defects. Also, it was not possible to reproduce the shape of the platinum distributions by numerical simulations with the parameters for point defect and platinum diffusion given by Zimmermann and Ryssel [3].

Simulations revealed that platinum diffusion in FZ wafers at the low temperatures of this investigation is dominated by the Frank-Turnbull mechanism. One of the main reasons for the major discrepancy between measured profiles and simulations is that Zimmermann and Ryssel underestimated drastically the Frank-Turnbull reaction constant because this parameter was not as important for their investigations. Besides this reaction constant, simulations were found to be sensitive predominantly to the initial concentration of vacancies, the product of interstitial platinum diffusion

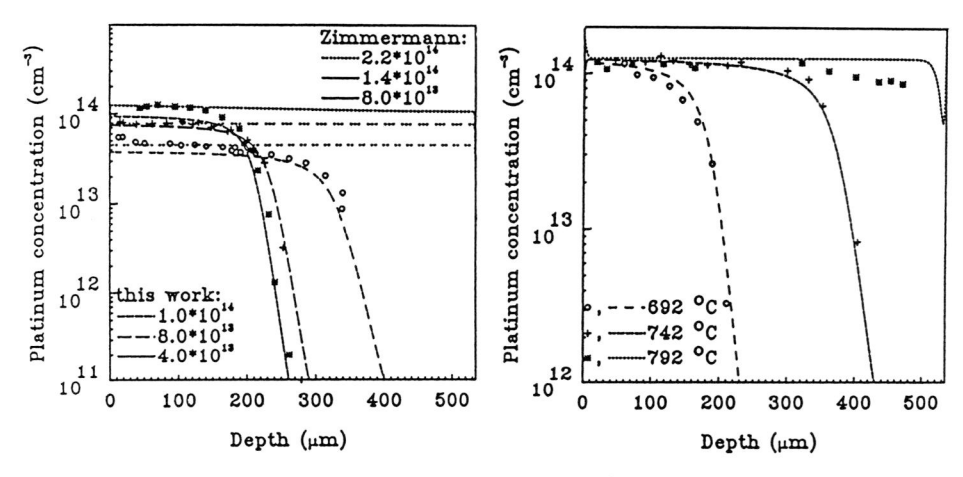

Fig. 1 One-sided diffusion in three different FZ wafers at 730 °C for 20 min

Fig. 2 One-sided diffusion for 1 h in FZ silicon

coefficient and interstitial platinum equilibrium concentration, and to the quotient of vacancy equilibrium concentration and substitutional platinum equilibrium concentration. The latter two parameters were determined from various experiments whereas the initial concentration of vacancies was assumed to vary from wafer to wafer. The Frank-Turnbull forward reaction was found to be diffusion limited with a reaction radius of 1 nm. The simulated results were obtained with the following parameters:

$k_{FT} = 1.89 \cdot 10^{-8} \exp(-0.604 \text{ eV}/kT) \text{ s}^{-1}$

$D_i \cdot C_i^{eq} = 6.62 \cdot 10^{19} \exp(-2.52 \text{ eV}/kT) \text{ cm}^{-1}\text{s}^{-1}$

$C_V^{eq}/C_s^{eq} = 3.28 \cdot 10^{-7} \exp(1.05 \text{ eV}/kT)$

Here and below, the symbols D and C^{eq} denote diffusion coefficients and equilibrium concentrations, k_{FT} stands for the Frank-Turnbull reaction constant, and the subscripts i, s, and V refer to interstitial and substitutional platinum and vacancies. From the profiles in Fig. 1, the initial concentrations of vacancies in the three investigated samples were estimated as $1 \cdot 10^{14}$ cm^{-3}, $8 \cdot 10^{13}$ cm^{-3}, and $4 \cdot 10^{13}$ cm^{-3}. Such different initial concentrations of vacancies can be explained easily by different thermal histories of the wafers during cooling after zone melting.

In Fig. 2, the influence of diffusion temperature on one-sided platinum diffusion is shown for processes of 1 h duration. All three samples were quarters from the same FZ wafer. The initial concentration of vacancies was estimated to be $1.3 \cdot 10^{14}$ cm^{-3} consistently. Due to the low temperature, the vacancies are nearly immobile. For steady state, because of the expected constant initial concentration of vacancies, a constant platinum concentration is expected which depends predominantly on the initial concentration of vacancies and only slightly on temperature. At the highest temperature, local equilibrium is nearly reached after 1 h.

In contrast to FZ wafers, platinum diffusion experiments at 730 °C and 780 °C with as-grown CZ silicon wafers resulted in much lower platinum concentrations with U-shaped depth profiles (Fig. 3, Fig. 4). Such profiles are typical for a dominance of the kick-out mechanism. The platinum concentration in the middle of the wafer was found to increase proportionally to the square root of diffusion time. Simulations at

730 °C shown in Fig. 3 used the product $C_s^{eq} \cdot D_I \cdot C_I^{eq} = 5.9 \cdot 10^{14}$ cm^{-4}s^{-1}, whereas at 780 °C, a value of $C_s^{eq} \cdot D_I \cdot C_I^{eq} = 5.6 \cdot 10^{16}$ cm^{-4}s^{-1} was used. The symbol I refers to self-interstitials. The dominance of the kick-out mechanism indicates that the initial concentration of vacancies did not influence the platinum diffusion in the investigated CZ wafers. A value of $C_V(t=0) = 3 \cdot 10^{10}$ cm^{-3} was used for the presented simulations. Therefore, the concentration of vacancies in these as-grown CZ silicon wafers was at least three orders of magnitude lower than in as-grown FZ silicon. Simulations with various values for the initial self-interstitial concentration showed a slight influence on the resulting platinum concentration in the bulk. The best fit was obtained by $C_I(t=0) = 3 \cdot 10^{10}$ cm^{-3}, but this value can only be interpreted as a rough estimation.

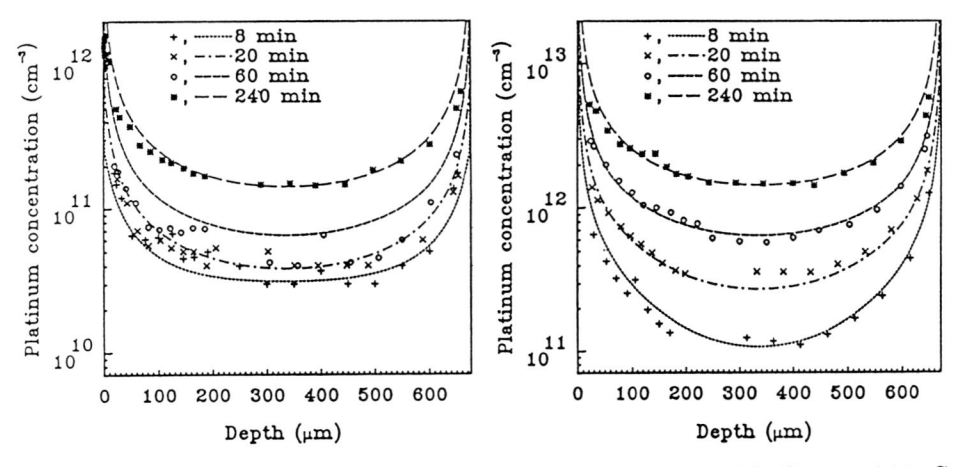

Fig. 3 Diffusion in CZ silicon at 730 °C Fig. 4 Diffusion in CZ silicon at 780 °C

4. Conclusions

Platinum diffusion in as-grown FZ and CZ silicon was investigated in the low temperature range from 700 °C to 800 °C. In FZ silicon it was found that the Frank-Turnbull mechanism dominates the platinum diffusion because of high initial vacancy concentrations in the range from $4 \cdot 10^{13}$ cm^{-3} to $2 \cdot 10^{14}$ cm^{-3}. In contrast, in CZ silicon wafers the vacancy concentrations were so low that the kick-out mechanism dominates the platinum diffusion. We conclude that the initial vacancy concentration is the key to understand the different diffusion behavior in FZ and CZ silicon.

References

[1] F. C. Frank, D. Turnbull, *Phys. Rev.* 104, 617 (1956)
[2] U. Gösele, W. Frank, A. Seeger, *Appl. Phys.* 23, 361 (1980)
[3] H. Zimmermann, H. Ryssel, *Appl. Phys. A* 55, 121 (1992)
[4] A. Prabhakar, T. C. McGill, M-A. Nicolet, *Appl. Phys. Lett.* 43, 1118, (1983)
[5] S. Mantovani, F. Nava, C. Nobili, G. Ottaviani, *Phys. Rev. B* 33, 5536 (1986)

Lattice Monte-Carlo Simulations of Vacancy-Mediated Diffusion and Implications for Continuum Models of Coupled Diffusion

S. T. Dunham and C. D. Wu

Department of Electrical, Computer and Systems Engineering
Boston University, Boston, MA 02215, USA

Abstract

In this paper, we analyze the interactions of dopants with vacancies using Lattice Monte-Carlo simulations and find that the assumptions underlying pair diffusion models lead to several corrections to the standard continuum models for coupled dopant/defect diffusion. Specifically, we find that at high doping levels, both the effective pair diffusivity as well self-diffusion due to vacancies increase rapidly with doping level due to the interactions of vacancies with multiple dopants. In addition, we find that pair diffusion theories overestimate the dopant flux resulting from gradients in the vacancy concentration.

Pair diffusion models have been very effective in modeling the coupled diffusion of dopants and point defects in silicon [1, 2, 3, 4]. However, the structure of the silicon lattice and the fact that the vacancy and the dopant move in opposite directions during exchanges, leads to a violation of the basic underlying assumption – that a dopant and a vacancy move together as a tightly-coupled pair. Lattice Monte-Carlo (LMC) simulations as used in this work provide a powerful tool for investigating vacancy-mediated dopant diffusion and provide information on how to modify pair diffusion models to better account for the underlying atomistic behavior.

The LMC simulations involve the hopping of vacancy atoms on a doped silicon lattice, with site exchanges with dopant atoms resulting in dopant diffusion. The vacancy hopping probabilities are biased by the dopant/vacancy interaction potential leading to the formation and diffusion of dopant/vacancy pairs. Since diffusion of a dopant/vacancy pair on the diamond lattice requires the vacancy to move away to at least a third-nearest-neighbor (3NN) distance, we consider a dopant/vacancy interaction out to 3NN sites. For the first-nearest neighbor binding energies, we used the experimental lower bounds for arsenic (1.23 eV) or phosphorus (1.04 eV) from Hirata et al. [5]. Since data is unavailable for the second and third-nearest neighbor binding energies, we simply used 2/3 and 1/3 of the nearest neighbor energy, respectively.

At very high doping levels, a vacancy is likely to interact with more than one dopant at a time. The net result is to reduce the activation energy for pair diffusion associated with third to second-nearest neighbor transitions, thereby increasing the dopant diffusivity. This behavior has been observed experimentally for group IV and V dopants in phosphorus-doped silicon by Nylandsted Larsen et al. [6]. Figure 1 illustrates how

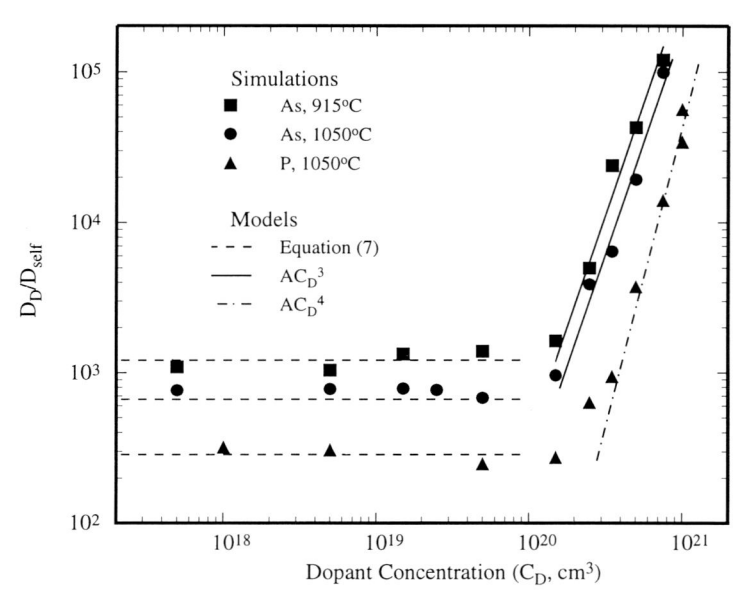

Figure 1: *Normalized diffusivity versus doping density using the pair binding energy for arsenic and phosphorus [5]. At high concentrations, the normalized diffusivity is approximately proportional to C_D^n with $3 < n < 4$ as indicated. Also shown for comparison is the analytic predictions for diffusivity in moderately doped material [7].*

dopant diffusivity as obtained from the LMC simulations varies as a function of doping level. The normalized diffusivity is nearly uniform at low and moderate doping levels and then rises rapidly for doping levels above about $2 \times 10^{20} \mathrm{cm}^{-3}$, manifesting an approximate third or fourth power dependence on the doping level. The results of the simulations agree well with the experimental observations of Nylandsted Larsen *et al.* [6], with a good match to both the doping level at which the onset of enhanced pair diffusion is observed and the dependence of diffusivity with increasing concentration (4th or 5th power dependence on doping once Fermi level effects are included).

The interaction of vacancies with multiple dopants can also be expected to increase self-diffusion via vacancies ($D_V C_V^*$) beyond just Fermi level effects, since vacancies can potentially travel long distances by transferring from one dopant to another. We examined vacancy density and displacement during LMC simulations and found that indeed the increase in self-diffusion via vacancies with doping level substantially exceeds that due simply to pair diffusion. Fig. 2 plots the components of self-diffusion via vacancies versus doping level as derived from the simulation results. The LMC simulations do not distinguish between paired and unpaired vacancies (indeed, vacancies switch during the course of the simulation), so within the pair diffusion formalism, we can break up the vacancy displacement into two parts, one part which is associated with the motion of dopant/vacancy pairs, and a second part which is due to the increase in loosely-bound vacancies which can move from the neighborhood of one dopant to another (equivalent to reduction in the quasi-vacancy formation energy [1]). The initial increase in vacancy displacement is primarily due to the displacement of loosely-bound vacancies, with the diffusion of pairs dominating at very high doping levels.

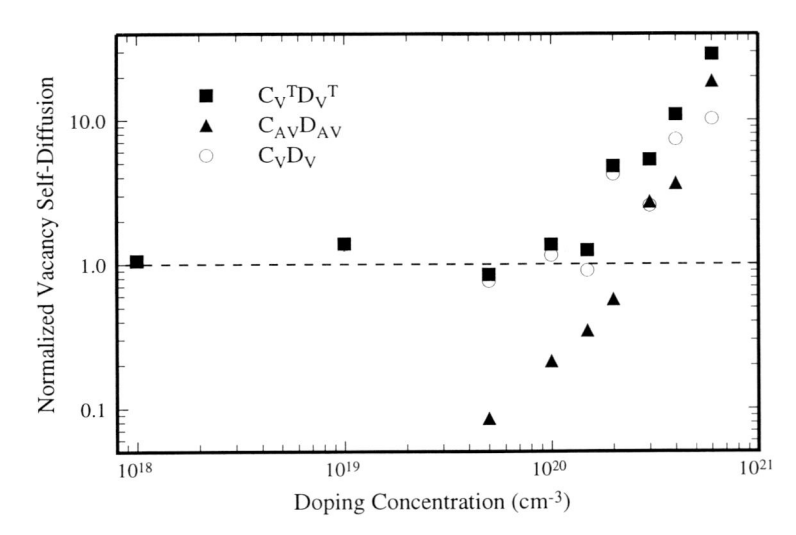

Figure 2: Vacancy self-diffusion normalized by its value in intrinsic material, with Fermi-level effects ignored, as a function of doping concentration for arsenic at $1050°C$ (or phosphorus at $850°C$). Plotted are the total vacancy displacement, the vacancy displacement due to pair diffusion and the vacancy displacement due to diffusion of unpaired vacancies, which is calculated from the difference of the first two quantities.

The formation of a high-concentration plateau in phosphorus diffusion profiles has been previously been accounted for by both enhanced pair diffusion [2, 3, 8, 9] and enhanced self-diffusion via vacancies [1]. This work shows that both effects can in fact be expected to operate in such systems. The combination of the two effects leads to a more pronounced plateau and sharper kink than with either of the effects acting alone.

In addition to examining diffusion in homogeneous systems, we also investigated the effect of a vacancy gradient on the motion of dopants via LMC. Pair diffusion models predict that the flux of dopants due diffusion via vacancies is given by:

$$J_A^V = -D_{AV} K_{A/V} \left(C_V \nabla C_A + C_A \nabla C_V \right), \qquad (1)$$

while simple vacancy diffusion theories [10] imply that the the sign of the second term is reversed since a dopant/vacancy exchange results in the dopant and vacancy moving in opposite directions. A vacancy gradient was sustained by adding an energy discontinuity along the $z = 0$ plane. The LMC simulations show that

$$J_A^V = -D_{AV} K_{A/V} \left(C_V \nabla C_A + \gamma C_A \nabla C_V \right), \qquad (2)$$

where γ is a function of the dopant/vacancy interaction potential. For the binding energies used in this work, we find $\gamma \cong 1/3$ as illustrated in Fig. 3, consistent with the value of 0.55 calculated by List et al. [11] for a stronger 3NN interaction.

In summary, Lattice Monte-Carlo simulations show that the coupled diffusion of vacancies and dopants deviates from ideal pair-diffusion behavior. At high doping levels, the interactions of vacancies with multiple dopants leads to rapid increases of both dopant diffusion and self-diffusion via vacancies at high doping levels. These effects

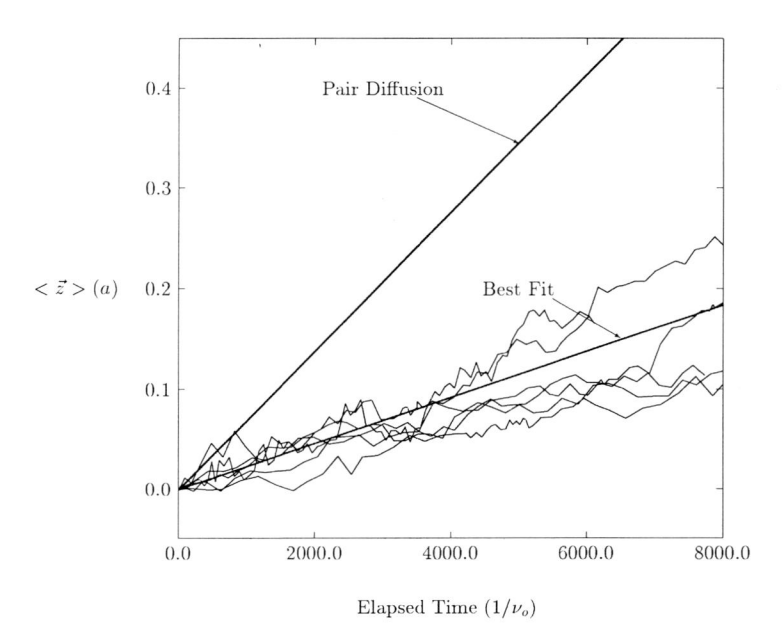

Figure 3: *Average dopant displacement versus time in the presence of a vacancy gradient.*

lead to the plateau observed in the high concentration region of phosphorus diffusion profiles. At moderate doping levels, the fact that vacancies and dopants cannot diffuse together as a tightly coupled pair leads to a reduction in the flux of dopants in a vacancy gradient relative to that predicted by pair-diffusion theories.

This work was supported by SRC/SEMATECH and NSF.

References

[1] M. Yoshida, *Jap. J. Appl. Phys.* **22**, 1404 (1983).
[2] D. Mathiot and J. C. Pfister, *J. Appl. Phys.* **55**, 3518 (1984).
[3] S. T. Dunham, *J. Electrochem. Soc.* **139**, 2628 (1992).
[4] P. M. Fahey, P. B. Griffin and J. D. Plummer, *Rev. Mod. Phys.* **61**, 289 (1989).
[5] M. Hirata, M. Hirata and H. Saito, *J. Phys. Soc. Jap.* **27**, 405 (1969).
[6] A. Nylandsted Larsen, K. Kyllesbech Larsen, P. E. Andersen, B. G. Svensson, *J. Appl. Phys.* **73**, 691 (1993).
[7] S. T. Dunham and C. D. Wu, **NUPAD V Proceedings**, 101 (1994).
[8] D. Mathiot and J. C. Pfister, *J. Phys. Lett.* **43**, L-453 (1982).
[9] D. Mathiot and J. C. Pfister, *J. Appl. Phys.* **66**, 1970 (1989).
[10] K. Maser, *Expt. Tech. Phys.* **39**, 169 (1991).
[11] S. List, P. Pichler and H. Ryssel, *J. Appl. Phys.* **76**, 223 (1994).

A New Hydrodynamic Equation for Ion-Implantation Simulation

Shiroo Kamohara, Megumi Kawakami

Semiconductor Development Center,
Semiconductor & Integrated Circuit Div., Hitachi, Ltd.,
1-280, Higashi-koigakubo Kokubunji, Tokyo 185, Japan

Abstract

In this paper, we propose a new hydrodynamic equation that reduces the simulation CPU time of the ion implantation to the complicated multi-layer structures. To simplify introduction of the hydrodynamic equation we introduced a BTE that includes the Fokker-Planch collision term and transformed this BTE into a hierarchical structure by moment expansion. We successfully shortened the CPU time by a factor of more than 100.

1. Introduction

As device structures become more complicated, implanted ions have become subject to more complicated physical phenomena, in which these ions pass through complicated multi-layer structures at an arbitrary angle and energy. To simulate these phenomena in two dimension (2D), it is necessary to change the simulation method from an analytical approach [1] using the Boltzmann transport equation (BTE) to other approaches. One is the Monte Carlo approach[2,3] and the other is solving numerically the five dimensional BTE [4,5,6]. However, both of these approaches require a prohibitive amount of the CPU time. Depending on circumstances, the CPU time for ion implantation can dominate the total process simulation.

In this paper, we propose a new hydrodynamic equation for analyzing the complicated physical phenomena that also reduces the CPU time for ion implantation simulation. Using this equation, we successfully shortened the CPU time by a factor of more than 100 by reducing the analysis space from a four-dimensional phase space to a two-dimensional coordinate space for 2D simulation.

2. BTE Including the Fokker-Plank Collision Term

Rosenbluth and his co-workers introduced the BTE for a two-body collision problem, in which two particles interact with the coulomb potential [7]. Based on this method, a two-body collision problem with the Moliere potential interaction is expressed as

$$\frac{\partial f}{\partial t} = (L_x + L_v) f, \qquad (1)$$

$$L_x = -v_i \frac{\partial}{\partial x_i}, \qquad (2)$$

$$L_v = \Gamma_a N_b \frac{m_a}{m_{ab}} \frac{\partial}{\partial v_i} \frac{v_i}{g^3} + \frac{\Gamma_a N_b}{2} \frac{\partial}{\partial v_i \partial v_j} \left(\frac{g^2 \delta_{i,j} - v_i v_j}{g^3} \right), \qquad (3)$$

where m_{ab}, g, and G_a are expressed as

$$m_{ab} = \frac{m_a m_b}{m_a + m_b}, \qquad (4)$$

$$g = \sqrt{\sum v_i^2}, \qquad (5)$$

$$\Gamma_a = \frac{8\pi Z_a^2 Z_b^2 e^2}{m_a^2} \int \phi(s) ds. \qquad (6)$$

Here, f is the phase space distribution function of the implanted ion, v is the velocity component, N is the concentration of the atom, m is the mass, Z is the atomic number, e is the elementary charge, ϕ is the collision function originating from the Moliere potential [8], t is the time, and x is the space component. Here, the subscripts a and b express the implanted ion and the atom in the target, respectively, and the subscript i, which obeys the Einstein rule, expresses the direction. While the L_v term of Eq. (3) describes only nuclear collisions, implanted ions also suffer electronic collisions. An electronic collision term can be added to Eq. (3) using the same analogy as the nuclear one. Assuming electronic collisions are not extremely dominant during the ion movement, g - which corresponds to the energy - changes slower than each of the velocity components. Under this assumption, we can place g outside of the differential operator. Then, we can approximately transform Eq. (3) into Eqs. (7) expressed as

$$L_v \approx \gamma \frac{\partial}{\partial v_i} v_i + \gamma_n k \frac{\partial}{\partial v_i \partial v_j} \left(g^2 \delta_{i,j} - v_i v_j \right), \qquad (7)$$

where

$$k = \frac{1}{2} \frac{m_{ab}}{m_a}, \qquad (8)$$

$$\gamma = \gamma_n + \gamma_e, \qquad (9)$$

$$\gamma_n = \Gamma_a N_b \frac{m_a}{m_{ab}} \frac{1}{g^3} = \frac{m_b}{m_{ab} m_a} \frac{S_n(g)}{g}, \qquad (10)$$

$$\gamma_e = \frac{m_b}{m_{ab} m_a} \frac{S_e(g)}{g}. \qquad (11)$$

Here γ is the collision cofficient, S is the stopping power, and δ is the Kronecker delta function. The subscript e and n, respectively denote an electronic collision and a nuclear collision. Using Eqs. (1), (2), and (7), we can analyze the time evolution of the implanted ion in the phase space. This BTE which describes the implanted ion dynamics makes the moment expansion the same as that generally used in fluid dynamics.

3. Hydrodynamic Equation for Ion Implantation

To reduce the analysis time of our BTE, we expand the BTE by the moments. The zero order moment is obtained from the v_1 and v_2 integration of Eq. (1). The 1st order moment of v_1 is obtained from the v_1 and v_2 integration of Eq. (1) after multiplying v_1 by both sides of Eq. (1). The higher order moments are also obtained in the same way. The integration does not operate on γ and γ_n, because these parameters change more slowly than the change of the phase space distribution function under our assumption. By these moment expansions, we obtained the simultaneous differential equation expressed as,

$$\frac{\partial C_{0,0}}{\partial t} = -\frac{\partial C_{1,0}}{\partial x_1} - \frac{\partial C_{0,1}}{\partial x_2}, \tag{12}$$

$$\frac{\partial C_{1,0}}{\partial t} = -\gamma C_{1,0} - \frac{\partial C_{2,0}}{\partial x_1} - \frac{\partial C_{1,1}}{\partial x_2}, \tag{13}$$

$$\frac{\partial C_{0,1}}{\partial t} = -\gamma C_{0,1} - \frac{\partial C_{1,1}}{\partial x_1} - \frac{\partial C_{0,2}}{\partial x_2}, \tag{14}$$

$$\frac{\partial C_{1,1}}{\partial t} = -2\gamma C_{1,1} - 2\gamma_n k C_{1,1} - \frac{\partial C_{0,1} C_{2,0}/C_{0,0}}{\partial x_1} - \frac{\partial C_{1,0} C_{0,2}/C_{0,0}}{\partial x_2}, \tag{15}$$

$$\frac{\partial C_{2,0}}{\partial t} = -2\gamma C_{2,0} + 2\gamma_n k C_{0,2} - \frac{\partial C_{1,0} C_{2,0}/C_{0,0}}{\partial x_1} - \frac{\partial C_{0,1} C_{2,0}/C_{0,0}}{\partial x_2}, \tag{16}$$

$$\frac{\partial C_{0,2}}{\partial t} = -2\gamma C_{0,2} + 2\gamma_n k C_{2,0} - \frac{\partial C_{1,0} C_{0,2}/C_{0,0}}{\partial x_1} - \frac{\partial C_{0,1} C_{0,2}/C_{0,0}}{\partial x_2}, \tag{17}$$

where $C_{1,0}=<v_1>$, $C_{0,1}=<v_2>$, $C_{2,0}=<v_1^2>$, $C_{0,2}=<v_2^2>$ and, $C_{1,1}=<v_1 v_2>$. Here, subscripts 1 and 2 denote two directions in the two dimensional coordinate space. We approximate the 3rd order moment by using $C_{2,1} = C_{0,1} C_{2,0}/C_{0,0}$, $C_{1,2} = C_{1,0} C_{0,2}/C_{0,0}$, $C_{3,0} = C_{1,0} C_{2,0}/C_{0,0}$, and $C_{0,3} = C_{0,1} C_{0,2}/C_{0,0}$, where $C_{2,1}=<v_1 v_1 v_2>$, $C_{1,2}=<v_1 v_2 v_2>$, $C_{3,0}=<v_1 v_1 v_1>$ and $C_{0,3}=<v_2 v_2 v_2>$. This approximation means the fluctuation of the 3rd order moment is negligible. We verified the accuracy of this assumption by checking the distribution function in velocity space using our in-house simulator [4]. Eq. (12) corresponds to the transport of the particle density. Eq. (13) and Eq. (14) correspond to that of the momentum component density. Eq. (15), Eq. (16) and Eq. (17) correspond to that of the energy density. This new hydrodynamic equation is numerically solved using the CGS method.

4. Simulation results

To verify the accuracy of our hydrodynamic equation, we compared the simulation and experimental results for when a 2.0×10^{16} cm^{-2} dose of arsenic was implanted into a silicon substrate at 30 keV and 50 keV, respectively. Good agreement is obtained as shown in Fig. 1. The CPU time for these simulations with 1000 mesh points is 6.6 sec when using the HITACHI M880 mainframe computer. Figure 2 shows the calculation results in which a 3.0×10^{15} cm^{-2} dose of boron is implanted into a trench-shaped silicon substrate at 100 keV. The CPU time of this simulation for 2500 mesh points is 134 sec. Table 1 lists the CPU time of the previous approaches and our approach. This confirms that our approach is 100 times faster than the conventional approach by reducing the analysis space from a four-dimensional phase space to a two-dimensional coordinate space.

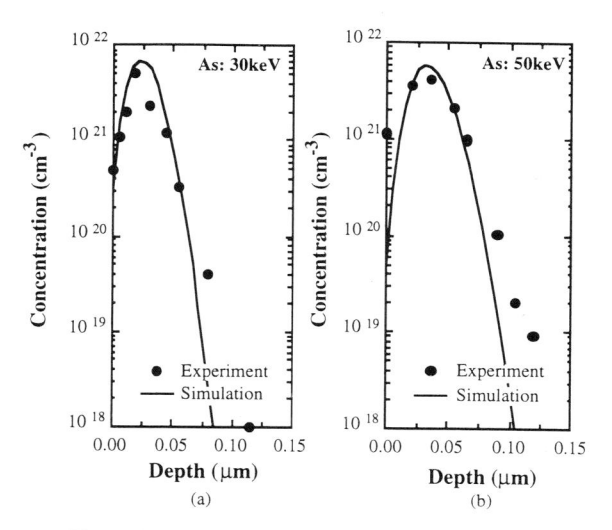

Fig. 1. Comparisons of the simulation and the experimental results for a 2.0×10^{16} cm^{-2} dose of As.

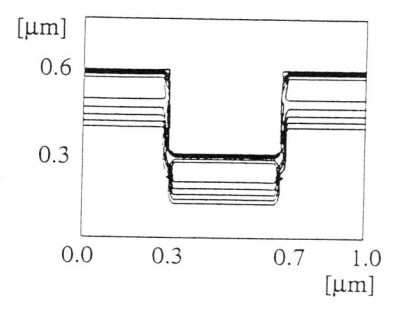

Fig. 2. Contour plot of the 3.0×10^{15} cm^{-2} dose of B implanted into a trench-shaped silicon substrate.

Table 1. CPU-time[*1] comparisons for ion implantation into a planar silicon substrate.

Method	CPU-time (min.)
This Work	0.1
Monte Carlo[*2]	20
Numerical[*3]	10

*1 : HITACHI M880 mainframe computer
*2 : Our in-house simulator.
(The method is the same as that of ref. [3])
*3 : Our in-house simulator [5].

References

[1] J. Linhard, M. Scharff and H. E. Schiott, Vol. 33, No. 14, pp. 3-42, 1963.
[2] T. Ishitani, R. Shimizu and K. Murata, Jpn. J. Appl. Phys., Vol. 11, pp. 125-133, 1972.
[3] T. Ishitani, R. Shimizu and K. Murata, Phys. stat. sol., Vol. 50, pp. 681-690, 1972.
[4] I. Saitoh, and N. Natsuaki, VLSI Process/Device Modeling Workshop, pp. 102-103, 1990.
[5] L. A. Christel and J. F. Gibbons, S. Mylroie, J. Appl. Phys., Vol. 51, pp. 6176-6182, 1980.
[6] T. Takeda and A. Yoshii, IEEE Electron Device Letters, Vol. EDL-4, pp. 430-432, 1983.
[7] M. N. Rosenbluth, W. M. MacDonald, and D. L. Judo, Phys. Rev. Vol. 107, pp. 1-6, 1957.
[8] G. P. Mueller, Radiat. Eff. Lett., Vol. 50, pp.87-92, 1980.

Monte Carlo Simulation of Multiple-Species Ion Implantation and its Application to the Modeling of 0.1μ PMOS Devices

A. Simionescu[a], G. Hobler[a], F. Laub[b]

[a] Institut für Festkörperelektronik, Technische Universität Wien,
Gußhausstraße 25-29/E362, A-1040 Wien, AUSTRIA
[b] Siemens AG, Zentralabteilung Forschung und Entwicklung,
ZFE T ME 3, 81730 München, GERMANY

Abstract

New process concepts require multiple implantations into the same volume of Si with only one annealing step after all implantations have been performed. Moreover, there is a trend towards BF2 implantation for shallow p-doping. We have extended the Monte Carlo simulator IMSIL to consider for each implantation the accumulated lattice damage of all previous implantation steps, and we have determined the parameters of the stopping power and damage accumulation models for BF2 ions. The influence of the cummulative damage on the implantation profiles and on the threshold voltage behavior is discussed for a specific PMOS device.

1. Introduction

The fabrication of ultra-small devices by ion implantation and subsequent annealing requires minimization of implantation energy and thermal budget. BF2 is an attractive implant species for p-doping, since the B atom carries only 11/49 of the energy of the BF2 molecule. Moreover, the additional damage generated by the F atoms reduces the channeling tail of the B profile. Low thermal budgets, on the other hand, result in a trend to perform an annealing step only after the last implantation. Modeling of ion implantation is challenged by these considerations in two ways: (1) Simulation of BF2 implantation in crystalline Si requires to keep track of the damage generated by both the B and the F atoms. (2) Multiple implantations into the same volume of Si require for each implantation to consider the accumulated damage of all previous implantation steps. In this paper we present for the first time Monte Carlo simulations which take into account both effects. As an application, we investigate the influence of the additional damage produced by F during the BF2 implantation on the electrical behavior of an ultra-short PMOS transistor [1] by coupled process and device simulation.

2. Implantation modeling

For implantation modeling we use our Monte Carlo simulator IMSIL [2]. IMSIL has been demonstrated to successfully predict impurity profiles not only after tilted implantations but also after implantations in all major channeling directions for different doses, in a wide energy range, and for light (B, P) as well as for heavy (As) ions [2-5]. This has been achieved primarily by appropriate models for electronic stopping and damage accumulation. Damage is taken into account using the modified Kinchin-Pease model and multiplying the amount of generated damage with a factor *frec*, modeling defect recombination [4,5]. Moreover, damage saturation is considered for B implantations. Employing the modified Kinchin-Pease model is justified since it has been shown [5] that neglecting the recoil range has only a minor effect on the dose dependence of implantation profiles due to damage accumulation, even for the heavy ion species As. The influence of damage on subsequent trajectories is considered by performing amorphous collisions with a probability proportional to the local number of displaced atoms. The recombination factor *frec* has been determined for B, P, and As as 0.125, 1, 2, respectively [2,4,5]. In this work we determine the recombination factor *frec* to be 1.2 for BF2. This value is obtained by calibrating the model with the 65 keV, $5 \cdot 10^{15}$ cm^{-2} dose implantation profile shown in Fig. 1. This result is verified by simulating the other BF2 implantations from Fig. 1 and additional low-energy channeling implantations from the literature[6,7], providing very good agreement between our simulations and the experimental data. Notice, that *frec* = 1.2 for BF2 is considerably larger than for B and even larger than for P, which is heavier than both B and F. This may be explained by damage multiplication due to the overlapping of recoil cascades generated by simultaneously implanted B and F. BF2 implantation is simulated by calculating alternatively one B and two F trajectories and recording the damage generated by each of the ions. Moreover, IMSIL has been extended for the simulation of successive implantations. The damage produced by previous implantations is regarded in the simulation of the following implantation steps. The simulator also allows the computation of implantation into arbitrarily shaped 2-D structures [2]. It was found, however, that consideration of the ions scattered out of mask sidewalls is not significant in the process considered here.

Fig. 1: B concentration distribution after 65 keV, 7° tilted BF2 implantation into (100) Si. Continuous lines: experiments [6], histograms: simulations.

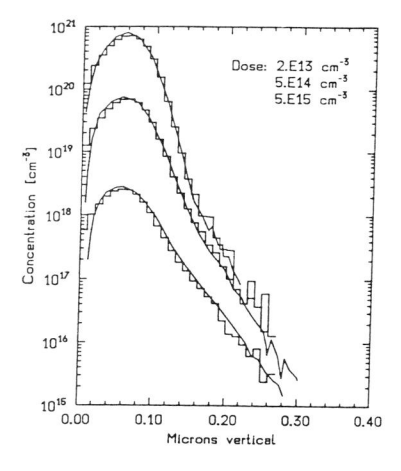

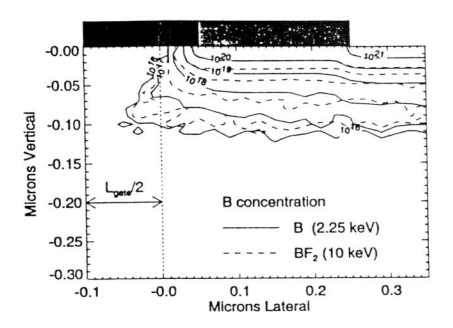

Fig.2: B profile resulting from B and BF2 S/D implantations for the 0.2μ PMOS transistor.

Fig. 3: Net doping of the 0.2μ PMOS transistor after all implantation steps have been performed. S/D implantations carried out with B (full line) or BF2 (dotted line).

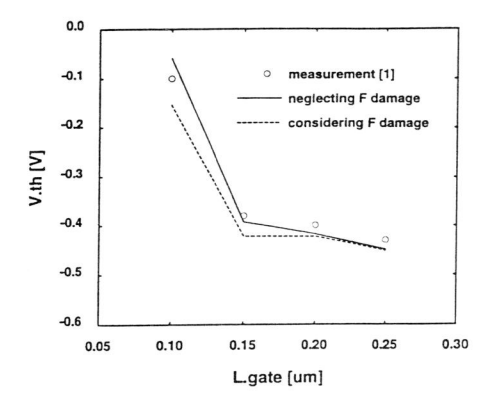

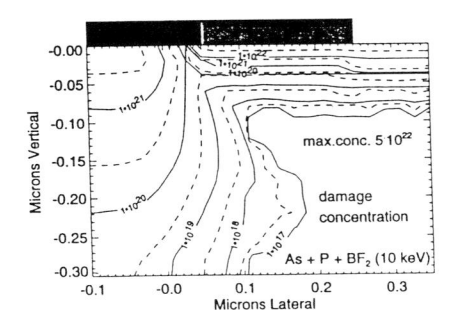

Fig. 4: Resulting threshold voltage over the gate length for the 0.1, 0.15, 0.2, and the 0.25μ PMOS transistors.

Fig. 5: Damage distribution for the 0.2μ PMOS transistor after all implantation steps have been performed. S/D implantation carried out with BF2.

3. Application to the Modeling of a 0.1μ PMOS Device

The implantation steps of the investigated process [1] have been simulated with the extended version of IMSIL. Fig. 2 shows the shape of the as-implanted B profile after the two S/D BF2 implantation steps (shallow junction: 0° tilt, $1 \cdot 10^{15}$ cm^{-2} dose, 10 keV energy, deep junction: 0° tilt, $5 \cdot 10^{15}$ cm^{-2} dose, 20 keV energy). For comparison, the B profile resulting from the similar B implantation steps with the energy reduced proportional to the mass ratio 11/49 is also shown. Notice that the B profile resulting from the BF2 implantation is shallower because of the increase in lattice damage induced by the F. Both implantation profiles show lateral channeling leading far below the gate. When looking at the net doping (Fig. 3), however, we observe that the As channel implantation ($1 \cdot 10^{13}$ cm^{-2} at 100 keV) and the P anti-punch implantation ($4 \cdot 10^{12}$ cm^{-2} at 120 keV) are chosen as to adjust the threshold voltage (V_{th}) of the 4 nm gate oxide device and to cover the B lateral channeling "nose". Regarding diffusion modeling, we mention that there is no widely accepted and confirmed model for diffusion of implanted boron at high concentrations. We obtained the information about the subdiffusion from V_{th} as a function of the gate length (L_{gate}) as measured in [1]. The time averaged boron motion during the RTA step at 1050° C for 10 sec was calibrated by a coupled process/device simulation (TSUPREM-4 [8], MINIMOS [9]) to give the same threshold roll-of as in [1] (Fig. 4). In addition these measurements show a delayed onset of the threshold roll-of. This reverse short channel effect (RSCE) was taken into account by fixed surface charges [10]. Fig. 4 shows the results for the V_{th} vs. L_{gate}. A difference of about 0.1 V can be observed for the 0.1 μ PMOS device depending on whether B or BF2 is used for the implantation. The difference is explained by the different amount of damage generated during the implantation. The damage distribution after all implantation steps (S/D implantation performed with BF2) is shown in Fig. 5.

4. Conclusions

BF2 implantations in crystalline silicon have been studied and the parameters for BF2 of the stopping power and damage accumulation models have been determined. The influence of the additional damage produced by F on the doping distribution and on the threshold behavior is discussed for a specific device. The simulation of BF2 implantation with the equivalent energy underestimates damage and therefore overestimates channeling.

Acknowledgement

The first author would like to acknowledge financial support by Siemens AG, Corporate Research and Technology, D-81730 Munich, Germany.

References

[1] K. Lee et al., *IEDM Techn. Dig.*, p. 131, 1993.
[2] G. Hobler, *Nucl. Instr. Meth.* B 96, p. 155, 1995.
[3] G. Hobler, and H. Pötzl, *Mat. Res. Soc. Symp. Proc.* 279, p. 165, 1993.
[4] G. Hobler et al., *J. Appl. Phys.* 77, 1995, in press.
[5] A. Simionescu et al., *Nucl. Instr. Meth.* B, in press.
[6] Al. F. Tasch et al., *J. Electrochem. Soc.* 136, p. 810, 1989.
[7] A. Walker et al., *J. Appl. Phys.* 73, p. 4048, 1993.
[8] TMA TSUPREM-4, Version 6, 1993.
[9] S. Selberherr, *Proc. VLSI Process/Dev. Model. Workshop*, p. 40-41, 1989.
[10] H. Jacobs et al., *IEDM Techn. Dig.*, p. 307, 1993.

Analytical Model for Phosphorus Large Angle Tilted Implantation

A. Burenkov[a], W. Bohmayr[b], J. Lorenz[a], H. Ryssel[a,c], and S. Selberherr[b]

[a]Fraunhofer-Institut für Integrierte Schaltungen,
Schottkystrasse 10, 91058 Erlangen, Germany
[b]Institute for Microelectronics, TU Vienna,
Gusshausstrasse 27-29, A-1040 Vienna, Austria
[c]Lehrstuhl für Elektronische Bauelemente, Universität Erlangen-Nürnberg,
Cauerstrasse 6, D-91058 Erlangen, Germany

Abstract

This paper describes a model for the simulation of large angle tilted implantation of phosphorus in silicon. To reduce the size of the precalculated look-up parameter table, the symmetry of the silicon crystal is exploited. Examples demonstrate the importance of ion channeling effects in LATID implantation.

1. Introduction

The large angle tilted implantation doping (LATID) technique provides an effective means to form the lateral dopant distributions under mask edges. A well known application of LATID is the formation of source/drain extensions in MOS transistors. Phosphorus implantation is mainly used to form the source/drain extensions of n-type LATID devices. The typical energy range is 20 to 80 keV, the dose depends on ion energy and is approximately $5 \cdot 10^{13} cm^{-2}$ or lower. Various tilt and rotation angles are used for this kind of devices. In LATID, channeling effects are more important than in other devices, because the ion beam may be oriented close to one of the main crystallographic axes. Therefore, the penetration depth and the lateral extension of the implanted profile depend on ion beam orientation relative to the silicon crystallographic directions. Since the tilt and rotation angles vary over a large range in LATID, the model for LATID implantation has to take into account several channeling directions, and not only one as in the conventional small tilt angie implantation models.

2. Model

In LATID application the tilt and rotation angles are defined relative to the wafer flat. Additionally for performing a simulation, the orientation of the simulation plane on the wafer and the silicon crystal orientation relative to the wafer flat must be known. These technology related parameters are used as input parameters for the suggested model. The angular variables build a relative large variety when we keep in mind that the critical channeling angles of phosphorus in silicon are about a few degrees. This means that in order to describe the angular dependence by a look-up table, an angular grid size of a few degrees is required. Taking a homogeneous grid size of 1 to 3 degree for each of the two independent angular variables, we would get several thousand of angular points for each energy. An unacceptable large amount of Monte Carlo calculations would be required to fill such a look-up table with parameters, therefore we suggest here another approach to store and to use the orientation dependent data.

We exploit that many combinations of tilt and rotation angles are crystallographically equivalent because of the symmetry of the silicon crystal. Table 1 shows some examples of the crystallographically equivalent directions. The rotation angle is zero, when the ion beam is normal to the flat of the wafer. The ion beam is parallel to a {110} type plane of the (100) oriented silicon wafer in this case. Rotation of the wafer is performed around the normal of the silicon wafer.

Table 1   Crystallographically equivalent ion beam directions for (100) silicon

Tilt	45°	45°	52.2°	52.2°	69.3°	69.3°
Rotation	15°	75°	18.4°	71.6°	4.1°	85.9°
Tilt	55°	55°	42.1°	42.1°	69.7°	69.7°
Rotation	20°	70°	13.9°	76.1°	7.3°	82.7°
Tilt	60°	60°	35.5°	35.5°	72.8°	72.8°
Rotation	25°	65°	14.4°	75.6°	13.4°	76.6°

All the combinations of tilt and rotation angles which build a horizontal row in the table are crystallographically equivalent for silicon. There are 6 crystallograpically equivalent directions per octant, 24 for the semi-sphere. The technologically relevant variety of tilt and rotation angles builds a semisphere, but only *1/48* of the sphere represents all non-equivalent directions. If we orient the z-axis normal to the wafer surface, this reduced sector can be defined by a spherical triangle build by the intersection of the planes *y=0, y=x,* and *x=z* with the unit sphere. We define the look-up table for the implantation parameters in this reduced sector of ion impact angles. To get the ion implantation parameters for a defined direction of the ion beam we proceed as follows: For the given beam direction, the crystallographically equivalent direction out of the reduced sector of non-equivalent directions is calculated. Subsequently, a linear two-dimensional three-point interpolation in the reduced angular sector is performed for all the model parameters which depend on the ion impact direction.

It is known that ion implantation profiles depend critically on the wafer orientation only for directions close to the main crystallographic directions [100], [110], and [111], but they vary smoothly in the areas beyond the critical channeling angles relative to main axes and planes. Taking advantage of this behavior, we use an inhomogeneous look-up table for the angular dependence of the implantation parameters. More parameter definition points are located near the main channeling directions and less in the areas between the main channeling direction. Moreover, we do not predefine the values of the angular variables in the look-up table. This means that new parameter description points can be added to the look-up table of the model without changing the code or the interface. The look-up table stored in an external file can be extended when new values for implantation parameters become available from Monte Carlo calculations or from experiments. To calculate the implantation distributions analytically, advanced models with the depth dependent lateral scattering [1] have been used. The vertical as well as the lateral distributions were approximated by Pearson distributions. The orientation dependent ion implantation parameters required for the analytical model were calculated with the Monte Carlo module MCIMPL of the VISTA framework [2].

3. Results

Figure 1 shows depth profiles of phosphorus implanted with an energy of 12keV into (100) silicon. The implantation dose per unit area normal to the ion beam amounted to $10^{14} cm^{-2}$. The profiles shown were calculated with different appoaches: crystalline based Monte Carlo [2] method, amorphous material analytic model [1], and using the model suggested in this paper. First, we should mention the large difference between the predictions of the crystalline based Monte Carlo calculation and the amorphous material model. The penetration depth at the $10^{16} cm^{-3}$ concentration level predicted by the amorphous material model is approximately 4 times smaller as the one calculated for crystalline silicon. This difference is due to strong channeling at low energies even for ion beam directions which deviate from the main channeling directions. In fact, the 7° tilt does not help to avoid channeling at these low energies, therefore only models which are based on a crystalline material appoach are capable to describe the deep penetration of impurities at low energies typical for modern MOS technology. The crystalline material based analytic model of this work satisfactorily reproduces the Monte Carlo profiles and is, therefore, able to predict the deep penetration of the ions associated with channeling. The difference between the profiles calculated at 7° and 45° with the new model is caused by two main factors: different ion beam impact directions and different point response distributions as a result of channeling. Figure 2 elucidates the effect of the point response modification in dependence of the ion impact direction. The figure shows different one-dimensional profiles of phosphorus (50 keV, $10^{13} cm^{-2}$), all implanted with the same tilt angle of 45°, but with different rotation angles. In amorphous material, the implantation profile would show no dependence on the rotation angle, since there is no orientation dependence of the point response function in amorphous material. With the new model, we observe a rather strong dependence on the rotation angle. At 0°, the ion beam direction coincides with the {110} plane of silicon, rotation of 9° does not correspond to any major channeling directions, and 36° is again close to the <110> axis and to the {100} type plane of the

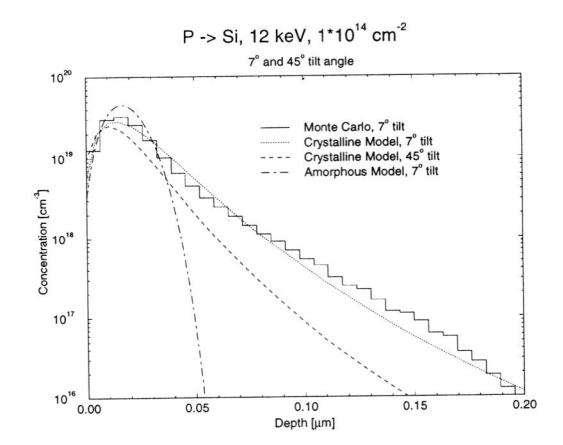

Fig. 1: Depth profiles of phosphorus ions

silicon. Figure 3 shows a two-dimensional post-implantation distribution of phosphorus in a test LATID structure simulated with the analytical model. The structure consists of a (100) silicon substrate and a polysilicon mask with a bevelled edge. The phosphorus ion beam was tilted counter-clockwise by 45° and rotated by 67° relative to the simulation plane. The implantation energy and dose are 20 keV and $5 \cdot 10^{13} cm^{-2}$, respectively. The simulation plane is parallel to the (100)-type plane of the silicon crystal. Concentration lines from 10^{17} to $10^{19} cm^{-3}$ are shown for the standard (solid line) and for the new (dotted line) model. The new crystalline based model predicts larger penetration depths as compared to conventional amorphous material based models and is sensitive to the ion beam impact direction.

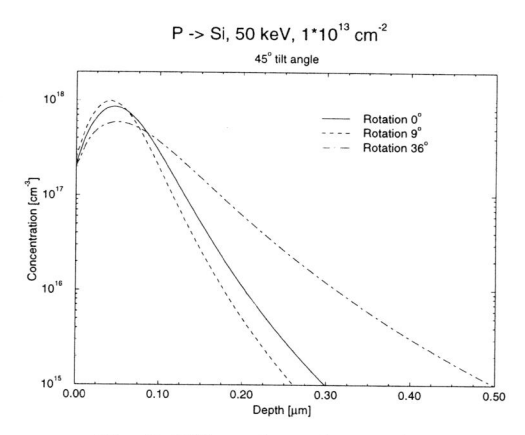

Fig. 2: Effect of rotation angle

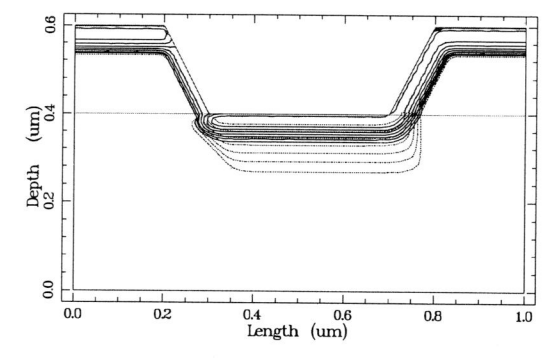

4. Acknowledgement

This work was carried out in cooperation between ADEQUAT (JESSI project BT11) and PROMPT (JESSI project BT8B) and has been funded by the EU as ESPRIT projects No. 8002 and 8150, respectively.

Fig. 3: Two-dimensional distributions of phosphorus in a LATID structure

References

[1] J. Lorenz, C. Hill, H. Jaouen, C. Lombardi, C. Lyden, K. de Meyer, J. Pelka, A. Poncet, M. Rudan, S. Solmi, *The STORM Technology CAD System*, Proceedings of the International Workshop on TCAD Systems (eds F. Fasching, S. Halama, and S. Selberherr), Springer Verlag, Wien, pp. 163-196, 1993.

[2] H. Stippel and S. Selberherr, *Three Dimensional Monte Carlo Simulation of Ion Implantation with Octree Based Point Location*, IEICE Transactions on Electronics, Vol. E77-C, No. 2, pp. 118-123, 1994.

Statistical Accuracy and CPU Time Characteristic of Three Trajectory Split Methods for Monte Carlo Simulation of Ion Implantation

W. Bohmayr[a], A. Burenkov[b], J. Lorenz[b], H. Ryssel[b,*], and S. Selberherr[a]

[a]Institute for Microelectronics, TU Vienna
Gusshausstrasse 27-29, A-1040 Vienna, Austria
[b]Fraunhofer-Institut für Integrierte Schaltungen,
Schottkystrasse 10, 91058 Erlangen, Germany
*Lehrstuhl für elektronische Bauelemente, Universität Erlangen-Nürnberg,
Cauerstrasse 6, 91058 Erlangen, Germany

Abstract

Three *trajectory split methods* [1] for the acceleration of two and three-dimensional Monte Carlo simulation of ion implantation into crystalline targets are presented. They ensure a much better statistical representation in regions with a dopant concentration several orders of magnitudes smaller than the maximum. As a result the time required to perform a simulation with comparable statistical accuracy is drastically reduced. The advantages of the new approaches have been confirmed by a thorough statistical analysis.

1. Introduction

Inspired by the results in [2] about the *rare event* approach implemented in the UT-MARLOWE code [3] for one-dimensional structures, we developed the *trajectory split method* [1] for the Monte Carlo simulation of ion implantation which drastically reduces the computational effort and is applicable for two and three-dimensional simulations. A similar method was first used in the work of Phillips and Price [4] to simulate hot electron transport.

2. The Trajectory Split Method

The traditional Monte Carlo approach for crystalline targets is based on the calculation of a large number of "distinct" ion trajectories, i.e. each trajectory is usually followed from the ion starting point at the surface of the target up to the stopping point of the ion. Since the majority of ion trajectories ends at the most probable penetration depth, the statistical noise of regions with a dopant concentration several orders of magnitudes smaller than the maximum (in the following we call these areas "peripheral") cannot be tolerated and we have to increase the **total** number of calculated ions.

The fundamental ideas of our new simulation approach are to locally increase the number of calculated ion trajectories in areas with large statistical uncertainty and to utilize the information we can derive from the flight-path of the ion up to a certain depth inside the target. For each ion, the local dopant concentration C_{loc} is checked at certain points of the flight-path (*checkpoints*). At each checkpoint we relate C_{loc} to the current maximum global concentration $C_{max,current}$ by calculating the ratio $C_{loc}/C_{max,current}$. The result is compared with given relative concentration levels (we define ten levels at $0.3, 0.09, 0.027, ..., 0.3^{10}$). Only if the current local concentration falls in an interval below the previous one, a *trajectory split point* is defined at this checkpoint. Therefore our approach is a self-adaptive algorithm because more split points are defined at areas with unsatisfying statistical accuracy. Additional trajectory branches are suppressed, if an ion moves from lower to higher local concentration levels. We store the position of the ion, its energy as well as the vector of velocity and use this data for virtual branches of ion trajectories starting at this split point. In this way, the peripheral areas of the dopant concentration are represented by a much higher number of ion trajectories and the statistical noise is reduced.

3. Three Split Strategies and their Statistical Characteristic

In our first attempt [1] we chose the simplest implementation of this method by suppressing recursive splits and by permitting only two virtual branches at each split point (Fig. 1). Such a virtual trajectory branch is calculated with the **same models and parameters** as a regular trajectory, but it starts **at the split point** with **initial conditions obtained from the regular ion**. To obtain the correct concentration, a weight is assigned to each branch. The different realizations of the virtual trajectories result from the thermal vibrations of the target atoms [5].

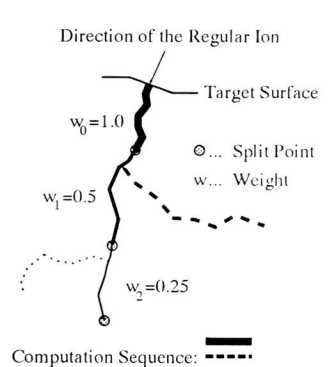

Figure 1: Topological structure of the non-recursive split method, the weight of each branch, and the sequence of its calculation

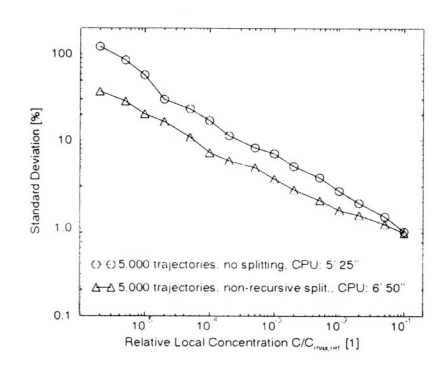

Figure 2: Two-dimensional point response of phosphorus implant, statistical accuracy and CPU time of non-recursive split method

In the present work we present two additional trajectory split methods: recursive <u>T</u>rajectory <u>R</u>elated <u>S</u>plit method (*TRS*, Fig. 3) and recursive <u>S</u>plit-level <u>R</u>elated <u>S</u>plit method (*SRS*, Fig. 5). As an important example, we perform a Monte Carlo simulation of a phosphorus implant at 50keV into (100) oriented single-crystal silicon covered by 2.5nm of oxide to obtain point response distributions. The required computational effort for such a simulation is about 70 minutes for the conventional approach and about 12 minutes for the new trajectory split methods using a HP 735/100 workstation. Implantation damage formation has been neglected.

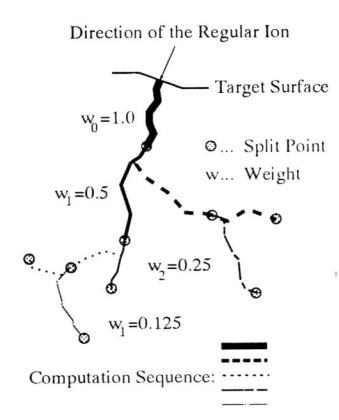

Figure 3: Topological structure of the trajectory related method (TRS), the weight of each branch, and the sequence of its calculation

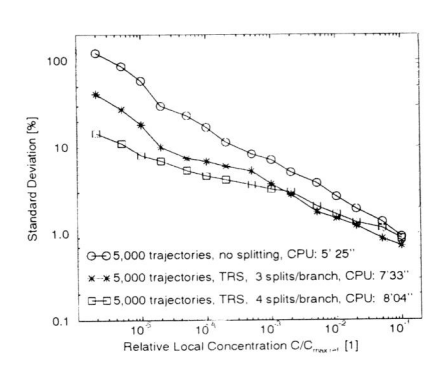

Figure 4: Two-dimensional point response of a phosphorus implant, statistical accuracy and CPU time of trajectory related split method (TRS)

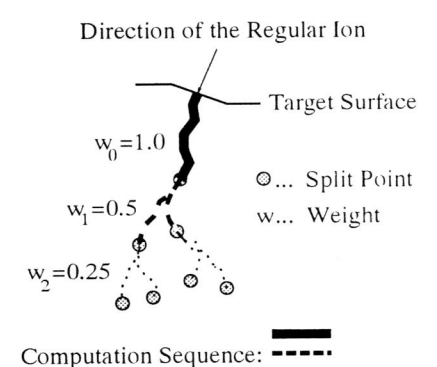

Figure 5: Topological structure of the split-level related split method (SRS), the weight of each branch, and the sequence of its calculation

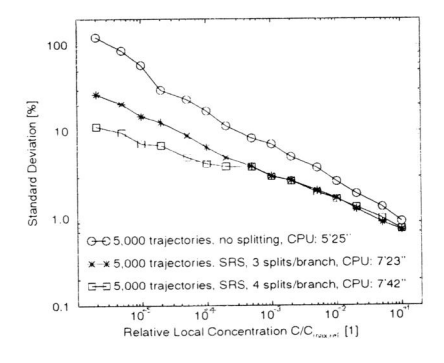

Figure 6: Two-dimensional point response of a phosphorus implant, statistical accuracy and CPU time of split-level related split method (SRS)

To assess the statistical accuracy of the results obtained from the conventional and from the trajectory split methods, we define a mean-square deviation from a reference distribution. For that reason we carry out a conventional simulation result with such a high number of ion trajectories (1,000,000) that statistical fluctuations are negligible in the concentration area considered.

4. Simulation Results

We present the deviation data for the non-recursive (Fig. 2), TRS (Fig. 4), and SRS method (Fig. 6) calculated with 5,000 distinct ion trajectories. The relative concentration in these figures is defined as the ratio of $C/C_{max,ref}$, where $C_{max,ref}$ means the maximum concentration of the reference distribution. It is evident that the relative standard deviation decreases for high local concentrations. Therefore areas with low local concentration are well suited to demonstrate the merits of the new simulation strategy. The differences in Fig. 4 and Fig. 6 result from slightly modified split algorithms. Our approaches show a deviation from the reference distribution which is up to ten times smaller compared to the deviation of the standard method. The computational effort is approximately proportional to the number of distinct ion trajectories and the additional overhead due to trajectory splits is only 25% to 35%. Further important advantages of the trajectory split method are its lower sensitivity to the local concentration and the opportunity to individualize its error behavior. Increasing the number of splits per branch (Fig. 4 and Fig. 6) and/or initalizing more than one virtual branch at each split point leads to a significantly smaller error in peripheral areas without effecting the statistic in other regions. In other words, there is a chance of optimizing the relation between CPU time and required statistical accuracy for a particular problem.

It should be mentioned that our new strategy is also well suited to compute the collision cascade of a displaced target atom ("recoil") because it offers the possibility to optimize the recoil statistic by a random deletion of recoil trajectories at places with a statistical "over-representation" and by splitting them at peripheral areas of the collision cascade. It is also expected that the *SRS* method will prove to be more profitable if implantation damage is considered.

5. Acknowledgement

Part of this work was carried out in cooperation between PROMPT (JESSI project BT8B) and ADEQUAT (JESSI project BT11) and has been funded by the EU as ESPRIT projects No. 8150 and 8002, respectively. TU Vienna wants to acknowledge important support by Digital Equipment Corporation, Hudson, USA.

References

[1] W. Bohmayr and S. Selberherr, "Trajectory Split Method for Monte Carlo Simulation of Ion Implantation Demonstrated by Three-Dimensional Poly-Buffered LOCOS Field Oxide Corners", *VLSI-TSA*, Taipei, Taiwan, 1995.

[2] S.-H. Yang, D. Lim, S. Morris, and A.F. Tasch, "A More Efficient Approach for Monte Carlo Simulation of Deeply-Channeled Implanted Profiles in Single-Crystal Silicon", *NUPAD*, pp. 97–100, 1994.

[3] K.M. Klein, C. Park, and A.F. Tasch, "Monte Carlo Simulation of Boron Implantation into Single-Crystal Silicon", *IEEE Trans. Electron Devices*, ED-39, pp. 1614–1621, 1992.

[4] A. Phillips and P.J. Price. "Monte Carlo Calculations on Hot Electron Energy Tails", *Applied Physics Letters*, vol. 30, no. 10, pp. 528–530, 1977.

[5] M. Jaraiz, J. Arias, E. Rubio, L.A. Marques, L. Pelaz, L. Bailon, and J. Barbolla, "Dechanneling by Thermal Vibrations in Silicon Ion Implantation", *X International Conference on Ion Implantation Technology*, Abstract P-2.19, Catania, Italy, 1994.

Author Index

Abe, T., 444
Abou-Elnour, A., 178
Abramo, A., 106
Altermatt, P. P., 348
Aluru, N. R., 86
Andronov, A., 325
Aoyama, K., 118
Arai, E., 444
Asenov, A., 226, 336
Axelrad, V., 10
Babiker, S., 226
Baccus, B., 460
Baltes, H., 1
Bang, D. S., 166
Barker, J. R., 226, 336
Baudelot, E., 364
Bauer, F., 332
Bauer, R., 151
Beaumont, S. P., 226
Benvenuti, A., 400
Bergunde, T., 328
Biesemans, S., 110
Bohmayr, W., 488, 492
Bonnaud, O., 428
Boulin, D. M., 424
Bozek, S., 460
Brennan, K. F., 274
Brisset, C., 26
Brown, A. R., 336
Bude, J., 106
Burenkov, A., 488, 492
Butel, Y., 360
Buturla, E. M., 163
Byers, J., 163
Carceller, J. E., 380
Carnevale, G. P., 286
Cea, S., 135, 139
Chen, T., 270
Cheng, M.-C., 202
Chung, H.-C., 298
Colalongo, L., 82
Colpani, P., 286
Curow, M., 250
Dauelsberg, M., 328
De Jaeger, J. C., 360
De Meyer, K., 110
Dollfus, P., 26
Dort, M. J. van, 372, 432

Dunham, S. T., 476
Durst, F., 258
Dutton, R. W., 66, 86, 114, 270, 424
Egorov, Yu., 328
Eicher, S., 332
Engl, W. L., 222
Falster, R., 472
Faricelli, J., 126
Feudel, T., 448
Fichtner, W., 332, 368, 400, 448, 456, 468
Fortin, B., 428
Frank, J., 22
Freund, D., 344
Fujii, K., 302
Fujinaga, M., 143
Fukuda, K., 90
Funk, J., 22
Gajewski, H., 234
Galjukov, A. O., 258
Gander, M. J., 66
Garretón, G., 468
Gärtner, K., 234, 456
Gergintschew, Z., 18
Gerlach, W., 262
Gerstenmaier, Y. C., 364
Ghetti, A., 388
Godoy, A., 380
Goldsman, N., 122
Górecki, K., 306
Gossner, H., 436
Griffin, P. B., 42
Gružinskis, V., 314, 318
Halama, S., 58
Hamaguchi, C., 420
Harrer, M., 396
Hazdra, P., 186
Hédoire, J., 360
Heinemann, B., 98
Heiser, G., 348
Hennacy, H., 122
Herzel, F., 98
Herzer, R., 290
Hesto, P., 26
Hirao, T., 143
Hobler, G., 484
Höfler, A., 448
Hofmann, F., 436
Hohmeyer, M., 166

Horino, Y., 302
Horio, K., 78
Howes, M. J., 352
Huang, C.-L., 126
Husain, A., 163
Hwang, D. M., 294
Hwang, H.-j., 214
Hwang, J. C. M., 294, 298
Iizuka, T., 384
Jacob, M., 472
Jaeger, J. C. De, 360
Jiménez-Tejada, J. A., 380
Johnson, J., 270
Jungemann, Chr., 222
Kadinski, L., 328
Kamohara, S., 194, 480
Kamon, K., 143
Kan, E. C., 66
Kataoka, Y., 448
Kato, H., 384
Kawakami, M., 480
Keith, S., 222
Kerr, D. C., 198
Kersch, A., 174
Khalil, N., 70, 126
Khrenov, G., 246
Kinomura, A., 302
Kishimoto, T., 302
Klaassen, D. B. M., 432
Klix, W., 230
Klös, A., 218
Kobayashi, T., 151
Kocsis, T., 238
Komatsu, Y., 151
Komori, S., 302
Korvink, J. G., 1
Kositza, J., 18
Kosina, H., 396
Kostka, A., 218, 344
Kotani, N., 143, 302
Koyama, K., 151
Krautschneider, W. H., 436
Krivokapic, Z., 166, 170
Krumbein, U., 400
Kulkova, E., 246
Kump, M., 163
Kunikiyo, T., 143
Kushner, M. J., 170
Kuznetsov, V. S., 322
Lades, M., 22
Lau, F., 436, 484
Laux, S. E., 404

Law, K. H., 86
Law, M. E., 46, 135, 139, 282
Ledl, Ch., 50
Lefebvre, M., 360
Leitner, E., 464
Leon, F. A., 388
Lhermite, H., 428
Liang, W-C., 122
Liebig, D., 74
Liegmann, A., 448, 456
Lin, C.-C., 282
Litsios, J., 348, 368
Lloyd, P., 163
López-Serrano, J., 62
Lorenz, J., 488, 492
Lucas, K. D., 14
Makarov, Yu. N., 258, 328
Manku, T., 94
Manukonda, R., 163
Marmiroli, A., 286
Masszi, F., 238, 340
Masuda, H., 408
Matsumoto, H., 384
Matsumoto, S., 444
Matsuo, H., 194
Mayergoyz, I. D., 122, 198
McVittie, J. P., 166, 170
Meinerzhagen, B., 222
Meyer, K. De, 110
Mijalković, S., 182
Miura, H., 147
Miura-Mattausch, M., 278
Miyoshi, H., 302
Mlekus, R., 50
Molzer, W., 102
Morton, C. G., 352
Mujtaba, S. A., 424
Mukai, M., 151
Musalem, F.-X., 26
Nakauchi, N., 151
Nasr, A., 126
Nathan, A., 30, 94
Nefedov, I., 325
Netzel, M., 290
Nishi, K., 90
Nishimura, T., 302
Niu, G. F., 190
Noell, M., 54
Nordlander, E., 340
Ohno, Y., 302
Oldiges, P., 310
Orlowski, M., 54, 159

Packan, P., 34
Palma, A., 380
Park, H.-s., 214
Paul, O., 1
Pichler, Ch., 70
Pichler, P., 472
Pigorsch, C., 230
Pinto, M. R., 106, 424
Poppe, A., 238
Prigge, O., 278
Prokaznikov, A. V., 322
Rafferty, C. S., 424
Rahm, A., 278
Rebora, A., 286
Reddy, K., 242
Reggiani, L., 314, 318
Reznik, D., 254
Rieger, G., 58, 151
Rottinger, M., 440
Roy, S., 336
Ruan, G., 190
Rudan, M., 82, 416
Rueda, H. A., 46
Runnels, S., 163
Ryssel, H., 472, 488, 492
Salaün, A.-C., 428
Salmer, G., 360
Sangiorgi, E., 106
Saraswat, K. C., 42, 166
Sasaki, N., 448
Satoh, K., 78
Sayama, H., 302
Schäfer, M., 174
Schäfer, M., 258, 328
Schäfer, H., 436
Scharfetter, D., 163
Schenk, A., 400
Schipanski, D., 18
Schmithüsen, B., 368
Schrom, G., 70
Schroeder, D., 155, 356
Schröter, M., 210
Schulz-Mirbach, C., 392
Schünemann, K., 178
Selberherr, S., 50, 58, 70, 151, 396, 440, 464, 488, 492
Senez, V., 460
Shigeta, K., 384
Shiktorov, P., 314, 318
Shimizu, T., 444
Simionescu, A., 484
Simlinger, T., 440

Singh, S., 122
Sittig, R., 266
Slotboom, J. W., 372
Smeys, P., 42
Smith, A. W., 274
Snowden, C. M., 352
Son, M.-s., 214
Sonoda, K., 302, 420
Starikov, E., 314, 318
Stenzel, R., 230
Stippel, H., 242
Strauch, G., 328
Strasser, E., 50
Strecker, N., 448, 468
Strojwas, A. J., 14, 62
Sugaya, M., 194
Suzuki, K., 448
Swart, N. R., 30
Takai, M., 302
Tanabe, H., 14
Tanaka, K., 384
Tang, T. A., 190
Taniguchi, K., 420
Tanizaki, Y., 147
Tarnay, K., 238
Tatsumi, T., 151
Taylor, W. J., 54, 159
Tixier, A., 286
Tkachenko, Y. A., 294, 298
Tolstikhin, V. I., 206
Troyanovsky, B., 114
Tsamis, C., 452
Tsoukalas, D., 452
Uchida, T., 143
Udal, A., 340
Valdinoci, M., 82
van Dort, M. J., 372, 432
Varani, L., 314, 318
Vecchi, M. C., 416
Velmre, E., 340
Venturi, F., 106, 388
Villablanca, L., 468
Vobecký, J., 186
Vogl, P., 396
Voinovich, P. A., 258
Wachutka, G., 22
Walkey, D. J., 210
Wang, X., 388
Wang, Z. Z., 460
Wei, C. J., 294, 298
Weyers, M., 328
Wiese, U., 262

Author Index

Wiesner, U., 266
Willander, M., 206
Wimmer, K., 54
Witkowski, U., 155, 356
Woerlee, P. H., 372
Wu, C. D., 476
Wu, Y-J., 122
Yamaji, M., 420
Yamada, T., 78
Yergeau, D. W., 66

Yoder, P. D., 400
Yoshida, M., 444
Yu, Z., 114
Yuan, C.-M., 14
Zaitsu, Y., 444
Zarebski, J., 306
Zheng, J., 170
Zhmakin, A. I., 258
Zimin, S. P., 322

S. Selberherr, H. Stippel, E. Strasser (eds.)

Simulation of Semiconductor Devices and Processes, Vol. 5

1993. 530 figures. XX, 504 pages.
Cloth DM 198,–, öS 1386,–
ISBN 3-211-82504-5

Prices are subject to change without notice

The SISDEP 93 conference proceedings present outstanding research and development results in the area of numerical process and device simulation. The miniaturization of today's semiconductor devices, the usage of new materials and advanced process steps in the development of new semiconductor technologies suggests the design of new computer programs. This trend towards more complex structures and increasingly sophisticated processes demands advanced simulators, such as fully three-dimensional tools for almost arbitrarily complicated geometries. With the increasing need for better models and improved understanding of physical effects, these proceedings support the simulation community and the process- and device engineers who need reliable numerical simulation tools for characterization, prediction, and development. This book covers the following topics: process simulation and equipment modeling, device modeling and simulation of complex structures, device simulation and parameter extraction for circuit models, integration of process, device and circuit simulation, practical applications of simulation, algorithms and software.

Springer-Verlag Wien NewYork

Sachsenplatz 4–6, P.O.Box 89, A-1201 Wien · 175 Fifth Avenue, New York, NY 10010, USA
Heidelberger Platz 3, D-14197 Berlin · 3-13, Hongo 3-chome, Bunkyo-ku, Tokyo 113, Japan

Springer-Verlag
and the Environment

WE AT SPRINGER-VERLAG FIRMLY BELIEVE THAT AN international science publisher has a special obligation to the environment, and our corporate policies consistently reflect this conviction.

WE ALSO EXPECT OUR BUSINESS PARTNERS – PRINTERS, paper mills, packaging manufacturers, etc. – to commit themselves to using environmentally friendly materials and production processes.

THE PAPER IN THIS BOOK IS MADE FROM NO-CHLORINE pulp and is acid free, in conformance with international standards for paper permanency.